AMERICAN IMAGINATION

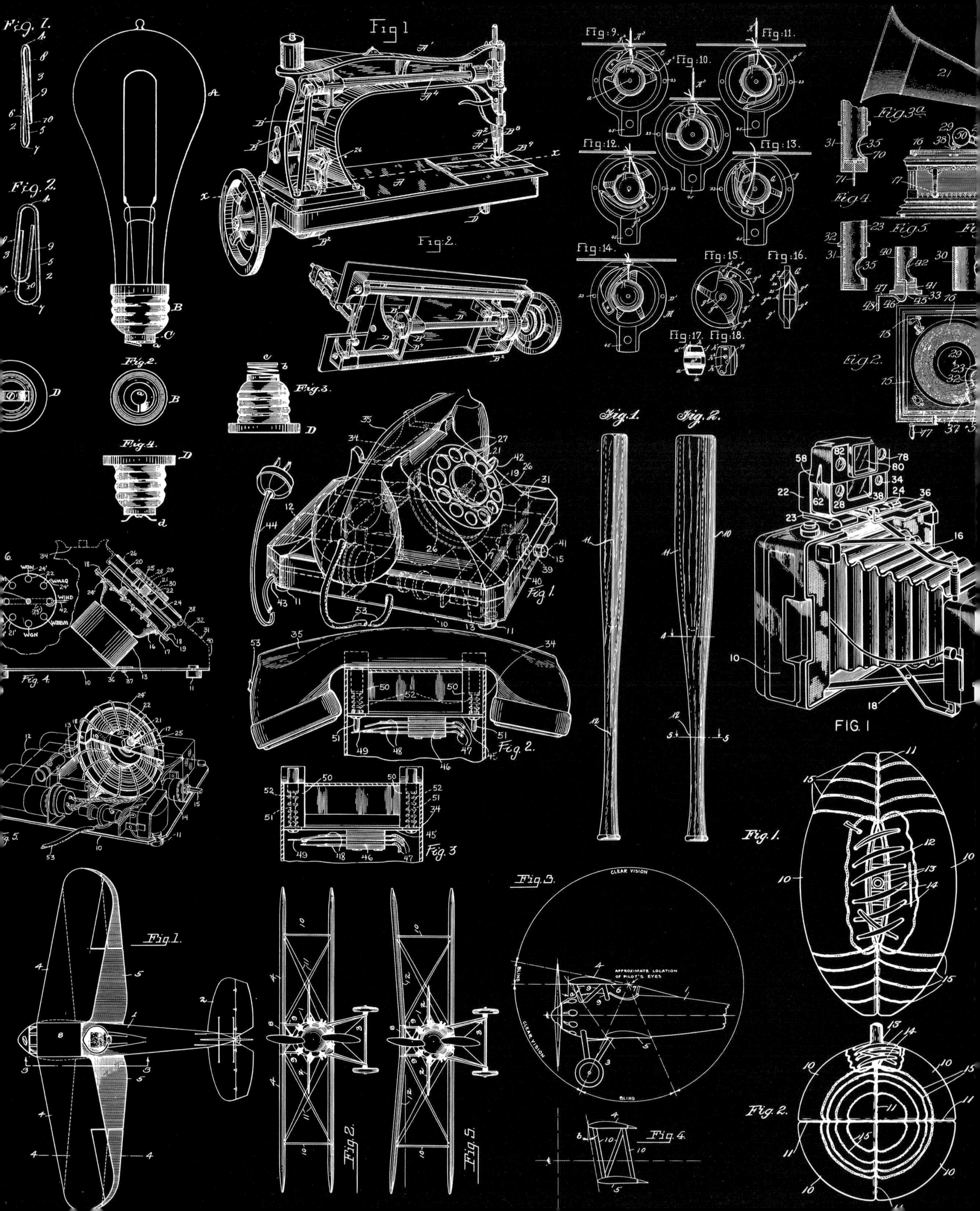

Fig. 1.
Fig. 2.
Fig.2.
Fig.3.
Fig.4.
Fig 1
Fig.2.
Fig:9.
Fig:10.
Fig:11.
Fig:12.
Fig:13.
Fig:14.
Fig:15.
Fig:16.
Fig:17.
Fig:18.
Fig.3a
Fig.4.
Fig.5.
Fig.2.
Fig. 4.
Fig.5.
Fig. 1.
Fig. 2.
Fig. 3
Fig.1.
Fig.2.
FIG. I
Fig. 1.
Fig.1.
Fig. 2.
Fig. 5.
Fig.3.
CLEAR VISION
APPROXIMATE LOCATION OF PILOT'S EYES
BLIND
Fig.4.
Fig. 2.

AMERICAN IMAGINATION

250 YEARS OF INVENTION, CREATIVITY, AND INNOVATION

JENNIFER NEWENS

weldonowen

CONTENTS

Subject Key

- AGRICULTURE
- ARCHITECTURE
- THE ARTS
- COMMUNICATION
- CULTURE
- FASHION
- FOOD AND DRINK
- GOVERNMENT AND MILITARY
- HEALTH AND MEDICINE
- IN THE HOME
- IN THE WORKPLACE
- POP CULTURE
- SCIENCE AND EXPLORATION
- SPORTS
- TECHNOLOGY
- TOYS AND GAMES
- TRANSPORTATION
- TV, MOVIES, AND THEATER

INTRODUCTION

The United States has always been a nation of builders, dreamers, and problem-solvers. From the moment the thirteen colonies declared independence in 1776, Americans have demonstrated a drive to innovate, create, and reshape the world around them. This restless energy has produced an extraordinary succession of inventions, innovations, and cultural breakthroughs that have fundamentally transformed not just American society, but nations all over the world.

The story of *American Imagination* spans 250 years of relentless ingenuity. It begins with the foundational documents and institutions that created a new form of government, moves through the industrial innovations that powered a growing nation, and continues through the digital revolution that has connected the globe. Each innovation emerged from a uniquely American context—a society built on the premise that ordinary people could achieve extraordinary things, where practical problem-solving met boundless ambition.

This book chronicles more than just inventions—it captures the evolution of American identity itself. The cotton gin and the steamboat enabled westward expansion. The telegraph and telephone shrank distances and connected communities. The brassiere and bloomers became important garments that allowed women a new form of expression. The assembly line and the skyscraper embodied industrial ambition. The personal computer and the Internet democratized information and reshaped human interaction. Each innovation reflects the values, challenges, and aspirations of its era while laying the groundwork for future breakthroughs.

What makes American innovation distinctive is not just its scope, but its accessibility. Many of the most transformative American inventions emerged from garages, workshops, laboratories, and kitchens run by individuals who saw problems and refused to accept that they couldn't be solved. From Benjamin Franklin's bifocal eyeglasses to Madam C. J. Walker's beauty line for African American hair to Apple's iPhone, American innovation has been driven by a democratic belief that good ideas can come from anywhere and change everything.

The diversity of innovations chronicled here—spanning the categories of government, arts and culture, architecture, sports, workplace and home, fashion, health and medicine, toys and games, science and exploration, food and drink, transportation, and technology—reveals the breadth of American creativity. These achievements represent countless hours of experimentation, frequent failures, and the persistence to keep trying. They embody the American conviction that progress is not only possible but inevitable when human ingenuity is unleashed.

Yet the story of *American Imagination* is not without complexity. Many innovations that improved lives also created new challenges. For example, the same industrial prowess that built prosperity also created environmental concerns. The technologies that connected people also raised questions about privacy and social cohesion. The economic systems that generated wealth also produced inequality. American innovation has always carried both promise and responsibility.

The innovations presented here were created by people of all backgrounds—immigrants and native-born Americans, men and women, people of different races and ethnicities—though historical barriers often limited recognition and opportunities. This collection acknowledges both the extraordinary achievements and the ongoing work to ensure that America's innovative capacity includes everyone.

As our country looks toward its future, this catalog of imagination serves as both celebration and inspiration. It demonstrates that the capacity for transformative innovation remains as vital today as it was in 1776. The challenges facing the nation and the world—climate change, disease, inequality, technological disruption—require the same spirit of creative problem-solving that has defined American innovation for two and a half centuries.

The American imagination will forever continue to evolve, but its essential character will always remain unchanged: the belief that human creativity, determination, and ingenuity can build a better tomorrow.

1776 THE DECLARATION OF INDEPENDENCE

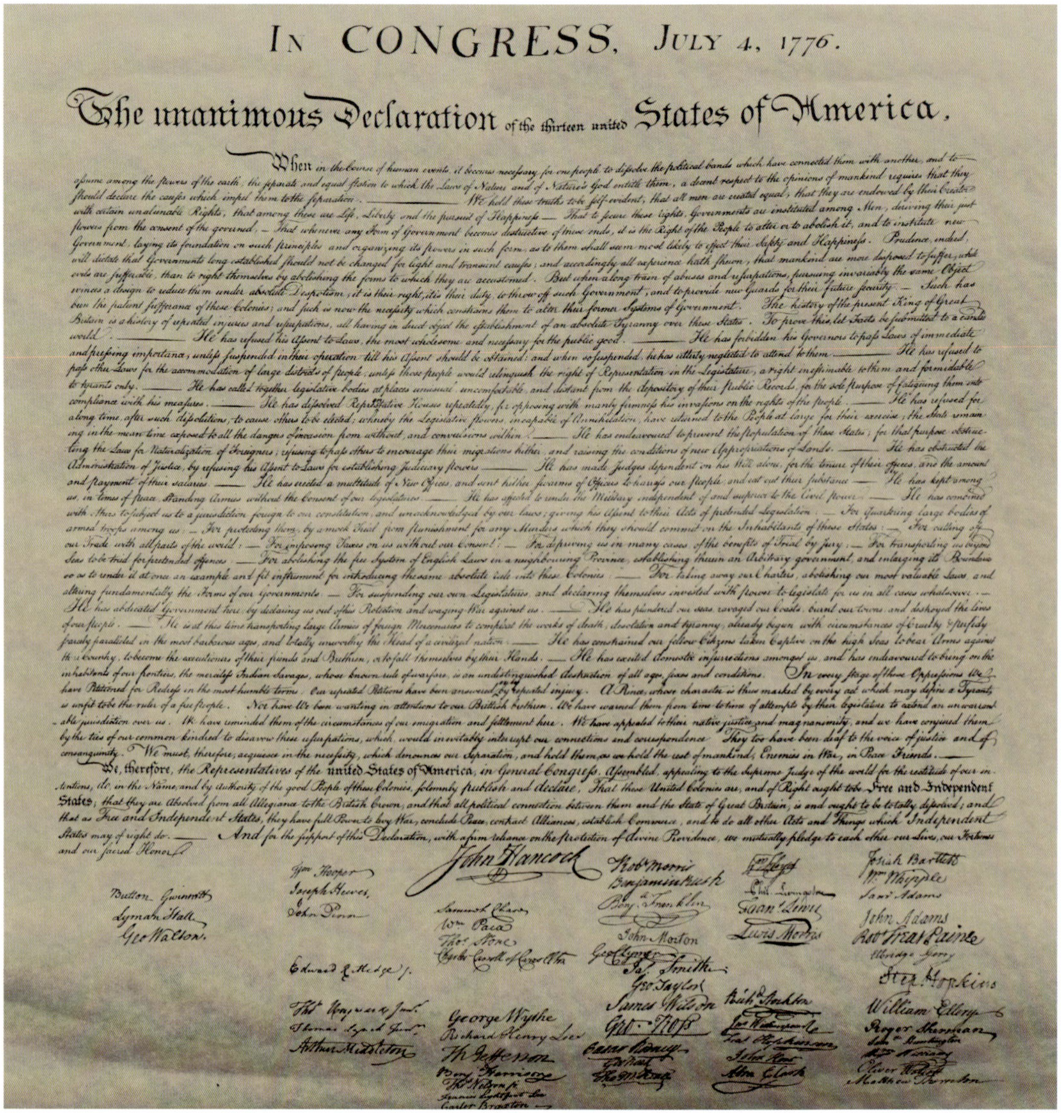
IN CONGRESS, JULY 4, 1776.

The unanimous Declaration of the thirteen united States of America,

When in the Course of human events, it becomes necessary for one people to dissolve the political bands which have connected them with another, and to assume among the powers of the earth, the separate and equal station to which the Laws of Nature and of Nature's God entitle them, a decent respect to the opinions of mankind requires that they should declare the causes which impel them to the separation. — We hold these truths to be self-evident, that all men are created equal, that they are endowed by their Creator with certain unalienable Rights, that among these are Life, Liberty and the pursuit of Happiness. — That to secure these rights, Governments are instituted among Men, deriving their just powers from the consent of the governed, — That whenever any Form of Government becomes destructive of these ends, it is the Right of the People to alter or to abolish it, and to institute new Government, laying its foundation on such principles and organizing its powers in such form, as to them shall seem most likely to effect their Safety and Happiness. Prudence, indeed, will dictate that Governments long established should not be changed for light and transient causes; and accordingly all experience hath shewn, that mankind are more disposed to suffer, while evils are sufferable, than to right themselves by abolishing the forms to which they are accustomed. But when a long train of abuses and usurpations, pursuing invariably the same Object evinces a design to reduce them under absolute Despotism, it is their right, it is their duty, to throw off such Government, and to provide new Guards for their future security. — Such has been the patient sufferance of these Colonies; and such is now the necessity which constrains them to alter their former Systems of Government. The history of the present King of Great Britain is a history of repeated injuries and usurpations, all having in direct object the establishment of an absolute Tyranny over these States. To prove this, let Facts be submitted to a candid world. — He has refused his Assent to Laws, the most wholesome and necessary for the public good. — He has forbidden his Governors to pass Laws of immediate and pressing importance, unless suspended in their operation till his Assent should be obtained; and when so suspended, he has utterly neglected to attend to them. — He has refused to pass other Laws for the accommodation of large districts of people, unless those people would relinquish the right of Representation in the Legislature, a right inestimable to them and formidable to tyrants only. — He has called together legislative bodies at places unusual, uncomfortable, and distant from the depository of their public Records, for the sole purpose of fatiguing them into compliance with his measures. — He has dissolved Representative Houses repeatedly, for opposing with manly firmness his invasions on the rights of the people. — He has refused for a long time, after such dissolutions, to cause others to be elected; whereby the Legislative powers, incapable of Annihilation, have returned to the People at large for their exercise; the State remaining in the mean time exposed to all the dangers of invasion from without, and convulsions within. — He has endeavoured to prevent the population of these States; for that purpose obstructing the Laws for Naturalization of Foreigners; refusing to pass others to encourage their migrations hither, and raising the conditions of new Appropriations of Lands. — He has obstructed the Administration of Justice, by refusing his Assent to Laws for establishing Judiciary powers. — He has made Judges dependent on his Will alone, for the tenure of their offices, and the amount and payment of their salaries. — He has erected a multitude of New Offices, and sent hither swarms of Officers to harrass our people, and eat out their substance. — He has kept among us, in times of peace, Standing Armies without the Consent of our legislatures. — He has affected to render the Military independent of and superior to the Civil power. — He has combined with others to subject us to a jurisdiction foreign to our constitution, and unacknowledged by our laws; giving his Assent to their Acts of pretended Legislation: — For Quartering large bodies of armed troops among us: — For protecting them, by a mock Trial, from punishment for any Murders which they should commit on the Inhabitants of these States: — For cutting off our Trade with all parts of the world: — For imposing Taxes on us without our Consent: — For depriving us in many cases, of the benefits of Trial by Jury: — For transporting us beyond Seas to be tried for pretended offences — For abolishing the free System of English Laws in a neighbouring Province, establishing therein an Arbitrary government, and enlarging its Boundaries so as to render it at once an example and fit instrument for introducing the same absolute rule into these Colonies: — For taking away our Charters, abolishing our most valuable Laws, and altering fundamentally the Forms of our Governments: — For suspending our own Legislatures, and declaring themselves invested with power to legislate for us in all cases whatsoever. — He has abdicated Government here, by declaring us out of his Protection and waging War against us. — He has plundered our seas, ravaged our Coasts, burnt our towns, and destroyed the lives of our people. — He is at this time transporting large Armies of foreign Mercenaries to compleat the works of death, desolation and tyranny, already begun with circumstances of Cruelty & perfidy scarcely paralleled in the most barbarous ages, and totally unworthy the Head of a civilized nation. — He has constrained our fellow Citizens taken Captive on the high Seas to bear Arms against their Country, to become the executioners of their friends and Brethren, or to fall themselves by their Hands. — He has excited domestic insurrections amongst us, and has endeavoured to bring on the inhabitants of our frontiers, the merciless Indian Savages, whose known rule of warfare, is an undistinguished destruction of all ages, sexes and conditions. In every stage of these Oppressions We have Petitioned for Redress in the most humble terms: Our repeated Petitions have been answered only by repeated injury. A Prince, whose character is thus marked by every act which may define a Tyrant, is unfit to be the ruler of a free people. Nor have We been wanting in attentions to our British brethren. We have warned them from time to time of attempts by their legislature to extend an unwarrantable jurisdiction over us. We have reminded them of the circumstances of our emigration and settlement here. We have appealed to their native justice and magnanimity, and we have conjured them by the ties of our common kindred to disavow these usurpations, which, would inevitably interrupt our connections and correspondence. They too have been deaf to the voice of justice and of consanguinity. We must, therefore, acquiesce in the necessity, which denounces our Separation, and hold them, as we hold the rest of mankind, Enemies in War, in Peace Friends. —

We, therefore, the Representatives of the united States of America, in General Congress, Assembled, appealing to the Supreme Judge of the world for the rectitude of our intentions, do, in the Name, and by Authority of the good People of these Colonies, solemnly publish and declare, That these United Colonies are, and of Right ought to be Free and Independent States; that they are Absolved from all Allegiance to the British Crown, and that all political connection between them and the State of Great Britain, is and ought to be totally dissolved; and that as Free and Independent States, they have full Power to levy War, conclude Peace, contract Alliances, establish Commerce, and to do all other Acts and Things which Independent States may of right do. — And for the support of this Declaration, with a firm reliance on the protection of divine Providence, we mutually pledge to each other our Lives, our Fortunes and our sacred Honor.

John Hancock

Arguably the most iconic American governmental document of all time—the one that started it all—the Declaration of Independence officially declared the American colonies' separation from Great Britain and set forth for the world to see exactly how they planned to do so. Ratified on July 4, 1776, the document, signed by Founding Fathers, including Benjamin Franklin (1706–1790), John Hancock (1737–1793), and Thomas Jefferson (1743–1826), outlines the three basic rights of the American people in its preamble: that all men are created equal; that all men have the right to life, liberty, and the pursuit of happiness; and that the people have the right to protest against their government. The United States declaring itself a sovereign nation paved the way for other governments to recognize it and form both commercial and political alliances. Today, the Declaration of Independence serves as the foundational text outlining American identity and inspires protests around freedom and equality.

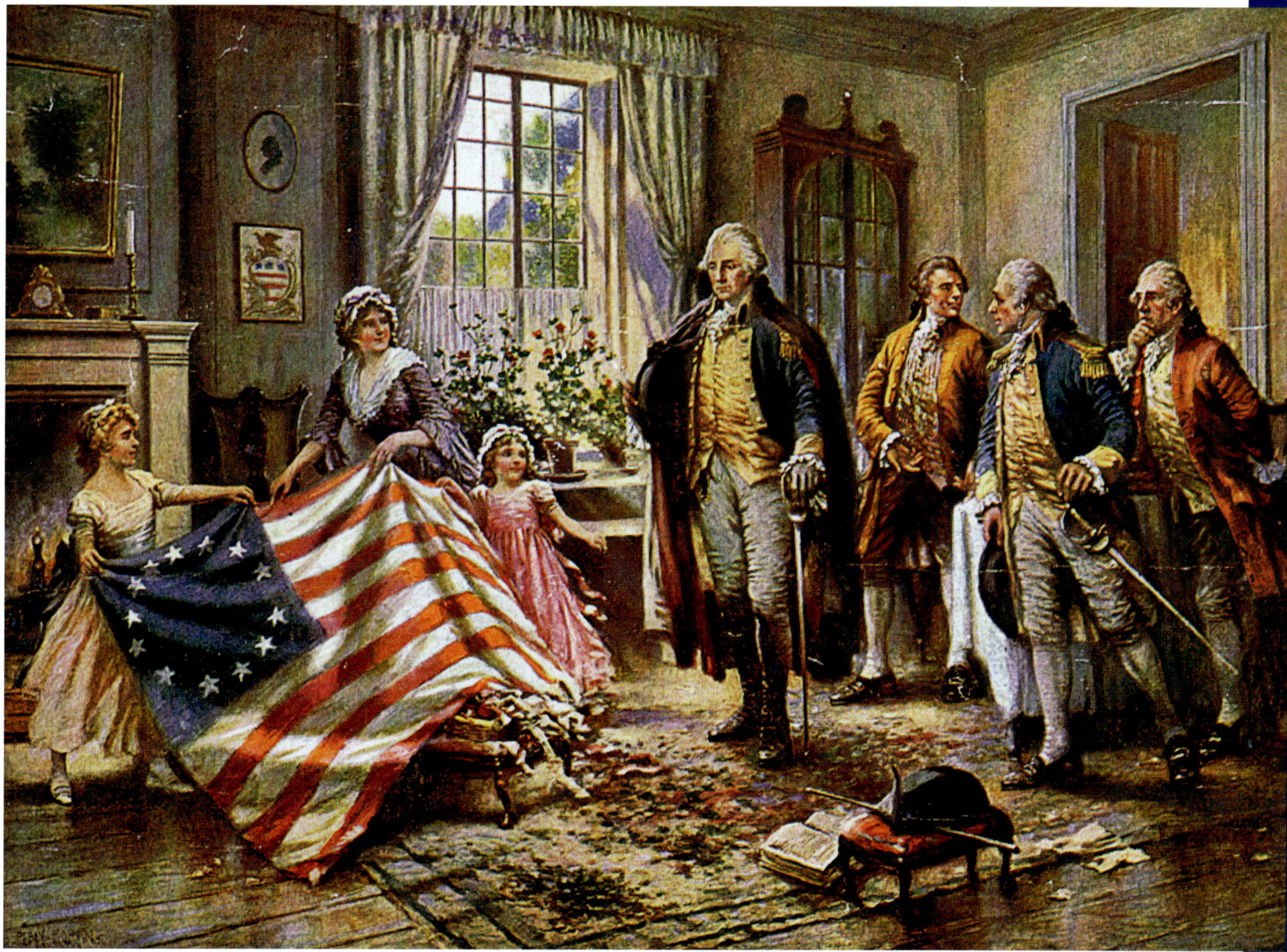

1777 THE AMERICAN FLAG

On June 14, 1777, the Flag Act was passed by the US Continental Congress, stating specific criteria for what the American flag must feature. Though it wasn't the first flag representing the United States, and the flag has evolved many times since that time, the United States now had their first official flag wholly original in design without borrowing symbolism from British heraldry. Whereas the Flag Act was specific about colors and shapes (thirteen red and white stripes, thirteen white stars on a blue background), it was not specific about the overall design. As such, there are many different legends around who was responsible for designing the early flag. One story purports that Betsy Ross, a Philadelphia seamstress, is responsible. Another tale puts forth that the design was inspired by George Washington's family coat of arms. The American flag has been iterated twenty-seven times to reflect the country's evolution. The current flag's thirteen red and white stripes represent the thirteen original colonies, and its fifty white stars on a blue background represent our fifty states. The Flag Act of 1777 became what we know of today as Flag Day, celebrated every year on June 14.

1778 THE FIRST MILITARY MANUAL

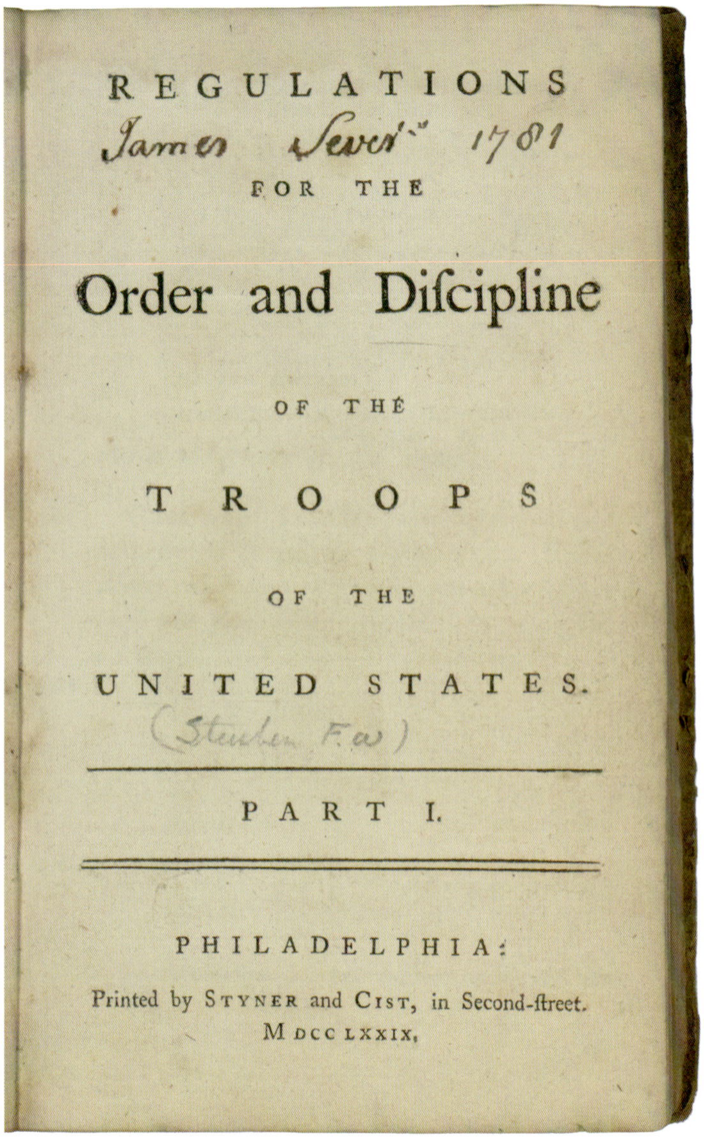
REGULATIONS
James Sever 1781
FOR THE
Order and Difcipline
OF THE
TROOPS
OF THE
UNITED STATES.
(Steuben F.W.)

PART I.

PHILADELPHIA:
Printed by STYNER and CIST, in Second-ftreet.
MDCCLXXIX.

Baron Friedrich Wilhelm Ludolf Gerhard Augustin von Steuben (1730–1794) was an aide to Frederick the Great, a monarch who made Prussia a major world power thanks to his military prowess. During the American Revolution, Steuben offered his services to George Washington and was assigned as inspector general of the Continental Army. Soon, Washington realized Steuben's military brilliance and asked him to create a standardized training handbook for the American troops. In 1778, work commenced on the *Regulations for the Order and Discipline of the Troops of the United States*, which became the first ever manual for drilling and regulating troops. Up until then, American troops had no real training. Steuben is credited with changing the course of how soldiers were instructed, impacting how they fought, and with helping the United States win the war against Great Britain. In addition to a standardized set of drills for soldiers, Steuben's manual also included rules around conduct, hygiene, and responsibilities by rank. In 1779, the US Congress officially adopted the manual's use for all US troops, and it remained in use until the War of 1812.

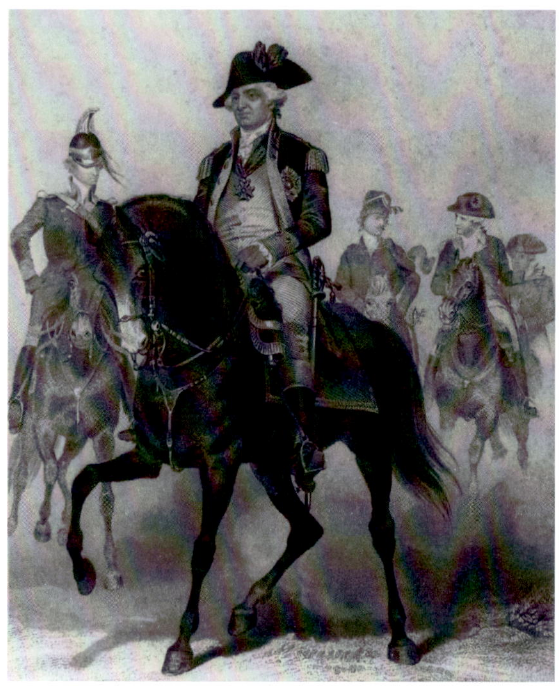

RIGHT: An engraving of Baron Steuben, c. 1857.

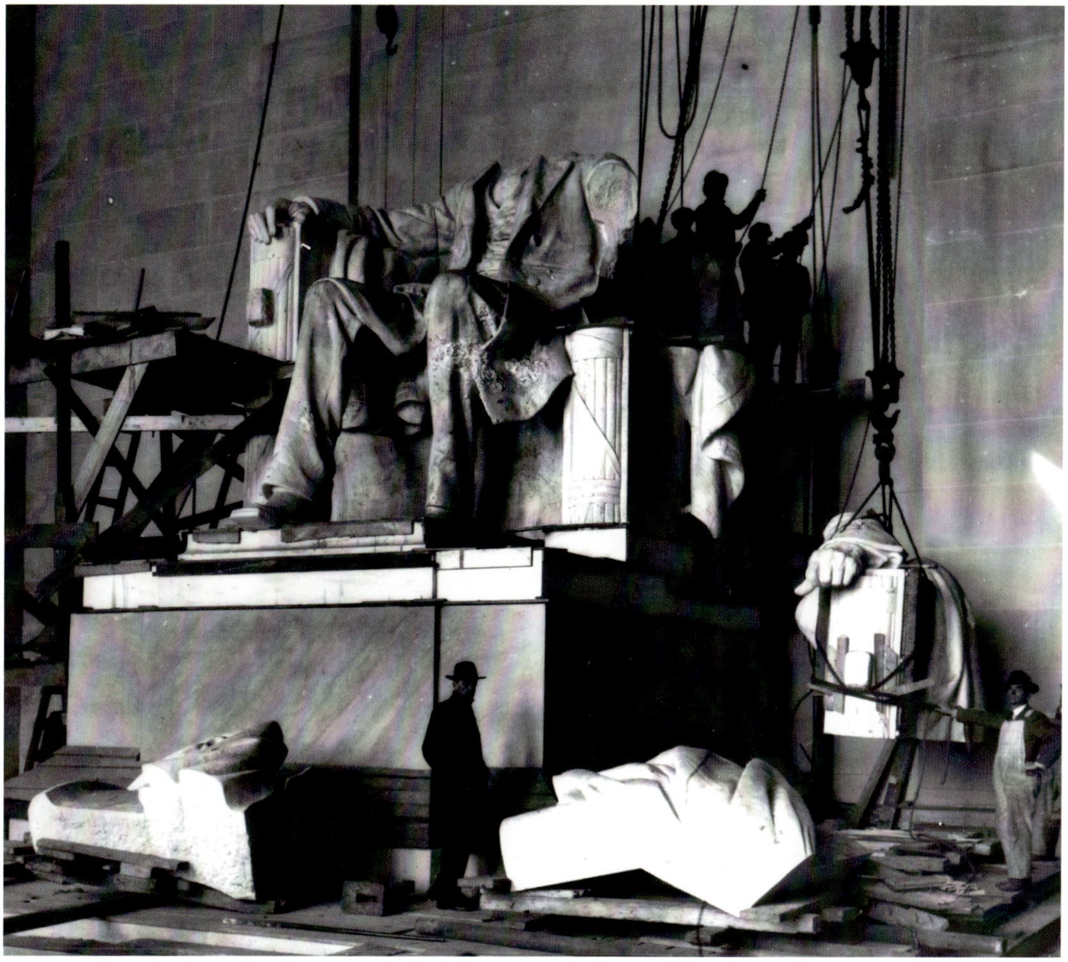

The Corps of Engineers building the Lincoln Memorial.

1779 THE US ARMY CORPS OF ENGINEERS

On March 11, 1779, the United States Congress established the US Army Corps of Engineers, an innovative way to help plan, design, and build fortifications during the Revolutionary War. In its original iteration, the Corps was composed of civilian workers and members of both the Continental and French armies. The work of the Corps was critical in securing a victory in the war. Although the Corps of Engineers disbanded after the Revolutionary War was over, and started briefly under a new name in 1794, it was reestablished as a permanent division of the US government in 1802. From 1802 to the present day, the Corps of Engineers' purpose remains to support the military in wartime, but a more enduring role is in civil construction projects, such as building the Washington Monument, the Lincoln Memorial, the base for the Statue of Liberty, and in reconstruction projects after natural disaster relief, such as the rebuilding of New Orleans after Hurricane Katrina. Perhaps a less glamorous but equally important role of the Corps lies in environmental preservation, such as coastal fortification; the improvement of national waterways; and in the construction and maintenance of structures such as canals, levees, dams, and roads.

1780 AMERICAN ACADEMY OF ARTS AND SCIENCES

The American Academy of Arts and Sciences was incorporated on May 4, 1780, in Massachusetts. Originated by Founding Fathers John Adams (1735–1826), John Hancock (1737–1793), and others who believed the new republic should honor accomplished citizens and engage them in progressing the public good, the academy was chartered "to cultivate every art and science which may tend to advance the interest, honor, dignity, and happiness of a free, independent, and virtuous people." From the onset, the academy distinguished itself by embracing both academic research and hands-on study while maintaining strict nonpartisanship and nondenominational thought. This approach allowed it to attract diverse intellectual talents and avoid the divisions that plagued other institutions of the era. The academy pioneered early efforts in data collection and analysis, particularly in medical research that focused on public health. Today, the academy continues its founding mission, serving as a prestigious honor society that convenes leaders to address national and global challenges, making it one of America's most enduring scholarly societies.

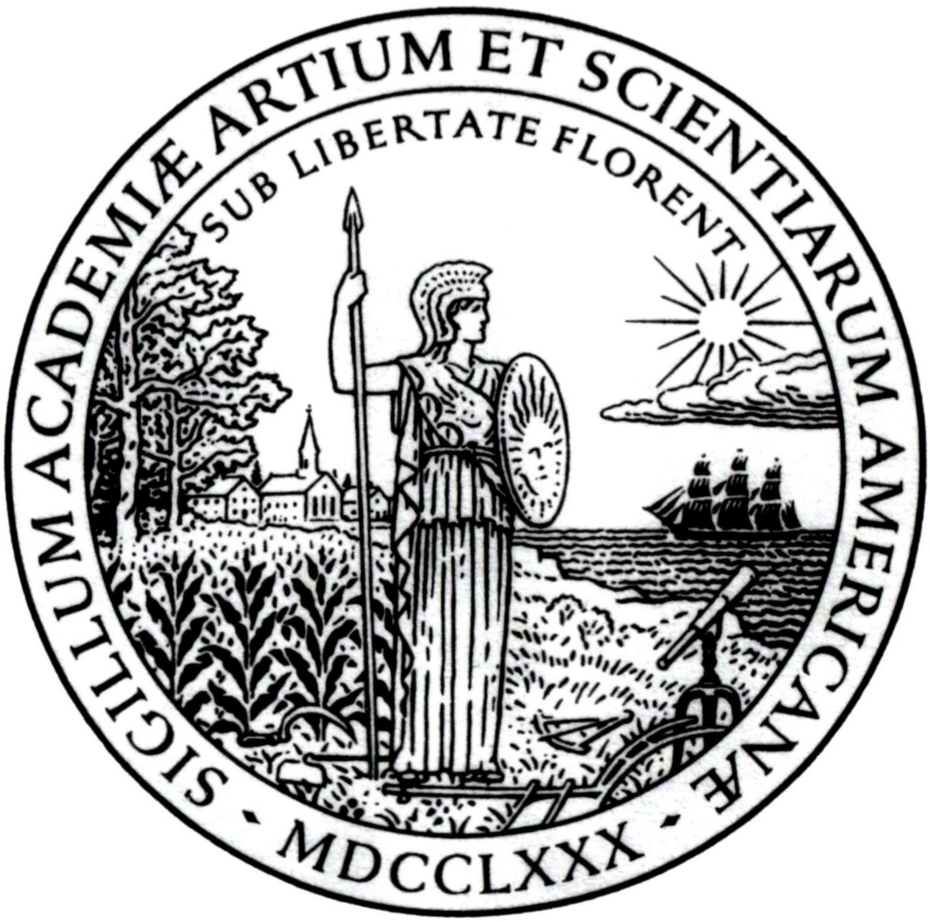

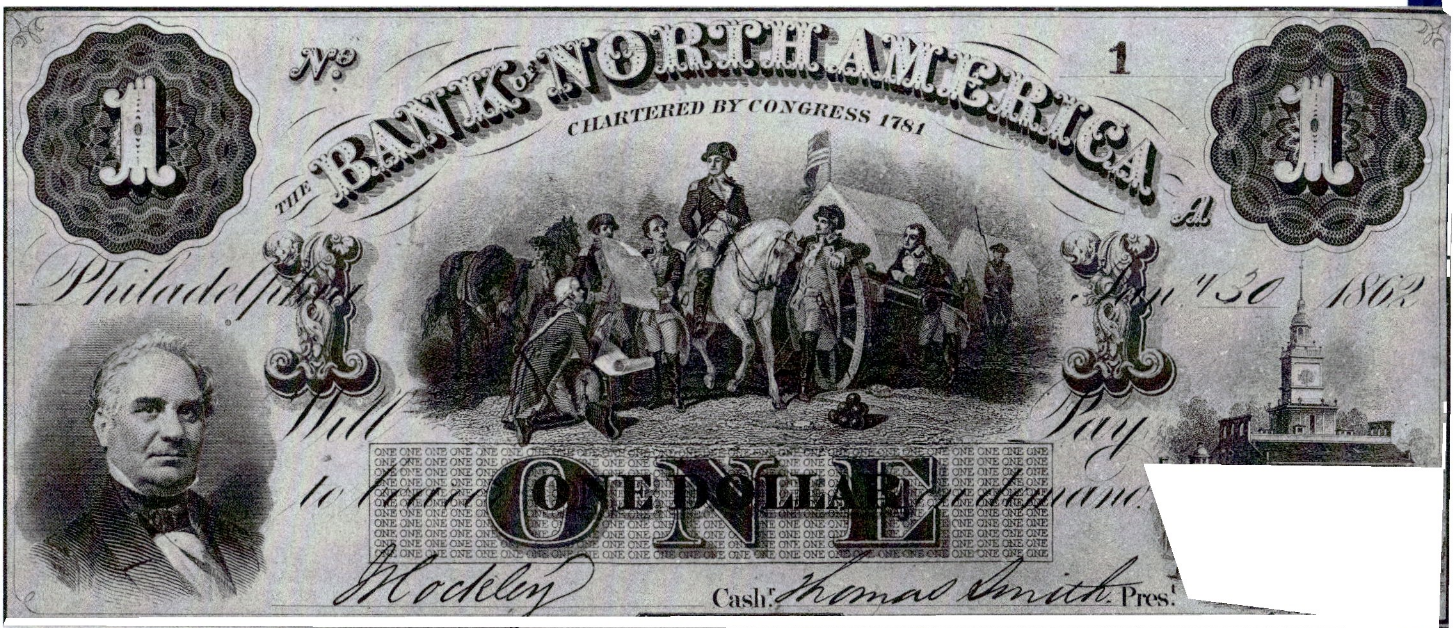

A one-dollar bill issued by the Bank of North America during the Civil War.

1781 BANK OF NORTH AMERICA

Chartered on May 26, 1781, the Bank of North America was the first commercial bank in the United States and served as the country's first de facto central bank. At the time of the bank's charter, the country's official currency was facing failure and the government struggled to pay soldiers who were fighting in the Revolutionary War. Robert Morris (1734–1806), then Superintendent of Finance, and Alexander Hamilton (1755–1804), Secretary of the Treasury, needed to find a way to attract wealthy investors to keep the economy afloat. Morris and Hamilton's groundbreaking plan to sell shares in the bank, then issue its own currency that competed with the country's troubled currency was risky but ultimately paid off—they attracted nearly one hundred Philadelphia residents along with Congressmen to buy shares in the bank. The Bank of North America opened in Philadelphia on January 7, 1782, and by 1783, several states passed laws allowing citizens to pay taxes with Bank of North America currency, essentially legitimizing it as legal tender.

1782 THE FLATBOAT

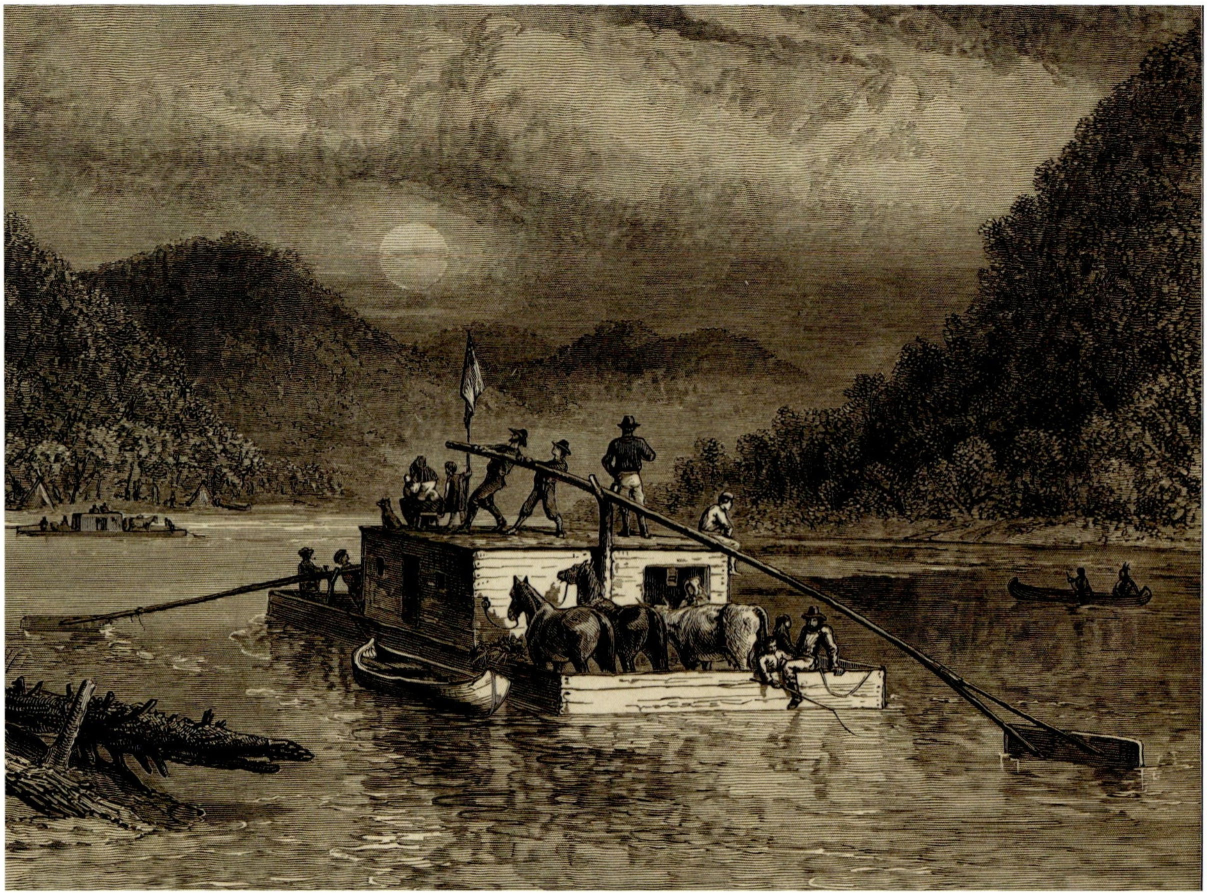

Jacob Yoder (1758–1832) is credited with building the first flatboat at Fort Red Stone, Pennsylvania, now known as Brownsville, situated on the Monongahela River, a tributary of the Ohio. Flatboats were heavy rectangular, flat-bottomed wooden structures without keels. In May 1782, Yoder piled his flatboat full of flour and successfully navigated it down the Ohio and Mississippi rivers to New Orleans, demonstrating how the waterways could be easily navigated using his boat-building prowess. Limited to about twenty feet in width and up to one hundred feet in length, flatboats were heavy and awkward to navigate, and had the disadvantage that they were unable to travel against the river current. But they were relatively easy to build and could be dismantled and salvaged for lumber once they reached their destination. Despite the flatboat's limitations, Yoder's voyage proved momentous and passenger, livestock, and cargo transportation by flatboat became one of the most important modes of water transportation of the era up until steamboats came into regular use in 1807.

1783 SELF-WINDING CLOCK

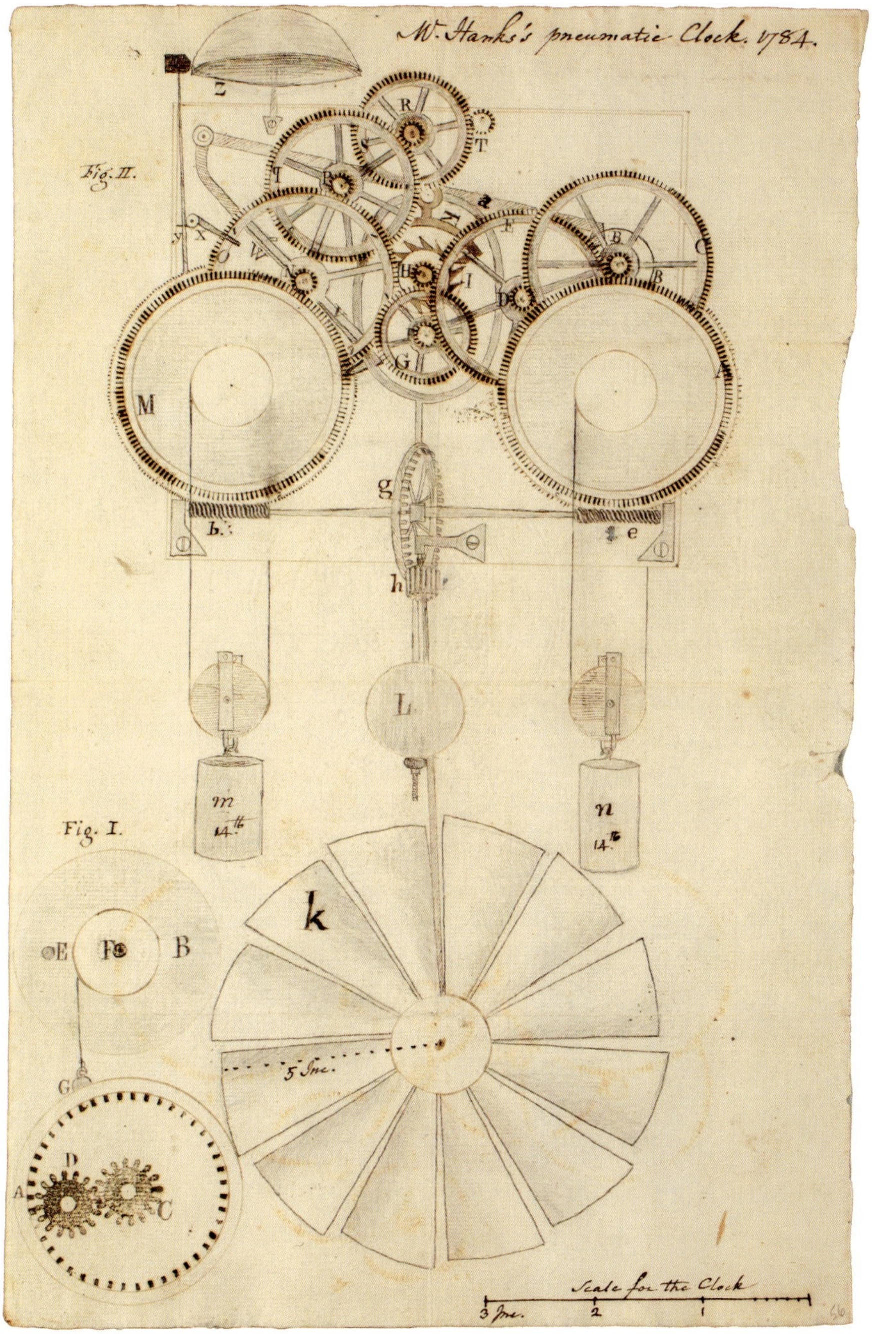

On October 6, 1783, clockmaker Benjamin Hanks (1755–1824) of Litchfield, Connecticut, obtained a fourteen-year intellectual rights patent for an apparatus that could wind itself and operate by using air pressure, essentially creating the first self-winding clock. This groundbreaking clock was said to automatically continue to wind itself up and operate until the mechanical parts wore out due to friction. Not just a technological breakthrough, self-winding clocks became an important part of the evolution of transportation and communication in United States history. The employment of these timepieces helped contribute to standard timekeeping across multiple cities and towns for railroads, manufacturing, government, and more. Though other manufacturers followed Hanks and perhaps gained more notoriety or credit for the technology (notably Charles Platt's Self-Winding Clock Company, formed in New York City in 1886), few can argue that the technology that allowed clocks to work without the need for manual winding helped launch the industrial revolution in the United States.

The original patent drawing for Benjamin Hanks's clock.

1784 DAILY NEWSPAPER

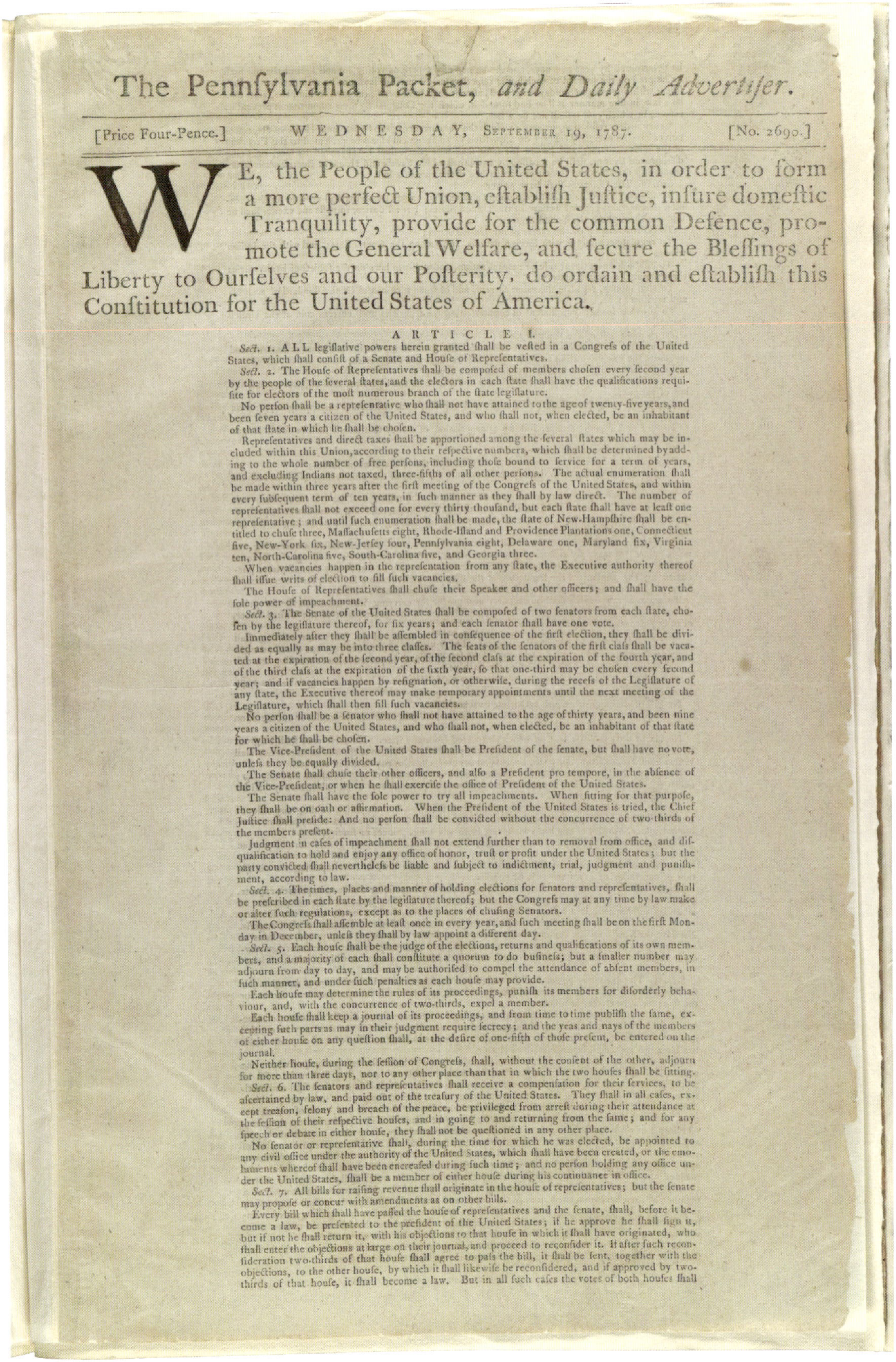

The Pennſylvania Packet, *and Daily Advertiſer.*

[Price Four-Pence.] WEDNESDAY, September 19, 1787. [No. 2690.]

WE, the People of the United States, in order to form a more perfect Union, eſtabliſh Juſtice, inſure domeſtic Tranquility, provide for the common Defence, promote the General Welfare, and ſecure the Bleſſings of Liberty to Ourſelves and our Poſterity, do ordain and eſtabliſh this Conſtitution for the United States of America.

ARTICLE I.

Sect. 1. ALL legiſlative powers herein granted ſhall be veſted in a Congreſs of the United States, which ſhall conſiſt of a Senate and Houſe of Repreſentatives.

Sect. 2. The Houſe of Repreſentatives ſhall be compoſed of members choſen every ſecond year by the people of the ſeveral ſtates, and the electors in each ſtate ſhall have the qualifications requiſite for electors of the moſt numerous branch of the ſtate legiſlature.

No perſon ſhall be a repreſentative who ſhall not have attained to the age of twenty-five years, and been ſeven years a citizen of the United States, and who ſhall not, when elected, be an inhabitant of that ſtate in which he ſhall be choſen.

Repreſentatives and direct taxes ſhall be apportioned among the ſeveral ſtates which may be included within this Union, according to their reſpective numbers, which ſhall be determined by adding to the whole number of free perſons, including thoſe bound to ſervice for a term of years, and excluding Indians not taxed, three-fifths of all other perſons. The actual enumeration ſhall be made within three years after the firſt meeting of the Congreſs of the United States, and within every ſubſequent term of ten years, in ſuch manner as they ſhall by law direct. The number of repreſentatives ſhall not exceed one for every thirty thouſand, but each ſtate ſhall have at leaſt one repreſentative; and until ſuch enumeration ſhall be made, the ſtate of New-Hampſhire ſhall be entitled to chuſe three, Maſſachuſetts eight, Rhode-Iſland and Providence Plantations one, Connecticut five, New-York ſix, New-Jerſey four, Pennſylvania eight, Delaware one, Maryland ſix, Virginia ten, North-Carolina five, South-Carolina five, and Georgia three.

When vacancies happen in the repreſentation from any ſtate, the Executive authority thereof ſhall iſſue writs of election to fill ſuch vacancies.

The Houſe of Repreſentatives ſhall chuſe their Speaker and other officers; and ſhall have the ſole power of impeachment.

Sect. 3. The Senate of the United States ſhall be compoſed of two ſenators from each ſtate, choſen by the legiſlature thereof, for ſix years; and each ſenator ſhall have one vote.

Immediately after they ſhall be aſſembled in conſequence of the firſt election, they ſhall be divided as equally as may be into three claſſes. The ſeats of the ſenators of the firſt claſs ſhall be vacated at the expiration of the ſecond year, of the ſecond claſs at the expiration of the fourth year, and of the third claſs at the expiration of the ſixth year, ſo that one-third may be choſen every ſecond year; and if vacancies happen by reſignation, or otherwiſe, during the receſs of the Legiſlature of any ſtate, the Executive thereof may make temporary appointments until the next meeting of the Legiſlature, which ſhall then fill ſuch vacancies.

No perſon ſhall be a ſenator who ſhall not have attained to the age of thirty years, and been nine years a citizen of the United States, and who ſhall not, when elected, be an inhabitant of that ſtate for which he ſhall be choſen.

The Vice-Preſident of the United States ſhall be Preſident of the ſenate, but ſhall have no vote, unleſs they be equally divided.

The Senate ſhall chuſe their other officers, and alſo a Preſident pro tempore, in the abſence of the Vice-Preſident, or when he ſhall exerciſe the office of Preſident of the United States.

The Senate ſhall have the ſole power to try all impeachments. When ſitting for that purpoſe, they ſhall be on oath or affirmation. When the Preſident of the United States is tried, the Chief Juſtice ſhall preſide: And no perſon ſhall be convicted without the concurrence of two-thirds of the members preſent.

Judgment in caſes of impeachment ſhall not extend further than to removal from office, and diſqualification to hold and enjoy any office of honor, truſt or profit under the United States; but the party convicted ſhall nevertheleſs be liable and ſubject to indictment, trial, judgment and puniſhment, according to law.

Sect. 4. The times, places and manner of holding elections for ſenators and repreſentatives, ſhall be preſcribed in each ſtate by the legiſlature thereof; but the Congreſs may at any time by law make or alter ſuch regulations, except as to the places of chuſing Senators.

The Congreſs ſhall aſſemble at leaſt once in every year, and ſuch meeting ſhall be on the firſt Monday in December, unleſs they ſhall by law appoint a different day.

Sect. 5. Each houſe ſhall be the judge of the elections, returns and qualifications of its own members, and a majority of each ſhall conſtitute a quorum to do buſineſs; but a ſmaller number may adjourn from day to day, and may be authoriſed to compel the attendance of abſent members, in ſuch manner, and under ſuch penalties as each houſe may provide.

Each houſe may determine the rules of its proceedings, puniſh its members for diſorderly behaviour, and, with the concurrence of two-thirds, expel a member.

Each houſe ſhall keep a journal of its proceedings, and from time to time publiſh the ſame, excepting ſuch parts as may in their judgment require ſecrecy; and the yeas and nays of the members of either houſe on any queſtion ſhall, at the deſire of one-fifth of thoſe preſent, be entered on the journal.

Neither houſe, during the ſeſſion of Congreſs, ſhall, without the conſent of the other, adjourn for more than three days, nor to any other place than that in which the two houſes ſhall be ſitting.

Sect. 6. The ſenators and repreſentatives ſhall receive a compenſation for their ſervices, to be aſcertained by law, and paid out of the treaſury of the United States. They ſhall in all caſes, except treaſon, felony and breach of the peace, be privileged from arreſt during their attendance at the ſeſſion of their reſpective houſes, and in going to and returning from the ſame; and for any ſpeech or debate in either houſe, they ſhall not be queſtioned in any other place.

No ſenator or repreſentative ſhall, during the time for which he was elected, be appointed to any civil office under the authority of the United States, which ſhall have been created, or the emoluments whereof ſhall have been encreaſed during ſuch time; and no perſon holding any office under the United States, ſhall be a member of either houſe during his continuance in office.

Sect. 7. All bills for raiſing revenue ſhall originate in the houſe of repreſentatives; but the ſenate may propoſe or concur with amendments as on other bills.

Every bill which ſhall have paſſed the houſe of repreſentatives and the ſenate, ſhall, before it become a law, be preſented to the preſident of the United States; if he approve he ſhall ſign it, but if not he ſhall return it, with his objections to that houſe in which it ſhall have originated, who ſhall enter the objections at large on their journal, and proceed to reconſider it. If after ſuch reconſideration two-thirds of that houſe ſhall agree to paſs the bill, it ſhall be ſent, together with the objections, to the other houſe, by which it ſhall likewiſe be reconſidered, and if approved by two-thirds of that houſe, it ſhall become a law. But in all ſuch caſes the votes of both houſes ſhall

According to the Library of Congress, the *Pennsylvania Packet and General Advertiser* was the first-ever daily newspaper published in America in Philadelphia, from September 1784 to 1790. The paper was originally founded by John Dunlap (1747–1812) as a weekly paper in 1771 to expose colonial life in America. David C. Claypoole (1757–1849) became a partner in the venture, which helped spur it into its daily incarnation. The newspaper played a pivotal role in broadcasting information during the Revolutionary War and beyond. It covered political events, debates, and announcements and served as a platform for the exchange of ideas during a crucial period in American history. You can still find facsimiles of old issues that listed enslaved people for sale in the Packet's classifieds. Despite that dark history, the *Pennsylvania Packet* is often considered a foundation of American journalism, reflecting the times and contributing to the formation of public opinion in the fledgling nation. One of the distinguishing facts is that, on September 19, 1787, the *Pennsylvania Packet and General Advertiser* devoted its entire issue to reprinting the newly created Constitution of the United States of America.

1784 Bifocal Lenses

Founding Father Benjamin Franklin (1706-1790) is credited with inventing "double spectacles," known today as bifocals. He described his invention, which allowed him to see both up close and at a distance using the same pair of glasses, in a letter to a friend. To accomplish this, legend has it that Franklin asked his optician to slice the lenses of two pairs of glasses in half, fuse them in the middle, then mount them in a single glasses frame. Bifocal lenses are just one of Franklin's long list of innovations for which he sought no monetary gain.

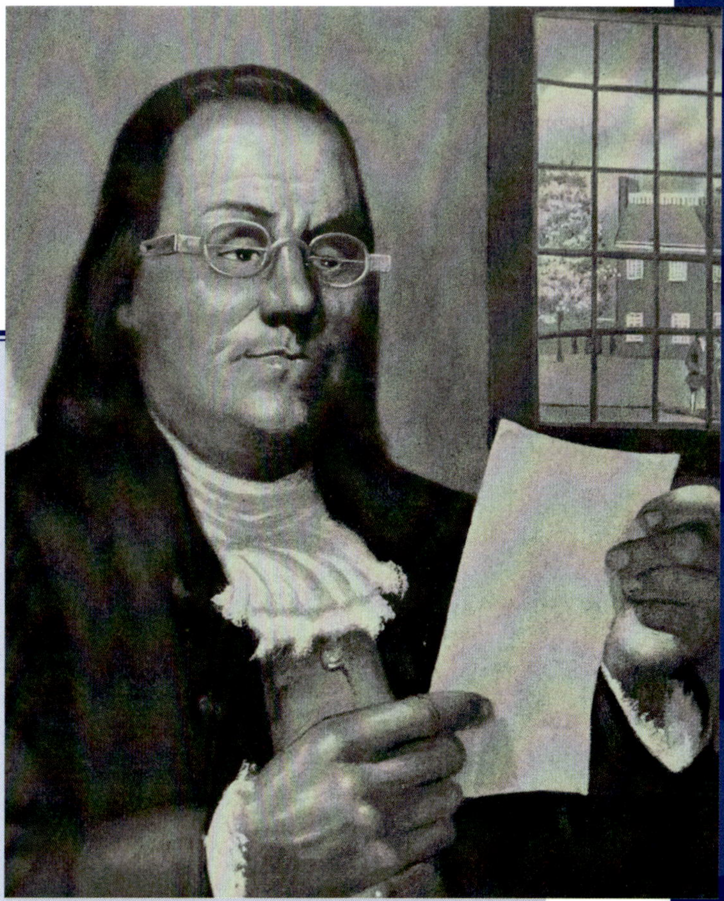

1785 FEDERAL STYLE ARCHITECTURE

Developed from Georgian architecture with elements drawn from Roman monuments, this style of architecture was an American innovation prominent after the Revolutionary War from 1785 to around 1820. Key figures associated with the movement include Founding Father Thomas Jefferson (1743–1826) and architect Benjamin Latrobe (1764–1820). Marked by balance and symmetry in its design, this style features elements such as rotundas, decorative columns, brick laid in a Flemish bond pattern, wrought iron handrails, and low stoops. Associated with the fledgling United States, the Federal style of architecture had philosophical ties to the concept of Rome as a powerful republic that the new country sought to emulate.

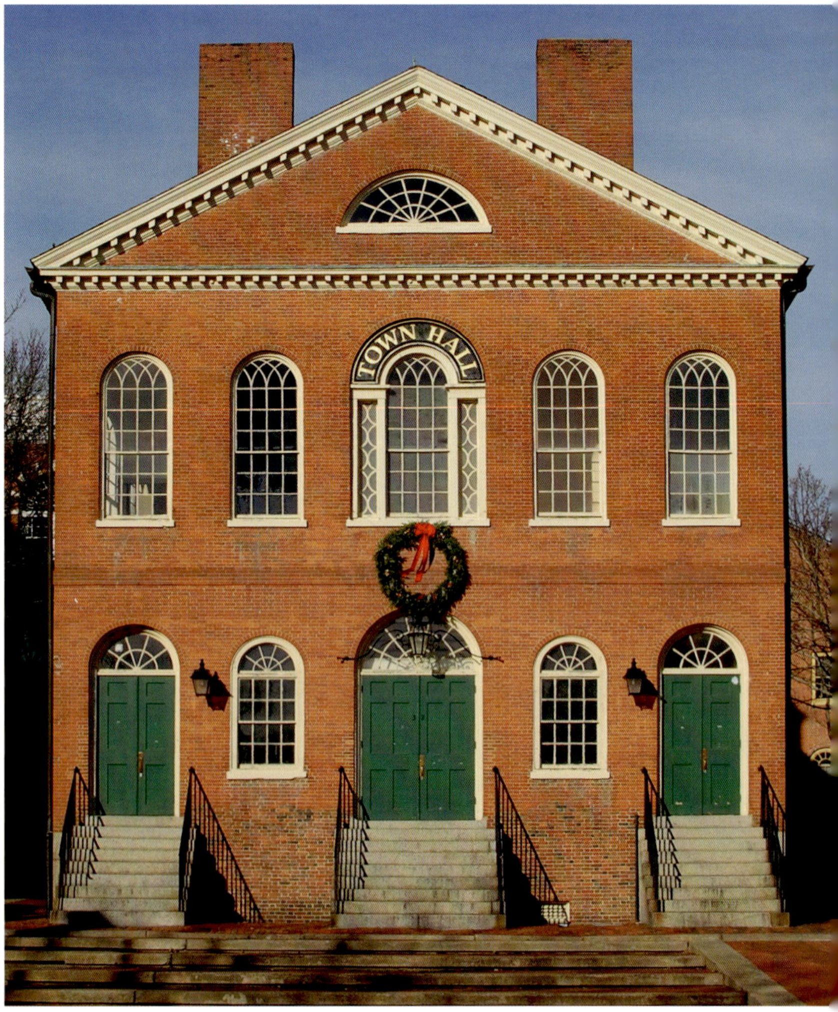

The Old Town Hall in Derby Square, Salem, Massachusetts.

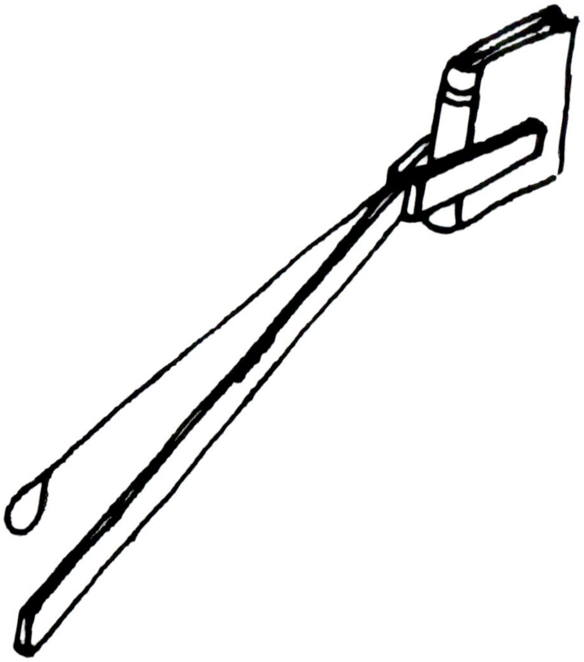

1786 LONG ARM

What was a book lover to do in the late 1700s when the book they desired to read was located on a shelf high above their head? Prolific inventor and Founding Father Benjamin Franklin (1706–1790) solved that problem with his "Long Arm." This eight-foot-long piece of wood with two grabbers on the end that were attached via a cord to a grip in the handle. Rather than mounting a bulky and dangerous ladder, a person could now use this low-profile device to select the desired book and bring it down from or return it to a high shelf with little effort. Franklin invented many things to better people's lives—but he never sought a patent for a single one. Instead, Franklin considered his innovations gifts to the public.

1787 THE FEDERALIST PAPERS

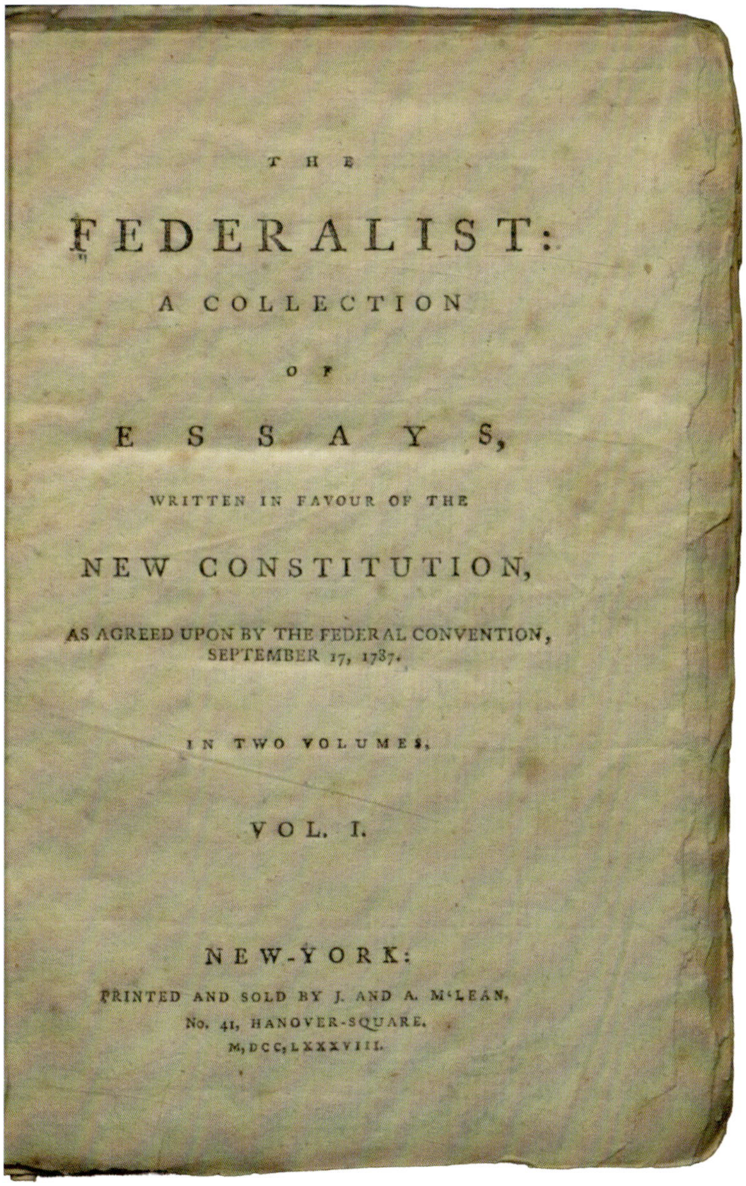

THE

FEDERALIST:

A COLLECTION

OF

ESSAYS,

WRITTEN IN FAVOUR OF THE

NEW CONSTITUTION,

AS AGREED UPON BY THE FEDERAL CONVENTION, SEPTEMBER 17, 1787.

IN TWO VOLUMES.

VOL. I.

NEW-YORK:

PRINTED AND SOLD BY J. AND A. M'LEAN,
No. 41, HANOVER-SQUARE,
M,DCC,LXXXVIII.

Second perhaps only to the Declaration of Independence, the US Constitution is arguably the most important American innovation in government in the history of the country, but it was not an easy route to ratification. The Federalist Papers, a series of eighty-five essays pressing the citizens of New York to ratify the new Constitution, are considered one of the most important sources for interpreting and understanding the original intent of the document. The Federalist Papers were published under the pen name "Publius," but the real authors were Alexander Hamilton (1755–1804), James Madison (1751–1836), and John Jay (1745–1829). The essays appeared in two New York newspapers between 1787 and 1788. The brilliance of these papers lies in their excellence in explaining parts of the Constitution in detail, how it would affect the American people, and why it was superior to the existing Articles of Confederation. Because they were written by two members of the Constitutional Convention, The Federalist Papers are still used today to understand and interpret the original intention of the constitution.

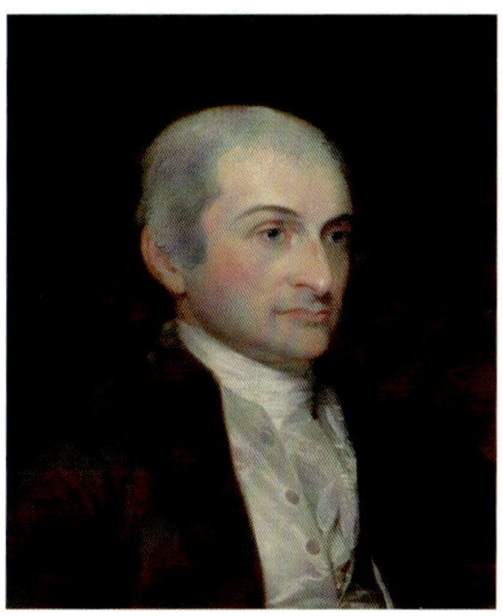
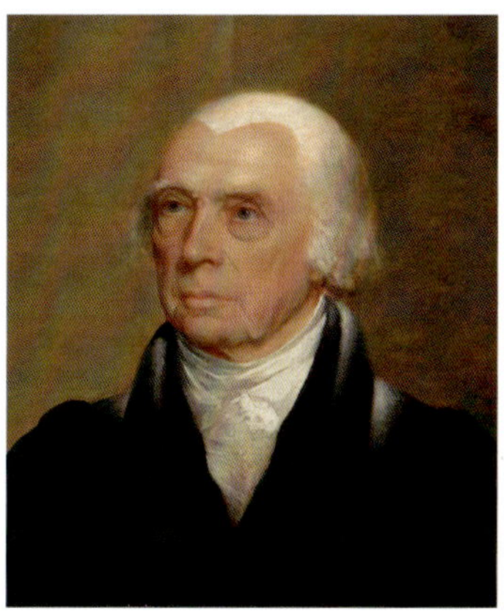
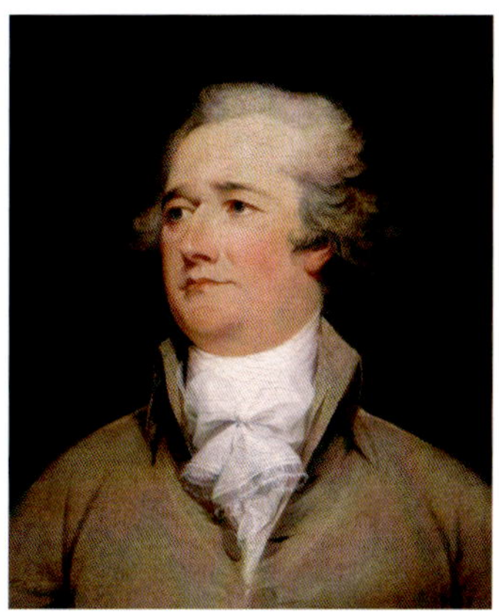

Authors of the papers (l-r): John Jay, James Madison, and Alexander Hamilton.

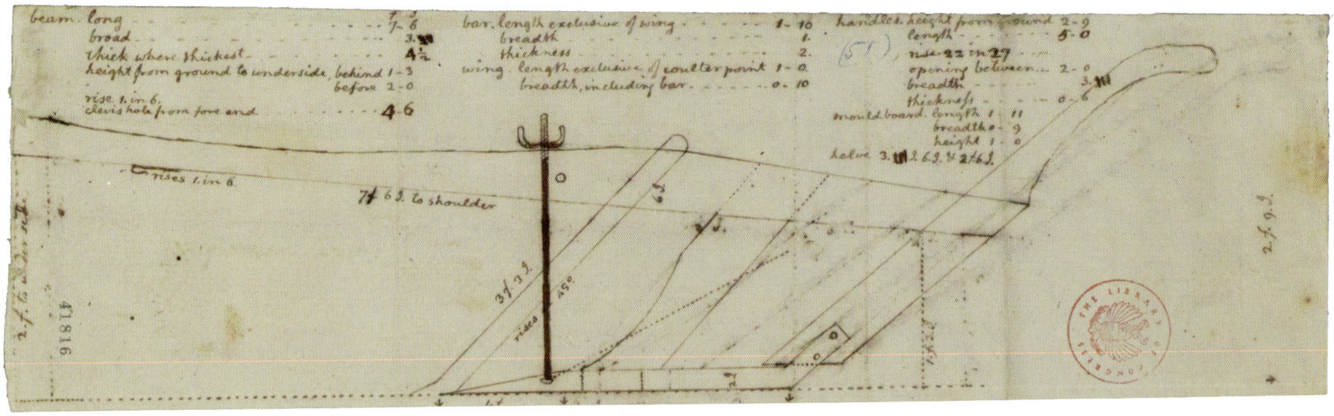

Thomas Jefferson's original sketch for the plow.

1788 MOLDBOARD PLOW

Not only was Thomas Jefferson (1743–1826) a Founding Father of our nation, but he was also responsible for numerous innovations that have stood the test of time; the moldboard plow is among his most important in the agriculture sector. While serving as minister to France, Jefferson had the opportunity to observe some European farm technology, including their plows. This inspired him to, in 1788, write up plans for a superior plow design for farming that incorporated a curved plate called a moldboard. This moldboard would, when run through the soil, lift the earth and turn it 180 degrees, inverting and aerating it and helping improve drainage. Jefferson's design was also considered brilliant in that it required a minimum amount of force from the farmer who was wielding it. The design was easy to duplicate without a lot of skill. Jefferson seemed exceedingly proud of his design—records show that he often boasted about its features and effectiveness—though he never applied for a patent.

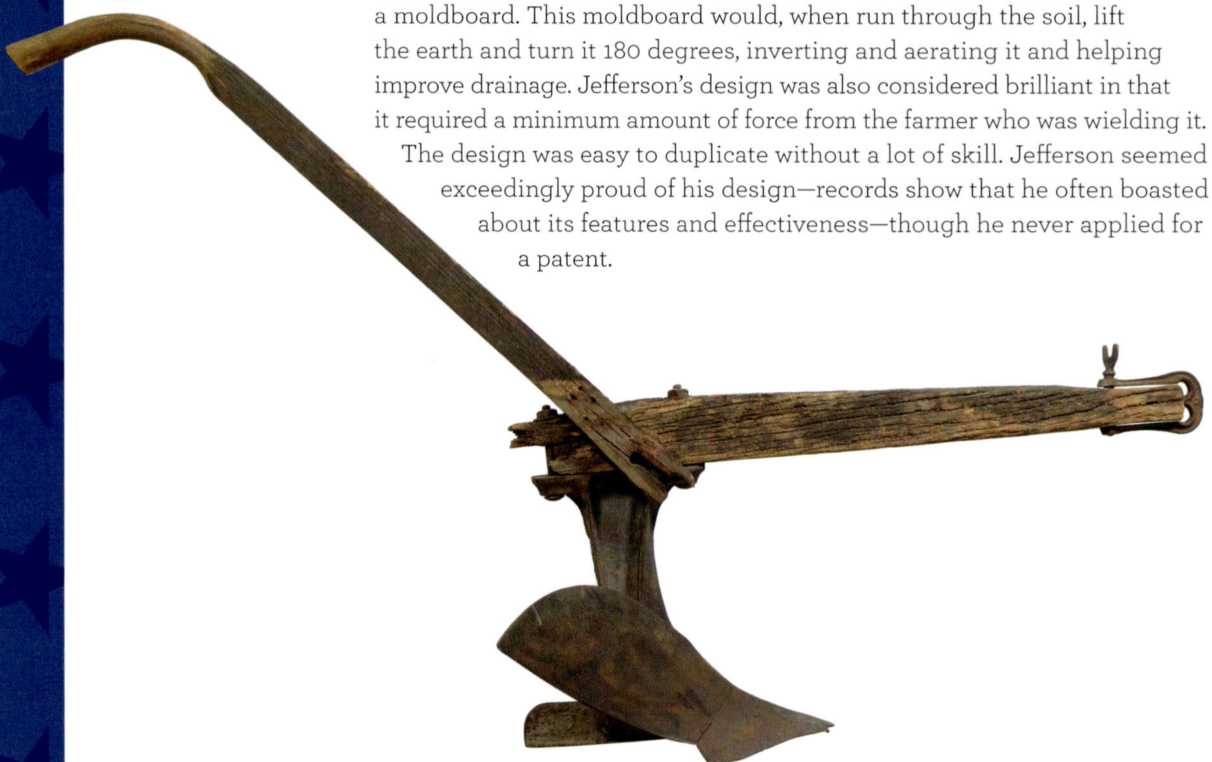

1789 THE FIRST AMERICAN NOVEL

Published in 1789, *The Power of Sympathy* by William Hill Brown (1765–1793) is widely considered to be America's first novel. It launched while America was establishing its cultural identity separate from British influence. This work, told through a series of letters, follows the tragic story of Harrington and Harriot, whose forbidden love leads to devastating consequences when they discover that they are half-siblings. Brown wrote this cautionary tale that warned against seduction and the dangers of unchecked passion, reflecting the moral concerns of post–Revolutionary War society. The work drew inspiration from contemporary scandals, particularly the Whitman family tragedy in Massachusetts. *The Power of Sympathy* established themes that would echo through centuries of American fiction: moral directives, social criticism, and the exploration of virtue versus vice. The novel's epistolary format and themes of seduction influenced later American novels, helping to create a distinctly American literary tradition that balanced entertainment with moral education.

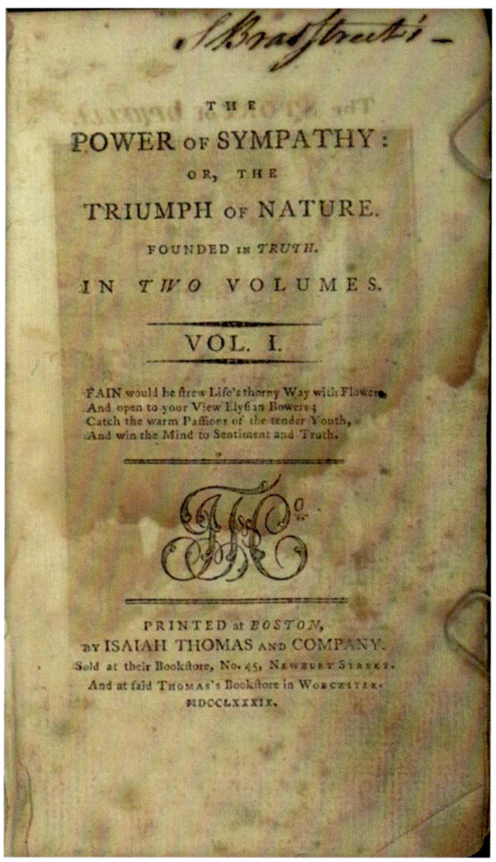

S. Bradstreet

THE POWER OF SYMPATHY: OR, THE TRIUMPH OF NATURE.

FOUNDED IN *TRUTH.*

IN *TWO* VOLUMES.

VOL. I.

FAIN would he ſtrew Life's thorny Way with Flowers,
And open to your View Elyſian Bowers;
Catch the warm Paſſions of the tender Youth,
And win the Mind to Sentiment and Truth.

PRINTED at *BOSTON,*
BY ISAIAH THOMAS AND COMPANY.
Sold at their Bookſtore, No. 45, NEWBURY STREET.
And at ſaid THOMAS's Bookſtore in WORCESTER.
MDCCLXXXIX.

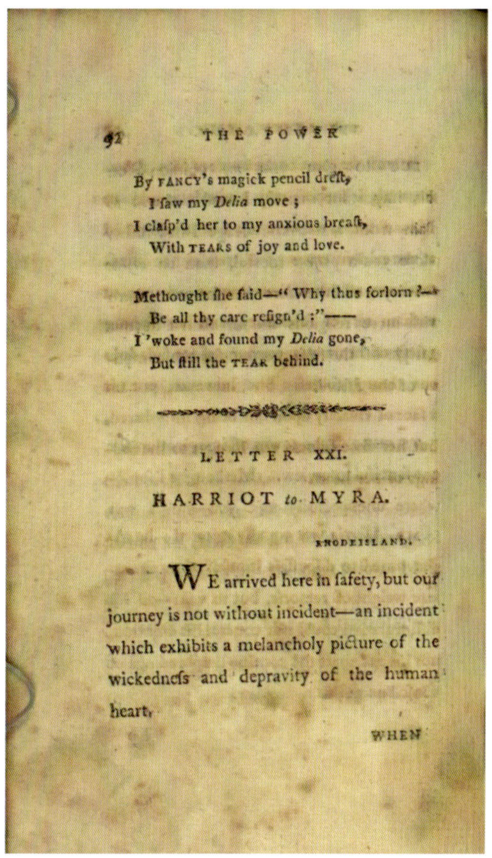

92 THE POWER

By FANCY's magick pencil dreſt,
I ſaw my *Delia* move;
I claſp'd her to my anxious breaſt,
With TEARS of joy and love.

Methought ſhe ſaid—"Why thus forlorn?—
Be all thy care reſign'd:"——
I 'woke and found my *Delia* gone,
But ſtill the TEAR behind.

LETTER XXI.

HARRIOT *to* MYRA.

RHODEISLAND.

WE arrived here in ſafety, but our journey is not without incident—an incident which exhibits a melancholy picture of the wickedneſs and depravity of the human heart.

WHEN

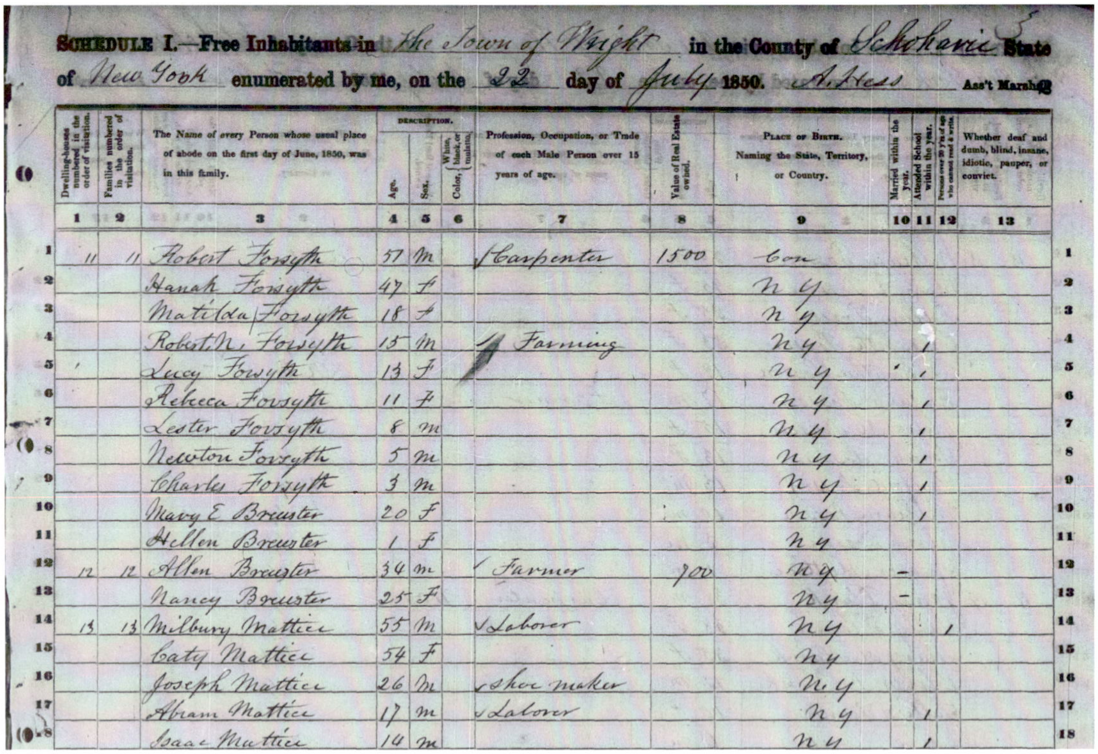

SCHEDULE I.—Free Inhabitants in The Town of Wright in the County of Schoharie State of New York enumerated by me, on the 22 day of July 1850. A. Hess Ass't Marshal.

Dwelling-houses numbered in the order of visitation.	Families numbered in the order of visitation.	The Name of every Person whose usual place of abode on the first day of June, 1850, was in this family.	Age.	Sex.	Color, White, black, or mulatto.	Profession, Occupation, or Trade of each Male Person over 15 years of age.	Value of Real Estate owned.	Place of Birth. Naming the State, Territory, or Country.	Married within the year.	Attended School within the year.	Persons over 20 y'rs of age who cannot read & write.	Whether deaf and dumb, blind, insane, idiotic, pauper, or convict.	
1	2	3	4	5	6	7	8	9	10	11	12	13	
11	11	Robert Forsyth	57	M		Carpenter	1500	Con					1
		Hanah Forsyth	47	F				N Y					2
		Matilda Forsyth	18	F				N Y					3
		Robert N. Forsyth	15	M		Farming		N Y		/			4
		Lucy Forsyth	13	F				N Y		/			5
		Rebecca Forsyth	11	F				N Y		/			6
		Lester Forsyth	8	M				N Y		/			7
		Newton Forsyth	5	M				N Y		/			8
		Charles Forsyth	3	M				N Y		/			9
		Mary E Brewster	20	F				N Y		/			10
		Hellen Brewster	1	F				N Y					11
12	12	Allen Brewster	34	M		Farmer	700	N Y	-				12
		Nancy Brewster	25	F				N Y	-				13
13	13	Milbury Mattice	55	M		Laborer		N Y			/		14
		Caty Mattice	54	F				N Y					15
		Joseph Mattice	26	M		shoe maker		N Y					16
		Abram Mattice	17	M		Laborer		N Y		/			17
		Isaac Mattice	14	M				N Y		/			18

1790 THE US CENSUS

August 2, 1790, was the first official counting of the population of the United States of America, then just the thirteen original colonies. This census, mandated in the newly ratified Constitution, was conducted by US marshals and considered one of the most important accomplishments of George Washington's first term as president. The attempt to get an accurate count of the country's population was seen as an innovation at the time, as European colonist nations shied away from the practice due to power struggles between the nobles and the central governments. In the beginning, only the heads of households and a general accounting of others, including enslaved people, was recorded by census takers. As time elapsed, the data became more and more detailed and eventually mechanized. The purpose of the census also changed over time. The original intention was to determine how many governmental representatives a state was entitled to according to its population. Though that is still a key purpose of the census in modern times, today it is also used for distribution of aid and drawing voting districts in municipalities.

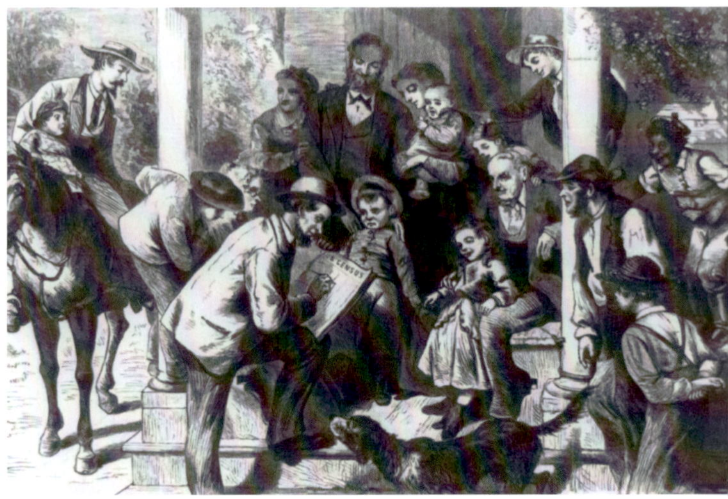

TOP: Detail of an 1850 census record from the town of Wright, New York. **RIGHT:** An illustration of a census taker doing his job from *Harpers Weekly*, 1870.

1791 US BILL OF RIGHTS IS RATIFIED

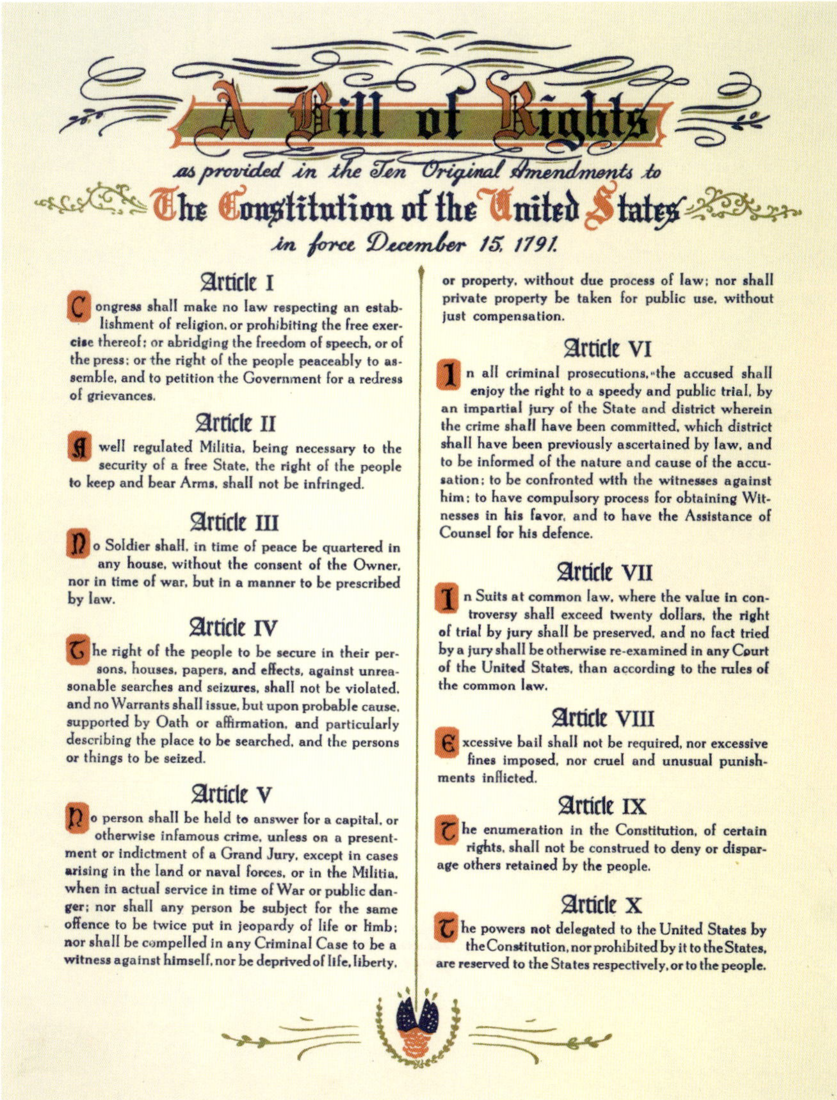

A Bill of Rights

as provided in the Ten Original Amendments to

The Constitution of the United States

in force December 15, 1791.

Article I

Congress shall make no law respecting an establishment of religion, or prohibiting the free exercise thereof; or abridging the freedom of speech, or of the press; or the right of the people peaceably to assemble, and to petition the Government for a redress of grievances.

Article II

A well regulated Militia, being necessary to the security of a free State, the right of the people to keep and bear Arms, shall not be infringed.

Article III

No Soldier shall, in time of peace be quartered in any house, without the consent of the Owner, nor in time of war, but in a manner to be prescribed by law.

Article IV

The right of the people to be secure in their persons, houses, papers, and effects, against unreasonable searches and seizures, shall not be violated, and no Warrants shall issue, but upon probable cause, supported by Oath or affirmation, and particularly describing the place to be searched, and the persons or things to be seized.

Article V

No person shall be held to answer for a capital, or otherwise infamous crime, unless on a presentment or indictment of a Grand Jury, except in cases arising in the land or naval forces, or in the Militia, when in actual service in time of War or public danger; nor shall any person be subject for the same offence to be twice put in jeopardy of life or limb; nor shall be compelled in any Criminal Case to be a witness against himself, nor be deprived of life, liberty, or property, without due process of law; nor shall private property be taken for public use, without just compensation.

Article VI

In all criminal prosecutions, the accused shall enjoy the right to a speedy and public trial, by an impartial jury of the State and district wherein the crime shall have been committed, which district shall have been previously ascertained by law, and to be informed of the nature and cause of the accusation; to be confronted with the witnesses against him; to have compulsory process for obtaining Witnesses in his favor, and to have the Assistance of Counsel for his defence.

Article VII

In Suits at common law, where the value in controversy shall exceed twenty dollars, the right of trial by jury shall be preserved, and no fact tried by a jury shall be otherwise re-examined in any Court of the United States, than according to the rules of the common law.

Article VIII

Excessive bail shall not be required, nor excessive fines imposed, nor cruel and unusual punishments inflicted.

Article IX

The enumeration in the Constitution, of certain rights, shall not be construed to deny or disparage others retained by the people.

Article X

The powers not delegated to the United States by the Constitution, nor prohibited by it to the States, are reserved to the States respectively, or to the people.

A copy of the Bill of Rights, c. 1950.

Creating the most influential democracy in the history of the world was not a straightforward progression. The grueling process of debating and eventually approving the Constitution was just the first step in the process. What came next was to outline limits on governmental power and to protect the freedoms of the citizens of the fledgling country. Certain rights for citizens in relation to their government needed to be spelled out. James Madison (1751–1836) tackled much of the job of creating the first ten amendments to the Constitution, which became known as the Bill of Rights. These amendments guarantee certain essential rights and civil liberties to the American people, for example: the right to free speech, the right to bear arms, freedom from any government-established religion, the free exercise of one's religion, freedom of assembly, due process of law, and more. After much debate, the Bill of Rights was ratified on December 15, 1791 and after that date was considered part of the Constitution.

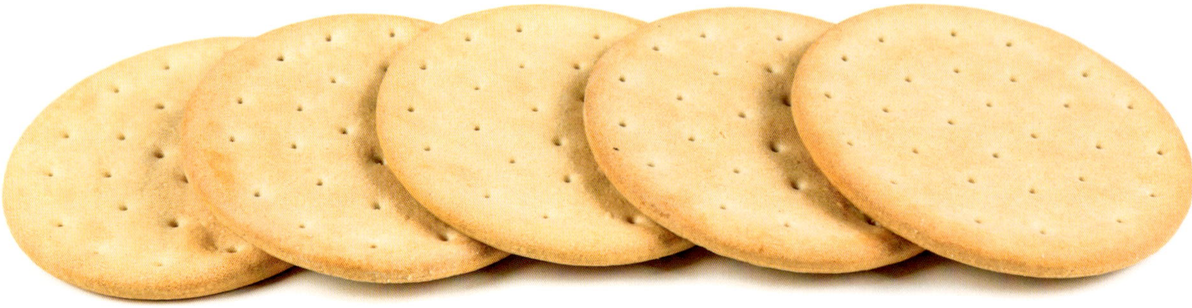

1792 A SNACK CRACKER EMPIRE IS BORN

John Pearson (birth and death dates unknown), of Pearson's Bakery, the first commercial bakery in America, invented pilot bread. Similar to hard tack, a cracker with a long shelf life specifically created for sailors to bring on their journeys, pilot bread had a more palatable flavor and a slightly softer texture. Pearson's bakery, located in Newburyport, Massachusetts, opened its doors in 1792. It quickly gained a reputation for delicious baked goods. Pearson's sons joined the business and helped expand it to open more shops locally and in other states and supplying hotels up and down the eastern seaboard. In 1889, the business, now called Pearson and Sons, was sold; its owners merged the bakery with six others they purchased to create the New York Biscuit Company. After another merger with a separate bakery conglomerate, the company ultimately became Nabisco (a portmanteau of the earlier name, National Biscuit Company), a company that is still around today. For a time, Nabisco marketed Pearson's original innovation under the name Crown Pilot Cracker.

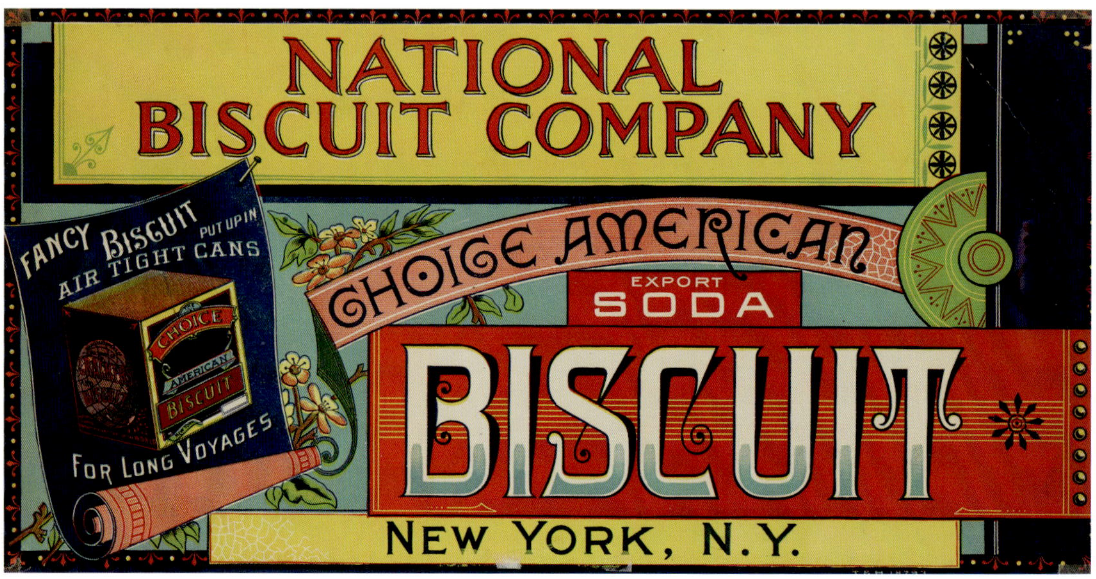

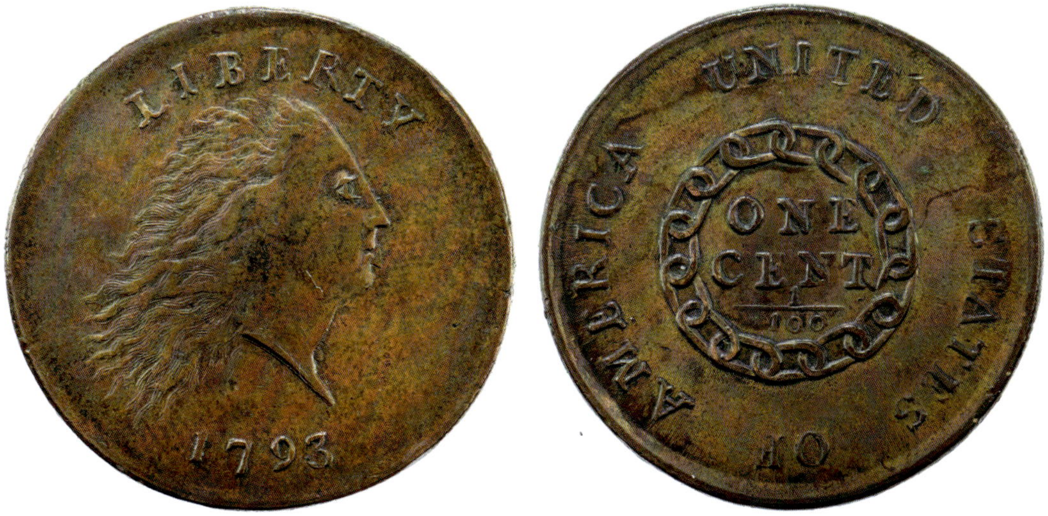

1793 FIRST US COINS IN CIRCULATION

After many years of circulating and accepting different currencies throughout the colonies—British, Spanish, German, and even colony-specific coins—the Coinage Act of 1792 established a national mint in Philadelphia and that a national currency would now be the standard. Precious metals were selected and denominations were determined as follows: copper for the half cent and cent; silver for the half dime, dime, quarter, half dollar, and dollar; and gold for the quarter eagle ($2.50), half eagle ($5), and eagle ($10). The mint delivered the nation's first circulating coins—just over 11,000 copper cents—on March 1, 1793, which was a momentous day for the young country. But the transition to a standard national currency was not an easy one. At first, it was difficult to manufacture enough coins to get into circulation. The high cost of materials was also a factor. During this time, foreign currency was still accepted until the Coinage Act of 1857, in which the US Congress finally banned it as legal tender for good.

1794 THE COTTON GIN

After visiting a friend in the American South, Eli Whitney (1765–1825) learned how labor intensive deseeding cotton could be. Eager to help, he invented a machine to comb and deseed bolls of cotton by pulling them through a set of wire teeth mounted on a revolving cylinder. He received a patent for the cotton gin in 1794. His invention revolutionized the cotton industry and the rise of it as the main cash crop in the South. Some scholars argue that Whitney's cotton gin has even further ramifications—that it helped institutionalize slavery in America. True, it made the work easier to do, but thanks to the cotton gin, cultivating cotton became so profitable that it exponentially increased the demand for labor and land to grow the crop. Despite Whitney's efforts to license his invention, it never made him wealthy. Resentful of the high licensing fees Whitney was asking, many farmers built counterfeit versions of their own cotton gins, ignoring patent law, which proved difficult for Whitney to fight in those days. Still, Whitney's cotton gin holds a place in history as one of the most significant inventions of all time.

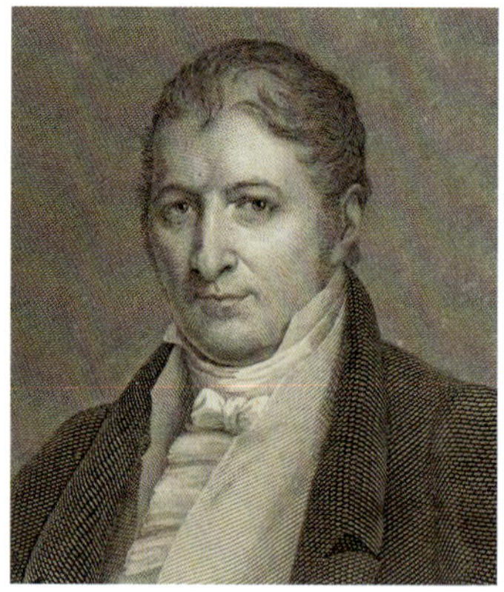

LEFT: Eli Whitney.
BELOW: The original gin.

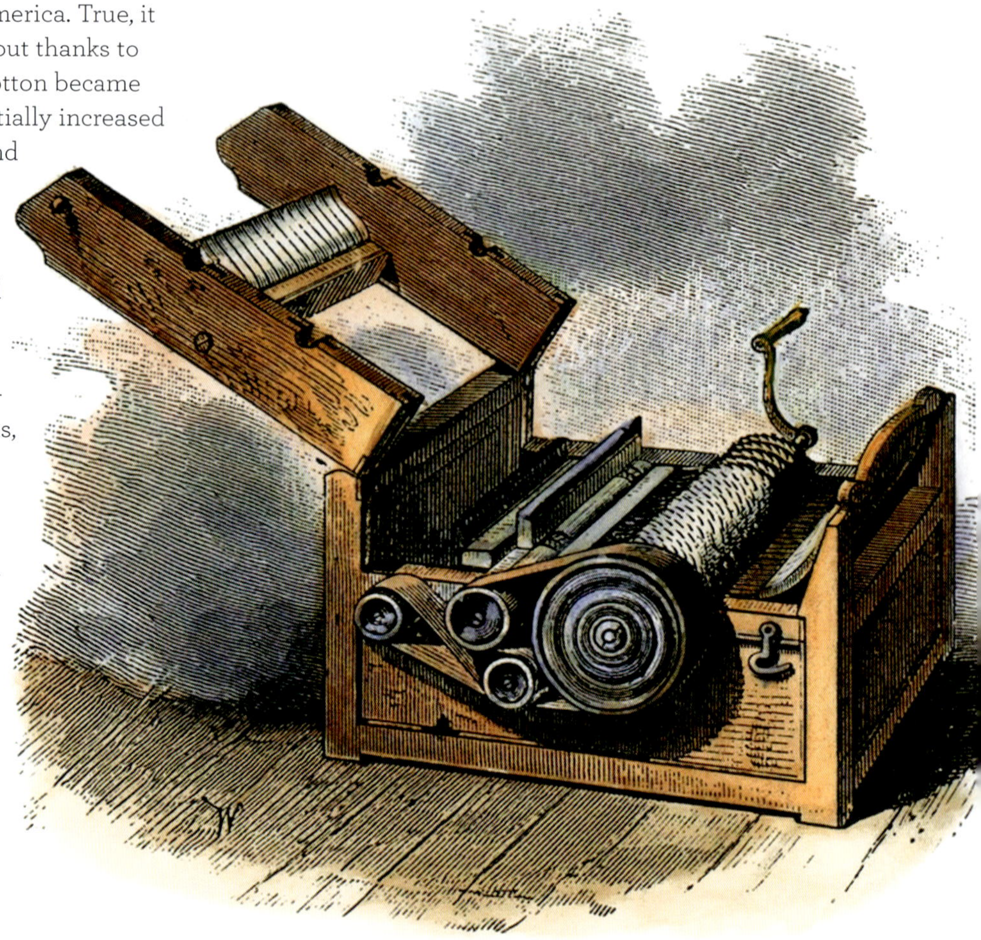

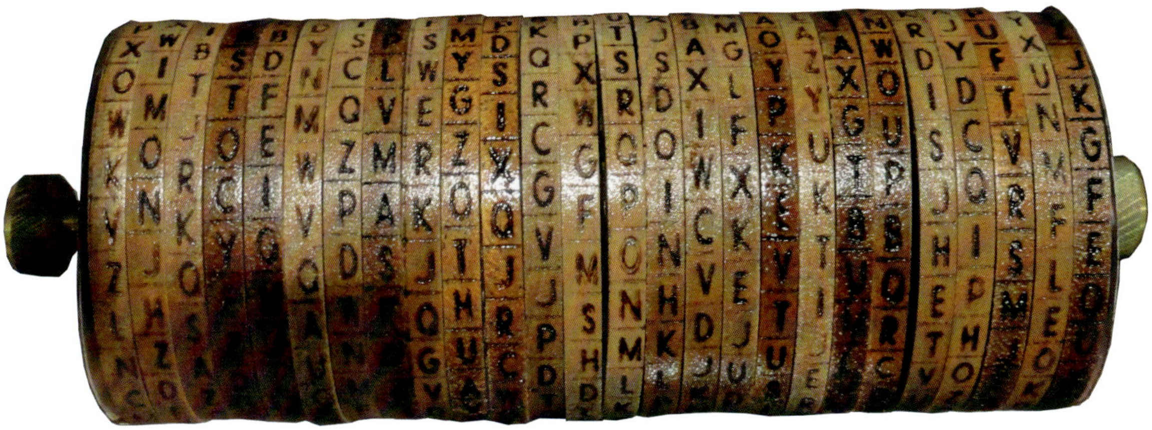

A replica of Jefferson's wheel cipher in the National Cryptologic Museum.

1795 WHEEL CIPHER

Another clever device from the mind of Thomas Jefferson (1743–1826) was the wheel cipher, which he envisioned to encode and decode messages while serving as George Washington's secretary of state. The inspiration behind the device is rumored to have come during Jefferson's service as America's minister to France from 1784 to 1789. Coded messages became a crucial part of communication during that time due to the animosity between countries. The wheel cipher apparatus involved a set of discs (the "wheels") that were stamped on their edges with letters in a random order. The discs were also labeled with a unique number. The discs were threaded on a spindle so that messages could be scrambled and unscrambled when the wheels were turned. Although there is some doubt if the apparatus was ever built during Jefferson's time, it was considered a secure and brilliant innovation in the post-Revolutionary War era.

1796 THE PRESIDENTIAL PORTRAIT

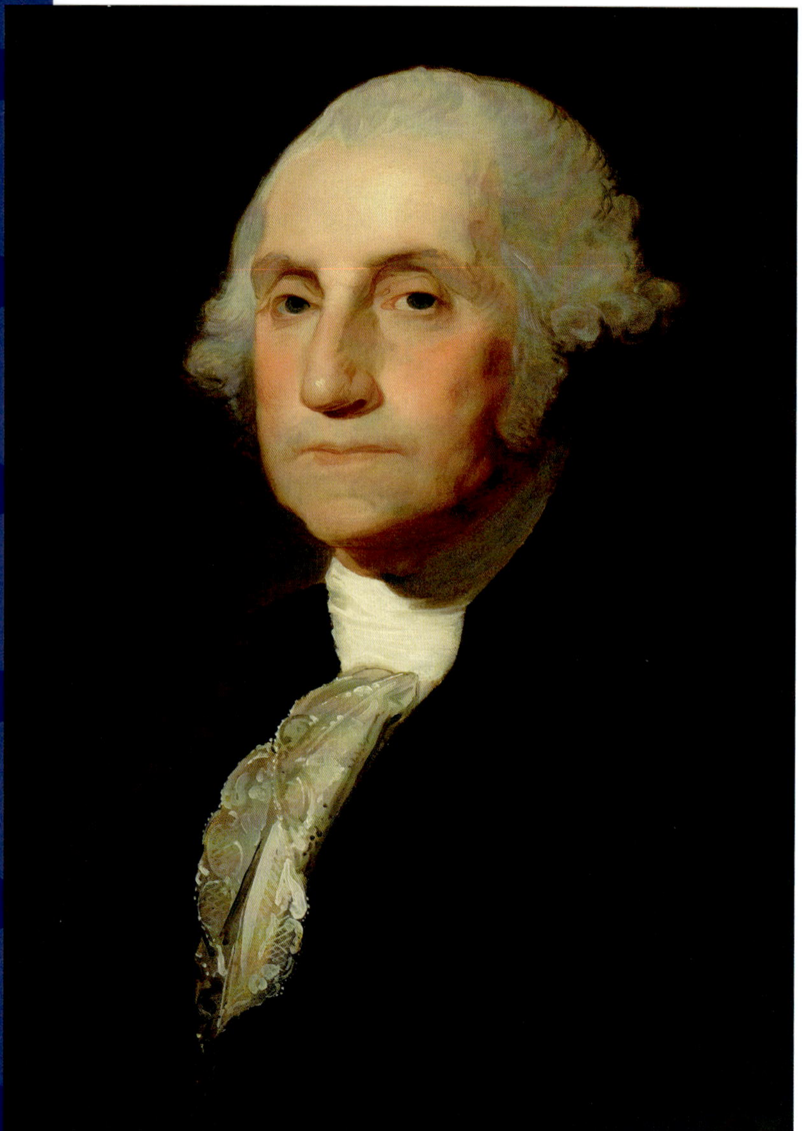

Fine artist Gilbert Stuart (1755–1828) of Rhode Island memorialized many prominent figures during his lifetime, including John Adams, John Quincy Adams, Thomas Jefferson, James Madison, and other dignitaries, but the portrait he rendered of the nation's first president, George Washington, is perhaps his most iconic. The pose that the artist chose for this first official presidential portrait is thought to be drawn from Washington's address to Congress at the end of 1795. Some sources claim that Washington's wife, Martha, commissioned the portrait, but Stuart decided he wanted to keep it for himself. Although Stuart never finished the portrait, Washington's regal visage in Stuart's portrait began to appear on the one-dollar bill in 1869. This work was dubbed the *Athenaeum Portrait* because the Boston Athenaeum acquired it after the artist's death and held it for 150 years. Although this work, and others of Stuart's, drew on European art traditions, Stuart's art style is considered to be a uniquely American portrait style; he had a particular aptitude for capturing his subjects' eyes. His portrait of George Washington started the tradition of creating portraits for each incoming president that still continues today.

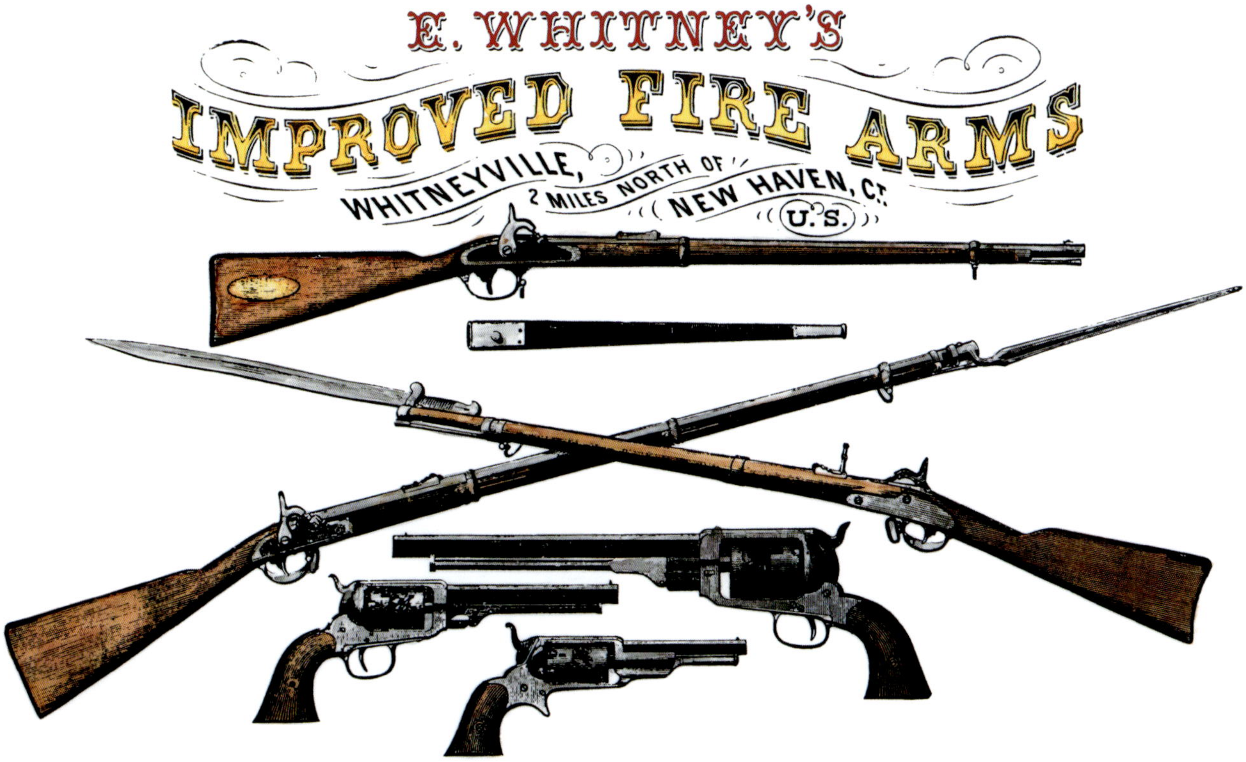

1797 INTERCHANGEABLE GUN PARTS

When Congress voted to appropriate funds to prepare the nation for war with France, inventor Eli Whitney (1765–1825) seized the opportunity to sell them on an idea that he thought would help him get rich—yet would have ramifications far more important for the country. Whitney's idea was to manufacture muskets with prefabricated, interchangeable parts, and thus make it possible to put large quantities of the weapons together with relatively unskilled workers. At the time, a single musket was made by a skilled craftsman using a very labor-intensive process. Whitney was successful at persuading Congress to offer him a contract to manufacture ten thousand of the weapons for the US Army and deliver them in just two years. Although the project was fraught with delays and widespread skepticism, Whitney eventually delivered the weapons about eight years late and sealed his place in history as a brilliant innovator. His thought leadership around interchangeable parts helped the United States enter the Industrial Revolution in earnest, wherein machines supplanted most of the manufacturing work from men, and factories propagated in cities and beyond.

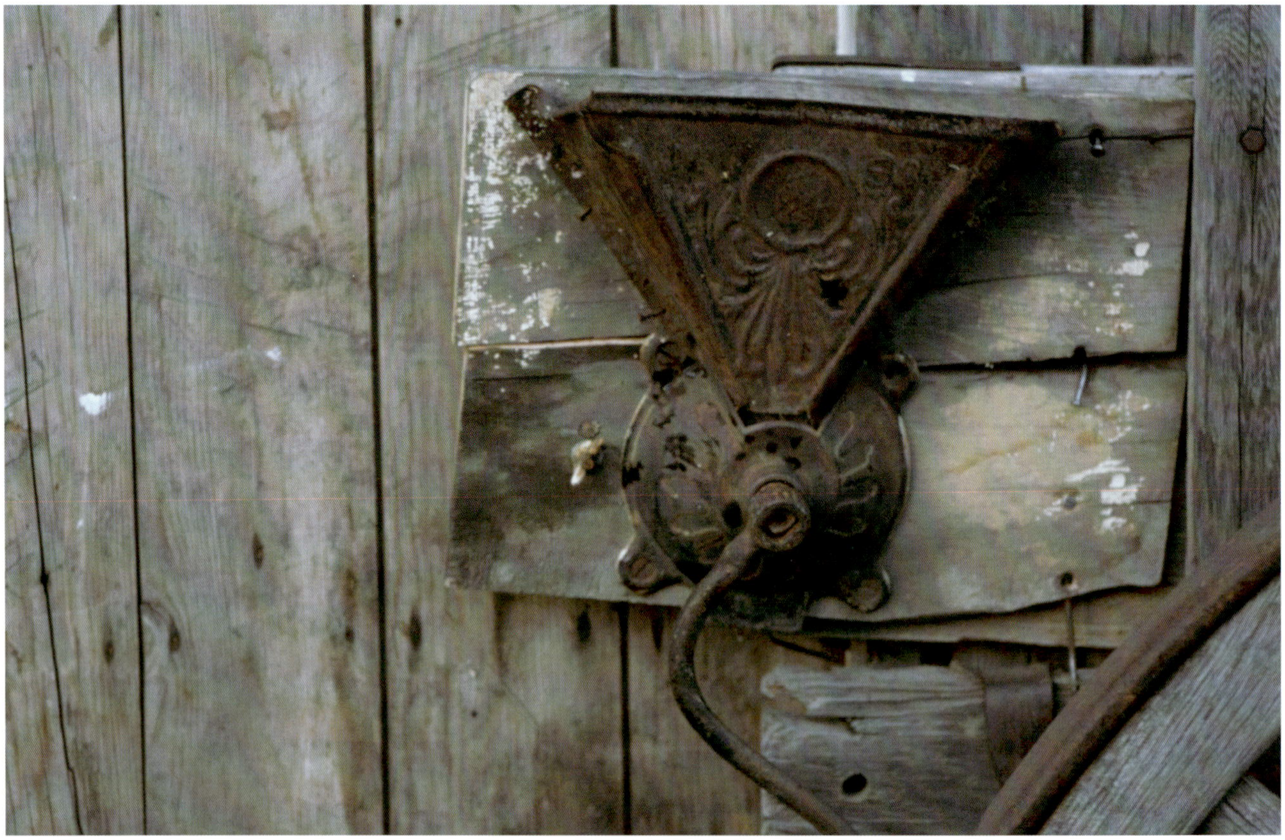

Coffee on the go: A Bruff coffee grinder attached to the side of an old chuck wagon.

1798 COFFEE GRINDER

Thomas Bruff Sr. (1648–1702), a dentist from Maryland—at one time Thomas Jefferson's personal dentist—revolutionized early American coffee culture when he received the first US patent for an improved coffee grinding mill in 1798, combining medical precision with mechanical innovation to create a device that would transform how Americans prepared their daily brew. Bruff's clever wall-mounted mill featured iron plates with both coarse and fine teeth that crushed the coffee beans between two circular grinding surfaces. This design represented a major advancement over the primitive mortar-and-pestle methods previously used for grinding the beans. His dental background likely influenced his understanding of how different textures and pressures could effectively break down materials. Bruff's basic principle—using burrlike surfaces to crush beans rather than slice them—remains the gold standard for quality coffee preparation today. Modern burr grinders still employ his core concept, though they are now powered by electricity rather than hand-cranking. Bruff's contribution laid the groundwork for America's evolving coffee culture, enabling consistently ground coffee that enhances flavor extraction and helping to establish coffee as the country's preferred morning beverage.

1799 SALT WORKS

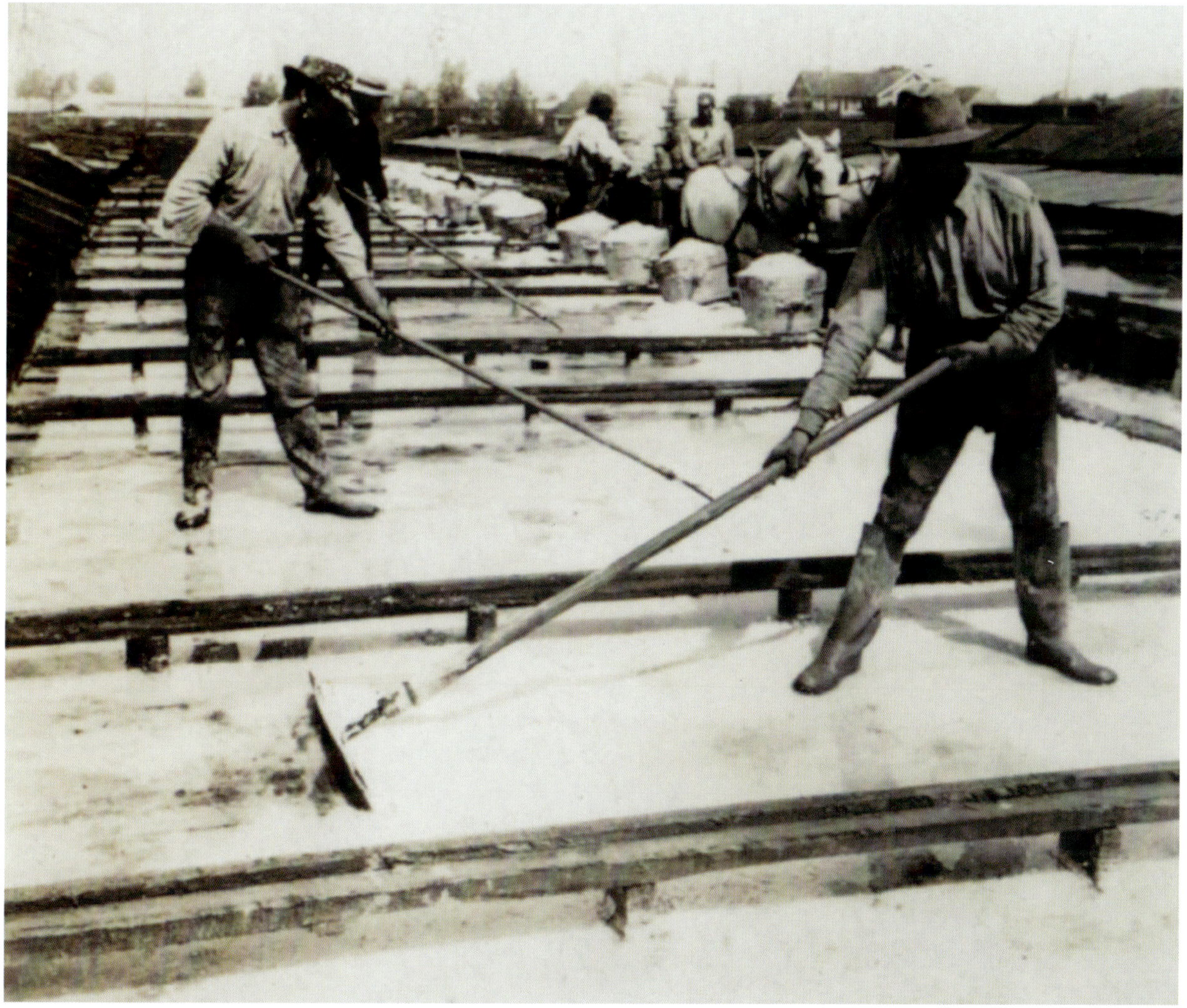

John Sears (1744–1817), a retired sea captain from Dennis, Massachusetts, revolutionized salt production when he obtained a patent in 1799 for his innovative portable roofs that covered salt vats, protecting them from rain. But the roofs were only a small part of Sears's salt-making process. During the American Revolution, Sears found a way to produce salt on a large scale, making Cape Cod's "white gold" salt until the middle of the nineteenth century. His salt-making method used solar evaporation, allowing seawater to naturally crystallize into salt in shallow vats through sun and wind exposure. This breakthrough was crucial for food preservation in the pre-refrigeration era. Sears generously granted patent rights to the roofs to others, including Nathaniel Freeman (1741–1827), a fellow local businessman, helping spread the expertise. By 1832, more than 880 salt works produced some 250,000 bushels of salt annually, and by 1837, there were an excess of 650 saltworks on Cape Cod alone. Thanks to John Sears's ingenuity, the industry thrived until cheaper alternatives from New York salt mines and products that arrived by railroad became available in the mid-1800s.

1800 THE WHITE HOUSE

The story of the White House dates back to 1791, when George Washington selected the location for the official presidential residence at 1600 Pennsylvania Avenue in Washington, DC. Then, in 1792, there was a public competition to choose a design for the building in the country's new capital. Irish American architect James Hoban (1755–1831) won the $500 prize and the commission to build arguably the most famous residence in the country. His design incorporated elements of Palladian and Neoclassical styles, which were popular architectural styles in the American colonies. That it was painted white when all of the other buildings around it remained unpainted redbrick also signaled the special significance of the structure. In the year 1800, the entire federal government moved from Philadelphia to Washington, DC. John Adams and his wife, Abigail, moved into the still-incomplete White House on November 1 of that year.

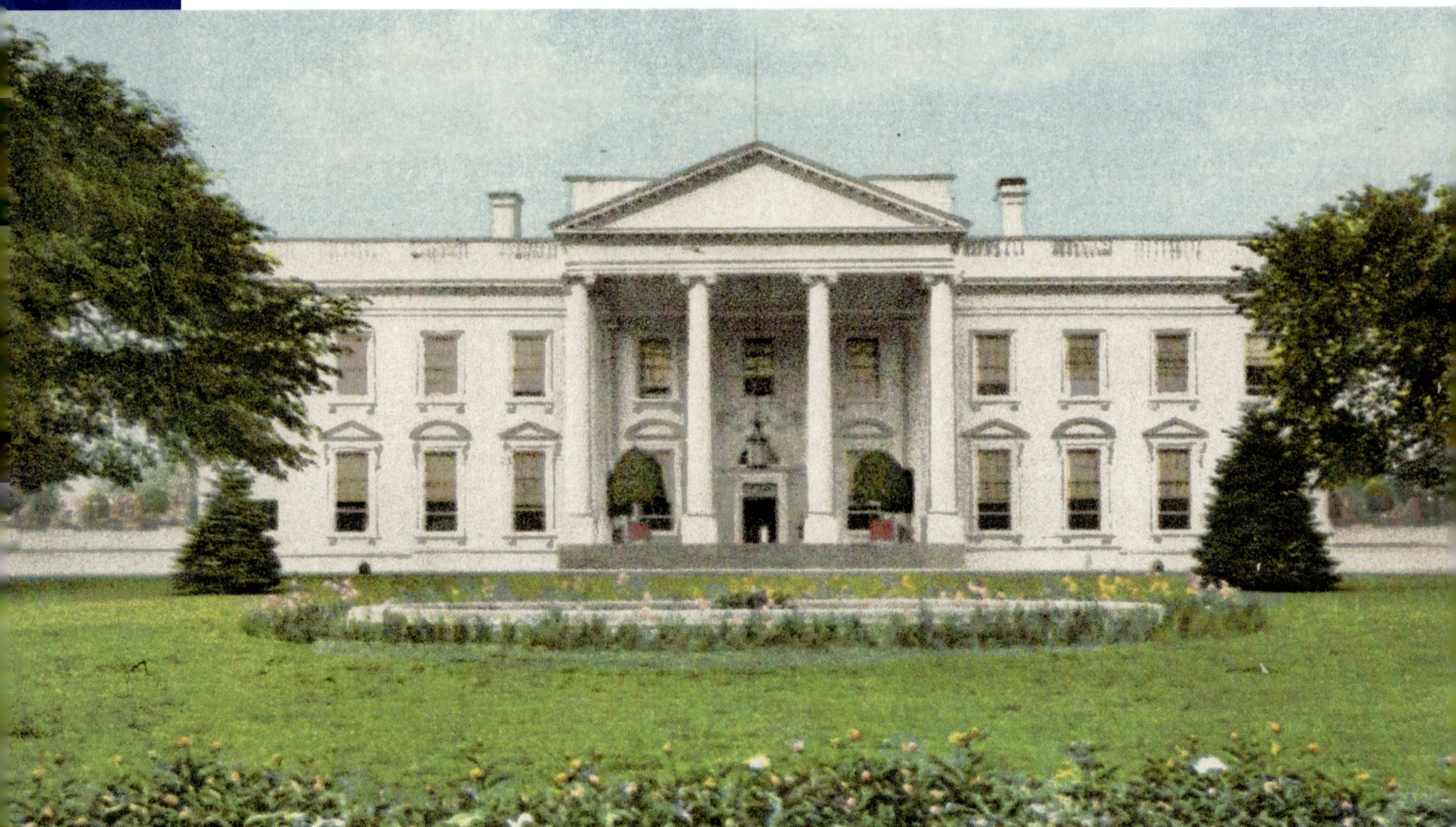

1801 FIRE HYDRANT

Walk down any city street today, and you're likely to see pillar-type fire hydrants on street corners painted in bright colors. Although it's been improved since 1801, the basic design behind these life-saving devices is often credited to Frederick Graff (1775–1847) of Philadelphia, who was a hydraulic engineer at Philadelphia's waterworks. Graff can't be absolutely confirmed as the inventor of the hydrant because it is believed that his patent was lost in the great Patent Office Fire of 1836. (Ironically, a device created to fight fires was perhaps a casualty of one.) The original "post-" or "pillar-type" fire hydrant stood aboveground and had a housing that allowed firefighters to connect it to a hose. Viewing it through modern eyes, the hydrant had a few disadvantages. Its "wet barrel" design (constant water flowing through it) meant that it was only efficient in locations where freezing wasn't an issue. Nevertheless, Graff's design paved the way for quick improvements by other innovators in subsequent years.

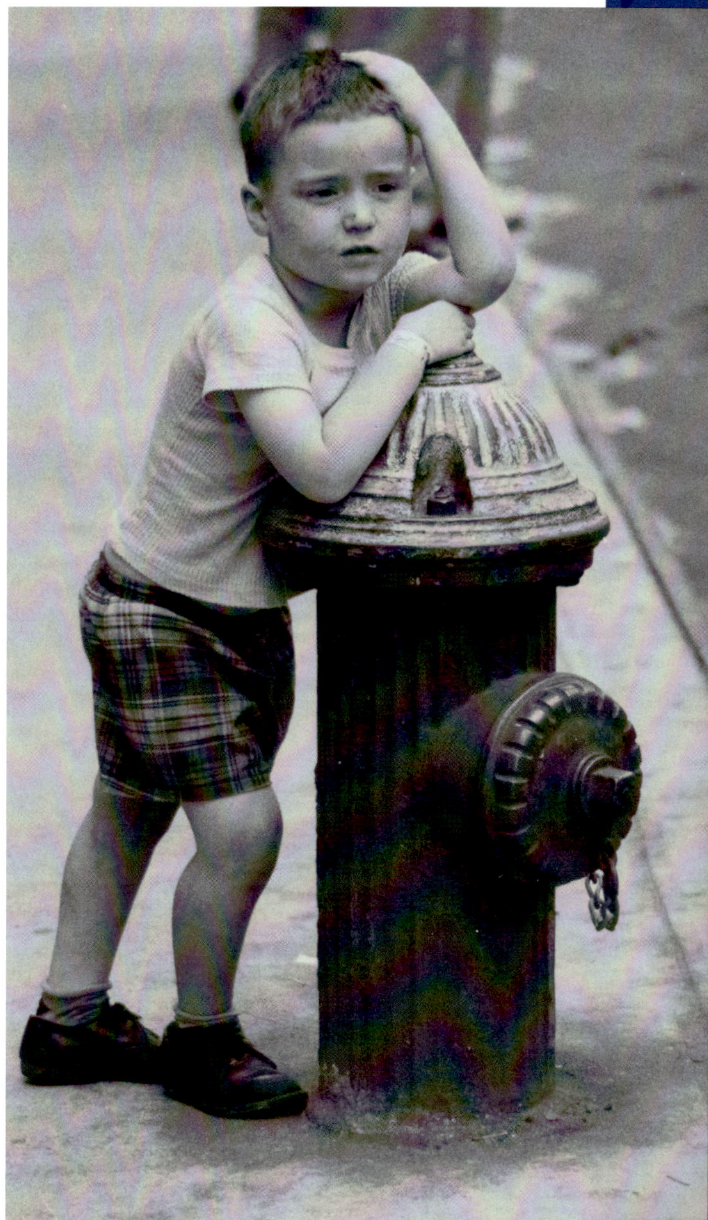

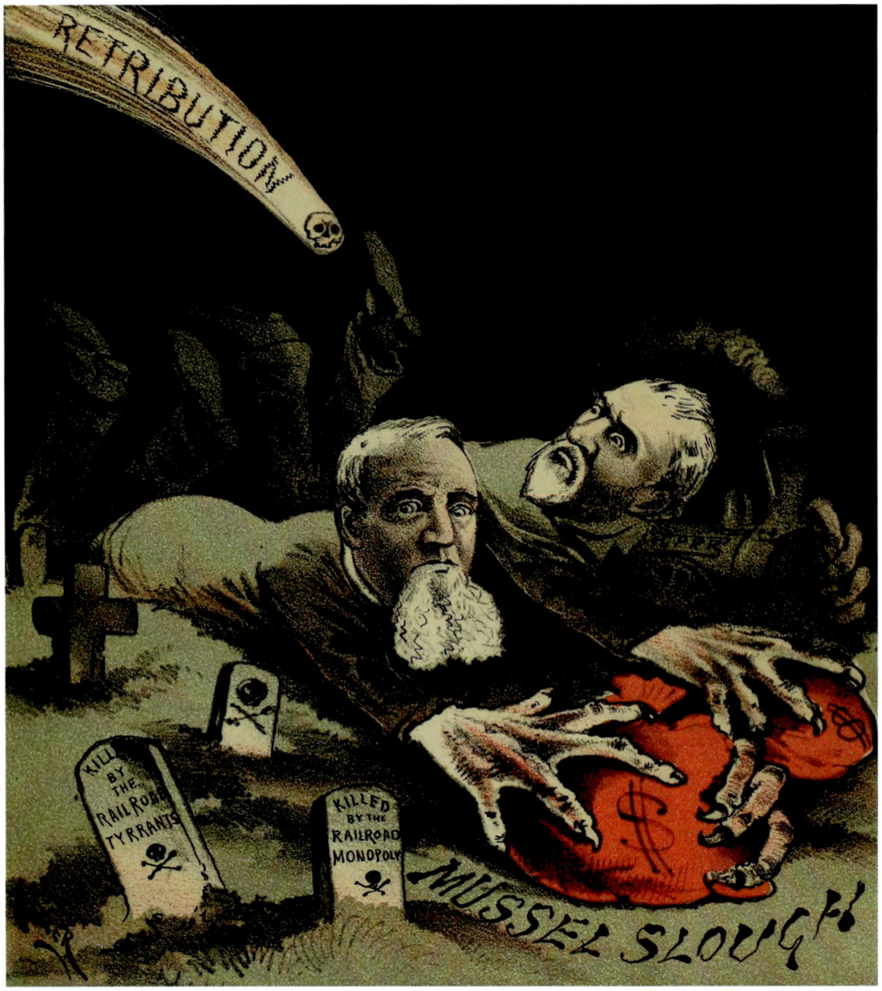

An editorial cartoon about the Mussel Slough Tragedy that appeared in an 1881 edition of *The Wasp*.

1802 THE FIRST US COMICS

Considered the first magazine of comics in the United States, *The Wasp* was first published on July 7, 1802. It was written, illustrated, and published by Robert Rusticoat, a pen name for political activist Harry Croswell (1778–1858) of New York. Just twenty-two years old at the time of publication, Croswell was a staunch supporter of John Adams and a sworn enemy of Thomas Jefferson. In response to a rival magazine called *The Bee*, which opposed Adams's federalist policies, Croswell used his platform not merely to write opinions about his rival but also to publish satirical illustrations that criticized Jefferson's Republican policies. Not only did Croswell have the audacity to criticize a sitting president, but he also did not hold back in his accusations. Among other insults, he publicly accused Jefferson of having an affair with a young woman he enslaved, Sally Hemings. The *People v. Croswell* became one of the early court cases that tested some of the amendments in the US Bill of Rights, including freedom of the press, state vs. federal rights, freedom of speech, and more.

1803 THE POLYGRAPH (AUTOPEN)

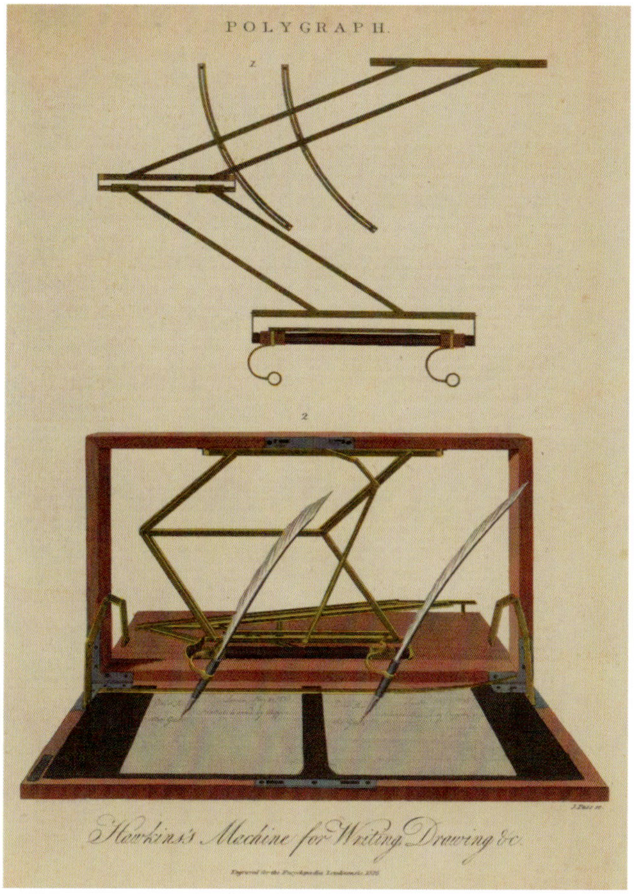

British American inventor John Isaac Hawkins (1772–1854) received a US patent in 1803 for his "polygraph," a mechanical copying device that served as the precursor to modern autopens. (Like many inventions of the era, the original patent document was lost in the Patent Office Fire in 1836.) The polygraph worked through synchronized pens—as the author moved one pen, the other moved simultaneously, producing an exact copy. Thomas Jefferson became an early adopter, acquiring the device in 1804 and calling it "the finest invention of the present age." Jefferson found it invaluable for his extensive correspondence, having written almost twenty thousand letters in his lifetime. The legacy of Hawkins's invention extends far beyond Jefferson's era. Modern autopens, though technologically advanced, operate on the same fundamental principle of the original autopens. The device has evolved from Jefferson's correspondence tool to today's presidential signing machines, demonstrating the lasting value of Hawkins's concept in administrative efficiency and document authentication.

ABOVE: Hawkins's original design. **RIGHT:** An autopen being used to sign checks at the Treasury Department, 1918.

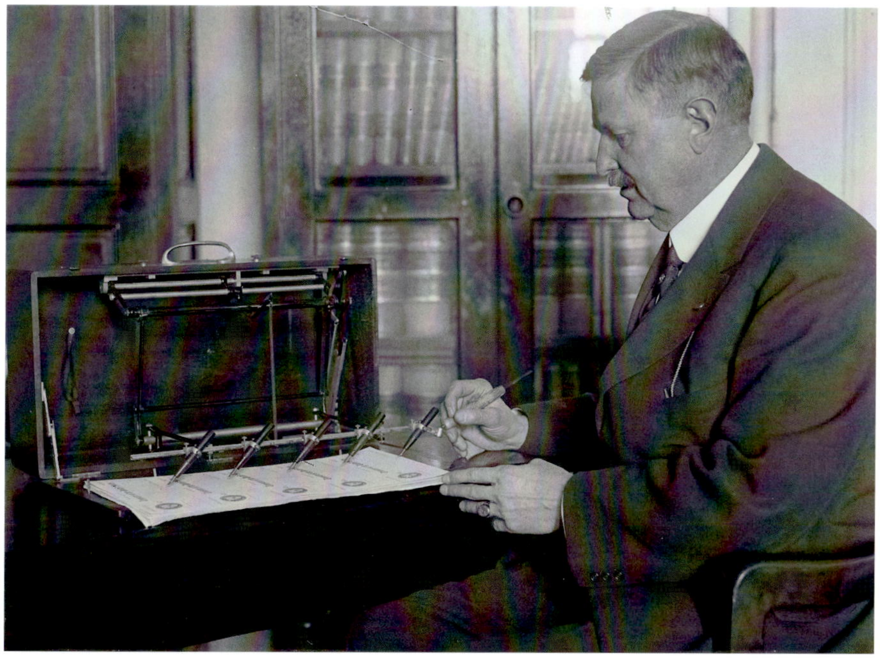

1804 LEWIS AND CLARK EXPEDITION

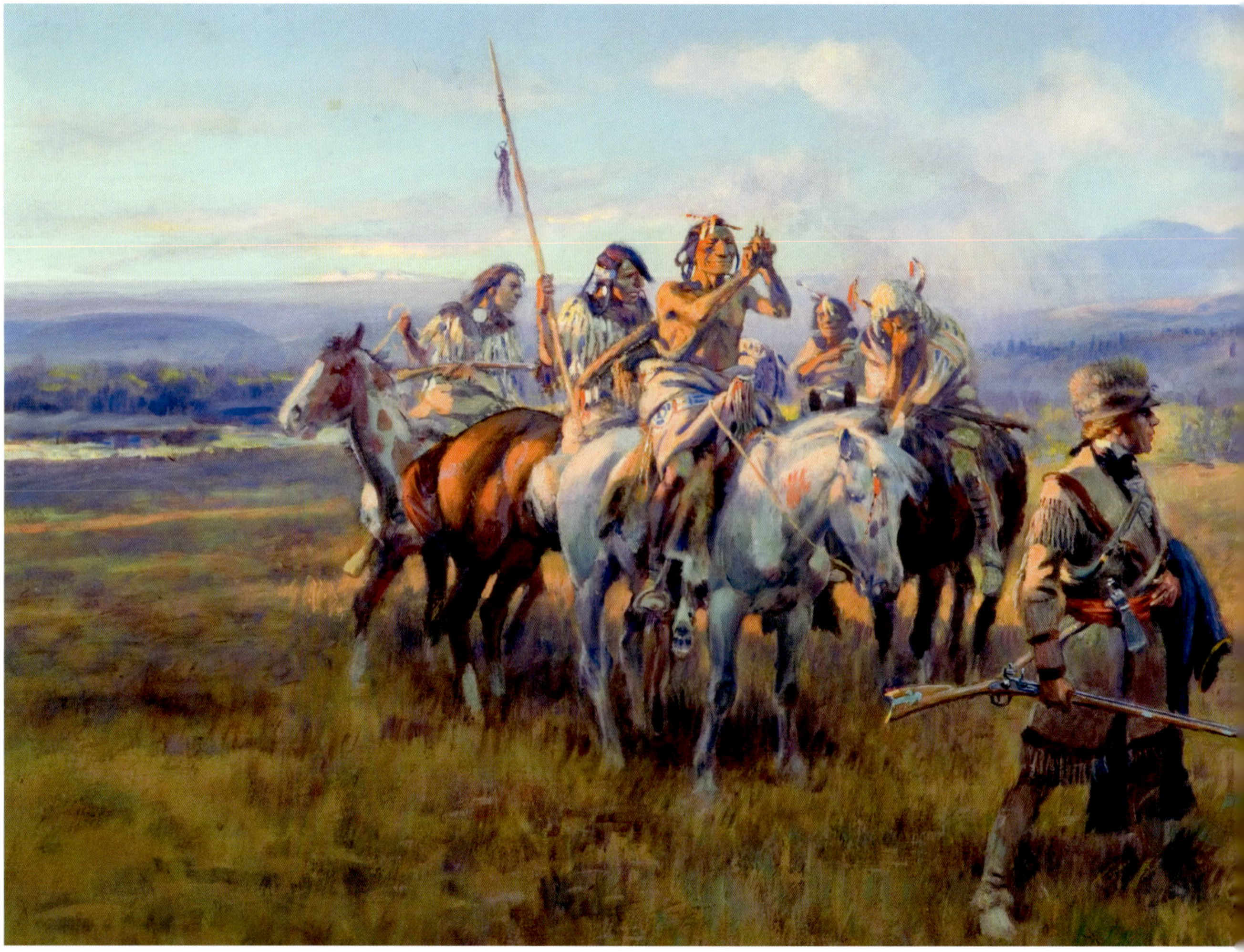

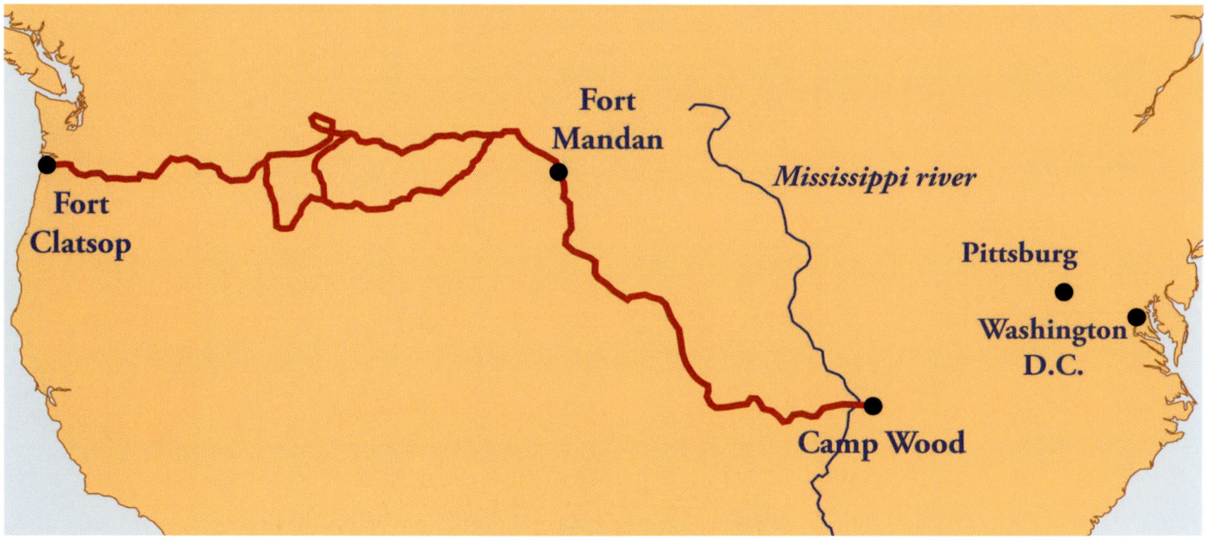

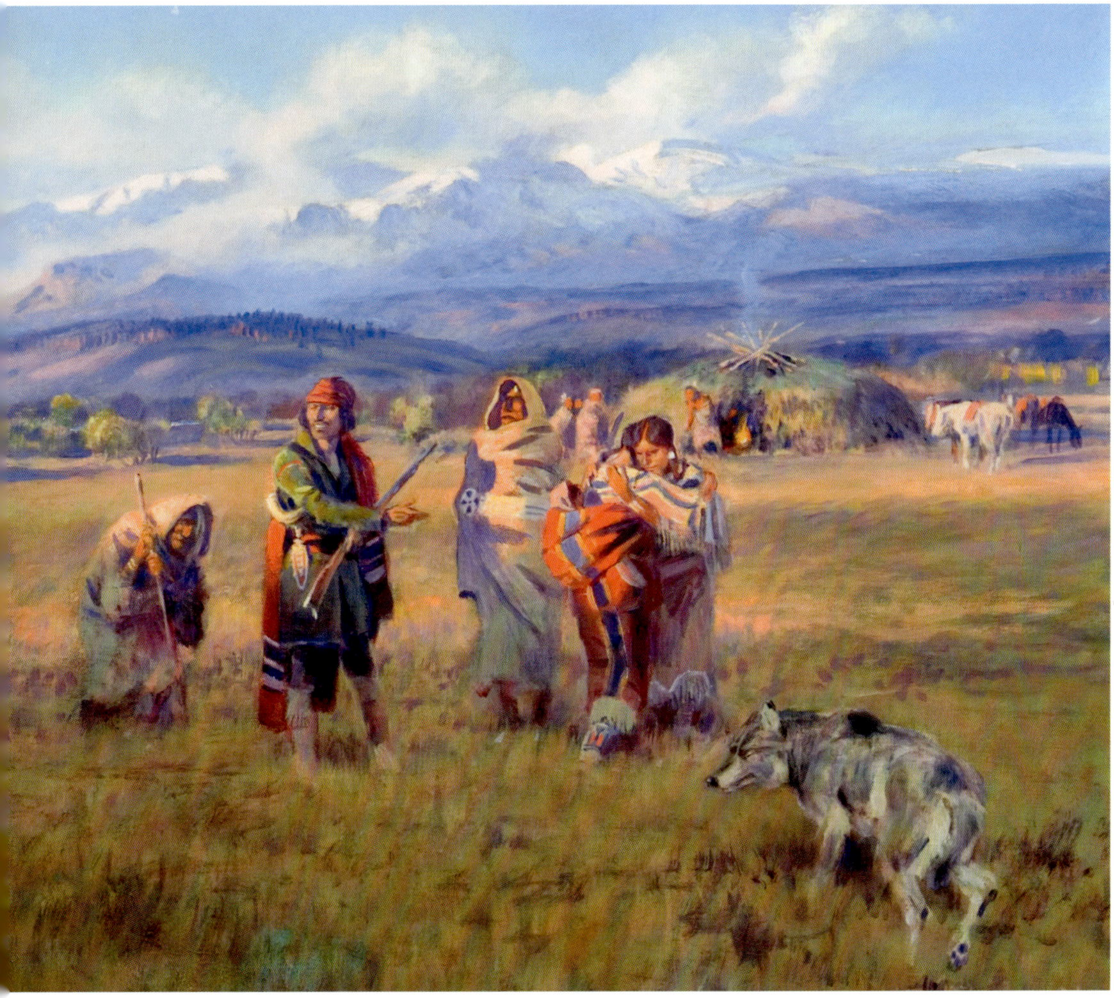

In May 1804, at the behest of then President Thomas Jefferson, Meriwether Lewis (1774–1809) and William Clark (1770–1838) set out to explore the lands west of the Mississippi River that were acquired in the Louisiana Purchase as well as the Pacific Northwest region. During the arduous two-year, eight-thousand-mile journey, the explorers and their team, which included the Native American interpreter and cultural liaison Sacagawea, documented geography (important waterways, mountains, and trails); ecology (soil, climate, flora, and fauna); and culture (mostly that of Indigenous peoples, with whom they had trade and diplomacy discussions). Their detailed notes and scientific findings were recorded in a series of journals. The expedition is the largest and most important in American history. In addition to knowledge, the Lewis and Clark Expedition contributed many innovations to the new country, including strengthening relations with Native peoples and expanding the fur trade.

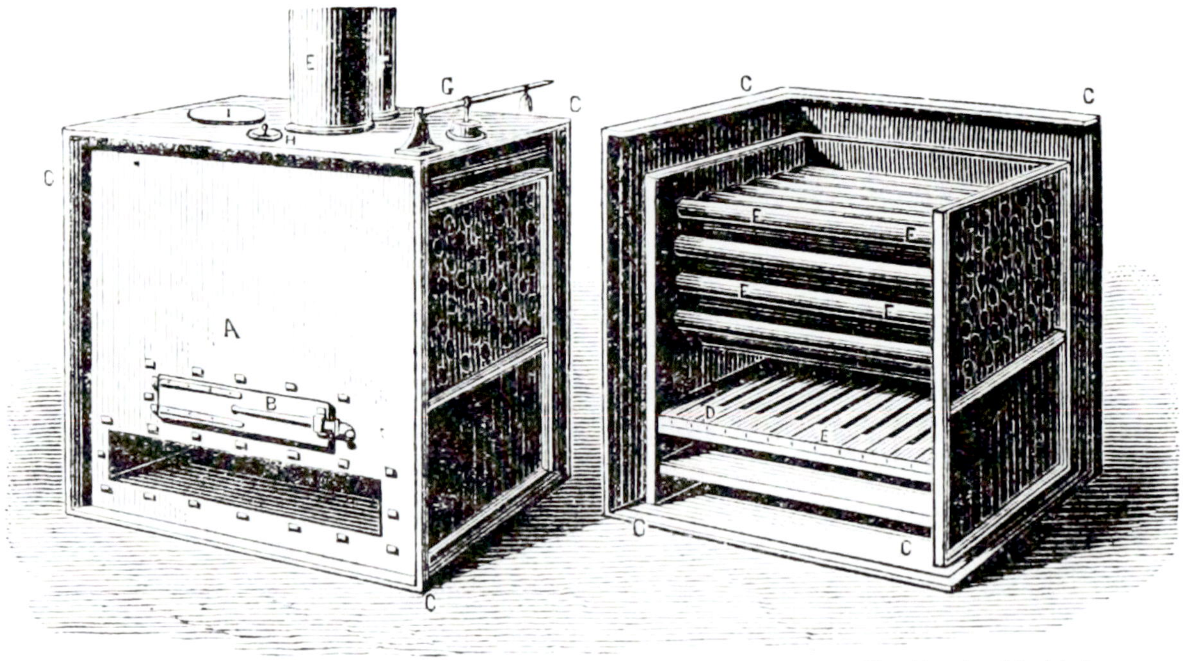

Oliver Evans's original design.

1805 VAPOR-COMPRESSION REFRIGERATION

In the early 1800s, food was preserved by drying, salting, pickling, or canning. Fresh food had to be eaten right away, or it would spoil. Enter Oliver Evans (1755–1819), an inventor from Philadelphia who developed vapor-compression refrigeration—a system that produced ice using a compressor and a condenser to remove heat from a vacuum-sealed environment and recycle a refrigerant solution through the closed system—in 1805. Although Evans created detailed drawings for his device, there is no evidence that he actually built a refrigerator. The credit for that goes to his colleague, Massachusetts-born Jacob Perkins (1766–1849), who built the devices based on Evans's original theories. It was Perkins who received the patents for the machines in 1834. The vapor-compression technology is considered one of the world's most significant mechanical innovations in history due to its widespread use in food preservation (refrigerators and freezers) and in air conditioners and heaters, especially when they were scaled up to the industrial level.

1806 COFFEE PERCOLATOR

Physicist and creative inventor Benjamin Thompson (1753–1814) was born in America, though he moved to England after the Revolutionary War. Unlike most residents in his newly adopted country, Thompson reportedly did not care for the flavor of tea. He was, however, a big fan of coffee, but not a big fan of the grounds that were left behind in the cup. To feed his coffee habit, he invented a coffee percolator. Though not like what we know as the modern coffee percolator with its upright tube, this coffee pot was inspired by the Turkish method of heating ground coffee and water in a constant cycle to make a strong brew, then pouring it out through a sieve to strain out the grounds. In England, Thompson was offered a countship and dubbed Count Rumford. When he invented his percolating coffee pot, it became known as the Rumford coffee maker. A prolific inventor, Thompson was known not only for his coffeepot, but also for designing warships, improving chimneys, and inventing the kitchen range and the double boiler.

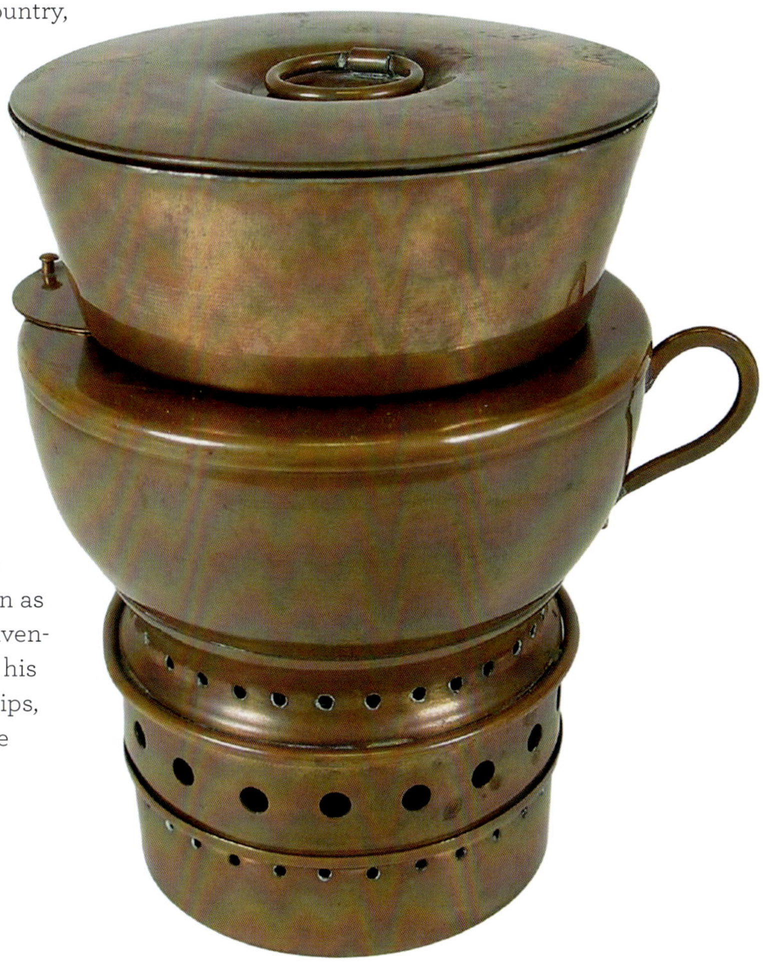

A model of Thompson's portable percolator.

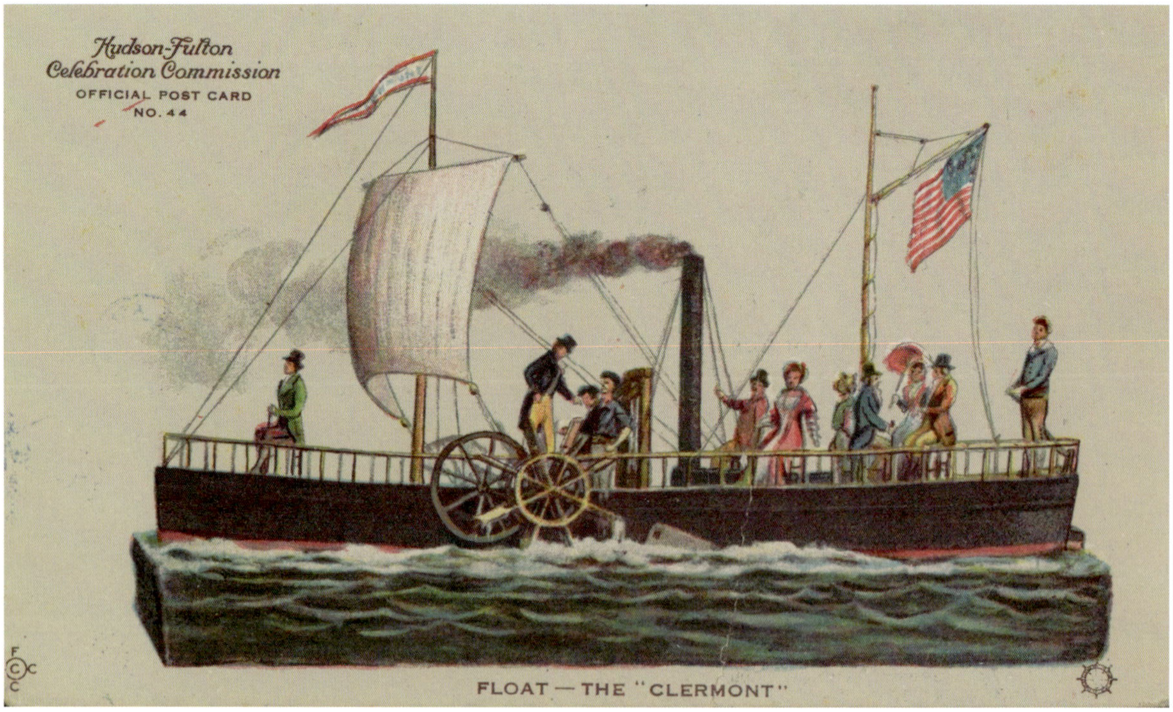

A 1909 postcard commemorating the *Clermont*.

1807 THE MODERN STEAMBOAT

In the early 1800s, much of American river travel occurred on sailboats, which often were hampered by high winds or rough waters. In 1807, American engineer, inventor, and artist Robert Fulton (1765–1815) modified the existing steamboat to be more practical and commercially viable. With its flat bottom, square stern, powerful engine, and two fifteen-foot paddle wheels, the *North River Steamboat*, colloquially dubbed the "*Clermont*," was able to withstand the rough river currents that plagued less-sturdy boats and traveled 150 miles on the Hudson river from New York City to Albany on August 17, 1807. Although it was a slow journey at only about five miles per hour, it was a significant one: The *Clermont* and steamboats that would follow in its wake would change river transportation and passenger travel in the United States in significant ways. Fulton is also credited with other American innovations, such as a submarine and a steam-powered warship.

1808 HEATING HOMES WITH COAL

On February 11, 1808, Judge Jesse Fell (1751–1830) of Wilkes-Barre, Pennsylvania, became the first person to successfully burn anthracite coal in his fireplace. Before Fell's experiment, anthracite was considered unsuitable for domestic heating because previous attempts had failed to sustain burning. His successful experiment on that day led to the widespread use of coal as the fuel source that helped to foster America's Industrial Revolution. Fell's breakthrough was transformative for American society in many ways. First, anthracite was cheaper and more fuel efficient than wood and could also be used for home heating and cooking fuel, revolutionizing American domestic life. Second, his discovery changed the Wyoming Valley from an agricultural area to an industrial superpower once coal mines were built there, soon followed by canals for transporting the anthracite. Next, the shift to anthracite decreased the country's dependence on scarce wood supplies and laid the foundation for coal-based industries. Finally, anthracite became crucial to America's industrialization and economic growth throughout the nineteenth century.

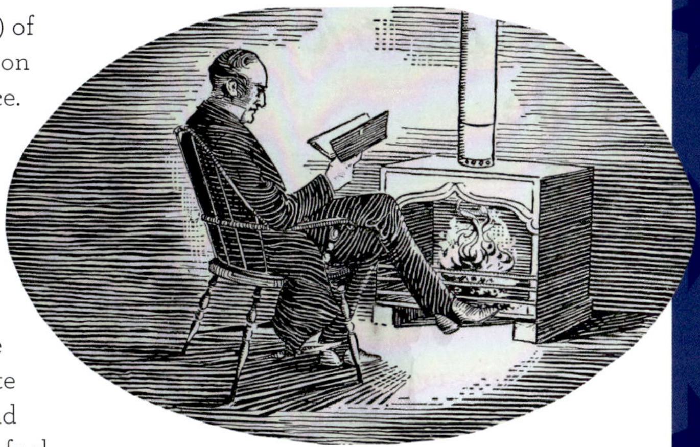

Etching depicting Jesse Fell enjoying the warmth of the anthracite coal.

1809 FASHION PATENT ASSIGNED TO A WOMAN

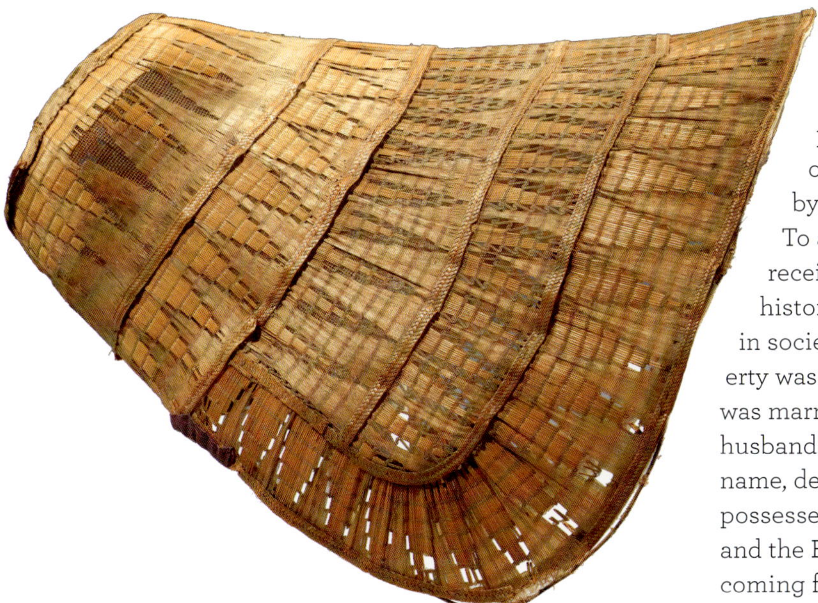

An example of intricately woven silk and straw, similar to what Kies may have patented, is shown in this 1835 poke bonnet.

Mary Dixon Kies (1752–1837) of Connecticut was among the first women to apply for and receive a US patent in her own name. Her patent, issued May 5, 1809, was for a process of weaving straw with silk that was adopted by the New England hat-making industry. To a woman in the early nineteenth century, receiving a patent was unheard of. At this time in history, women had absolutely no rights or power in society. If she were unmarried, a woman's property was considered that of her father; if a woman was married, her property was considered that of her husband. Kies's moxie in pursuing a patent in her own name, despite the times, displayed courage few women possessed at the time. Due to the Napoleonic Wars and the Embargo of 1807, fashion items were no longer coming from or going to Europe. Kies's innovation gave her (temporary) financial gain and praise from both President James Madison and his wife Dolley. Sadly, Kies's patent was destroyed in the fire that devastated the Patent Office in 1836.

1810 CONESTOGA WAGON

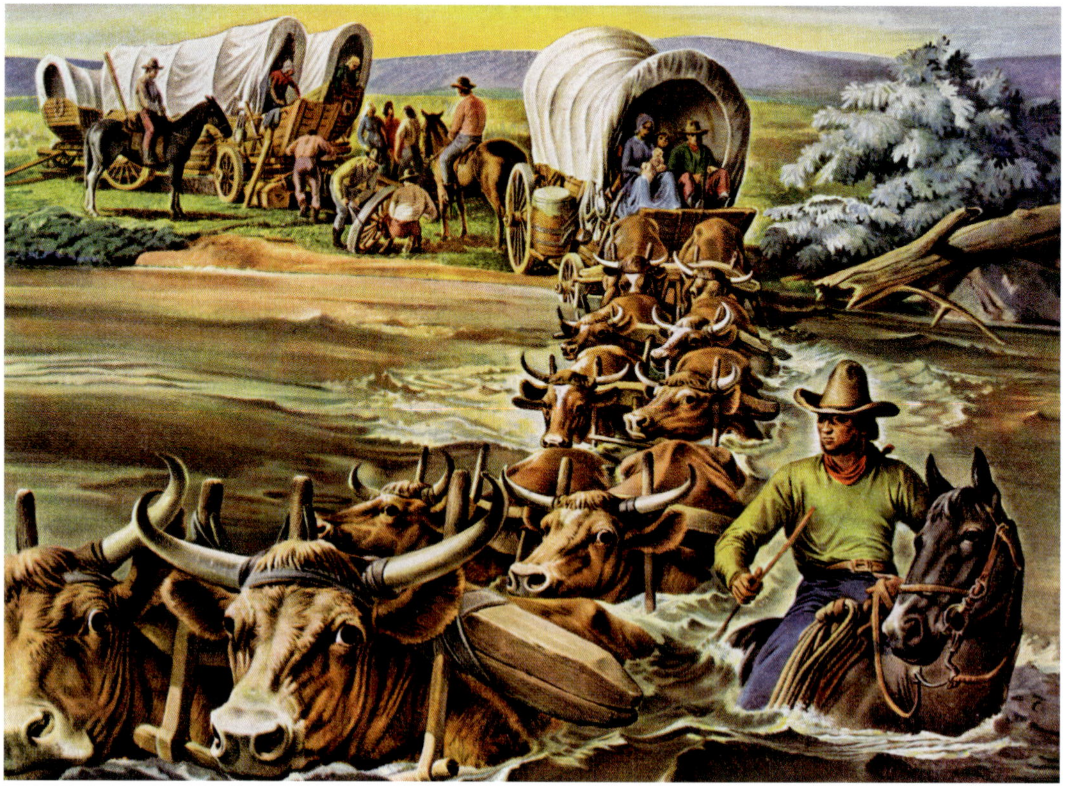

An illustration of a train of Conestoga wagons crossing the Platte River in Nebraska.

The Conestoga wagon stood as the backbone of American westward expansion by 1810, earning it a reputation as the "ship of the prairie." Originally developed in Pennsylvania's Conestoga Valley during the mid-eighteenth century, these massive wagons were indispensable for transporting goods across the rugged American frontier. Distinguished by their curved floors and white canvas tops, Conestoga wagons could haul up to six tons of freight. Their clever design prevented goods from shifting during travel over uneven terrain, and their broad wheels distributed weight effectively. Teams of four to six horses or oxen pulled these heavyweight wagons along the primitive roads and trails. By 1810, the Conestoga wagons had established crucial supply lines connecting eastern manufacturers with western settlements. They transported everything from household goods and farming equipment to whiskey and grain, facilitating the commerce that bound the expanding nation together. The Conestoga wagon represented the spirit of westward migration that would define nineteenth-century America, inspiring countless families to seek new opportunities beyond the frontier. Even as railroads eventually replaced wagon trains, the Conestoga remained an enduring symbol of American pioneering spirit.

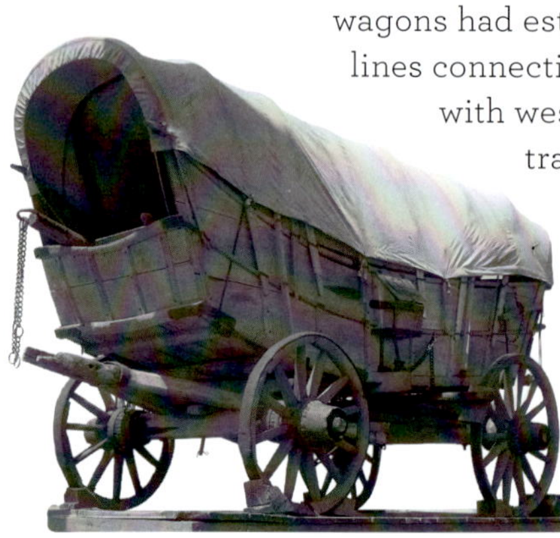

1811 THE NATIONAL ROAD

When construction began in 1811 on the National Road, America's first federally funded interstate highway, it marked a pivotal moment in the nation's transportation history. Stretching from Cumberland, Maryland, to Vandalia, Illinois, this ambitious project represented the federal government's first major investment in infrastructure, sparking debates about constitutional authority and federal versus state responsibilities that would echo throughout American history. Originally conceived to connect the Eastern Seaboard with western territories, the National Road facilitated westward expansion and economic development during the early nineteenth century. The highway enabled faster movement of people, goods, and ideas across the Appalachian Mountains, transforming isolated frontier communities into thriving commercial centers. Towns along the route experienced unprecedented growth as the road brought increased trade and settlement. The National Road established important precedents for federal involvement in transportation projects, laying the groundwork for future initiatives like the Interstate Highway System. Today, much of the original road follows US Route 40, and some of the original sections serve as historical landmarks, reminding Americans of this early experiment in federally sponsored infrastructure that helped bind a growing nation together.

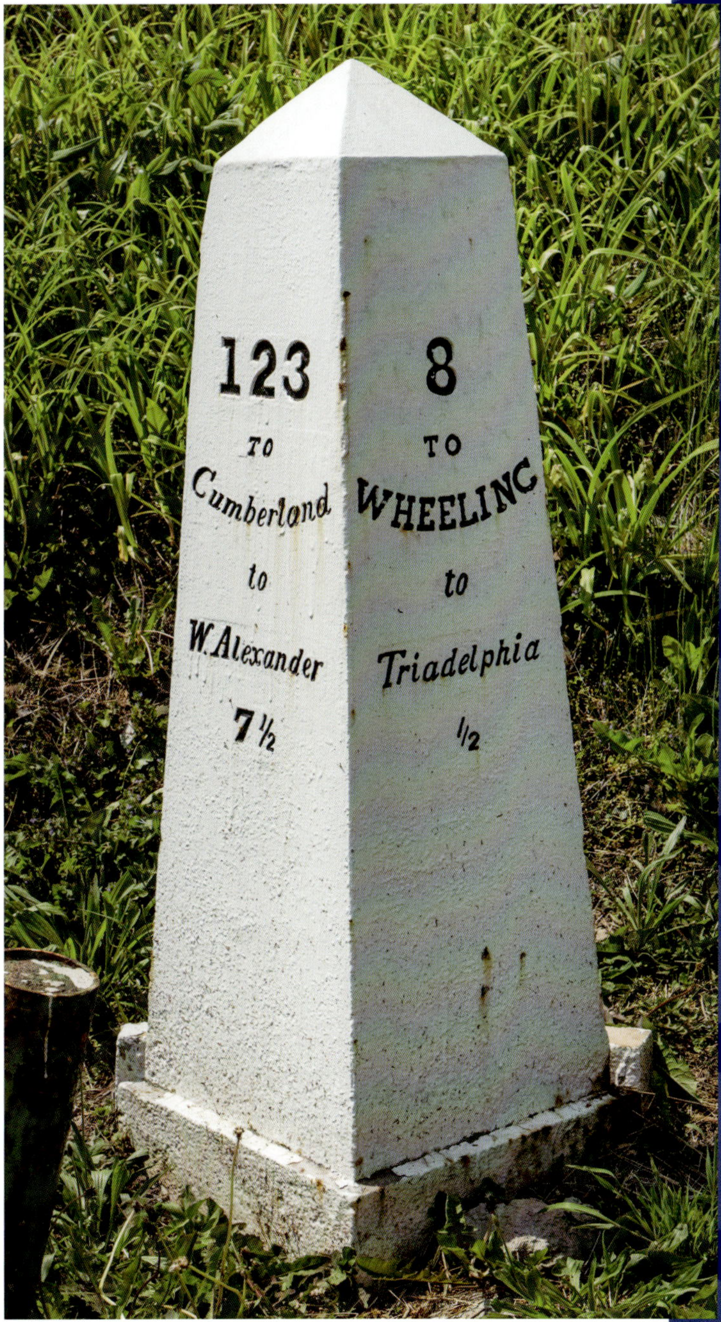

An original mile marker in Triadelphia, West Virginia. Congress mandated these stone markers be placed every mile to tell travelers how many miles they were from the starting point of the highway (Cumberland), as well as the distance to closer points.

1812 KENTUCKY BOURBON

Built in 1812 in Woodford County, Kentucky, by Elijah Pepper (1760–1830s), Old Oscar Pepper Distillery is one of the oldest standing distilleries in Kentucky. Bourbon is a distinctly American spirit with its historic origins in the state. Nearly all the United States bourbon was—and is—crafted in Kentucky. The Peppers ran their distillery for three generations, and its Old Crow brand still survives today, though its ownership has since changed. The Old Oscar Pepper Distillery is on the National Register of Historic Places. One of the early employees of the distillery, Dr. James Crow (1789–1856), a chemist, refined the typical "sour mash" technique of making the bourbon by bringing science into the process, ensuring a more consistent product. Crow also advocated using charred oak barrels for aging the bourbon, enhancing its flavor. These two innovations are still used to this day in bourbon-making.

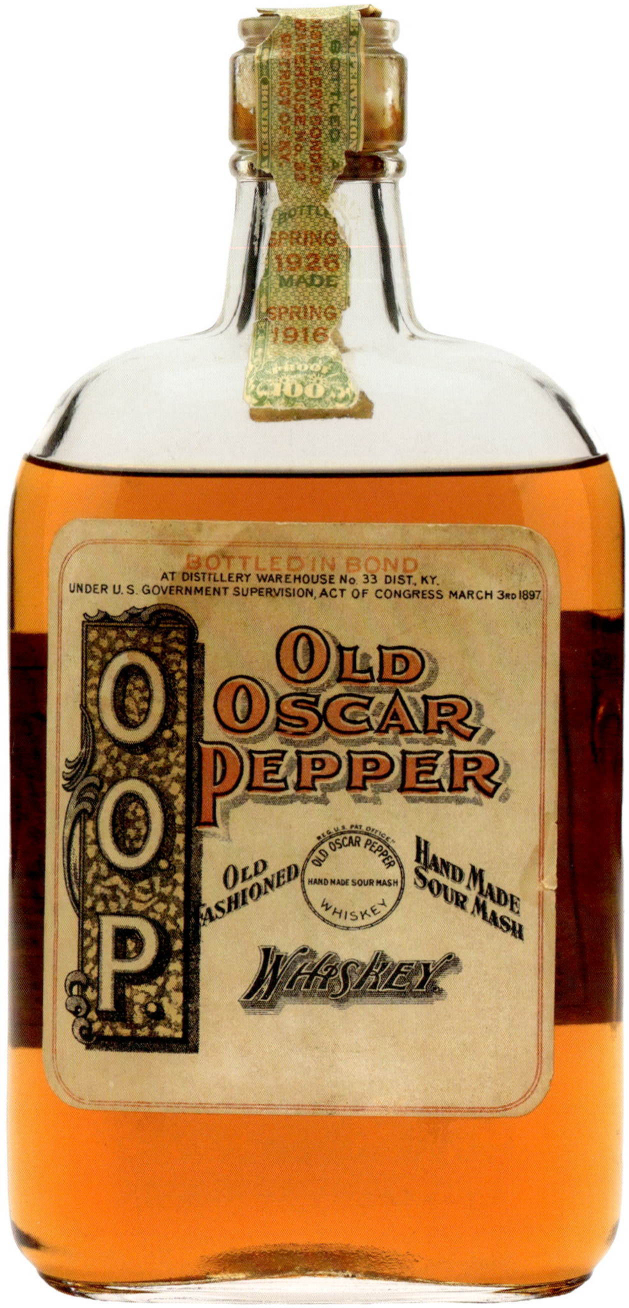

1813 CIRCULAR SAW

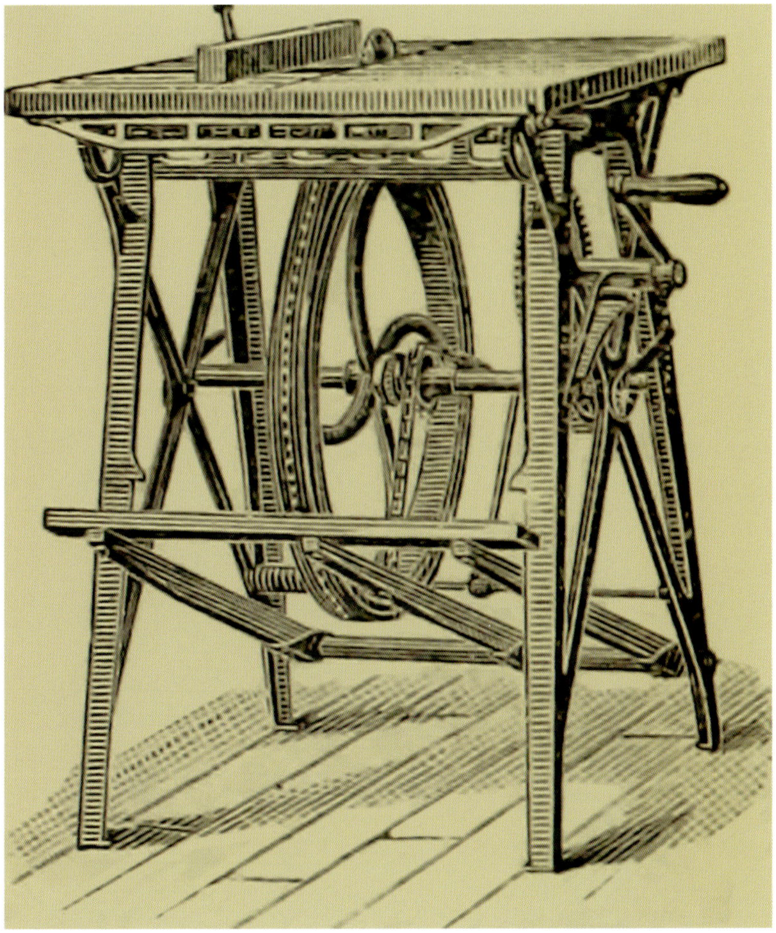

Sarah "Tabitha" Babbitt (1779–1853) was an American toolmaker and inventor. She was a member of the Shaker sect (a Protestant Christian sect that believed in, among other things, shared property). The Shakers had a tradition of furniture making that was popular even outside their ranks. Witnessing the labor it took to wield a whipsaw, which needed two men to operate, Babbitt came up with the idea for the circular saw, a much more efficient tool that employed a round blade attached to a power source—originally a pedal-operated spinning wheel, then a water-powered machine—that cut down on both the manpower as well as the effort. Despite the fact that the Shakers believed in gender equality, they forbade women from applying for patents, so Sarah was unable to monetize her achievement. Because of the lack of actual records, there remains some controversy over whether Sarah was actually the one to invent the circular saw. Indeed, the patent was recorded by two Frenchmen (some lore posits that they did so after reading about it in a Shaker newspaper). Whether or not you believe that this particular innovation belongs to Sarah, there is no doubt that she was a visionary. She is credited with other improvements or inventions relating to such diverse objects as false teeth, nails for building, and spinning wheels.

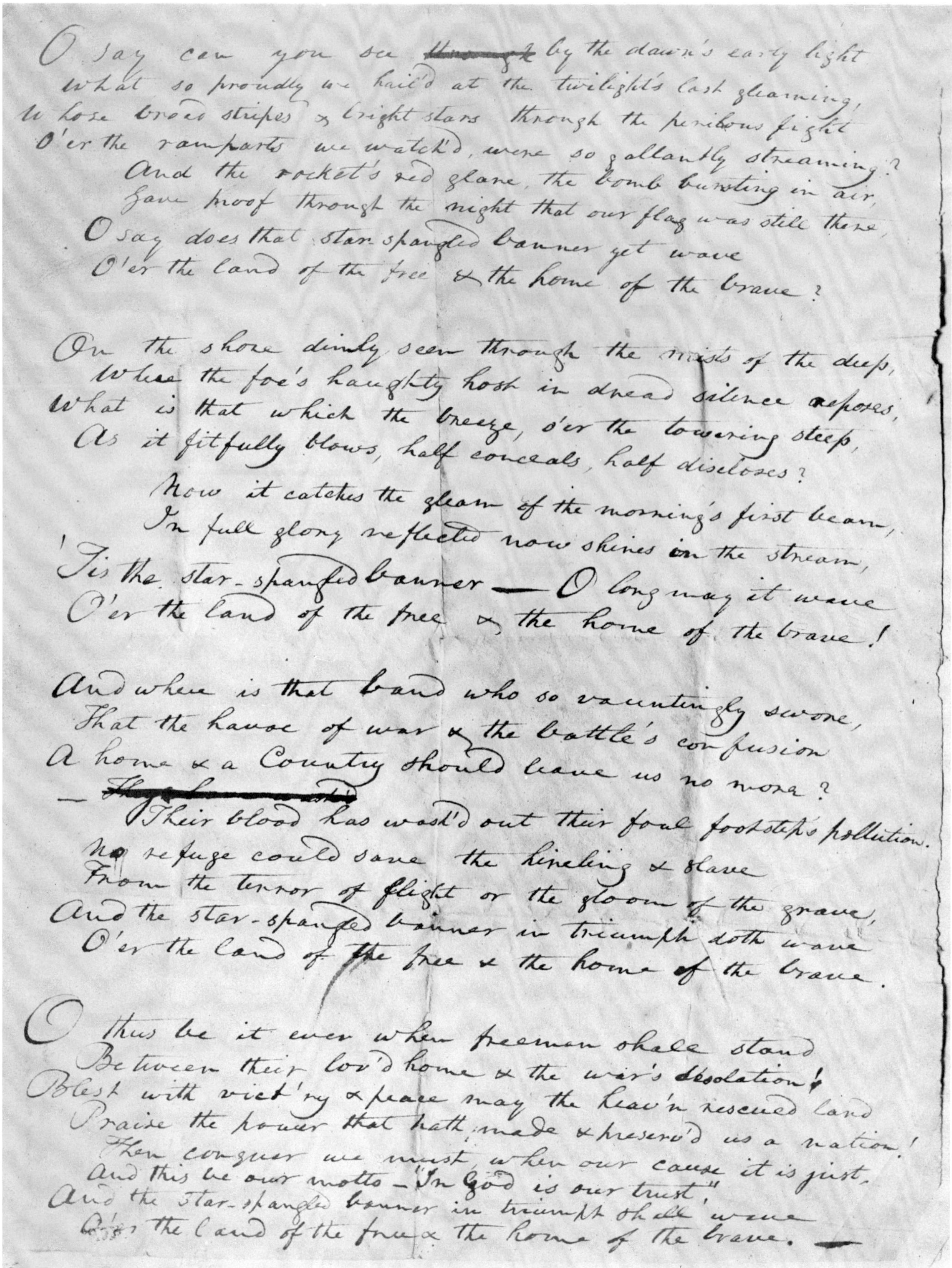

O say can you see ~~through~~ by the dawn's early light
What so proudly we hail'd at the twilight's last gleaming,
Whose broad stripes & bright stars through the perilous fight
O'er the ramparts we watch'd, were so gallantly streaming?
And the rocket's red glare, the bomb bursting in air,
Gave proof through the night that our flag was still there,
O say does that star spangled banner yet wave
O'er the land of the free & the home of the brave?

On the shore dimly seen through the mists of the deep,
Where the foe's haughty host in dread silence reposes,
What is that which the breeze, o'er the towering steep,
As it fitfully blows, half conceals, half discloses?
Now it catches the gleam of the morning's first beam,
In full glory reflected now shines in the stream,
'Tis the star-spangled banner — O long may it wave
O'er the land of the free & the home of the brave!

And where is that band who so vauntingly swore,
That the havoc of war & the battle's confusion
A home & a Country should leave us no more?
— ~~[illegible]~~
Their blood has wash'd out their foul footstep's pollution.
No refuge could save the hireling & slave
From the terror of flight or the gloom of the grave,
And the star-spangled banner in triumph doth wave
O'er the land of the free & the home of the brave.

O thus be it ever when freemen shall stand
Between their lov'd home & the war's desolation!
Blest with vict'ry & peace may the heav'n rescued land
Praise the power that hath made & preserv'd us a nation!
Then conquer we must, when our cause it is just,
And this be our motto — "In God is our trust,"
And the star-spangled banner in triumph shall wave
O'er the land of the free & the home of the brave. —

1814 "THE STAR-SPANGLED BANNER"

On September 13, 1814, American attorney Francis Scott Key (1779–1843) was imprisoned on a British warship throughout the bombardment of Fort McHenry, an attempt by the British Royal Navy to overtake the city of Baltimore during the war of 1812. At the end of the fracas, on September 14, Key saw a large American flag—a "star-spangled banner"—still waving, celebrating America's victory over the British. His experience inspired him to write a patriotic poem celebrating the nation's flag, which he set to the tune of a British song written by John Stafford Smith that was popular in America at the time. Soon, Key's song became a well-liked, contemporary patriotic song, and one that the fledgling country could be immensely proud of. Despite its popularity, it took some time before "The Star-Spangled Banner" was used in any official capacity. In 1889, the United States Navy first recognized the song for official use. Then, on March 3, 1931, Congress passed a joint resolution, which then President Herbert Hoover signed into law, making the song the official national anthem of the United States.

LEFT: Francis Scott Key's original manuscript for "The Star-Spangled Banner."

1815 DENTAL FLOSS

In 1815, a dentist in New Orleans named Levi Parmly (1790–1859) is credited with inventing dental floss. The first incarnation of the simple yet brilliant device that protects the health of teeth and gums came in the form of thin, wax-coated silk thread. Although Parmly did not hold the patent for the device (that was granted in 1898 to the Johnson & Johnson company), he was the first to recognize that there were oral health issues to be dealt with beyond simple brushing. In 1819, Parmly described his invention in a book called *A Practical Guide to the Management of the Teeth,* which was no doubt an influential guide for dental practitioners of the day. In World War II, silk became hard to procure, so other dental pioneers, including Charles C. Bass (1875–1975), stepped in to develop improvements to Parmly's original invention, including using nylon as a replacement for silk.

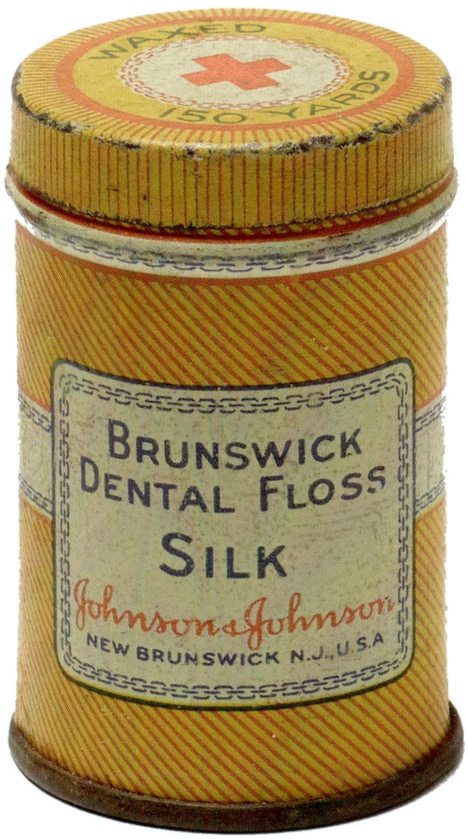

1816 COMMERCIALLY PRODUCED CRANBERRIES

In 1816, Revolutionary War hero Captain Henry Hall (1761–1850) discovered one of his cranberry plants in Cape Cod, Massachusetts thriving in the sand, despite his expectations. This sight galvanized Hall to become the first person to successfully cultivate cranberries. But to tell the full story, we need to back up in time. Sources tell us that wild cranberries have been grown by Native peoples in the sandy soil of coastal New England for thousands of years. The winter after Captain Hall returned to his farm after the War of 1812, and he had cut down several trees for firewood for warmth during the cold months, it is rumored that he returned to the site where his cranberries thrived on his land only to discover them entirely covered by sand. He thought his crops had been lost. But when spring emerged, he was delighted to discover that the cranberry plants were propagating, thriving more than ever before. Hall's discovery of sand as a natural fertilizer, insect deterrent, and temperature insulator helped lead to the thriving cultivated cranberry industry we know today.

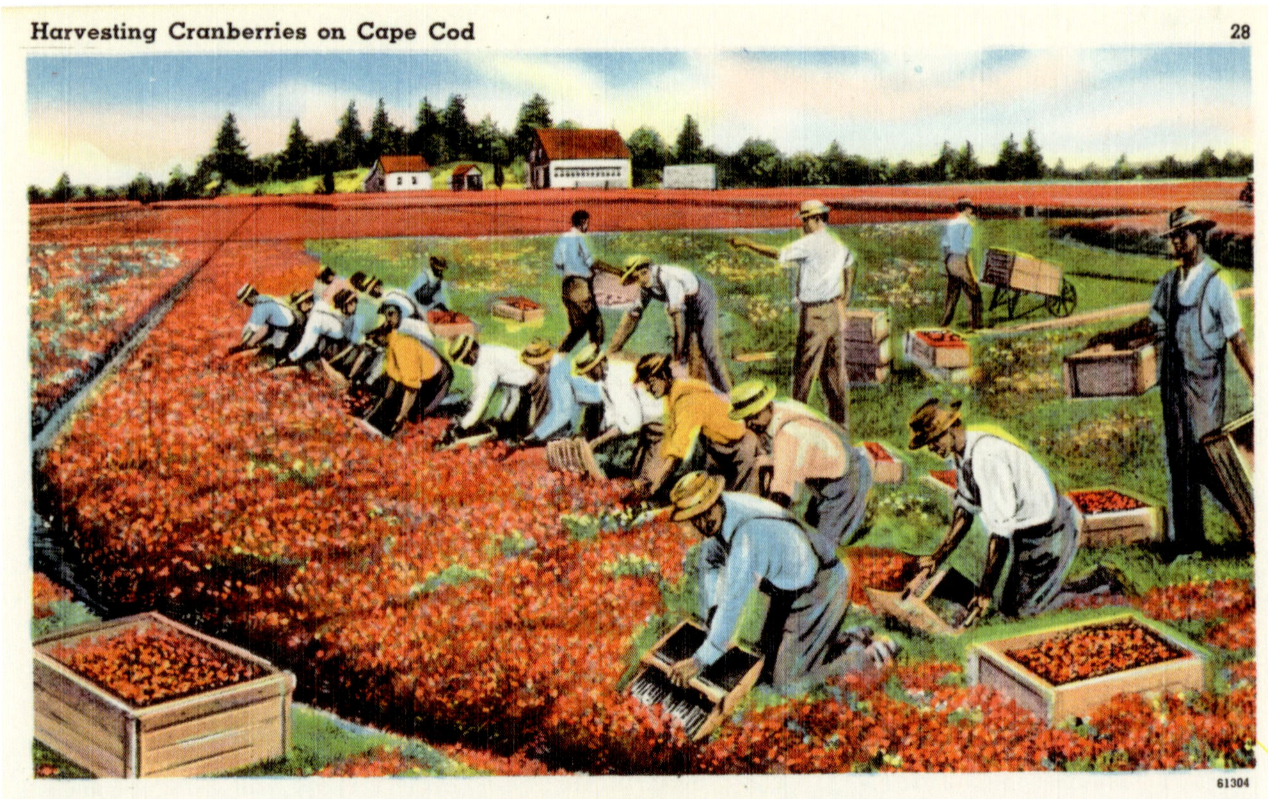

This postcard, c. 1940, depicts workers collecting cranberries using the water harvesting technique. The area is flooded with water, and because cranberries have air pockets, they float to the top. The workers then use rakes to loosen the berries from the vines, and scoops to gather them.

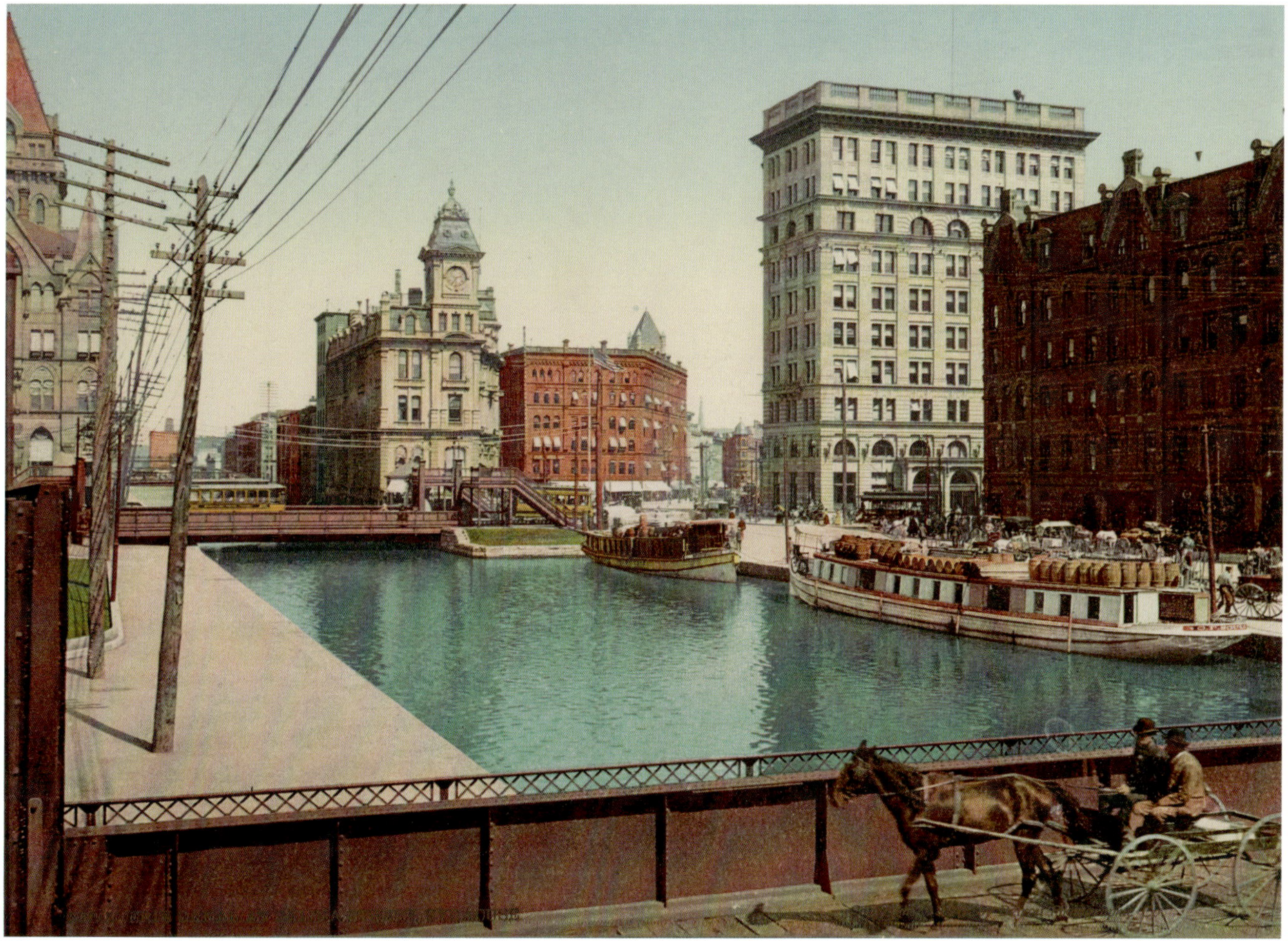

The Erie Canal at Salina Street in Syracuse, New York, c. 1900.

1817 ERIE CANAL

On July 4, 1817, workers broke ground on the Erie Canal at Rome, New York. Although it wouldn't be completed until 1825, the 363-mile-long waterway linking Lake Erie to the Hudson River was such an important engineering feat that some dubbed it at the time the Eighth Wonder of the World. Among the canal's impressive statistics: It was the first in the United States to connect inland rivers with the Atlantic Ocean; the engineers of the day figured out a lock system to move boats up and down through five hundred feet of elevation; eighteen aqueducts were constructed to carry the canal over existing bodies of water; Canvass White (1790–1834), an engineer on the project, discovered a cement that could firm up underwater, speeding the building process; and finally, the canal made it possible for New York City to become a major commercial port. DeWitt Clinton (1769–1828) was the visionary behind the project. A senator at the time he conceived the canal circa 1811, he was governor of the state of New York in 1817 when he worked with the state's legislature to approve the $7 million in funds needed (over $170 trillion in today's dollars) to build the important waterway.

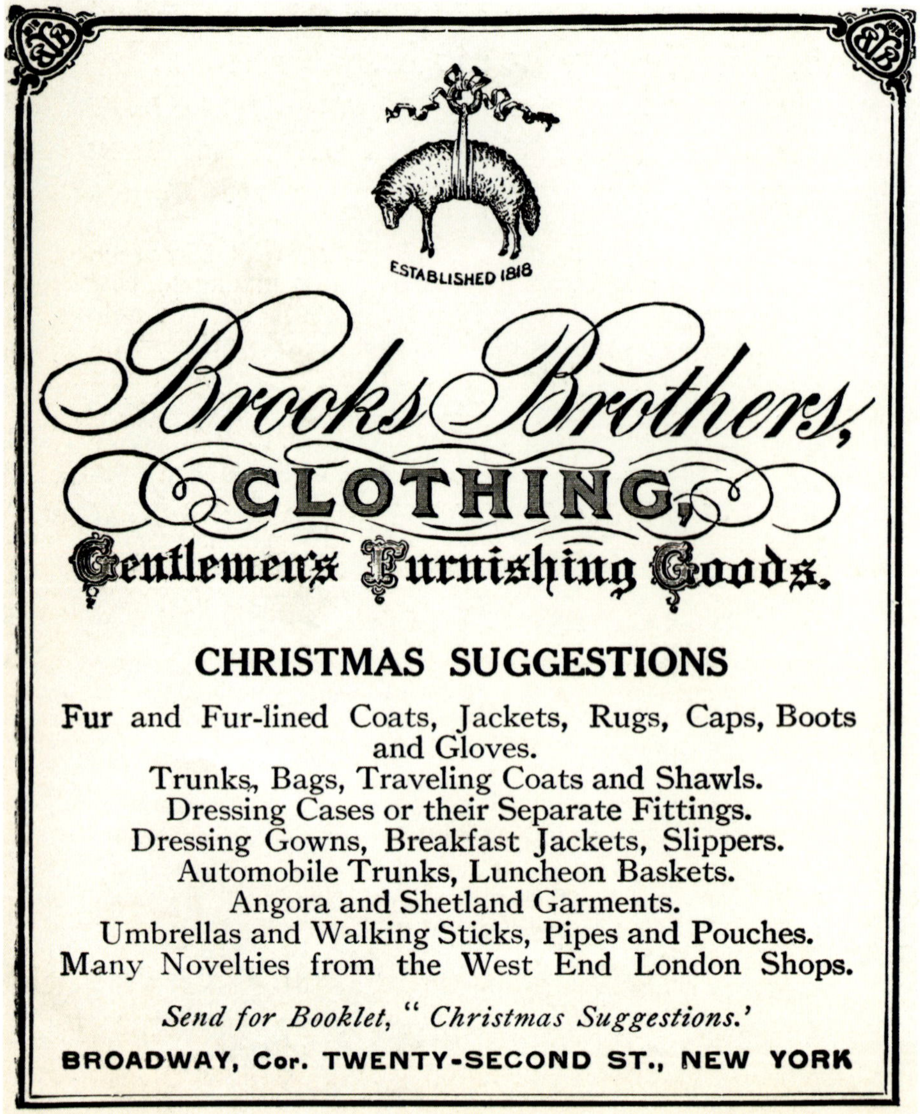

Gifts for the well-dressed man: A 1910 Brooks Brothers Christmas ad with the now-iconic golden fleece logo.

1818 BROOKS BROTHERS

Founded in 1818 in New York City by Henry Sands Brooks, Brooks Brothers is the oldest continuously operating clothing brand in the United States. The original name of the company was H. & D. H. Brooks & Co., but it changed to Brooks Brothers when Brooks's sons inherited the company in 1850 after his death. During the nineteenth century, Brooks Brothers was well-known for a couple of reasons: They were favored by presidents, including James Madison and Abraham Lincoln (Lincoln was wearing custom Brooks Brothers when he was assassinated) and also had a contract to outfit the Union Army during the Civil War. In later years, the company would become known for innovations, including launching the first ready-to-wear line of suiting in America, introducing the seersucker suit for warm-weather wear, and bringing out the first Oxford cloth button-down shirt. In 1850, Brooks Brothers introduced their iconic golden fleece logo—a lamb suspended by a ribbon. This trademark inspired by a European legend from an order of knights became known to loyal customers as a symbol of superior quality.

1819 THE FIRST AMERICAN HORROR STORY

The Sketch Book of Geoffrey Crayon, Gent., by Washington Irving (1783–1859), was published in installments between 1819 and 1820. The collection includes "The Legend of Sleepy Hollow" (among other famous stories), one of America's most famous ghost stories and a cornerstone of early American literature. Set in the Dutch settlement of Sleepy Hollow, New York, the tale follows superstitious schoolmaster Ichabod Crane and his terrifying encounter with the legendary Headless Horseman while pursuing the affections of Katrina Van Tassel. The story's portrayal of the Hudson Valley's fall landscape and its exploration of New World versus Old World values resonated deeply with readers of the day who were seeking cultural identity in the young country. The tale has inspired countless adaptations across every medium—from Tim Burton's film interpretation to the television series *Sleepy Hollow*. The story also started many Halloween traditions and gothic imagery that persist today. Irving's Headless Horseman character became an archetypal figure in the American horror genre. The story's themes of superstition versus rationality, outsider versus community, and the power of local legend continue to captivate readers and viewers alike, cementing its place in the American cultural imagination.

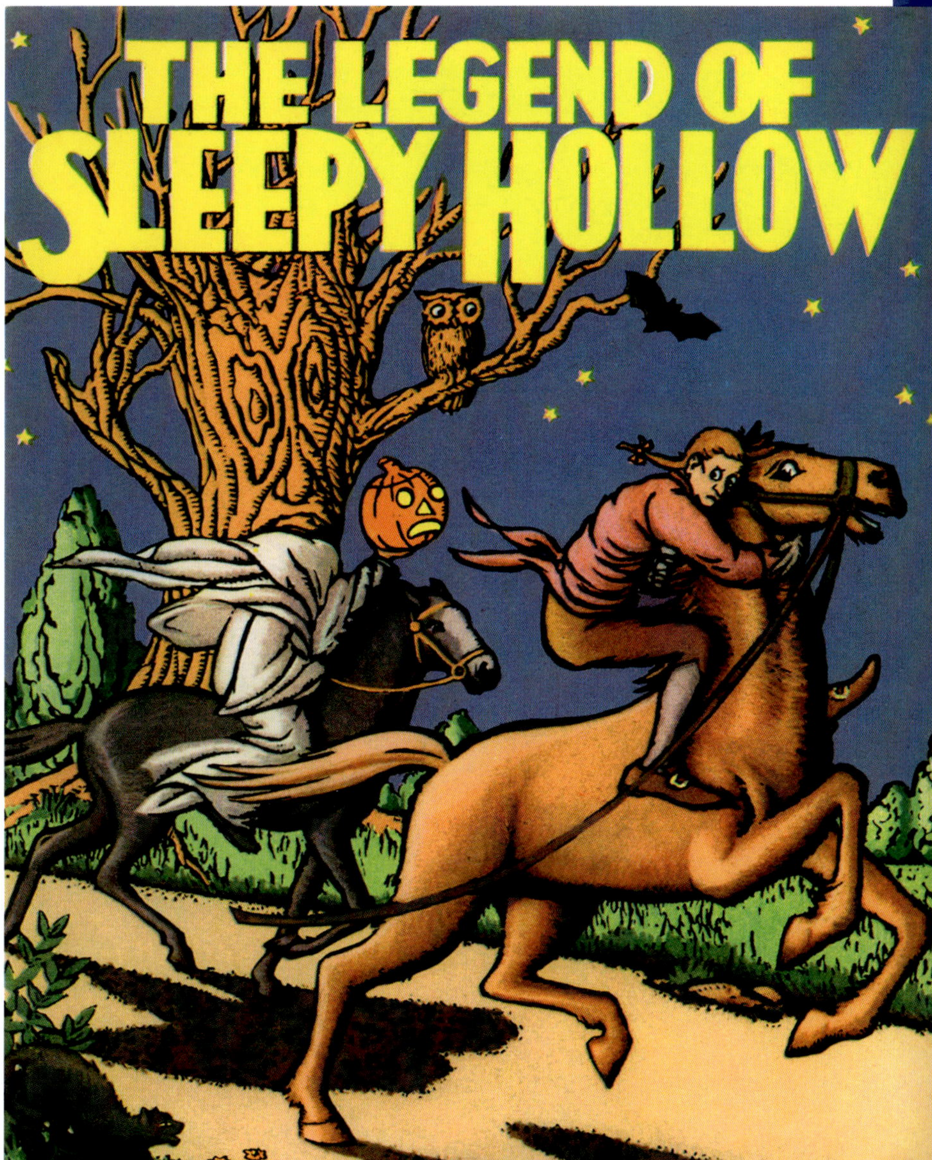

Cover illustration for the book by Frances Brundage, c. 1926.

1820 US BOTANIC GARDEN

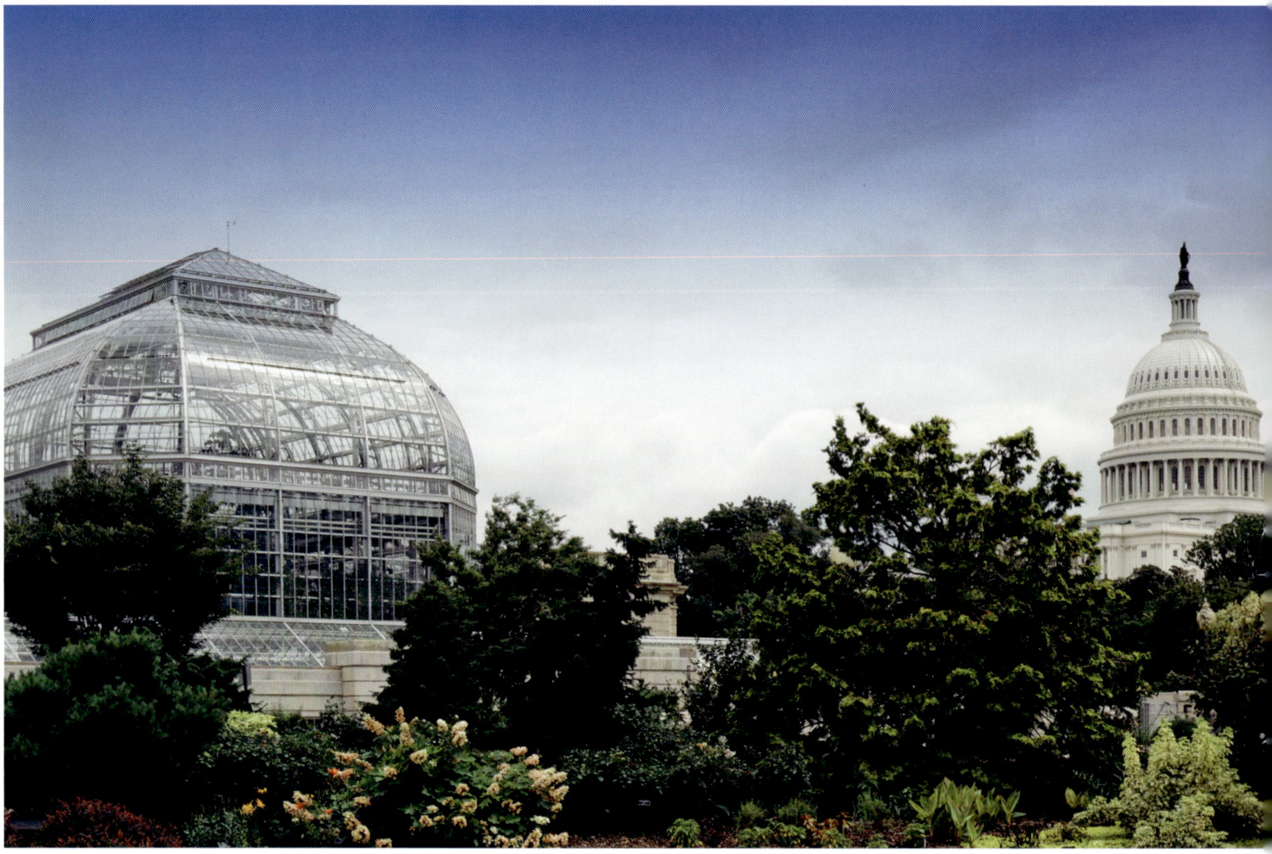

The United States Botanic Garden, located in Washington, DC, is significant not only because of its impressive array of plants from around the world and that it welcomes more than a million visitors a year to its grounds, but also because it is the oldest continuously operating public garden in the country. A national public garden was a longtime dream of George Washington during the late eighteenth century, so much so that he wrote a letter to Congress persuading them to designate land for his dream. On May 8, 1820, President James Madison signed that bill into being. The original garden was located just to the west of the Capitol grounds. This facility was open until 1837. In 1842, members of the Wilkes Expedition brought a number of previously unknown plant samples from its travels to Washington, DC, and the dream of a national botanic garden was rekindled. The Wilkes Expedition plants became the inaugural permanent collection in the garden's history. Additional expansions and relocations occurred over the years, and the Garden moved to its present location in 1933. Impressively, the Garden's collection still contains specimens believed to be directly related to the Wilkes Expedition plants.

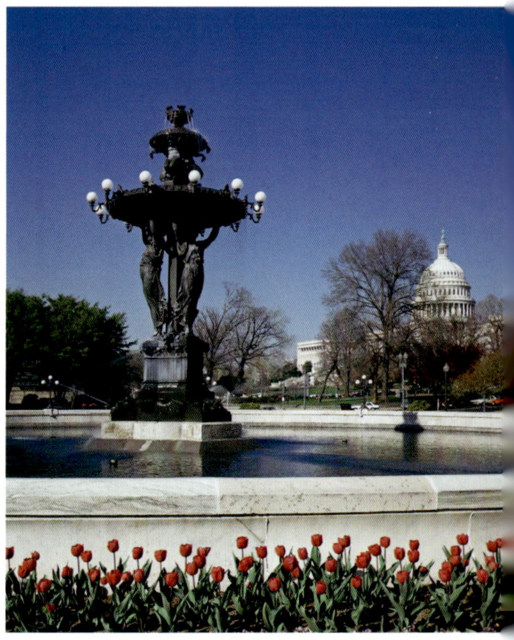

The Bartholdi Fountain at the Gardens.

1821 DRY SCOURING CLOTHES

Thomas Jennings (1791–1859), a Black man, was born free in New York City and thus was able to pursue a trade (he became a skilled tailor) and operate his own business. Jennings was dissatisfied with the methods available at the time for cleaning clothing and came up with a way to remove dirt and oil from garments without the use of water. Because Jennings was a free man, there was another important opportunity available to him: patenting his invention. He seized the opportunity in 1821 and was the first Black man to be granted a patent for his "dry scouring," a precursor to dry cleaning. Jennings used the earnings from his successful invention to fight for the abolition of slavery. He became an influential and vocal member of his community, and was active in the civil rights movement. Sadly, Jennings died six years before the Thirteenth Amendment officially abolished slavery in the United States. Another shadow on Jennings's legacy is that his patent was lost in a fire in 1836 along with ten thousand other "X patents" issued between 1790 and 1836.

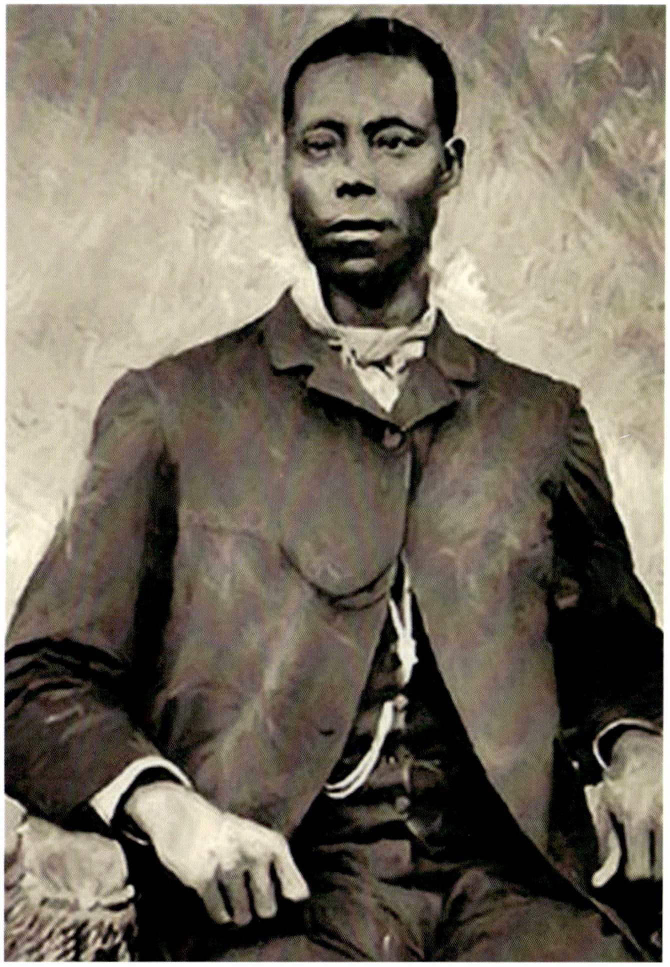

Thomas Jennings.

1822 FIRST PLAY BY A BLACK PLAYWRIGHT

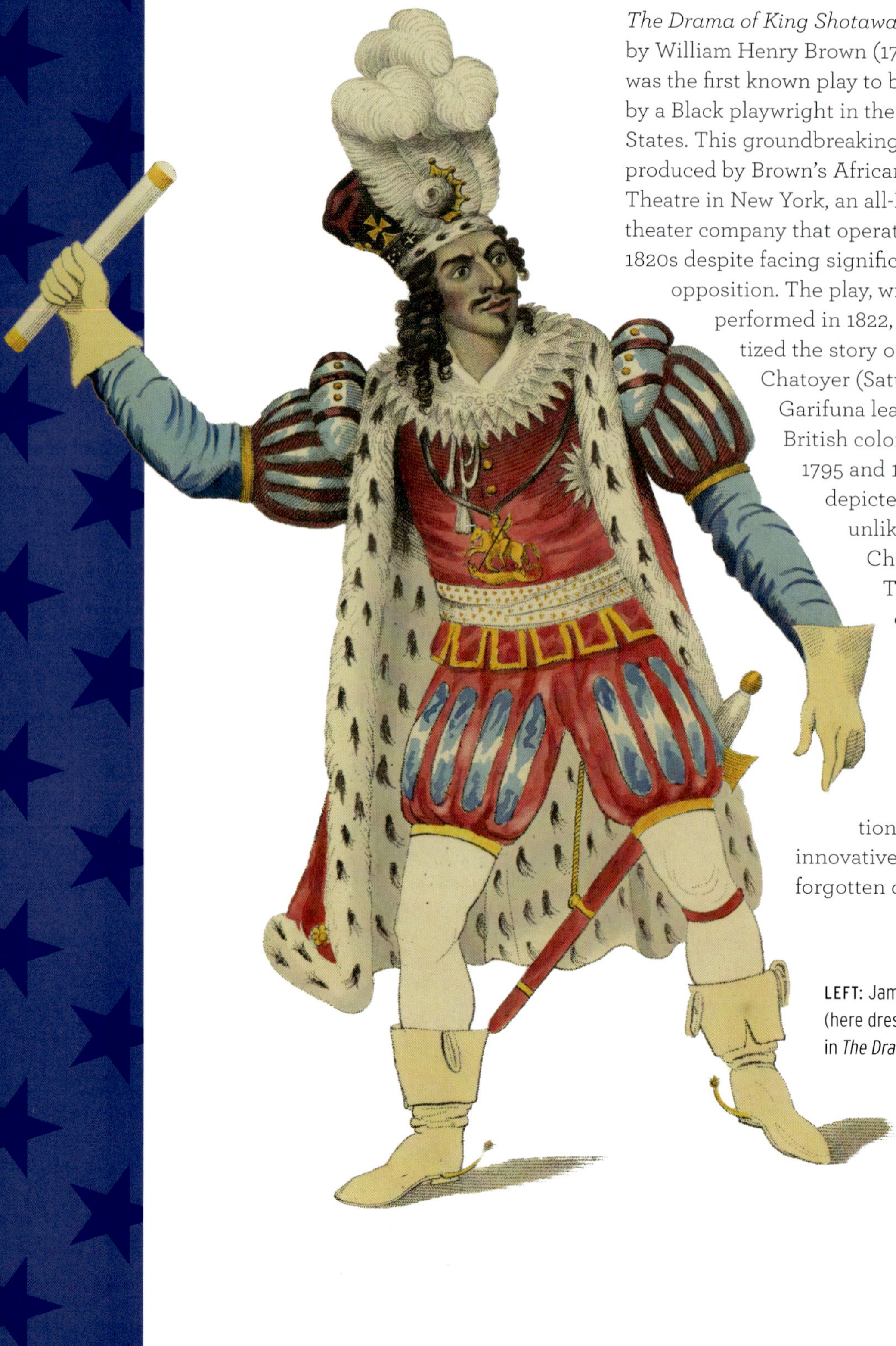

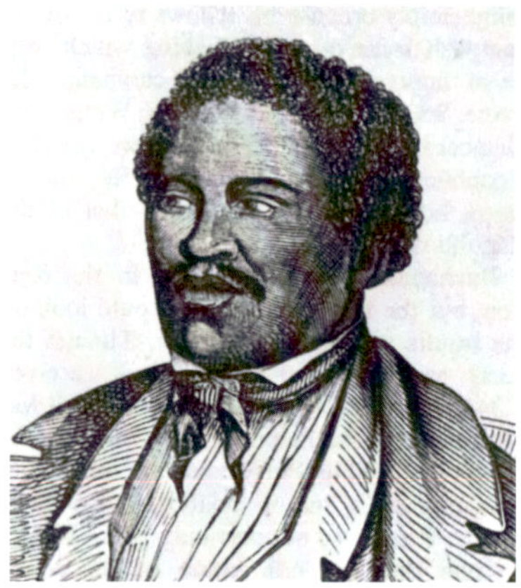

The Drama of King Shotaway, written by William Henry Brown (1790–1884), was the first known play to be written by a Black playwright in the United States. This groundbreaking work was produced by Brown's African Grove Theatre in New York, an all-Black theater company that operated in the 1820s despite facing significant local opposition. The play, written and performed in 1822, dramatized the story of Joseph Chatoyer (Satuye), a Garifuna leader who led the Black Carib resistance against British colonialism on the island of Saint Vincent between 1795 and 1796. Similar to *Hamilton!*, *King Shotaway* depicted a revolution against British tyranny, though unlike Hamilton's successful American Revolution, Chatoyer's rebellion was ultimately unsuccessful. Tragically, no known copies of the original play exist today. However, its legacy endures as a pioneering work that challenged theatrical conventions and gave voice to Black Caribbean resistance narratives. The play remains widely celebrated as a cultural landmark. Modern artists continue to honor the lost masterpiece through contemporary adaptations and stage readings, ensuring that Brown's innovative contribution to American theater is not forgotten despite the absence of the original text.

LEFT: James Hewlett, as the star actor of the African Grove Theatre (here dressed as Richard III), most likely portrayed Joseph Chatoyer in *The Drama of King Shotaway*. **ABOVE:** William Henry Brown.

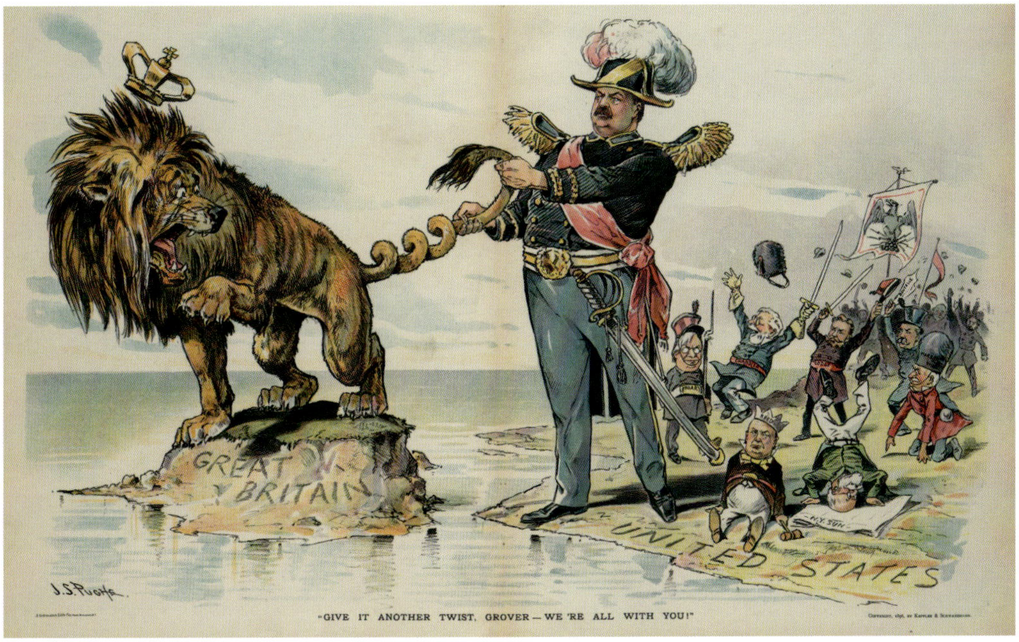

President Grover Cleveland twists the tail of the Lion (Britain) in this *Puck* cartoon.

1823 THE MONROE DOCTRINE

It can be said that the Founding Fathers of the United States of America were constantly innovating as they conceived the democracy of the new country. Considered the last Founding Father, James Monroe (1758–1831), the fifth US president, delivered a famous statement to Congress on December 2, 1823, that outlined the United States' policy toward European nations. Although the statement was a small part of a larger address to Congress, it had far-reaching and lasting effects. The main component of the doctrine was that Europe should not interfere in Western Hemisphere affairs—no colonization or attempts at placing monarchs in or near the country would be tolerated. At first, the European powers took little notice of Monroe's statement, but it had a profound and lasting effect. Since those early years, the Monroe Doctrine has been evoked both overtly and symbolically to defend nearby nations by presidents such as Theodore Roosevelt when defending the US intervention in Latin America (the Roosevelt Corollary), John F. Kennedy during the Cuban Missile Crisis, and more.

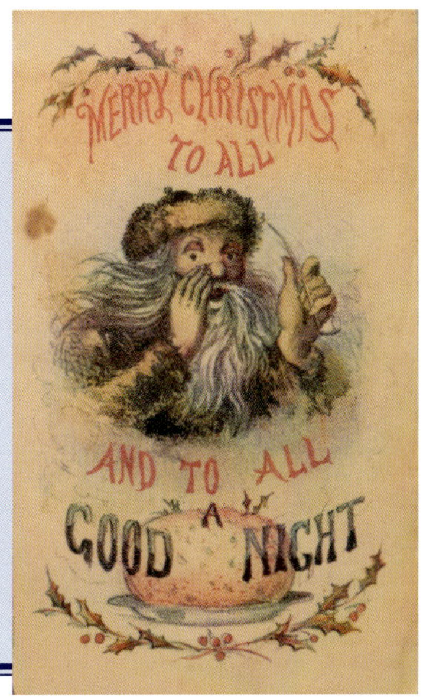

1823 "A Visit from St. Nicholas"

On December 23, 1823, a Christmas poem was published anonymously in the *Sentinel*, the local newspaper of Troy, New York. The poem, titled "Account of a Visit from St. Nicholas," is now known by its first line, "The Night Before Christmas," and is one of the most influential pieces of holiday writing ever created. Not only is it responsible for our modern vision of Santa Claus, but it also shaped our ideas of Christmas gift-giving. The poem is credited to Clement Clarke Moore; as the story goes, he wrote it one snowy day while on a shopping trip on a sleigh. Although there is some question as to whether Moore is the writer, there is no doubt that it is one of the best-known poems ever written, and a true holiday classic.

1824 RUBBER GALOSHES

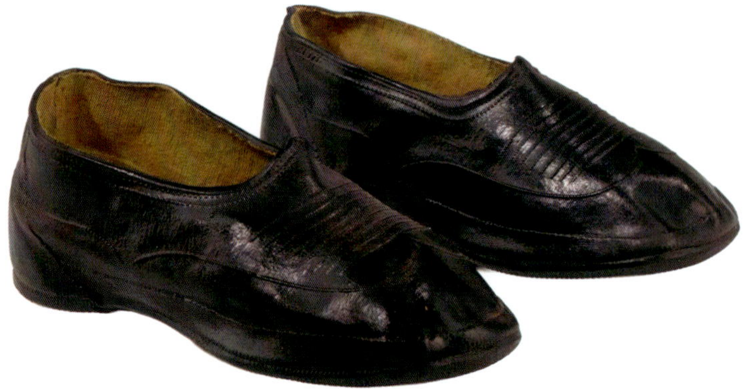

On February 4, 1824, J. W. Goodrich (birth and death date unknown) made a revolutionary contribution to footwear history by introducing rubber galoshes to the American public. Goodrich's galoshes transformed how people navigated inclement weather. Prior to this invention, people had limited options for keeping their shoes dry, often relying on primitive overshoes or simply accepting that their footwear would become damaged by moisture and mud. The introduction of the flexible, waterproof rubber galoshes provided an accessible and effective barrier against the elements, offering people a practical solution to protect their shoes from rain, snow, and mud. Modern galoshes, foul-weather boots, and even workwear evolved from Goodrich's foundational concept, becoming essential equipment across numerous industries including construction, food processing, agriculture, and health care. Today's versions feature advanced materials, steel toe protection, slip-resistant soles, and ergonomic designs. Goodrich's simple yet ingenious invention allowed people from all walks of life to maintain dry, protected feet regardless of environmental conditions, fundamentally changing how society approached outdoor activities during adverse weather.

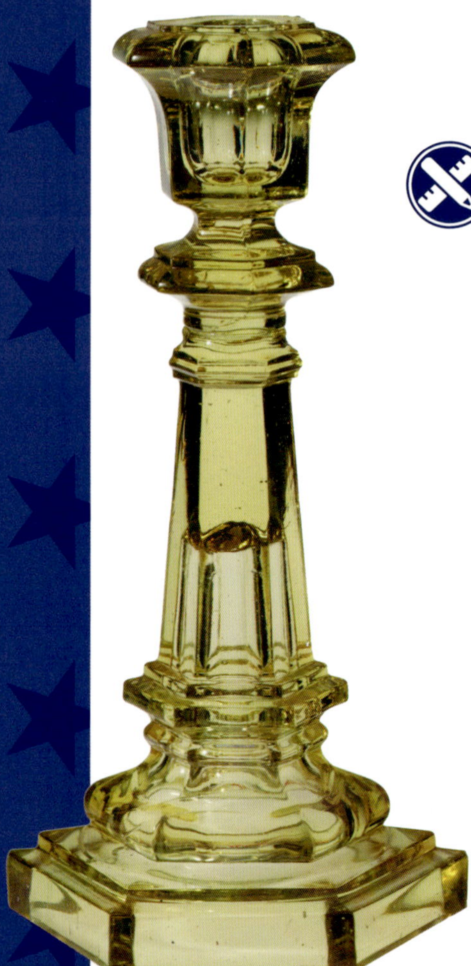

1825 BOSTON AND SANDWICH GLASS COMPANY

Deming Jarves (1790–1869), a former executive of the New England Glass Company, founded the Boston and Sandwich Glass Company in Sandwich, Massachusetts, in 1825. The company was known for its high-quality pressed glass products. Sandwich, the oldest town on Cape Cod, was an ideal location for a factory because it was situated near a harbor that could allow for easy delivery of the silica needed to create the glass and the wood needed to fuel the furnaces for heating the raw material. The factory's pressed glass proved very popular, and Jarves was granted several patents for improving the glass-pressing process. His prosperity helped create a thriving community in the town of Sandwich for many years to come. The company expanded to include a large array of home goods, including lamps, perfume bottles, and tableware in a range of colors and textures. Sandwich glass is still prized by collectors today.

1826 INTERNAL COMBUSTION ENGINE

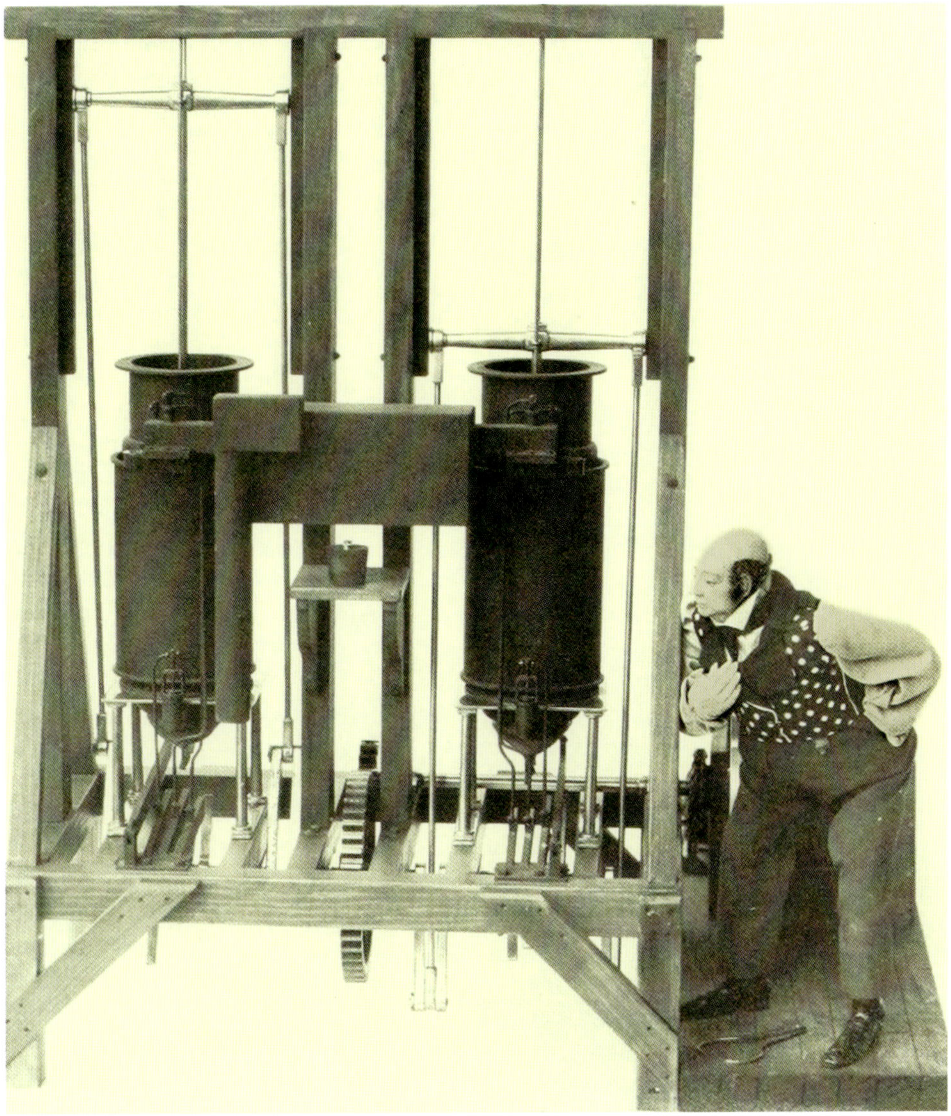

In 1826, American inventor Samuel Morey (1762–1843) received the first US patent for an internal combustion engine, a device that would eventually revolutionize the automotive industry. During his early experiments, Morey discovered that turpentine vapors exploded when exposed to air, and he harnessed this action to drive a piston in a vacuum-controlled environment. Although there were other engines in development at the time, Morey's device was also the first to employ a precursor to a modern carburetor separated from the liquid fuel by a piece of wire mesh. Morey successfully demonstrated the engine's use in front of live audiences both in a boat and a wagon in the years after the patent was granted, but he was unsuccessful in obtaining a buyer for the engine and was rumored to have retired in frustration. Sadly, his patent was also one of the many lost in the 1836 Patent Office Fire.

1827 FIRST TOURIST STUNT AT NIAGARA FALLS

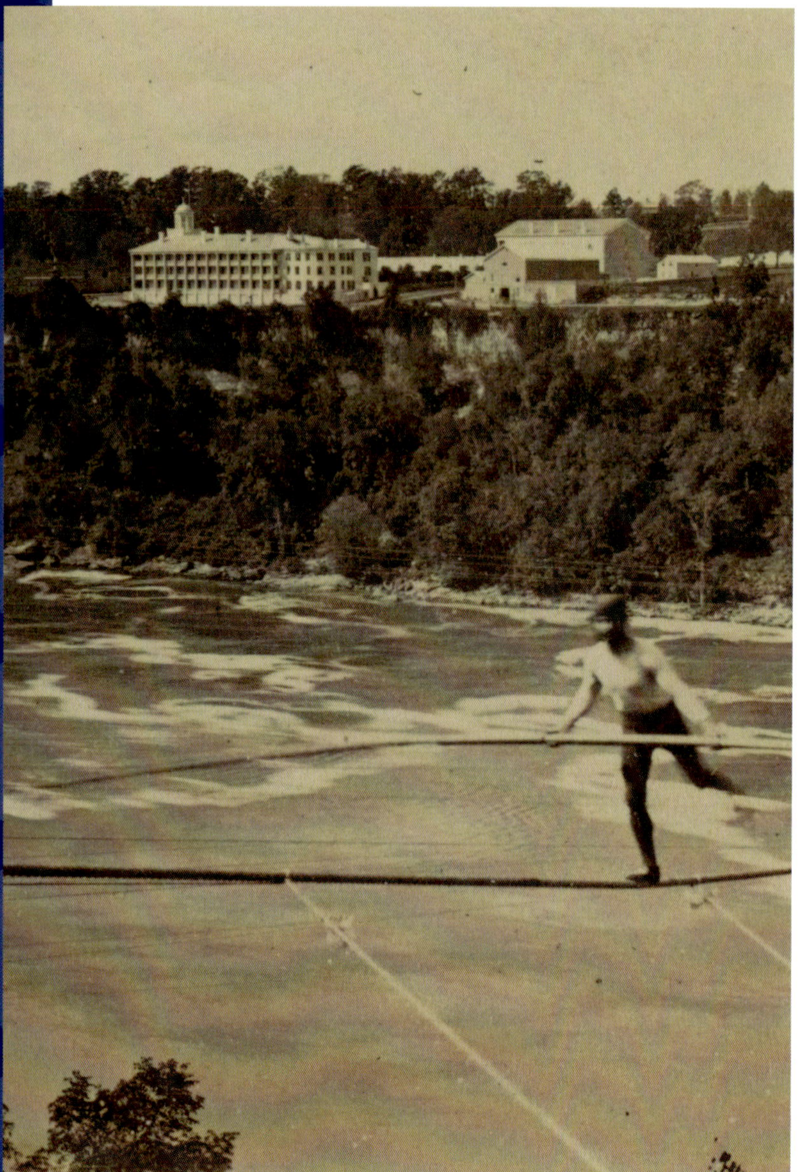

Niagara Falls has a long and storied history of stunts staged to attract tourists and their dollars to area hotels, gift shops, and bars and restaurants, but September 8, 1827, marks the day that the first one played out on the American side of those famous Falls. Parkhurst Whitney (1784–1862), owner of the Eagle Hotel on the New York side of the falls, conspired with two other local businessmen on the stunt. In an effort to draw people to the area, the promoters advertised the upcoming spectacle: an old boat filled with "ferocious" beasts, or "wild animals," would be sent over the falls. But on the day, the boat had been filled with more domestic creatures—reports vary, but small bears, foxes, a dog, geese, a raccoon, and an old buffalo have been mentioned. The crowd didn't disappoint—it was estimated to be around ten thousand people of mixed social classes. When the schooner arrived at the falls' edge with the animals aboard, eyewitnesses report that it split in two before plunging over the cascading water to cheers from the crowd, who apparently didn't much mind that it meant the demise of most, or all of, the animals. The spectacle was considered a resounding success, and descriptions of the day were reported in newspapers all across the globe. The events of September 8, 1827, solidified Niagara Falls as a tourist destination and paved the way for many more wacky stunts to come.

LEFT: A tightrope performer goes over the falls in the late nineteenth century. **RIGHT:** Annie Edson Taylor, the first human to go over the falls in a barrel and live.

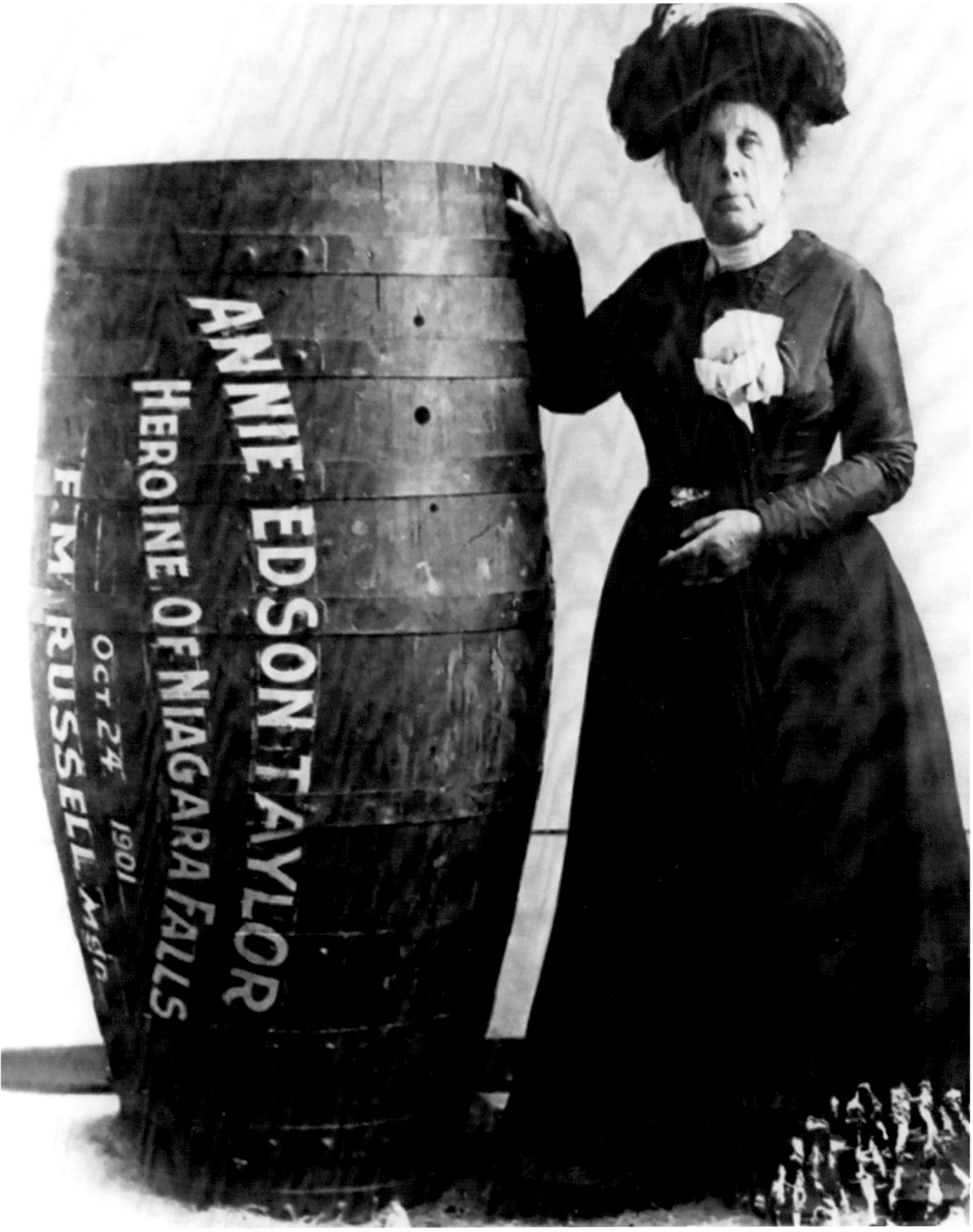
ANNIE EDSON TAYLOR
HEROINE OF NIAGARA FALLS
OCT 24 1901
F M RUSSELL MGR

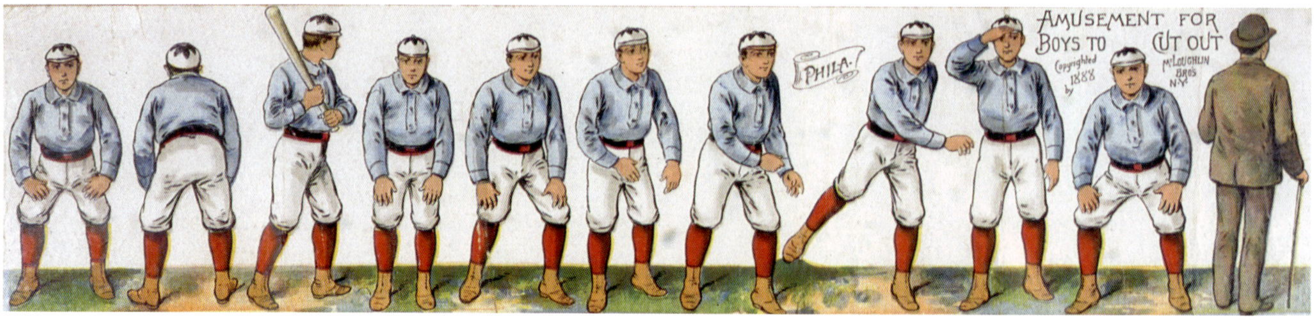

1828 PAPER DOLLS

In 1828, McLoughlin Brothers, a printing company located in New York City, was founded. The company became a prominent figure in the paper doll industry, printing their dolls from engraved wood blocks (prior to this, the dolls were made of thin paper without tabs, using sealing wax to affix the figures). This wood-block technology marked a significant development in the production and popularity of paper dolls. The McLoughlin Brothers' most popular doll models were alliteratively dubbed Dottie Dimple, Jenney June, and Lottie Love. McLoughlin Brothers was responsible for many innovations of its day, including publishing high-quality children's picture books (the company was a groundbreaker in color printing technology), puzzles, games, and board books. After a very successful run, in 1920, McLoughlin Brothers, founded by John McLoughlin Jr. (1827–1905) who was later joined by his brother Edmund McLoughlin (1833–1889), was sold to Milton Bradley & Company.

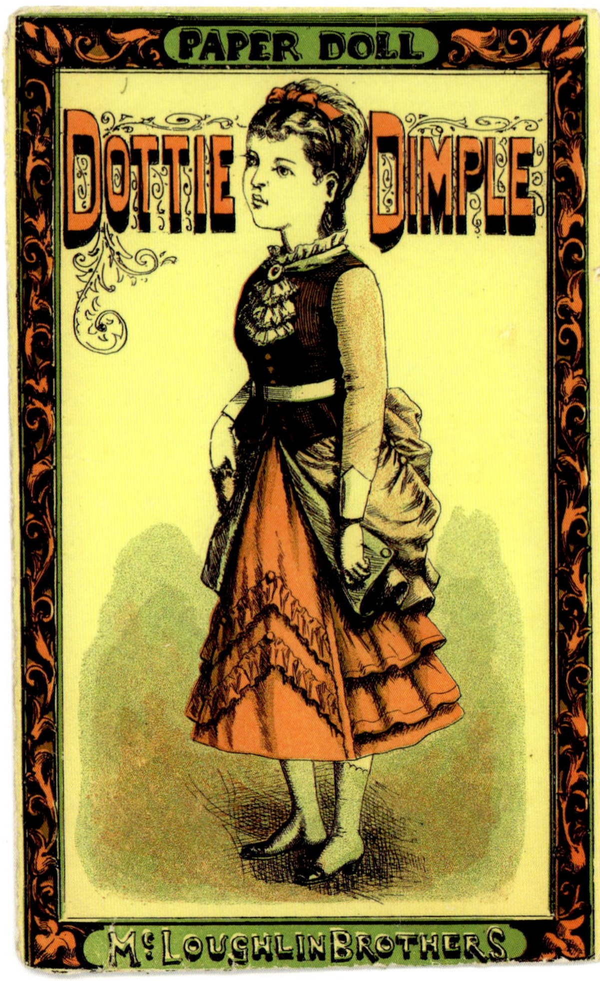

McLoughlin paper dolls: Dottie Dimple for girls, and baseball players for boys.

1829 GRAHAM CRACKERS

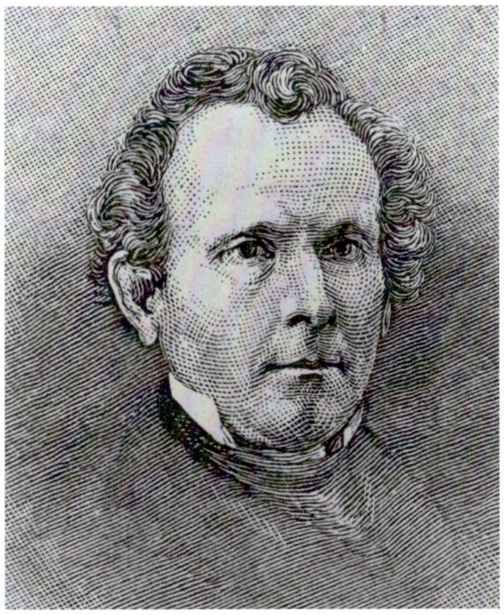

Sylvester Graham.

Known today as an ingredient in one of the ultimate campfire indulgences, s'mores, graham crackers didn't always have such a decadent reputation. The graham cracker was invented by the unconventional American Presbyterian minister Sylvester Graham (1794–1851) from Connecticut. Graham was a religious zealot who prescribed to his followers a rigorous lifestyle of abstinence and moral and spiritual purity. He was an early proponent of vegetarianism and an advocate of temperance and controlling sexual behavior. Graham developed his eponymous cracker as an offshoot of his diet, which recommended eating lots of fresh produce, whole wheat bread, cereal, and plenty of fiber and water. The original cracker was very bland, made from coarse whole wheat flour with no sugar or fat. Over time, Graham modified his recipe to include small quantities of oil, natural sweeteners, and salt for a more palatable cracker. Followers of Graham created a coarse whole wheat flour they dubbed "graham flour" to make their own crackers. In 1898, the National Biscuit Company started mass-producing graham crackers, which have been a popular staple in America ever since.

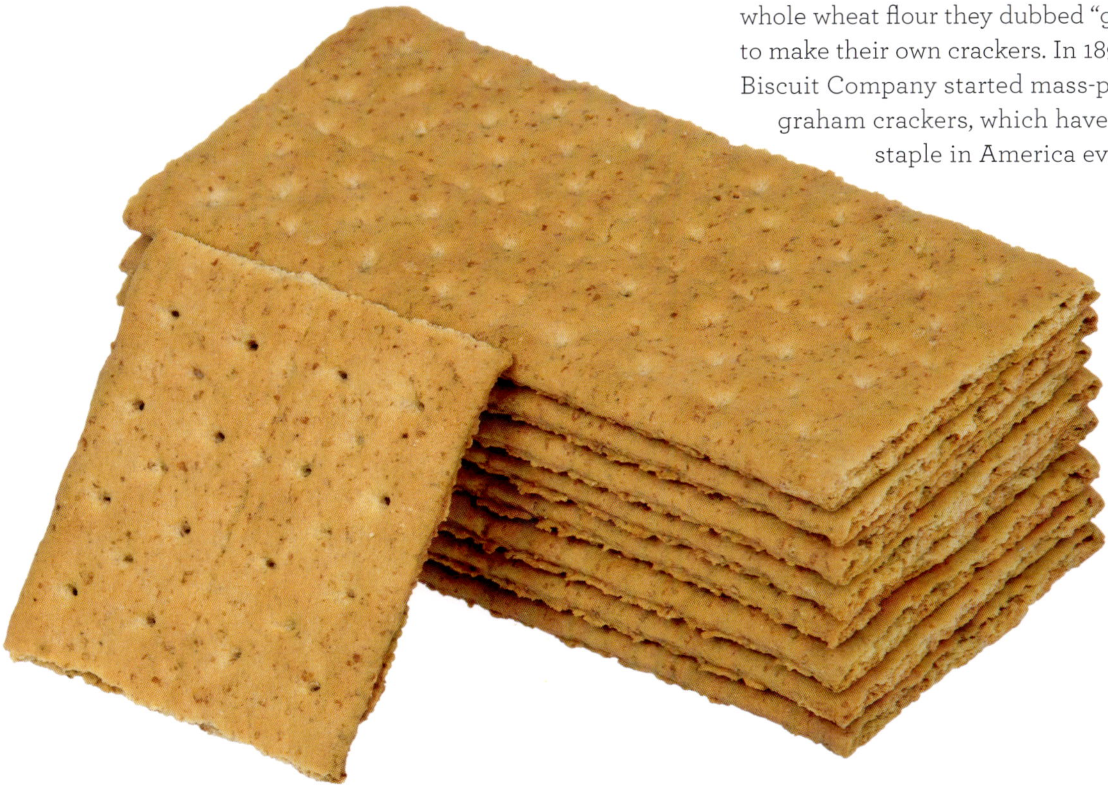

Exciting Trial of Speed between Mr. Peter Cooper's Locomotive, "Tom Thumb," and one of Stockton & Stokes's Horse-Cars.

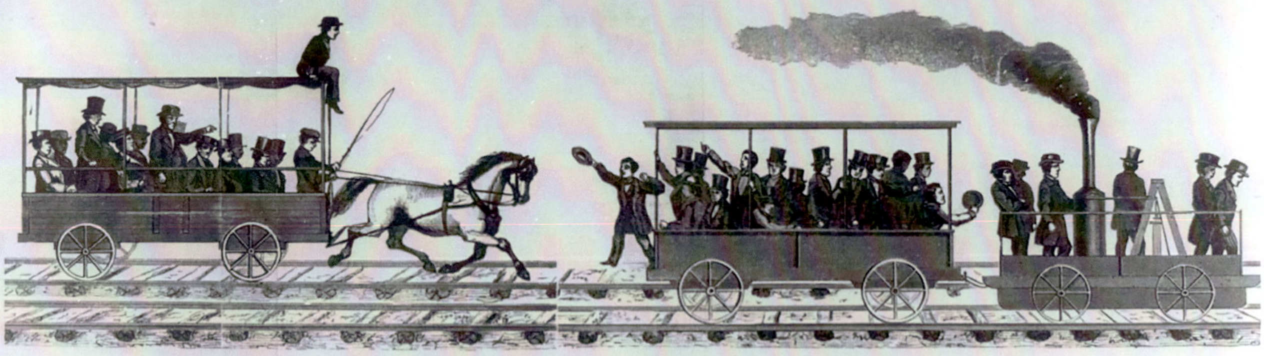

The trial took place on the Baltimore and Ohio Railroad, on the 28th August, 1830. The sketch represents the moment the Engine overtook and passed the Horse-Car, the passengers filled with excitement.

(See Mr. Latrobe's description, page 116.)

1830 TOM THUMB LOCOMOTIVE

Steam locomotives were invented in Britain in the beginning of the nineteenth century. Meanwhile, in America, the Baltimore and Ohio (B&O) Railroad was using horses to pull its trains. In 1830, inventor Peter Cooper (1791–1883) built a miniature steam locomotive that was powered by coal and could operate on the footprint of B&O's track, which at the time ran from Baltimore to what is now Ellicott City. On August 28 of that year, Cooper's locomotive brought the B&O board of directors and other passengers along the track to demonstrate its use. The locomotive was christened "Tom Thumb" due to its small size, and it became known as the first American steam locomotive. The Tom Thumb was in operation unofficially until 1831. Its success helped the B&O Railroad expand into steam-powered instead of horse-driven locomotives. The Tom Thumb remains a fond relic in the minds of Americans. A model of the train is part of the collections at the National Museum of American History, and its image appeared on a US postage stamp in 1952.

1831 MECHANICAL REAPER

Virginia native Cyrus McCormick (1809–1884) is credited with building the first mechanical grain reaper—one of the most important agricultural innovations ever created—after his father was unsuccessful at doing the same. At just twenty-two years old, he toiled in his father's workshop to fashion the implement—a cutting blade that vibrated, a reel that helped gather the grain, and a base that received the grain after it was cut. The device was mounted on two wheels that were connected via a harness to a team of horses in a chariot-like fashion. The reaper combined several steps that earlier harvesting machines had done separately and the new tool helped even the proverbial playing field for farmers who could not afford to hire a large labor force (or who had fewer enslaved people working for them). McCormick patented his machine in 1834 and began licensing it a few years later. Subsequently, he began manufacturing his reaper in a factory in Chicago and grew the business into a highly profitable one.

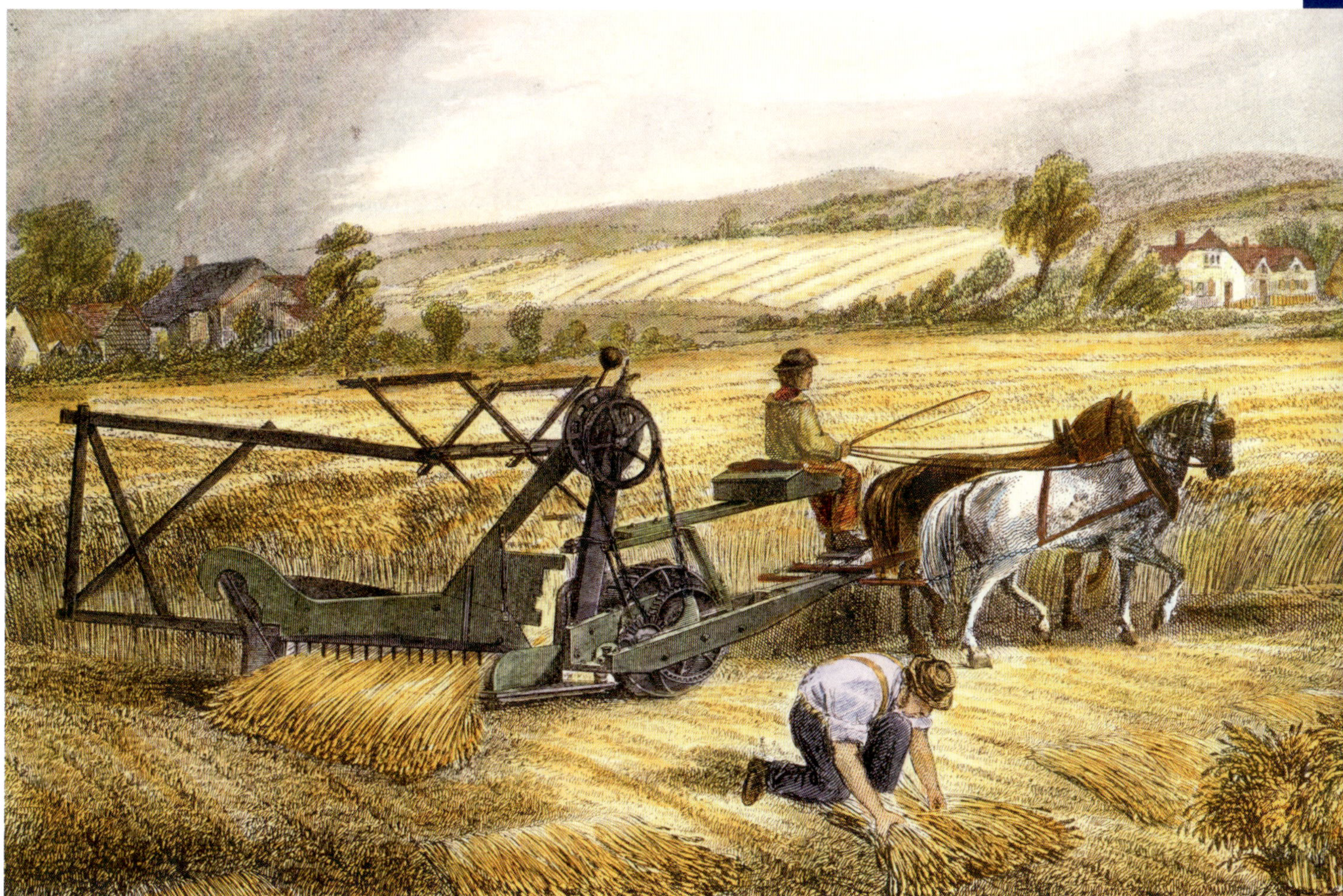

1832 SODA FOUNTAIN

Englishman John Matthews (1808–1870), who grew up with the naturally carbonated waters in his native England, immigrated to New York in 1832 with the goal of producing soda water for what he hoped would be a captive audience. Although soda fountains already existed, Matthews's new system made the process of carbonating water far more efficient and cost effective. Legend has it that marble dust was at one point part of Matthews's carbonation technique, and he procured his supply from construction remnants from the building of St. Patrick's Cathedral. Matthews patented his device in 1936 under the name "Genuine Matthews Apparatus." He sold his machine to soda fountain owners in New York City and beyond and became very successful due to the sustained popularity of the carbonated drinks accented with flavored syrups such as ginger, sarsaparilla, and assorted fruit flavors. After his death in 1870, records show that Matthews's son, John Matthews Jr., followed in his footsteps and continued to design and seek patents for other products relating to soda fountains.

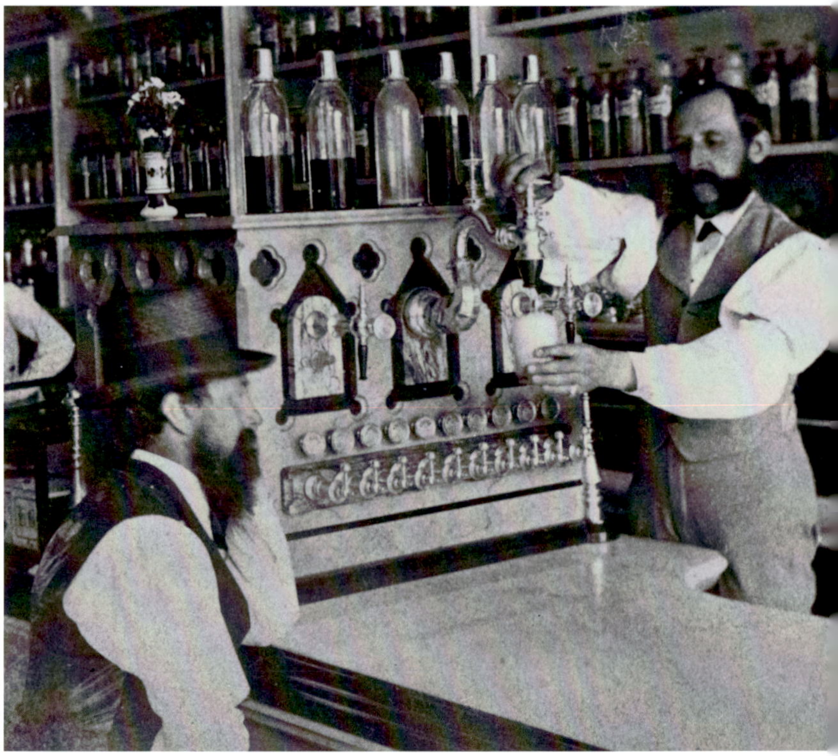

1833 FIRST FREE PUBLIC LIBRARY

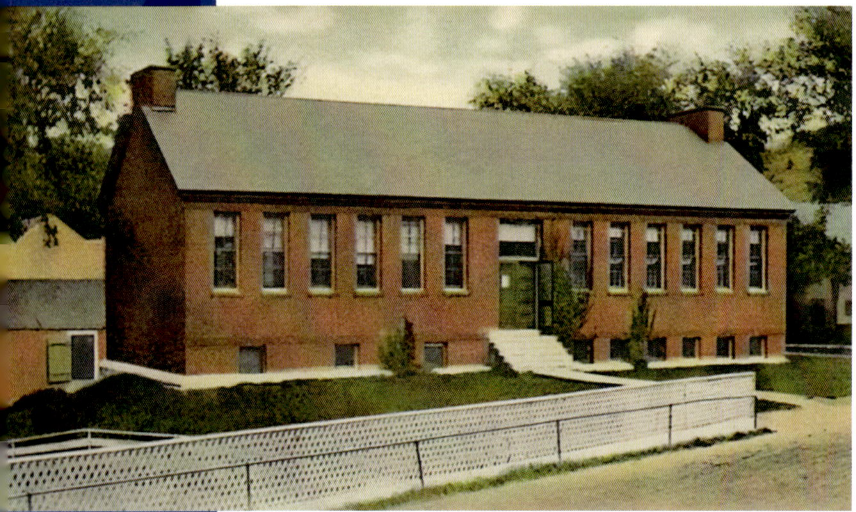

Originally housed in the town hall, the library moved into this new home, built by George S. Morison, in 1893.

The first truly free public library in the United States was established on April 9, 1833, in Peterborough, New Hampshire. This pioneering institution was the vision of Reverend Abiel Abbot (1765–1859), who advocated for a library funded entirely by public taxation rather than private donations or membership fees. The Peterborough Town Library represented a radical departure from existing libraries and subscription-based collections that served only paying members. This groundbreaking model demonstrated that knowledge and literacy were public goods that deserved community investment. The library's success inspired similar initiatives across New England and eventually nationwide, establishing the principle that access to information should not depend on economic status. The Peterborough library laid the foundation for America's public library system, and modern public libraries continue this democratic mission, providing free access to books, digital resources, Internet connectivity, and educational programming. The concept of tax-supported public libraries has become so fundamental to American civic life that it's difficult to imagine communities without them.

1834 CORN PLANTER

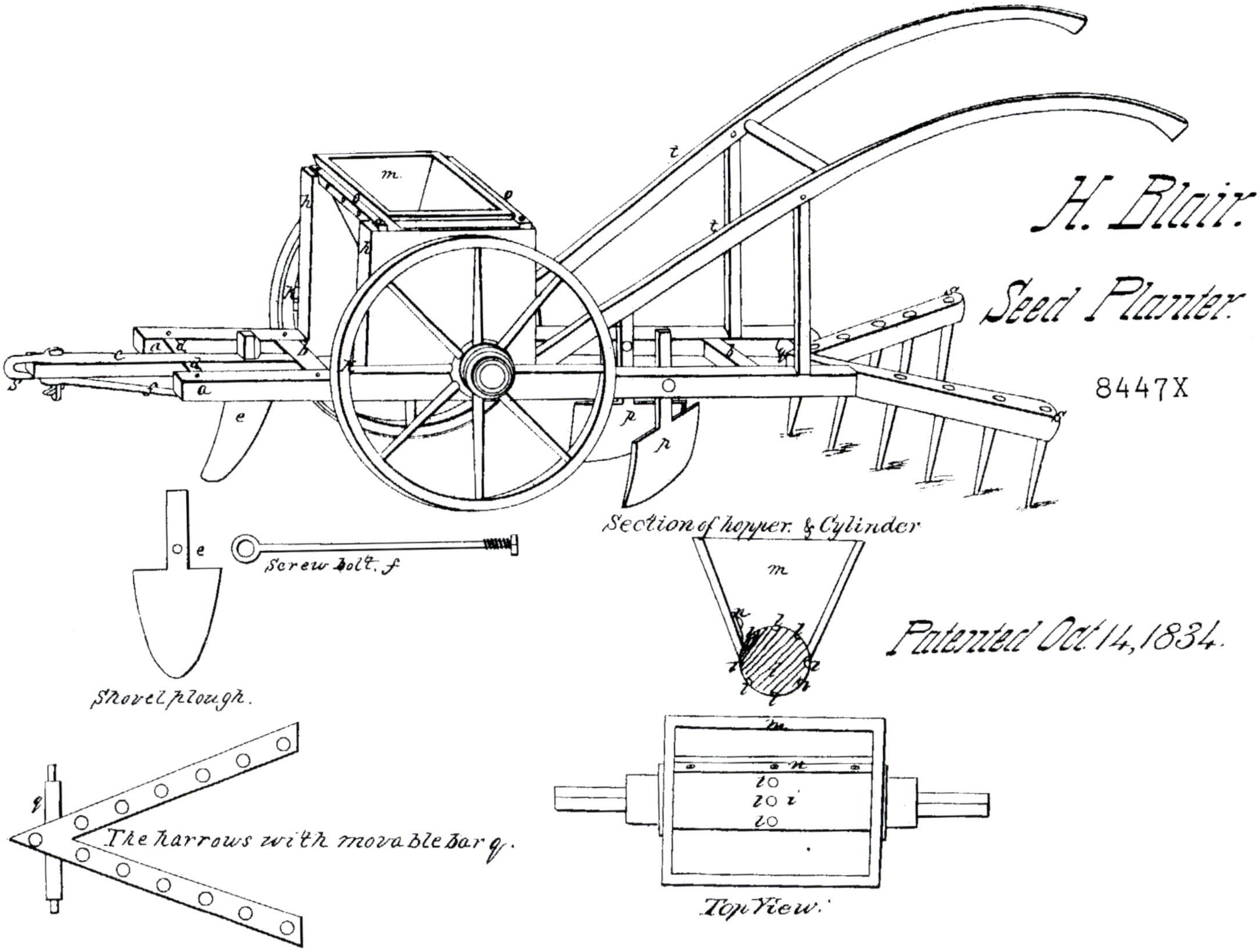

Henry Blair's patent drawing for his corn planter.

Henry Blair (1807–1860) was an African American farmer and inventor from Maryland in the nineteenth century. Although there is little information available about his personal life, it appears that he was not enslaved, so he was therefore able to run his own business during the era. On October 14, 1834, Blair received a patent—likely the second known patent received by a Black man—for what he called a corn planter. This wheelbarrow-like device had a compartment affixed to the base of it that spread seeds and large combs affixed to the back that worked the seeds into the soil as they were dispersed. This device helped farmers plant seed crops more efficiently and with a greater yield than before. Blair later received a patent for another agricultural invention that he called a cotton planter, a somewhat similar mechanism with modifications optimized specifically for cotton seeds. Though he was clearly an innovator, it is speculated that Blair was unable to read or write, as he signed his patents with an X.

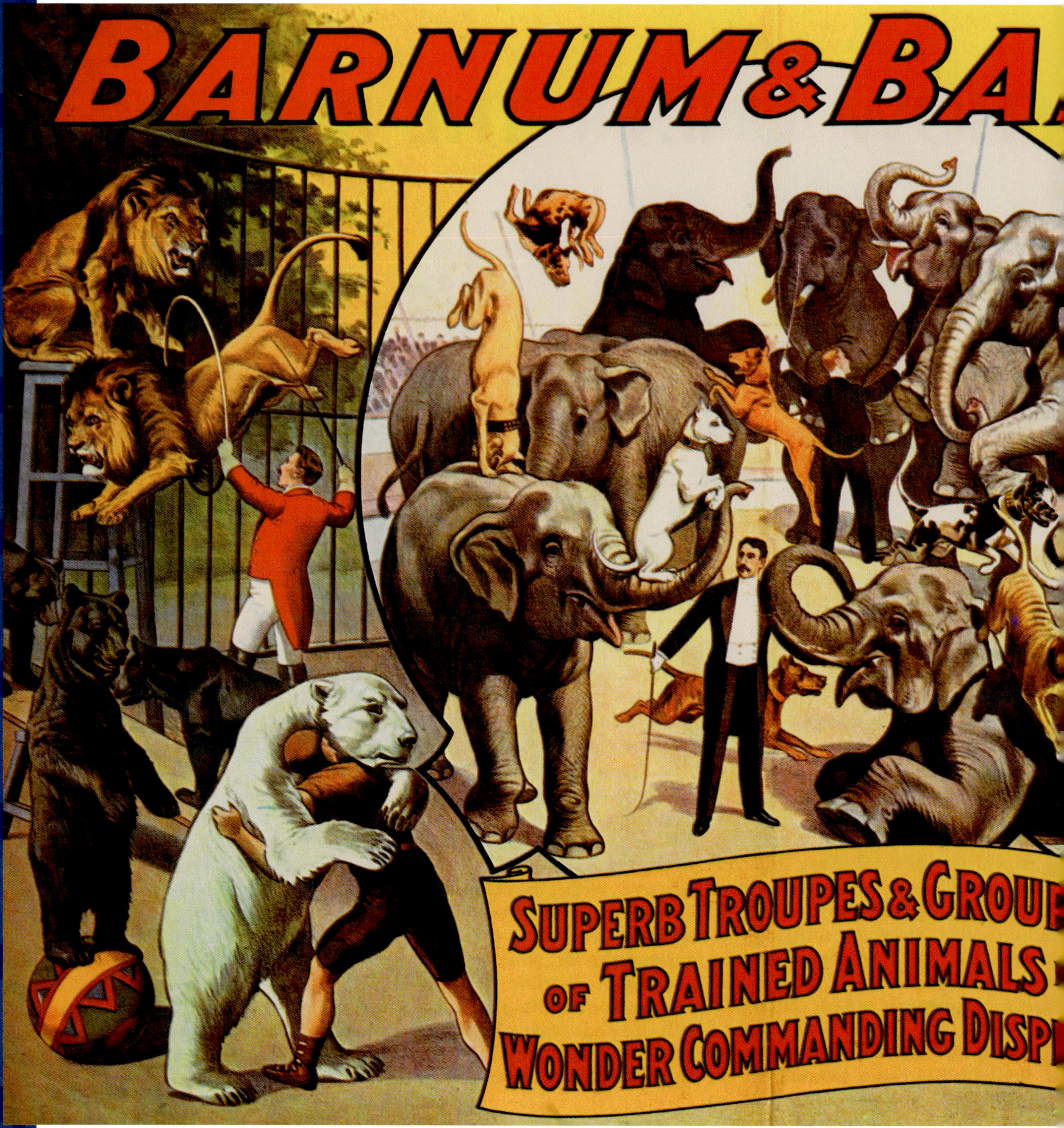
BARNUM&BA
SUPERB TROUPES & GROUP
OF TRAINED ANIMALS
WONDER COMMANDING DISP

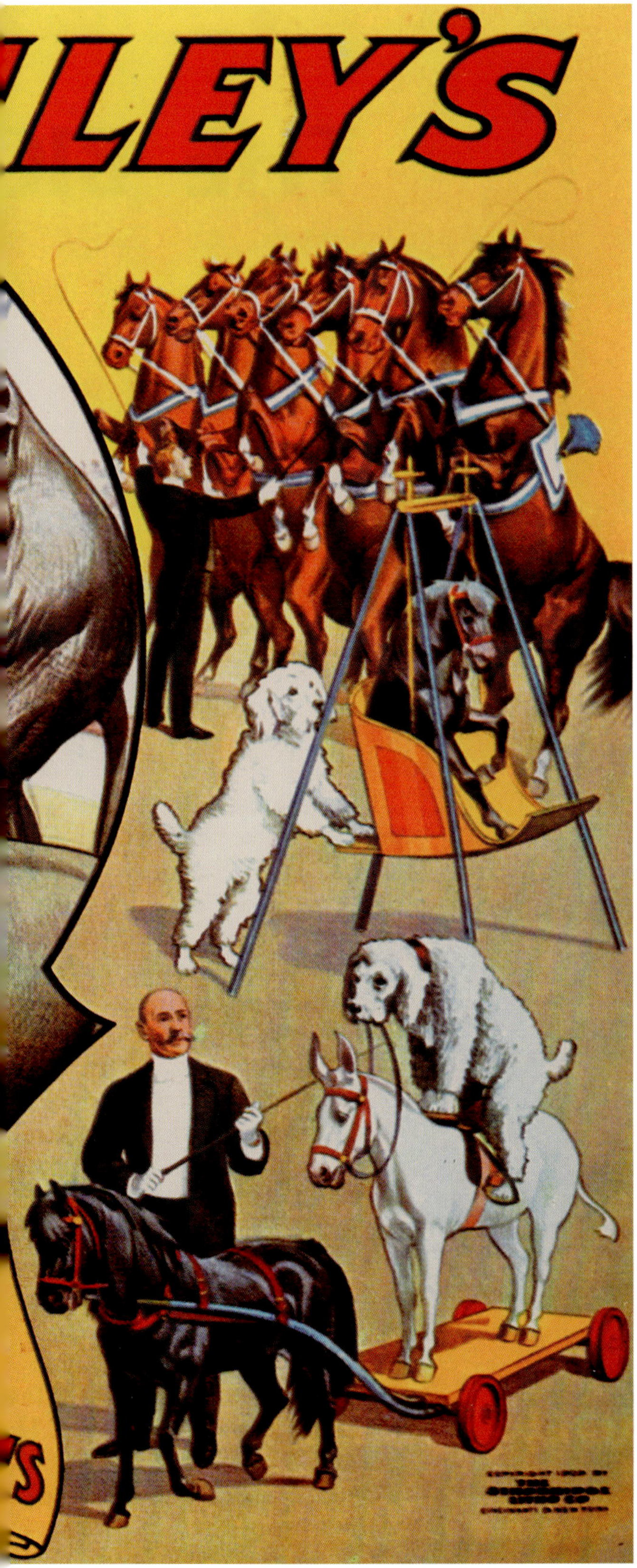

1835 BILLBOARDS

Some mistakenly believe that P.T. Barnum created billboards, but even though Barnum was certainly involved, credit for the achievement goes to a New York resident named Jared Bell (birth and death date unknown). Bell was a printer who originally created large-scale, highly colorful, sometimes political signs on special, durable paper that could withstand being outside. One story purports that Bell created a sign for a zoological society in 1835 that was spotted by P. T. Barnum. Ever the self-promotor, Barnum wanted a sign for himself advertising his forthcoming circus—only he wanted his billboard even bigger and better than the ones he had seen. Bell created a billboard more than fifty square feet in size advertising Barnum's clowns and tigers that would soon be arriving in town. Bell's pioneering work, using saturated colors and appealing figures rendered with depth and movement, became the impetus for outdoor advertising as we know it today.

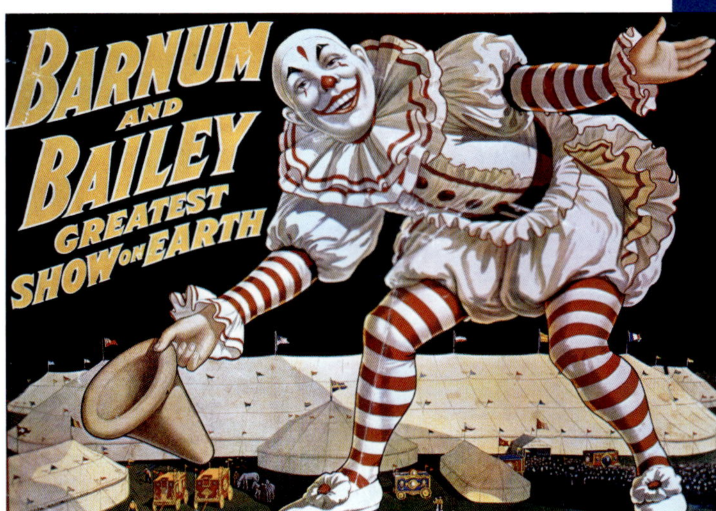

P. T. Barnum used billboards to great effect, as shown in these late nineteenth-century examples.

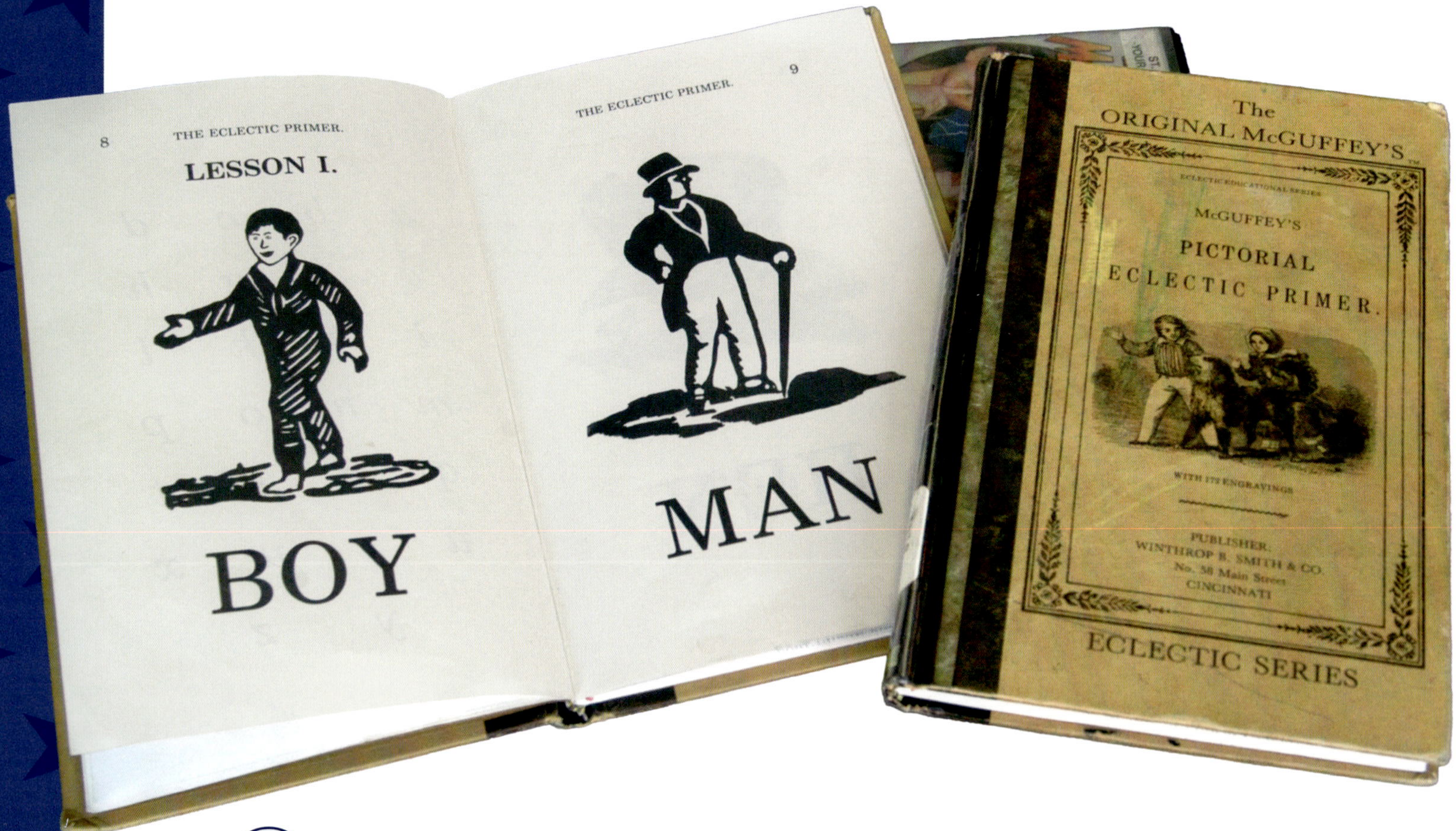

1836 THE MCGUFFEY READERS

Created by Ohio teacher William Holmes McGuffey (1800–1873), this series of primary school textbooks were widely used in American schools beginning in 1836. About 120 million copies of the readers were sold between 1836 and 1960, placing their sales in a category on par with the Bible. The reading guides were monumental in a number of ways: First, due to their widespread adoption, the series reflected the moral values of the country in the nineteenth century; second, due to their popularity, the readers helped standardize the English language for school children in the country at the time; and finally, the subject matter was designed to be age appropriate to the reader, something that had not been focused on before. It has been said that the McGuffey Readers—filled with charming illustrations to capture the attention of young learners—has helped more American school children read than any other book series. The McGuffey Readers are still used by homeschooling parents and some small school districts today.

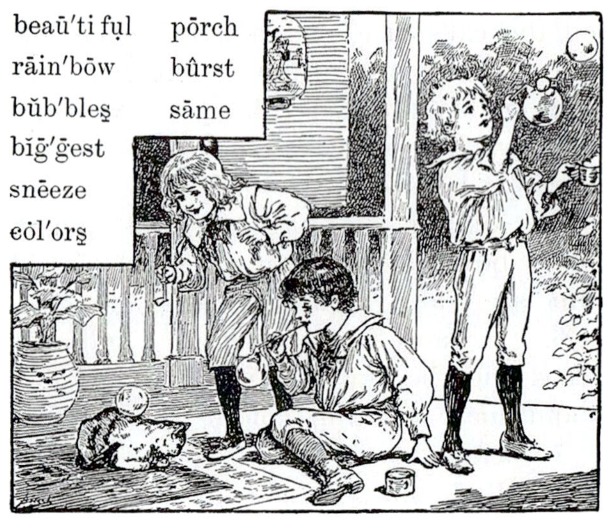

1837 JOHN DEERE PLOW

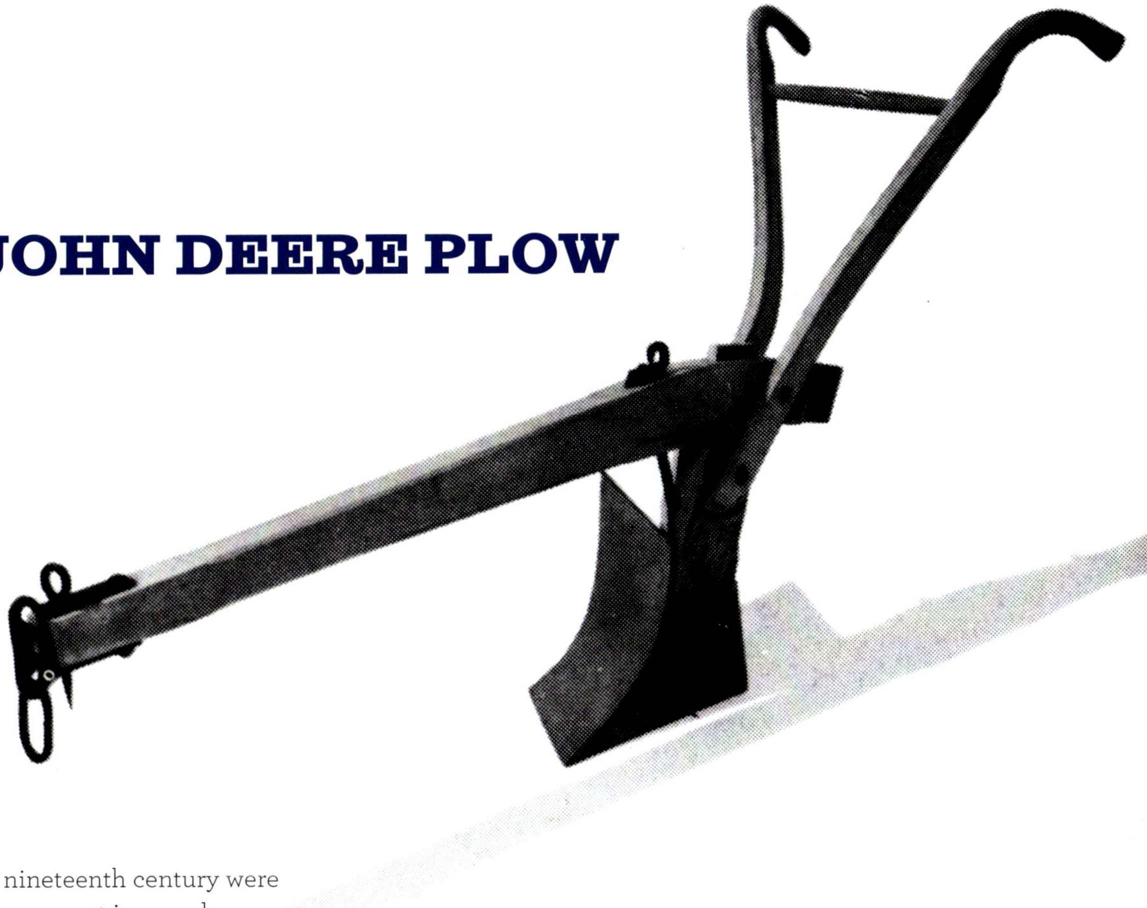

Plows in the nineteenth century were made from heavy cast iron and proved cumbersome and difficult to wield in the soil of the Illinois farmland. Enter John Deere (1804–1886), a blacksmith originally from Vermont, who relocated to Illinois when economic conditions forced him. In 1837, after learning from local farmers that the sticky prairie soil stuck to iron plow blades, Deere deduced that a plow from much lighter and highly polished cast steel would solve their problem. And he was right. The new "self-scouring" plow proved so successful that Deere was soon able to open his own plow company less than a hundred miles away in Moline, Illinois, next to the Mississippi River. The river was an ideal location for a factory, as it provided both power and loading docks for shipments. By 1856, Deere's company was one of America's great success stories, due to both his insistence on high quality and high personal and professional morals. Deere's legacy lives on today with a thriving company still located in Moline, Illinois.

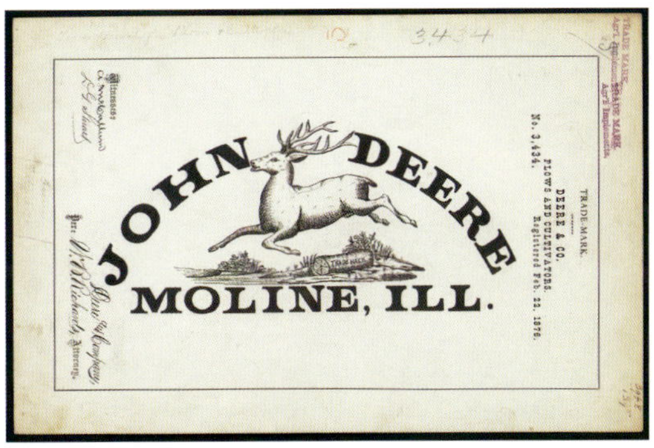

Deere's original plow, and the 1876 trademark application for the company.

1838 WILKES EXPEDITION

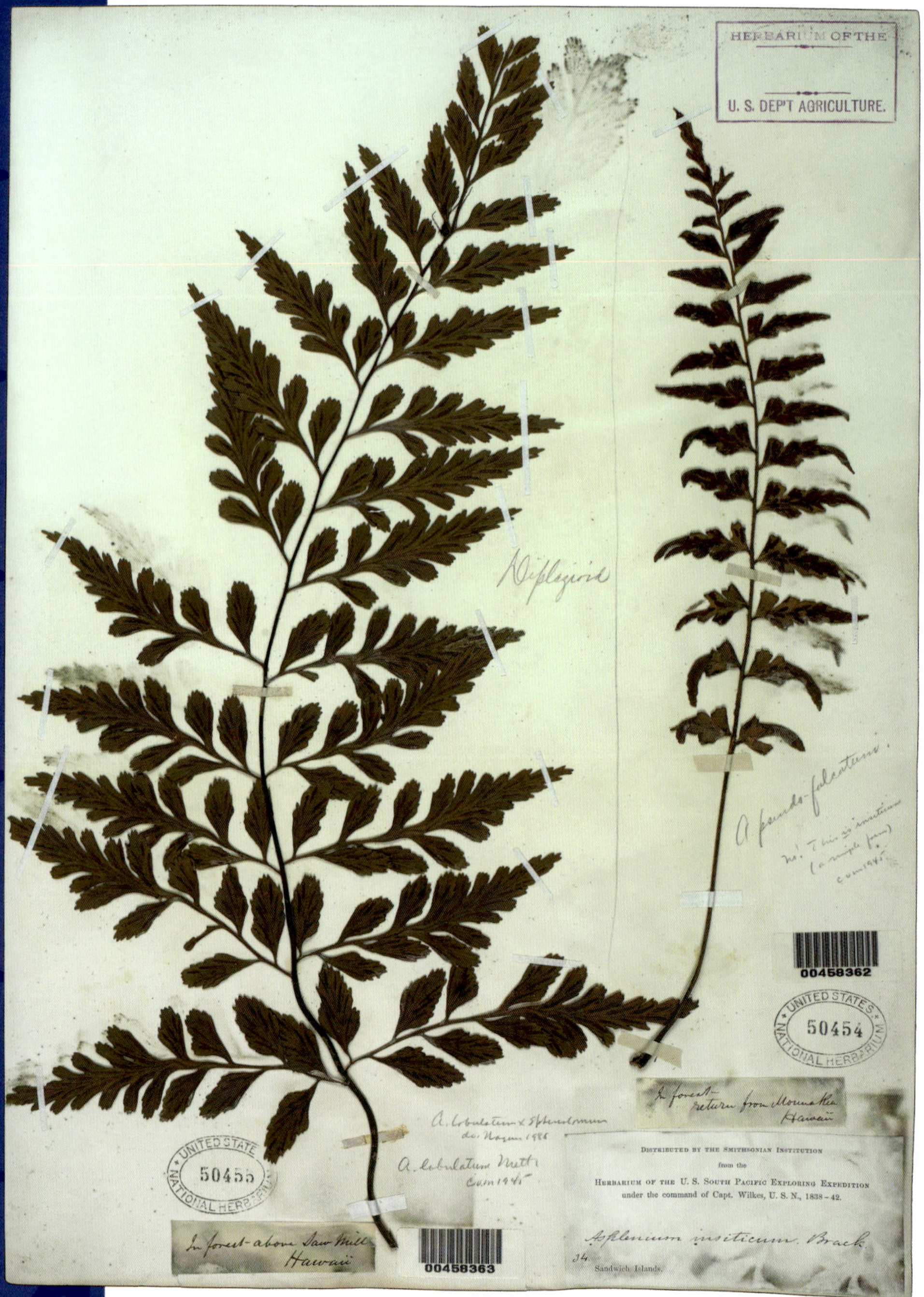

Authorized by Congress in 1836, the Wilkes Expedition launched from Hampton Roads, Virginia, on August 18, 1838, with hundreds of scientists aboard six US Navy vessels under the command of Lt. Charles Wilkes (1798–1877). Their mission was to explore and map the Pacific ocean, the northwest coast of the United States, and Antarctica. At face value, the expedition was an incredible demonstration of navigation prowess. But the results of the journey went far beyond that. The Wilkes Expedition not only broadened the knowledge of the world—its geology, botany, ethnographical data, and more—it also expanded American scientific knowledge and commerce. As a result, the Wilkes Expedition helped impress upon the rest of the world the United States' status as an economic leader. Although the expedition he led was monumental, Wilkes proved a controversial leader. Nevertheless, the expedition would go down in history as one of the most important information-gathering missions of the country's early years.

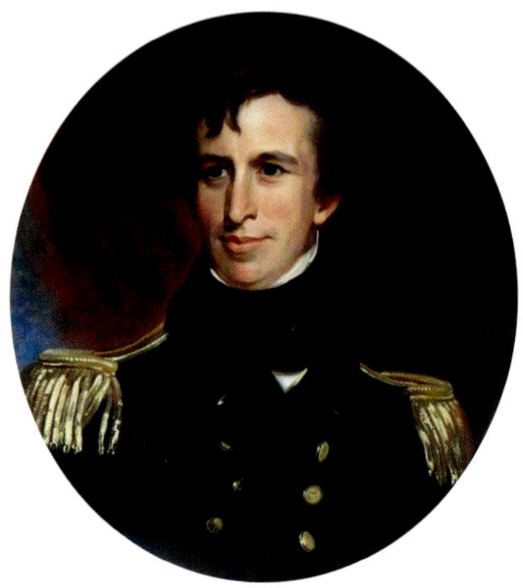

ABOVE: One of the plant specimens gathered during the expedition. **RIGHT:** Lt. Charles Wilkes.

1839 STEAM SHOVEL

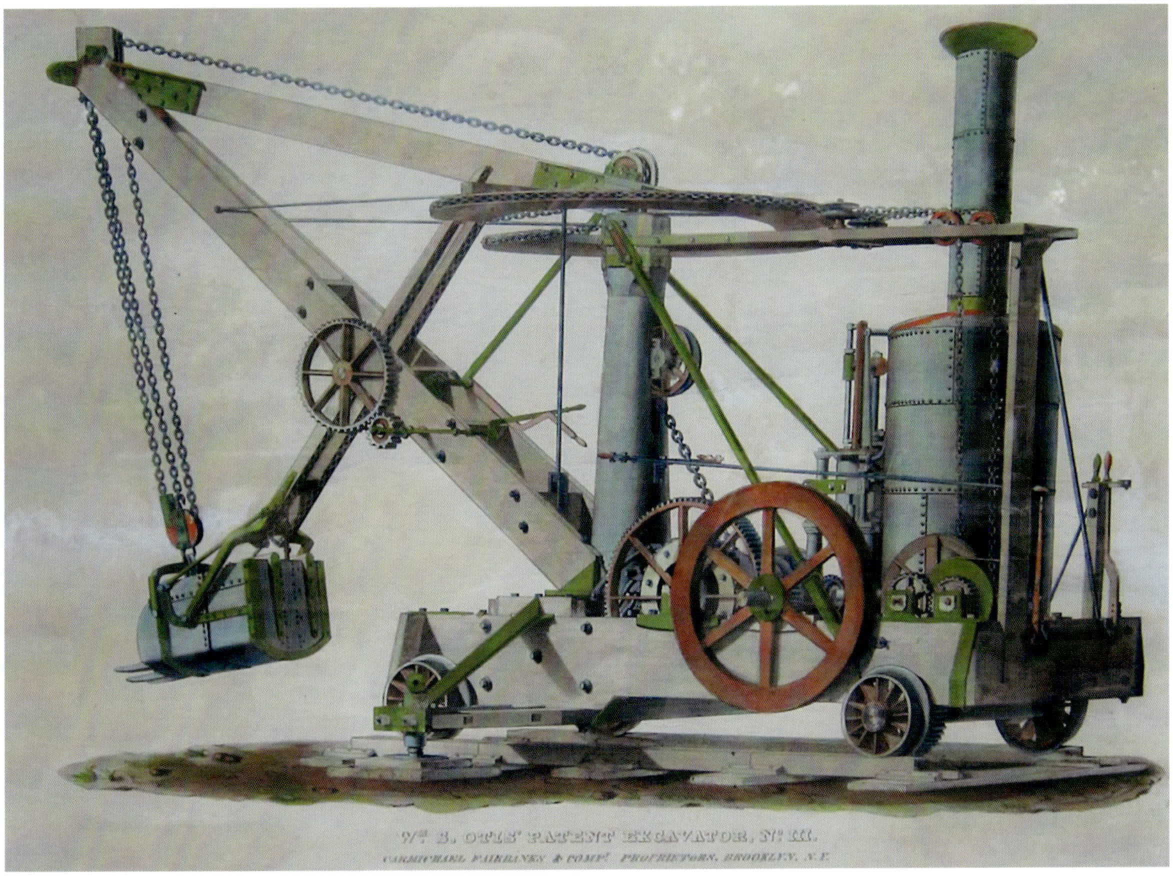

ABOVE: The patent drawing for Otis's excavator. BELOW: William Otis.

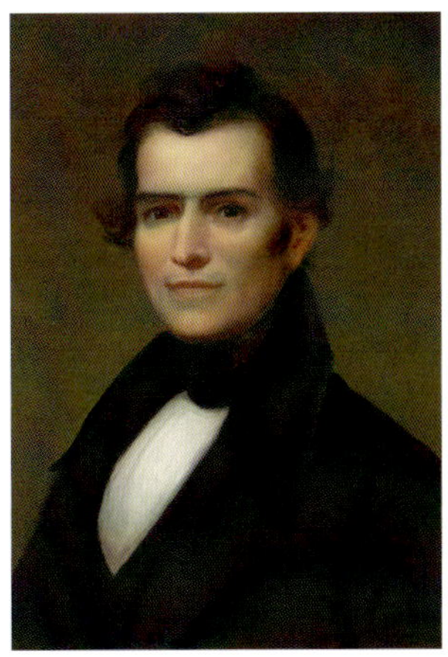

On February 24, 1839, Massachusetts-born William Otis (1813–1839) received a patent for the steam shovel, also known as the power shovel, a type of mechanical excavator. The steam-powered machine could lift loads of dirt by the bucketful, raise them on a crane that was mounted on a railroad car, and then move the dirt via a pulley system to dump it into an empty railcar. This innovation helped increase digging production exponentially over human-powered labor. Otis died several months after receiving his patent at the young age of twenty-six, but the firm for which he worked at the time—railroad construction company Carmichael and Fairbanks, which was owned by Otis's relatives—continued to develop the technology for even more improvements. Because it competed with inexpensive immigrant labor, the steam shovel was not an overnight success. However, it eventually became a major component of the United States' western expansion far beyond just the railroad, including canals, dams, mines, and bridges. William must have had inventing in his blood, because his cousin Elisha invented the passenger elevator a few years later.

1840 PAINT TUBES

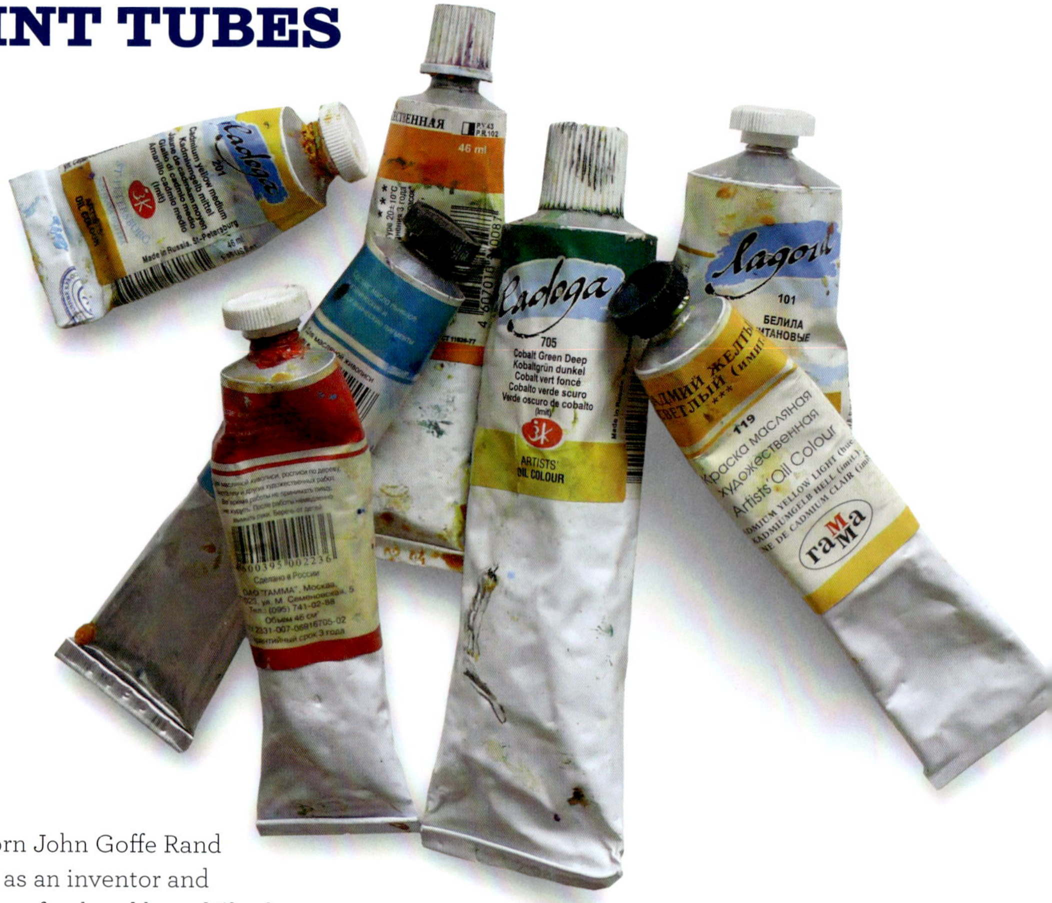

New Hampshire-born John Goffe Rand (1801–1873) worked as an inventor and portraitist, so he knew firsthand how difficult it was to store the pigments that artists used for oil painting. In that era, mixed paints waiting to be used were commonly stored in a pig's bladder tied with string, which would need to be punctured in order to retrieve the paint. Records show that, while on a trip to England, Rand invented "metal rolls for paint" or what we know of today as paint tubes. These were collapsable tubes made from metal—tin or another inactive metal—and tightly sealed with a cap. These tubes allowed the paints to be stored much longer without drying out and easily carried to different venues. Rand's paint tubes revolutionized art in the day, in particular impressionism, whose artists were able to leave their studios and get close to nature, one of the movement's favorite subjects. Rand received a patent for the paint tubes from the US Patent Office in September 1841. Famous artists have been quoted as saying that their art would not have been possible without the invention of paint tubes, among them Pierre-Auguste Renoir.

1841 THE STAPLER

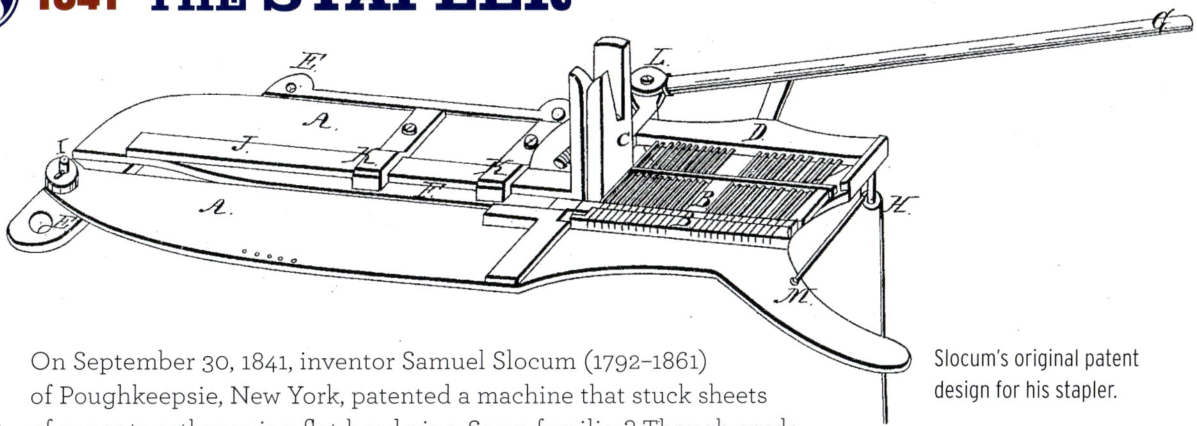

Slocum's original patent design for his stapler.

On September 30, 1841, inventor Samuel Slocum (1792–1861) of Poughkeepsie, New York, patented a machine that stuck sheets of paper together using flat-head pins. Seem familiar? Though crude and soon improved to more closely resemble the bent metal fastener we know today, Slocum's innovation is believed by many to be the world's first commercial stapler. Slocum was born in Rhode Island, moved to London to become a pin manufacturer, and later moved back to the United States where he invented his fastener, patented as a "Machine for Sticking Pins into Paper." Although the device was likely not envisioned as the ubiquitous office tool it is today—it was conceived as a method for packaging pins—Slocum's innovative mind had far-reaching results. Several more improvements would occur and patents would be issued in the stapler realm before the version we know today became standard.

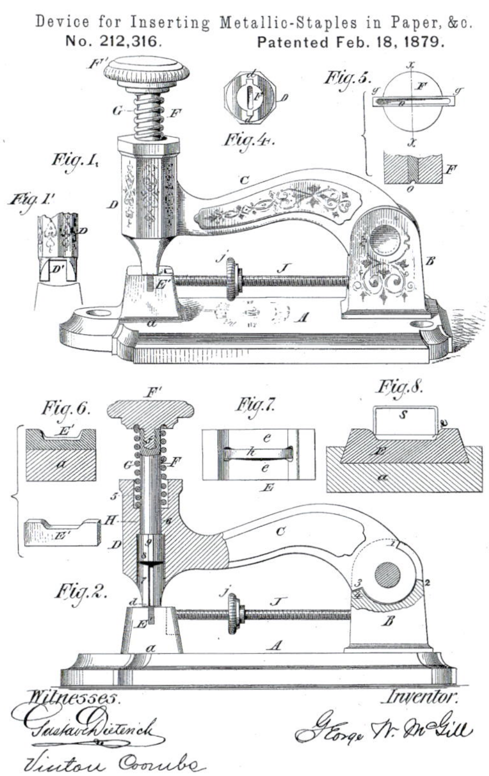

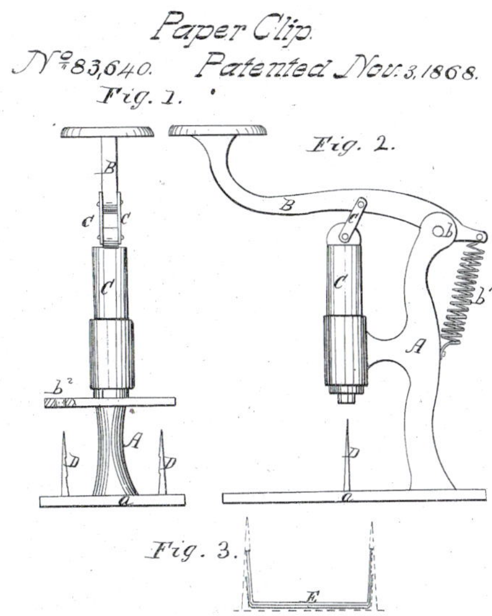

Two of the many "improvements" on the stapler idea: Albert J. Kletzker of St. Louis's 1868 "paper-fastener" and New Yorker George W. McGill's 1879 "staple-fastener."

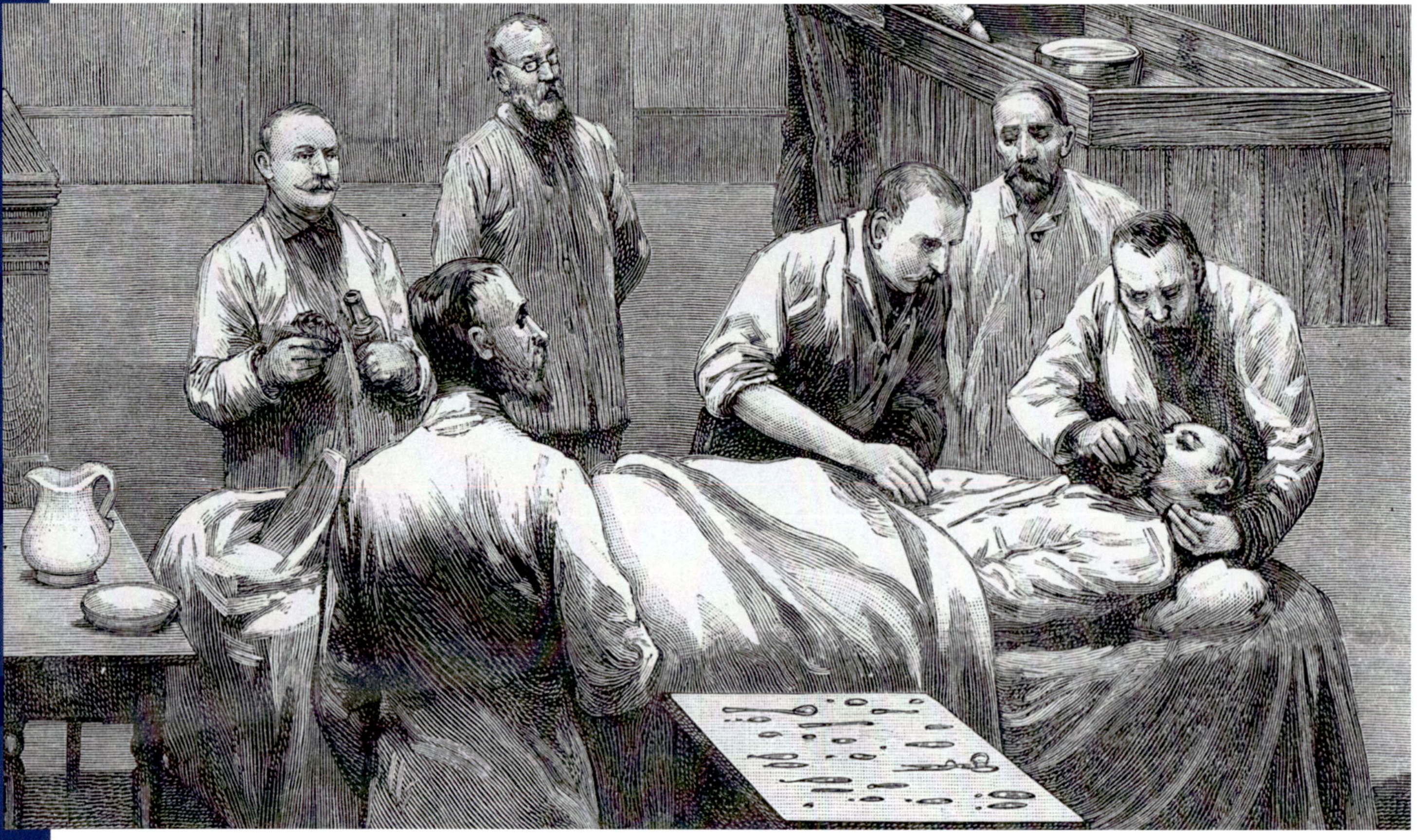

1842 ETHER

Crawford Williamson Long (1815–1878) was a Georgia-based surgeon who observed that people intoxicated by inhaling ether recreationally (dubbed "ether frolics") experienced little pain. This discovery led to his theory that ether could be used as an anesthetic to perform major surgeries without discomfort. Long first proved his theory on March 30, 1842, by administering ether to a patient in advance of removing a neck tumor. The patient reported experiencing no pain during the procedure. Long continued to use ether in subsequent surgeries with repeated success. A humble man, Long did not immediately report his findings, therefore others claimed the discovery of ether first, notably William T. G. Morton who, in 1846, demonstrated the anesthetic's properties in a public operating theater. Because Long didn't publish his findings in a timely manner, scholars were skeptical of his claims. But Long explained that he wanted to perform numerous experiments on the substance before claiming its efficacy in order to assure its safety. Thanks to Long's brilliance and careful experimentation, the use of ether revolutionized public health, and it could potentially do so again. According to a 2015 scientific article, "the widespread use of ether anesthesia in developing countries would improve patient safety, cut costs, and help struggling health care systems become more self-sufficient."

1843 VULCANIZED RUBBER

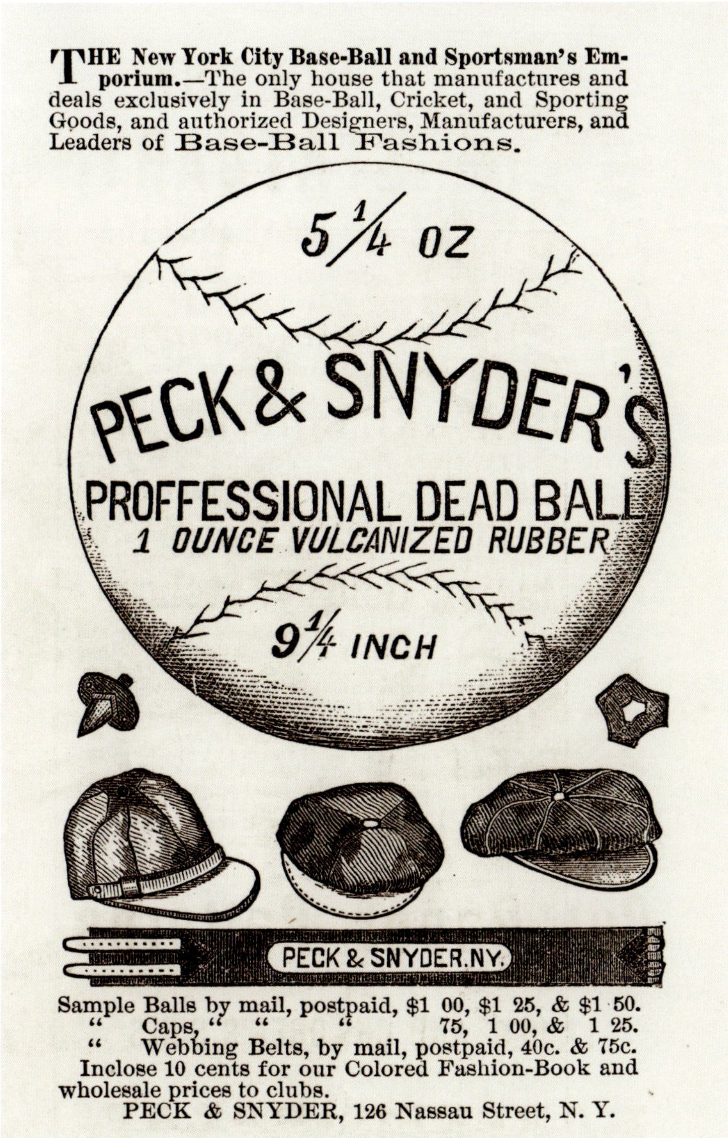

In 1839, Charles Goodyear (1800–1860) discovered the process of "vulcanizing" rubber, or heating it with sulfur and adding other substances in order to improve its strength and texture, but it took until 1843 to perfect the process. Goodyear then patented vulcanization, named after the Roman god of fire, the following year. Prior to vulcanization, natural rubber, a coagulated, viscous sap from a tropical tree, was prone to melting in hot weather or cracking at cold temperatures—even sticking or losing its shape in ambient environments. Goodyear was arguably obsessed with rubber and performed experiments on it for years, adding different substances to try to improve it and going into debt in the process. After Goodyear's success, he licensed his proprietary method to manufacturers to make such items as shoes, umbrellas, rafts, and eventually tires and other automotive products. Although Goodyear never made much money from his innovation, his contribution will go down in history as one of the most important technological achievements in American history.

Vulcanized rubber even found its way into sports, as seen in this 1874 ad.

1844 THE MODERN TELEGRAPH

Although the telegraph already existed at this time, it is arguably Samuel Morse (1791–1872) who gets the credit for its rapid growth and wide acceptance throughout the world. The improvements were based on Morse Code, a system of dots and dashes that he invented along with Leonard Gale (1800–1883), that was paired with letters that could be strung together to form messages without the need for decoding. This new telegraph system worked by transmitting electrical currents by electromagnetism into dots and dashes via indentations on a paper tape. The system was first used on May 24, 1844 to send the inaugural inter-city message, from Washington, DC, to Baltimore, saying "What hath God wrought?" a quote from the Bible. It is said that telegraph lines spread more quickly across the country than had the railroad (although the telegraph often followed the railroad routes so they did not need to be built from scratch). Morse's innovation in telegraphy meant that information could be spread as fast as electricity; people no longer needed to wait for ships or horses to deliver news, good or bad.

Fig. 3225.

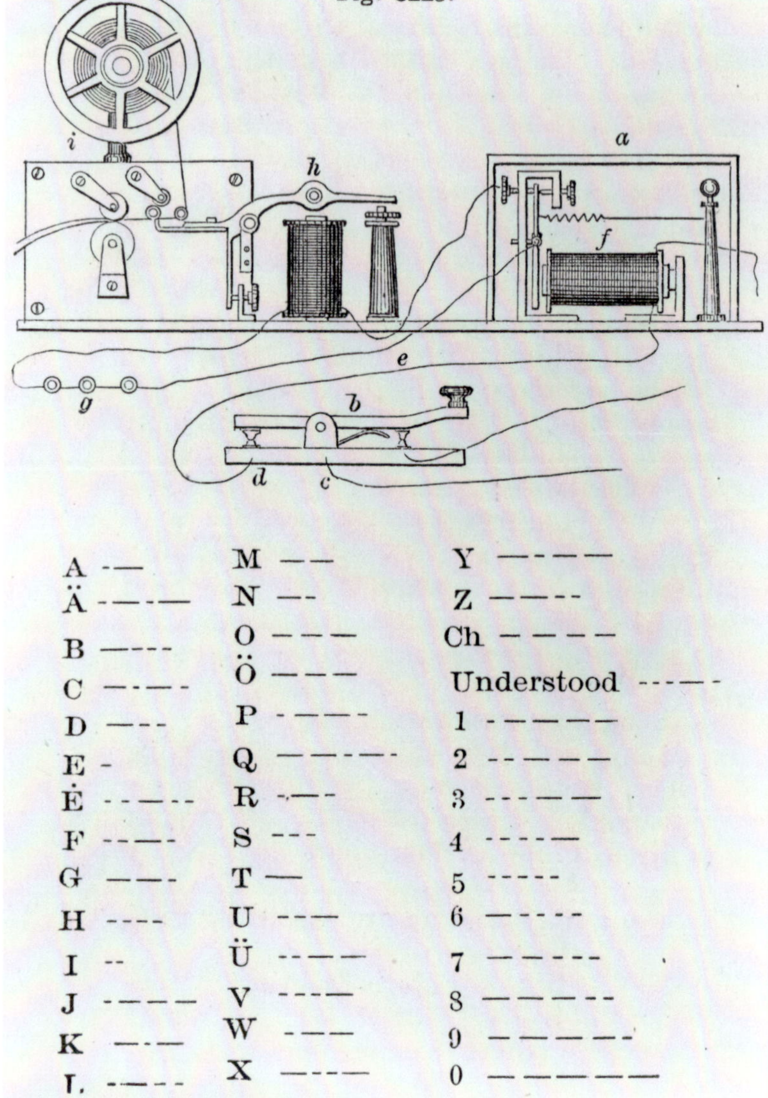

Morse Apparatus and Alphabet.

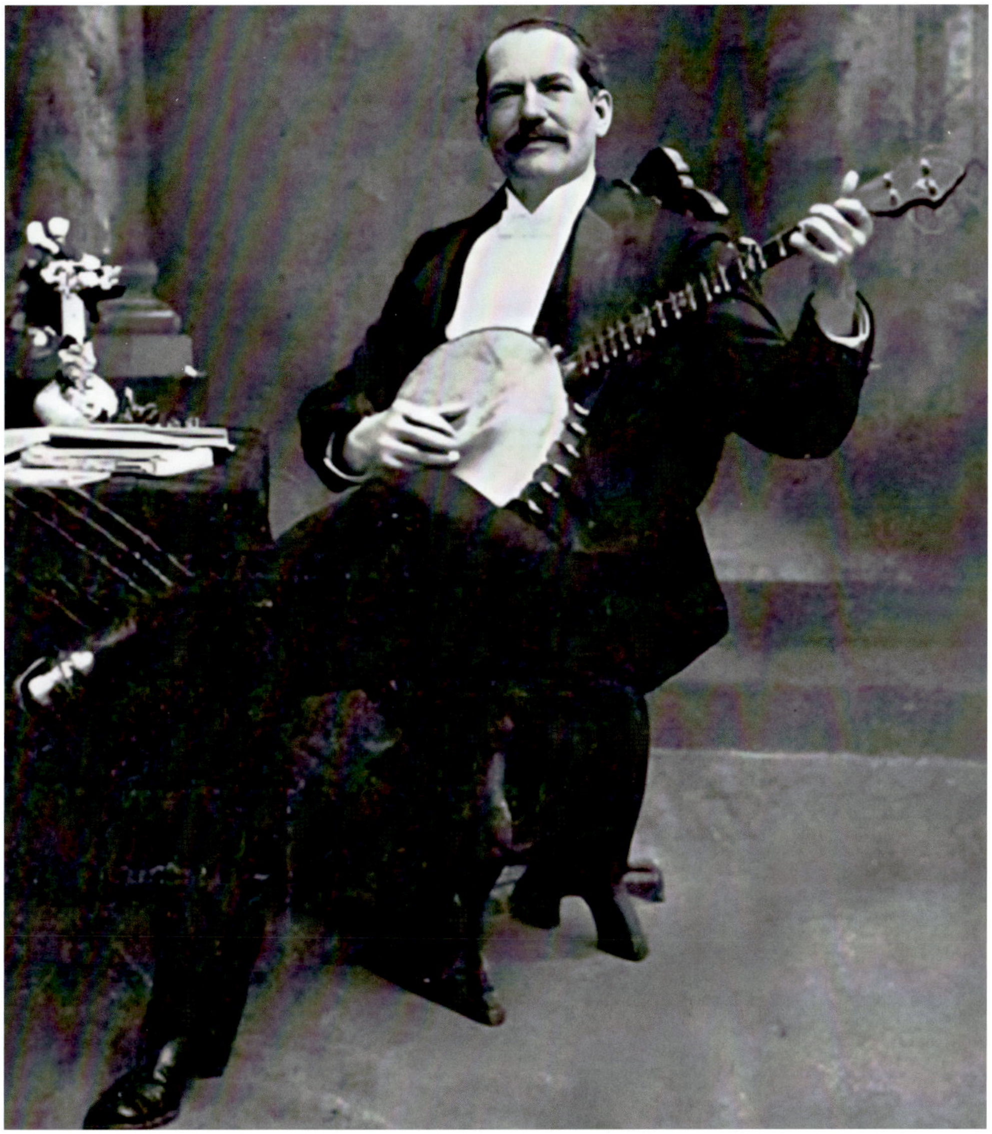

One of the most popular banjo players of the late nineteenth century, Polk Miller.

1845 THE BANJO

Drum maker and musical instrument dealer William Boucher (1822–1899) of Baltimore, Maryland was the first commercial maker of the banjo, building his first instrument in his shop in 1845, according to records. With roots in Africa, the original banjo was a handcrafted rural instrument made from a gourd, turtle shell, or wood. Boucher replaced the gourd body with a bent-wood rim borrowed from a drumhead, which could be produced at scale with greater consistency attached to a fretless neck. Boucher's banjos can still be found in fine art museums around the country and are highly prized by musical instrument collectors. During the mid-1800s, the minstrel show was one of America's most popular forms of entertainment, and the five-string banjo was a central element of these performances. Although race and racism played a big part in minstrel shows, and they are controversial in the modern era, it is impossible to mention banjos without a reference to the then massively popular entertainment—they likely introduced nineteenth-century Americans to banjos as a musical instrument. Five-string banjos, as Boucher conceived them, remain immensely popular in folk and bluegrass music today.

1846 FIRST OFFICIAL BASEBALL GAME

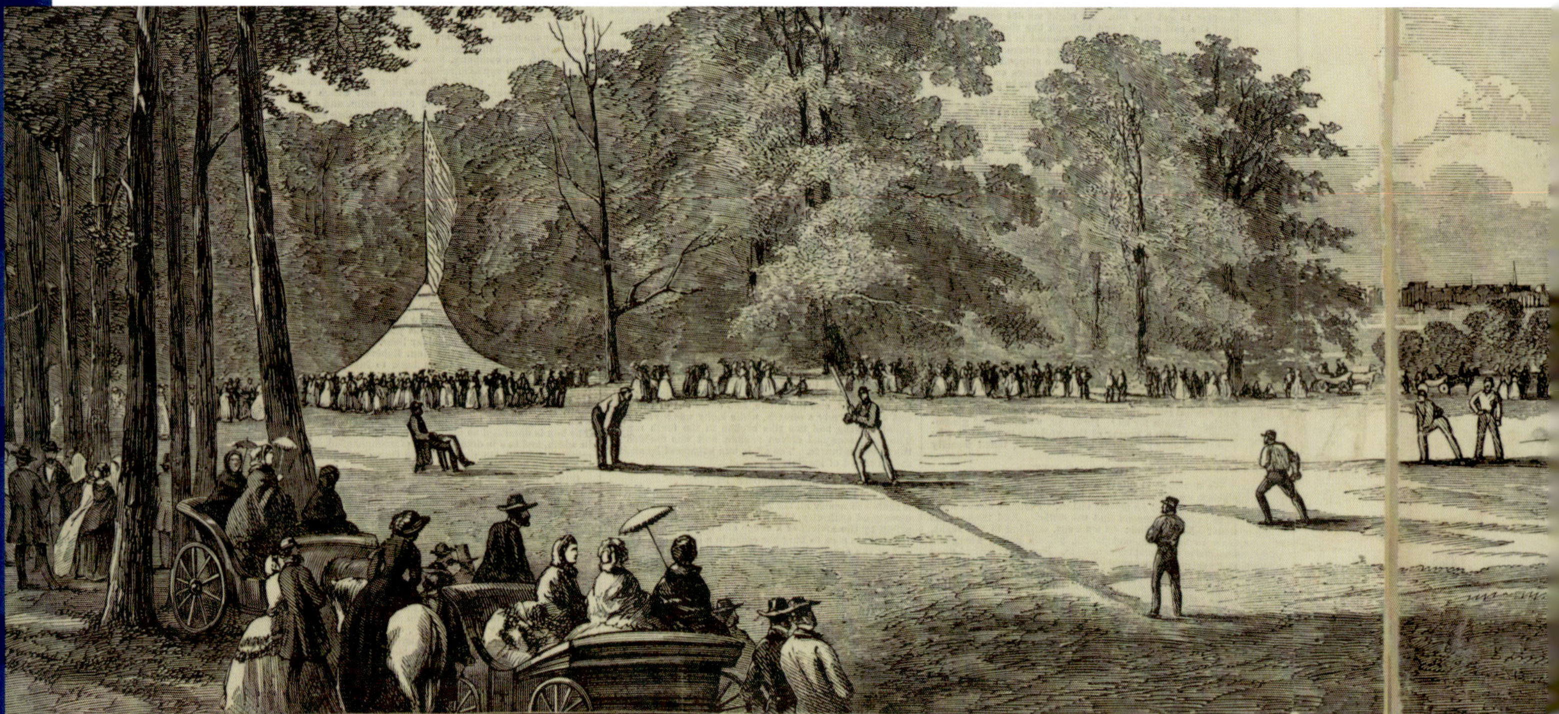

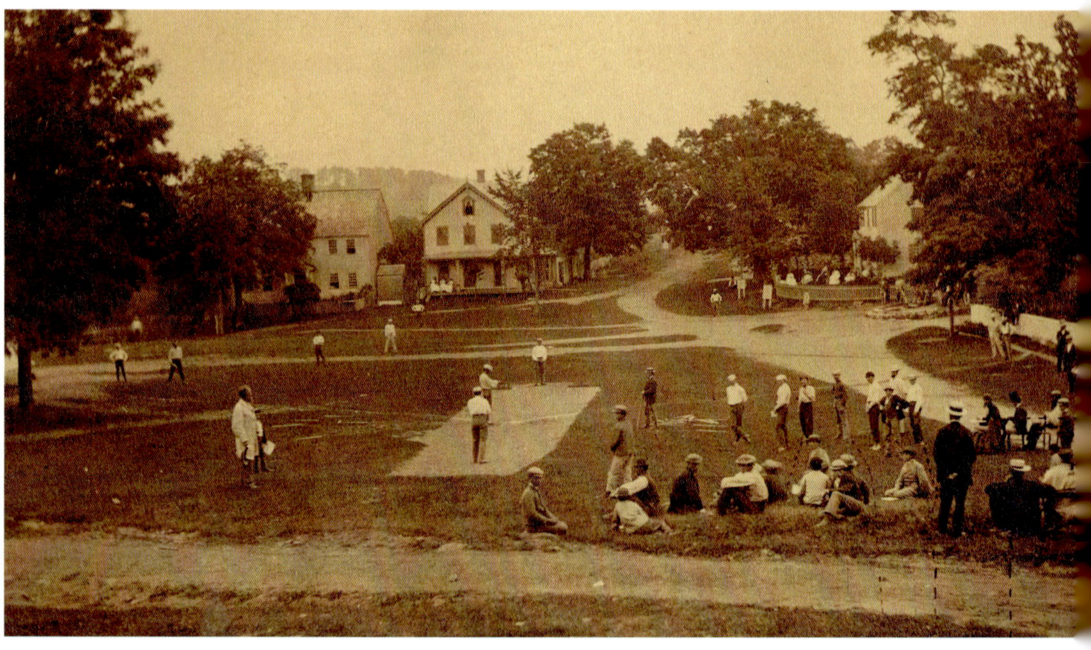

ABOVE: A drawing of the game at Elysian Fields in Hoboken, New Jersey.
RIGHT: One of the first known photographs of a baseball game, taken on the Washington Green in Connecticut, August 4, 1869.

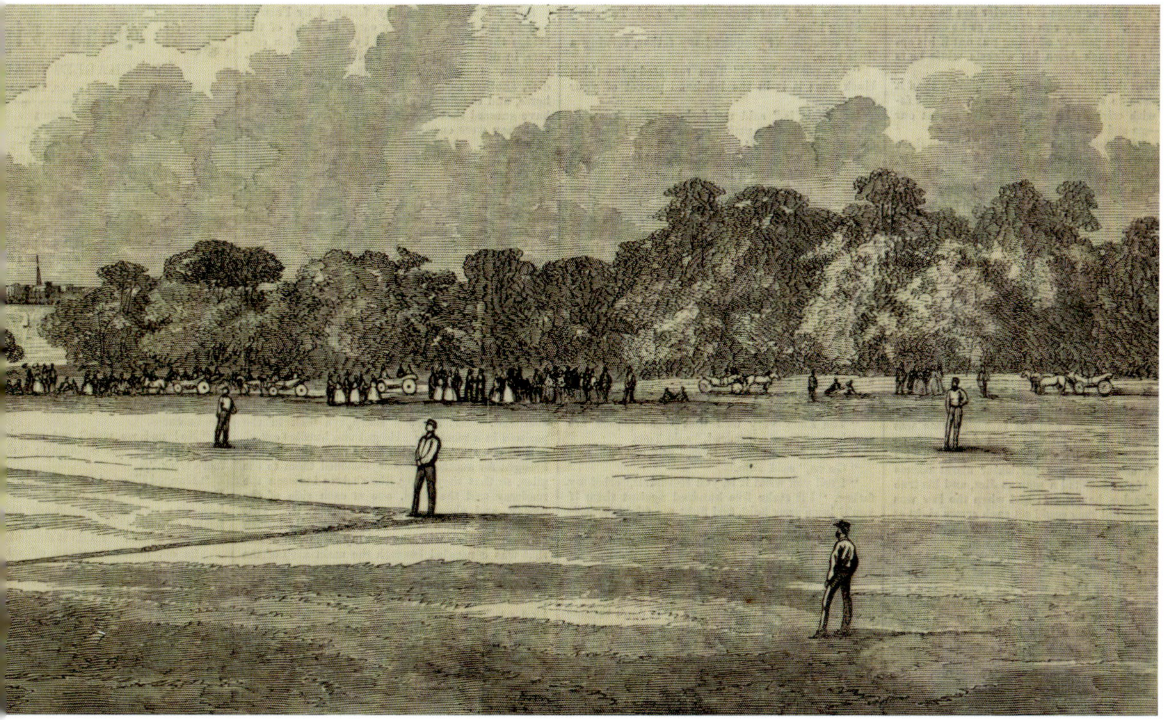

The game of baseball can be traced to the 1800s, but no official organized games were recorded until the mid-1800s. In the ensuing years, men began to develop rules of the game, guidelines for how the field is laid out, and field etiquette. On June 19, 1846, the first official organized baseball game in history was played in Hoboken, New Jersey, at Elysian Fields (there is now a historical marker at the site) between the New York Knickerbockers Base Ball Club and the New York Nine (or New York Mutuals, depending on the source). The rules for that game were devised by members of the Knickerbockers and became the foundation for the modern game of baseball. Frequently credited as the father of American baseball, Alexander Cartwright (1820–1892) was then a member of the Knickerbockers. He is credited with inventing the modern baseball field the year prior and was part of the committee to develop the rules for the game played that day. Cartwright also served as referee at the landmark game, which the Knickerbockers lost 1–23. Cartwright was elected to the Baseball Hall of Fame in 1938. The innovations created by Cartwright and his teammates paved the way for the sport as we know it today.

1847 ROTARY PRINTING PRESS

Richard Hoe (1812–1886), an American inventor born in New York City, revolutionized the printing industry in 1847 when he patented the steam-powered rotary printing press. At the age of fifteen, Hoe started working in his father's printing press company, R. Hoe and Company, which at the time used only flat-bed printers. After the younger Hoe took over the company, he developed improvements to the printing presses, including mounting the type on a rotating cylinder instead of a flat bed. This "lightning" press, which used one cylinder to carry the type and multiple other cylinders to hold the paper on long rolls, dramatically increased the number of sheets that could be printed in a given time frame. The new printing technology was first used by the newspaper the *Philadelphia Public Ledger* that same year, printing 8,000 sheets per hour. Richard Hoe's advanced printing technology helped newspaper publishers respond to increasing demand from consumers. Hoe continued to innovate his technology and has gone down in history as one of the fathers of mass media.

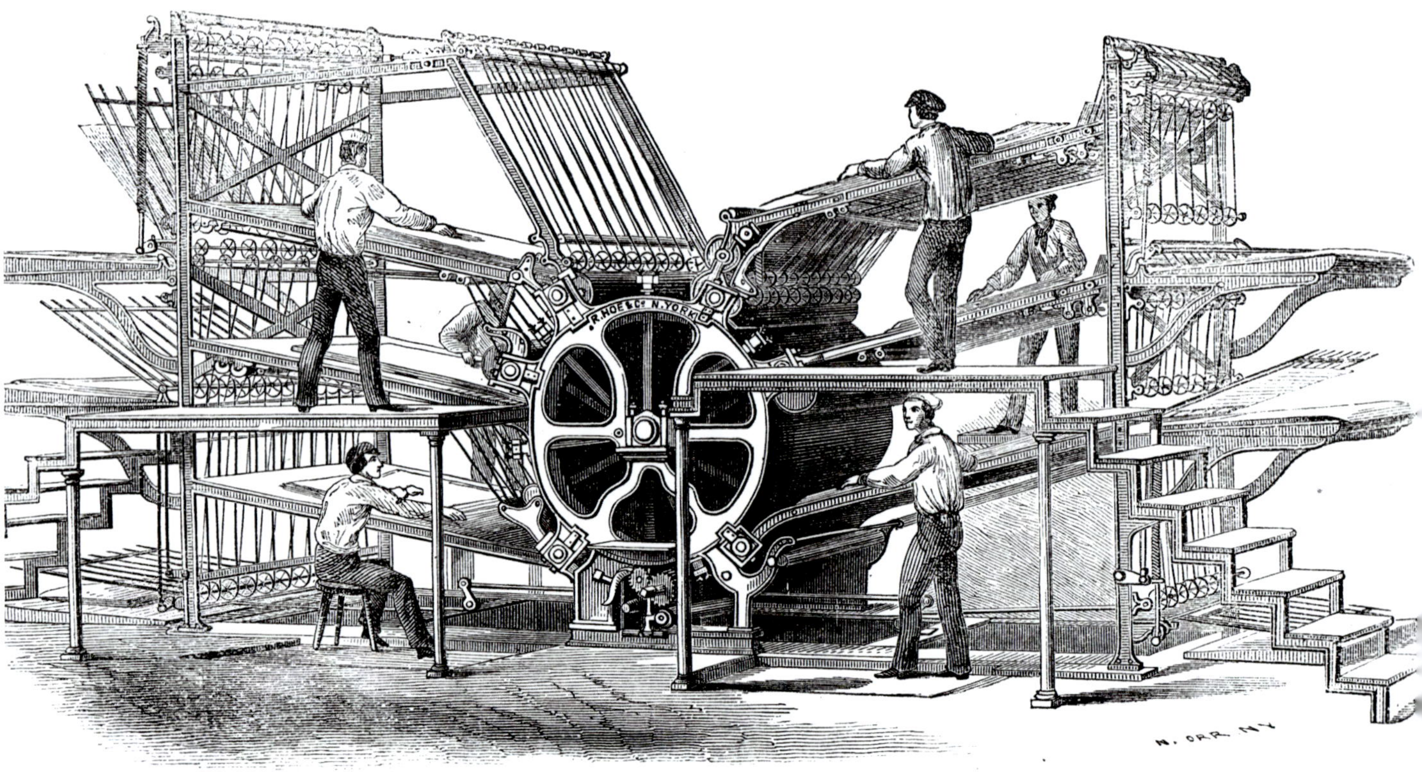

HOE'S SIX CYLINDER PRINTING PRESS.

1848 THE LONG TOM

During the Gold Rush, any device that helped speed the processing of gold was a boon to the prospectors who labored to find "The Motherload." Such a device was the Long Tom, also called a Broad Tom. A primitive sluice box, a Long Tom featured a trough that was up to twenty feet long with "riffles" (grooves) along its surface. It was positioned at an angle on the side of a slope. Dirt was shoveled into the upstream end of the apparatus and then a steady stream of water was applied to it. The flow of the water acted to rinse away the lighter dirt and leave the heavier gold behind in the riffles below. Allegedly, twenty times more earth could be processed after the implementation of the Long Tom, transforming the industry for many hard-toiling prospectors searching for treasure.

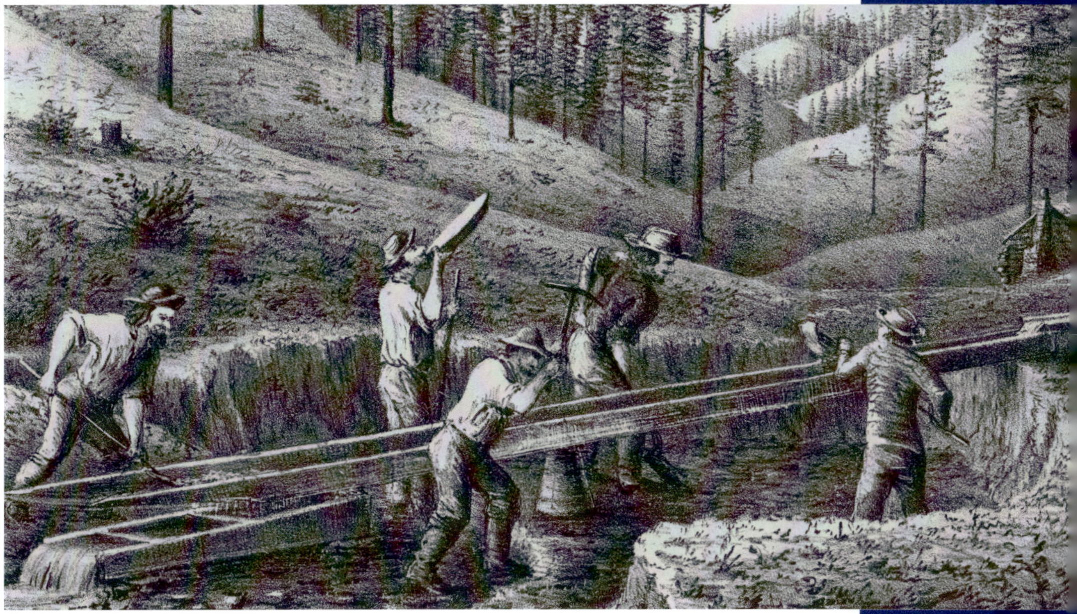

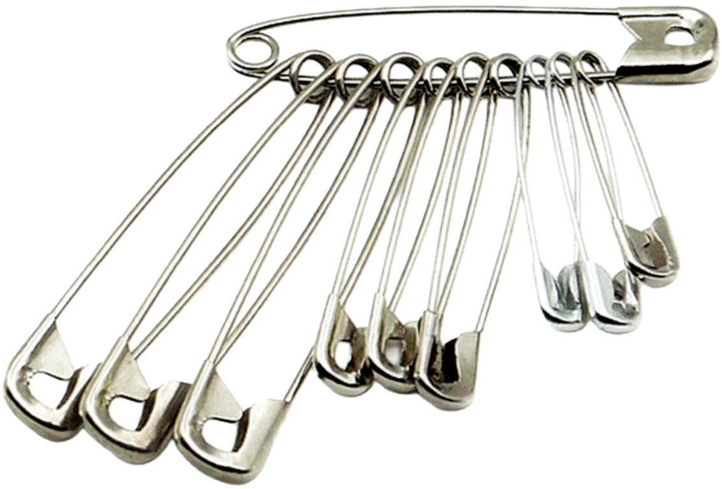

1849 SAFETY PIN

Mechanic and inventor Walter Hunt (1796–1859) of New York City patented the safety pin on April 10, 1849. Legend has it that Hunt owed a man $15 and was desperate to pay his debt. While ruminating over how he would pay his bill, he twisted a piece of metal wire into a circular shape that acted as a spring and added a clasp that held the ends together. Thus, the safety pin was born. Hunt sold his patent rights for about $400 and solved his debt problems.

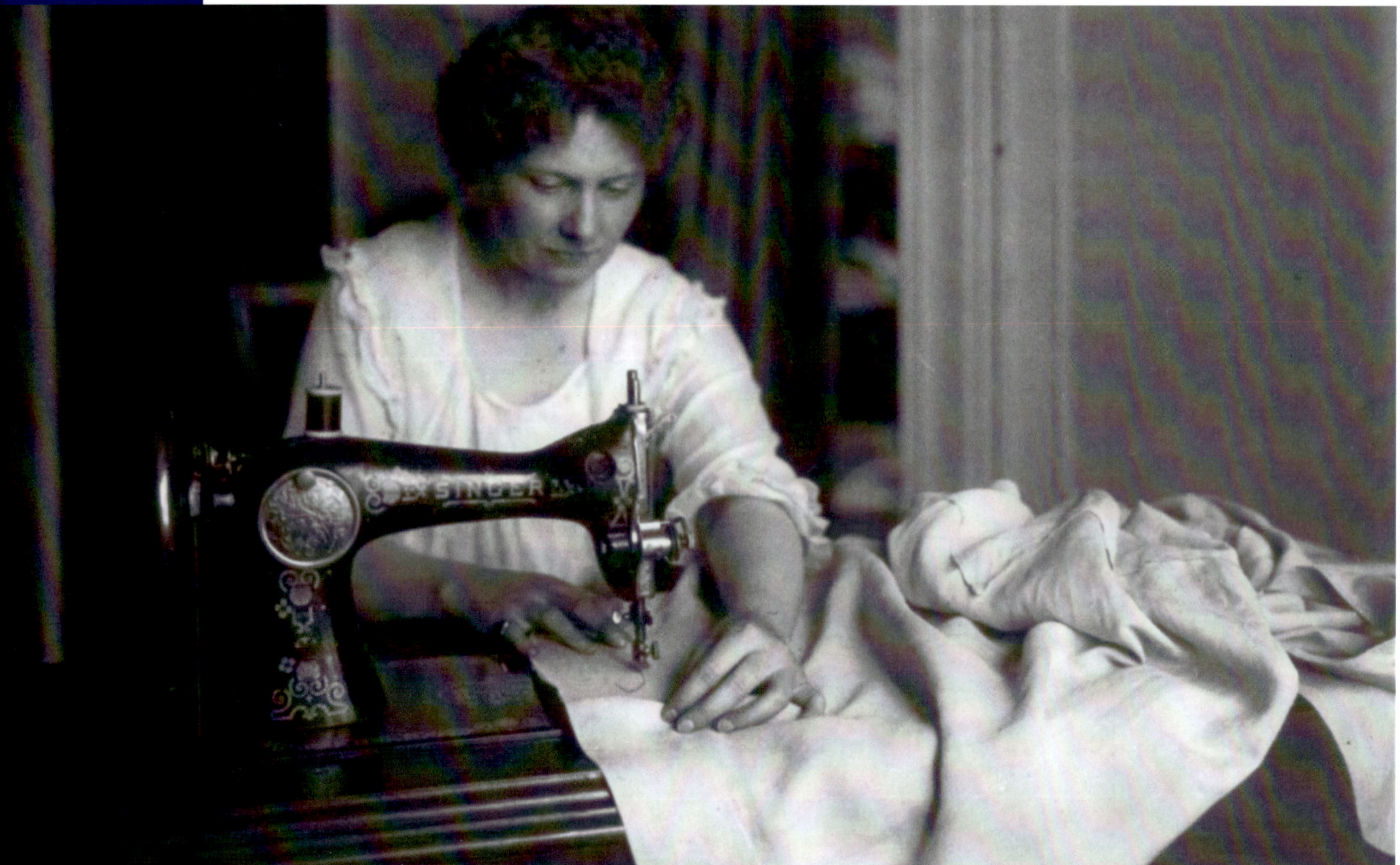

A woman using a Singer, c. 1900.

1850 SINGER SEWING MACHINE

Isaac Singer (1811–1875) was an American inventor whose last name became synonymous with sewing machines. In 1850, he created a partnership with two other businessmen to create the first reliable sewing machine for use in the home. He then established I. M. Singer & Co. in Boston in 1851, later moving operations to New York City in 1853. The new sewing machine, with its eye-pointed needle and lock stitch design (challenged in patent court by Elias Howe) was revolutionary at the time, allowing garments to be made in just a fraction of the time as before. In 1855, the Singer sewing machine won first prize at the Paris World's Fair—many believe it to be one of the most important inventions of the nineteenth century. Singer's business partner, Edward Clark (1811–1882), came up with some innovations of his own, including the first-ever installment plans by a major American company, which helped sell the expensive sewing machines to American housewives. He also helped the company become among the first to be successful internationally. By 1860, the Singer company was the largest manufacturer of sewing machines throughout the world.

1851 BLOOMERS

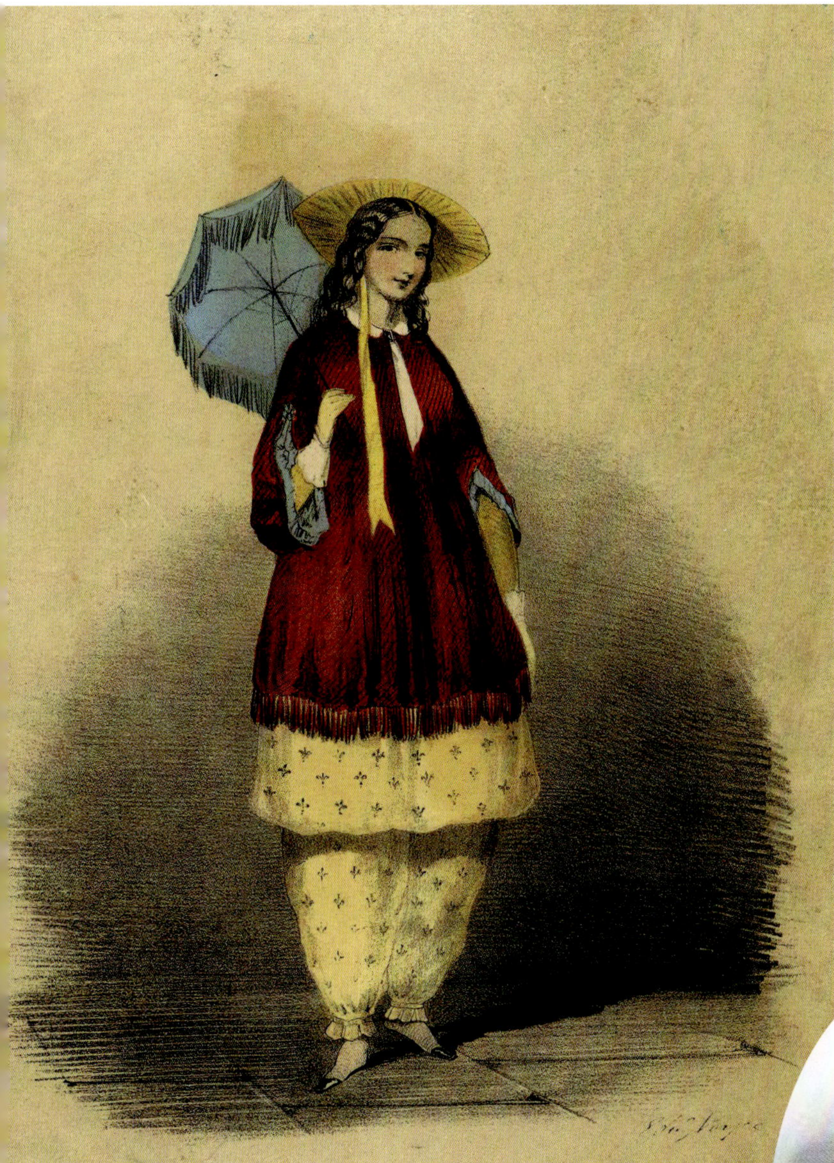

In mid-1800s America, women's clothing was tight (the tiny waists idealized in the day called for squeezing into a corset), heavy (full-length skirts and multiple petticoats to fill out the shape were the norm), and restrictive (the combined constriction and weight could inhibit movement). Enter Amelia Bloomer (1818–1894) from Seneca Falls, New York, who in 1851 made popular an outfit consisting of a knee-length dress paired with a pair of loose, Turkish-style pantaloons, or "bloomers." Bloomer was the editor of *The Lily*, known as the first women's newspaper. Starting in the April 1851 edition, Bloomer used her position to advocate for the outfit, for dress reform, and for women's rights. Although Bloomer was not the inventor of the pantaloons, articles and illustrations in *The Lily* and her habit of wearing the costume led to folks calling the pantaloons "bloomers." The symbolism of bloomers as an icon of feminist rebellion earns them a spot as an iconic innovation in history that can't be ignored.

ABOVE: The fashion that started a rebellion. **RIGHT:** Amelia Bloomer.

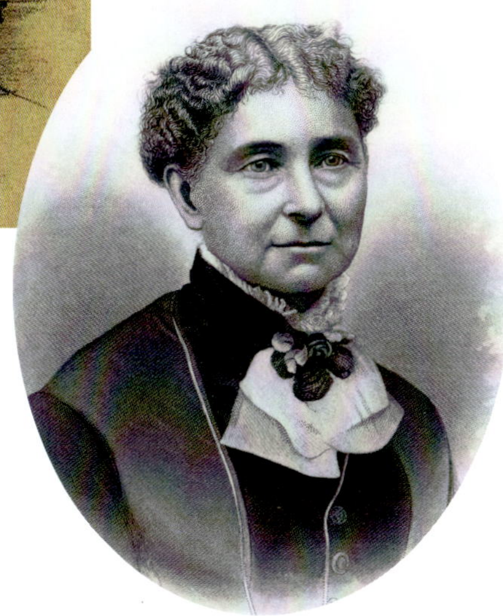

1852 BIRTH OF POP MUSIC

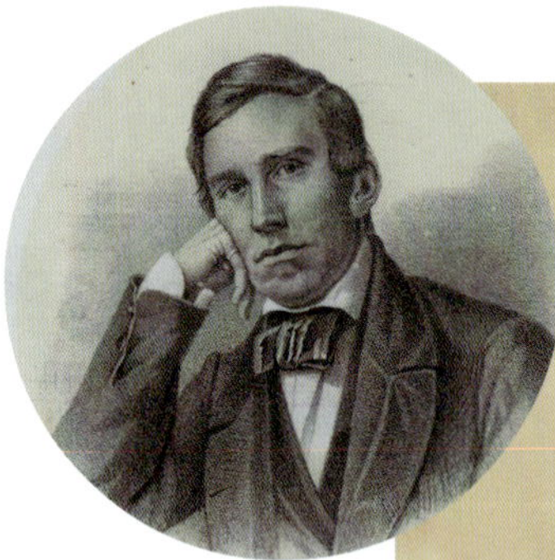

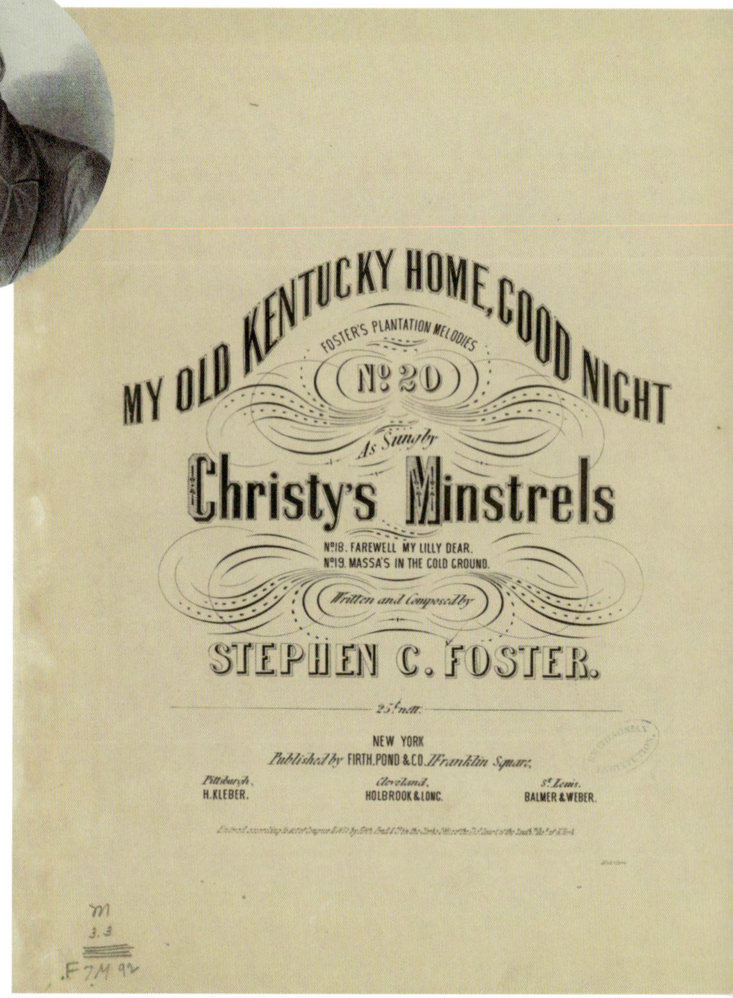

Stephen Foster, c. 1859, and the sheet music for his most famous song.

Many consider Stephen Foster (1826–1864) to be one of America's most beloved melodists as well as the country's first great professional songwriter. Records show that he composed more than two hundred songs in his short life, including favorites such as "Oh! Susanna," "Camptown Races," and "Beautiful Dreamer." Foster's songs gave the young country a collection of melodies so popular, they could readily be considered our national songbook long before radio existed. More than simple folk songs, Foster's compositions drew on diverse traditions. One of Foster's most important works was "My Old Kentucky Home," originally called "My Old Kentucky Home, Good-Night!" Despite a racial slur in the original lyrics, it was written as a condemnation of the state's enslavers. Indeed, he wrote the song in 1852 after reading the abolitionist novel *Uncle Tom's Cabin* by Harriet Beecher Stowe. Foster took care to minimize the offensive caricatures of Black people of the day in his music and portray empathy for enslaved workers. Today, "My Old Kentucky Home" is the official song of the state of Kentucky, and Foster's songbook is still well-known among Americans, showing his enduring power as the original king of American pop music.

1853 ENVELOPE-FOLDING MACHINE

In the 1840s, a nationwide postal reform lowered the price of sending a piece of mail, drastically increasing the number of letters being sent and effectively launching the envelope industry. However, making envelopes was still done by hand; stationery clerks using tin forms could only make about twenty-five envelopes per hour. But on June 21, 1853, that all changed when Russell L. Hawes (1823–1867) of Worcester, Massachusetts, received a patent for an envelope-folding machine. This innovative device could produce 10,000 to 12,500 envelopes daily, marking a dramatic leap in manufacturing efficiency. Hawes's design built upon previous inventions by incorporating a self-feeding mechanism that automatically picked up blank sheets of paper, significantly reducing the need for human aid. This automation was a crucial step toward modern mass production. Hawes's 1853 patent helped transform envelope-making from a labor-intensive craft into an industrial process, contributing to the democratization of written communication. By dramatically reducing production costs and increasing output, Hawes's machine made envelopes more accessible to ordinary citizens, supporting the expansion of both personal and business correspondence that would help further the industrial age.

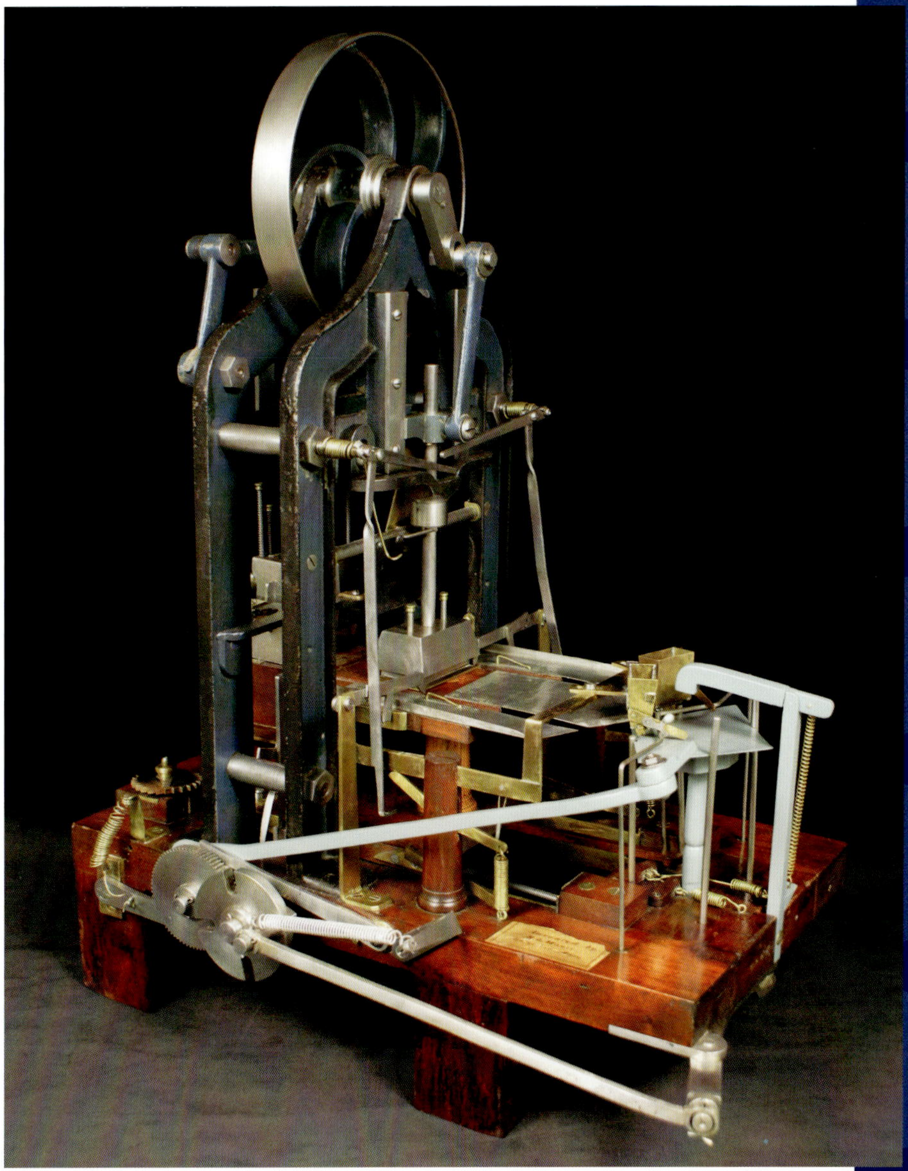

1854 *WALDEN*

The philosophical memoir *Walden* was published in 1854, seven years after author Henry David Thoreau (1817–1862) ended his two-year experiment of living in a small cabin near Walden Pond in Concord, Massachusetts. The book was compiled from the extensive journals Thoreau kept from 1845 to 1847, which he carefully developed into a cohesive narrative structured around the cycle of seasons. During Thoreau's lifetime, *Walden* received modest attention and sales were disappointing. However, its influence grew dramatically in the twentieth century, becoming one of America's most celebrated works of literature and philosophy. The book explored the art of simple living, self-reliance, and humanity's relationship with nature. Thoreau's detailed observations of the natural world, combined with his philosophical reflections on materialism, conformity, and individual purpose, established him as a foundational voice in American transcendentalism and environmental literature. *Walden* greatly influenced later environmental movements, inspiring figures like John Muir and contemporary conservationists. Its themes of deliberate living, civil disobedience, and finding meaning through direct experience with nature continue to resonate with readers seeking alternatives to modern consumer culture and a deeper connection to the natural world.

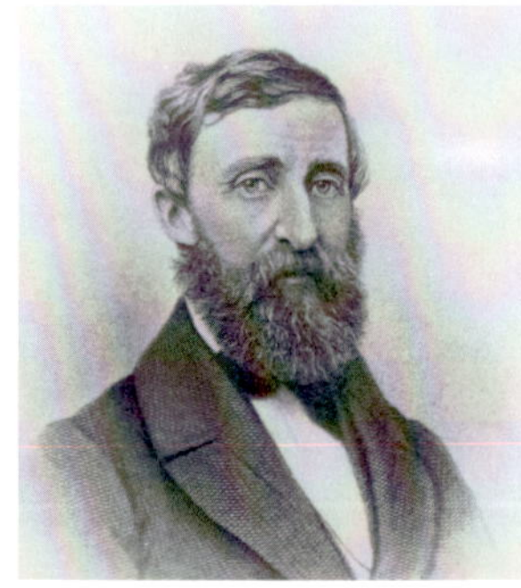

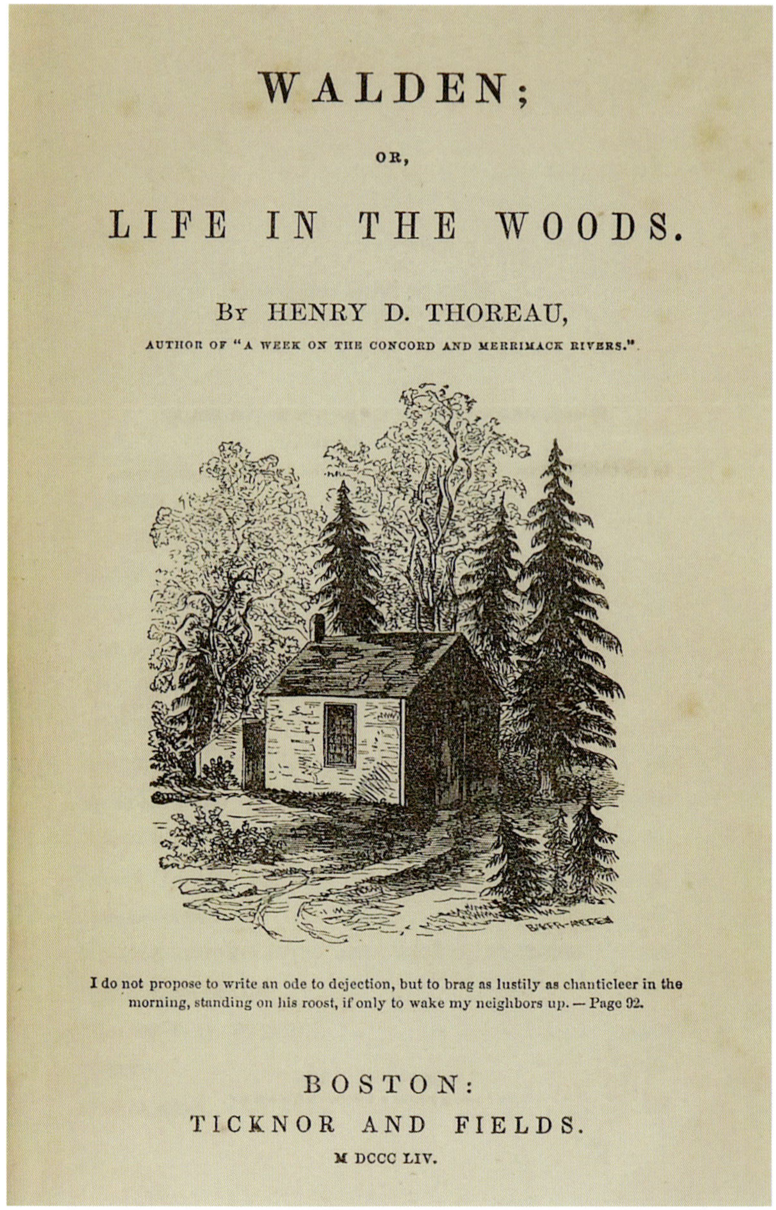

WALDEN;

OR,

LIFE IN THE WOODS.

BY HENRY D. THOREAU,

AUTHOR OF "A WEEK ON THE CONCORD AND MERRIMACK RIVERS."

I do not propose to write an ode to dejection, but to brag as lustily as chanticleer in the morning, standing on his roost, if only to wake my neighbors up. — Page 92.

BOSTON:

TICKNOR AND FIELDS.

M DCCC LIV.

TOP: Henry David Thoreau. **RIGHT:** The title page for the first edition of the book, showing Thoreau's own cabin near Walden Pond.

1855 *LEAVES OF GRASS*

Leaves of Grass was first published anonymously in 1855 by Walt Whitman (1819–1892) as a collection of twelve poems, which he revised and expanded many times during his lifetime. Whitman's verse was innovative in four ways: First, it was among the first poems to feature free verse instead of borrowing from techniques of the European poets. Second, the poems celebrate both the individual and the collective, hailing democratic concepts. Next, Whitman supported sexual tolerance and freedom in his poems. (He was rumored to be gay or bisexual and was for that reason a controversial figure.) Finally, the poems in *Leaves of Grass* glorified the regenerative power of nature. As is true of many great works of art, *Leaves of Grass* was little appreciated during Whitman's lifetime—it was critically panned and often banned from public libraries. Whitman was even fired from a federal job over objections to the content of his work. But gradually, over time, the brilliance of the work revealed itself. Writers such as Ralph Waldo Emerson and Oscar Wilde helped the collection gain notoriety. For all its innovations and Whitman's courage in writing a work of astonishing originality, *Leaves of Grass* is arguably the most important American poetry collection ever written.

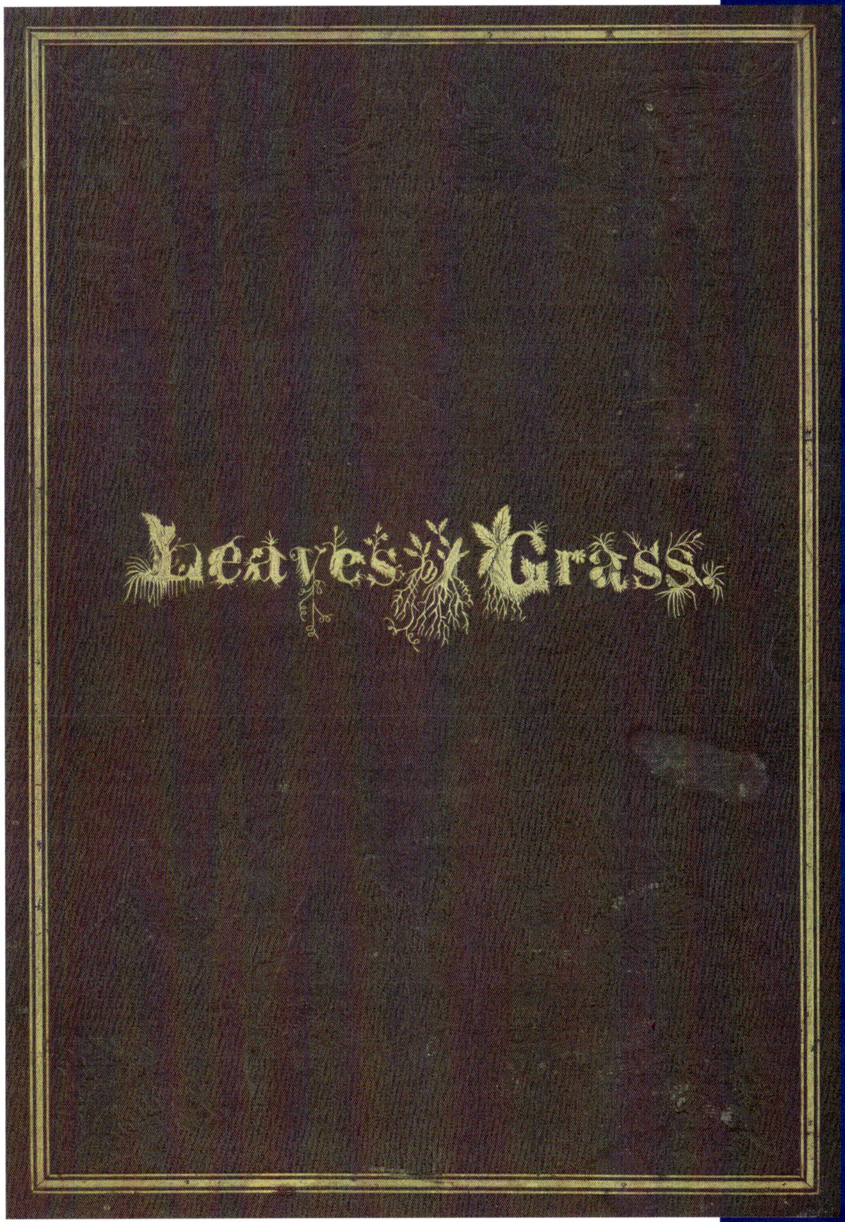

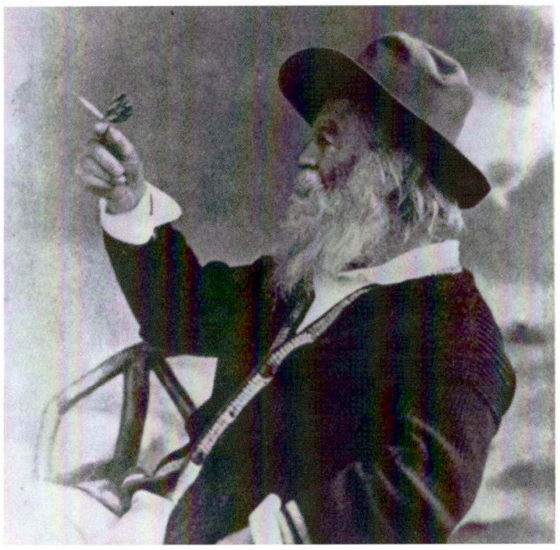

LEFT: This portrait of Walt Whitman holding a butterfly, taken in 1877, was one of his favorites. **ABOVE:** A sample proof of *Leaves of Grass* from 1891 shows that he intended for the photo to be used as the frontispiece for the book.

1856 CONDENSED MILK

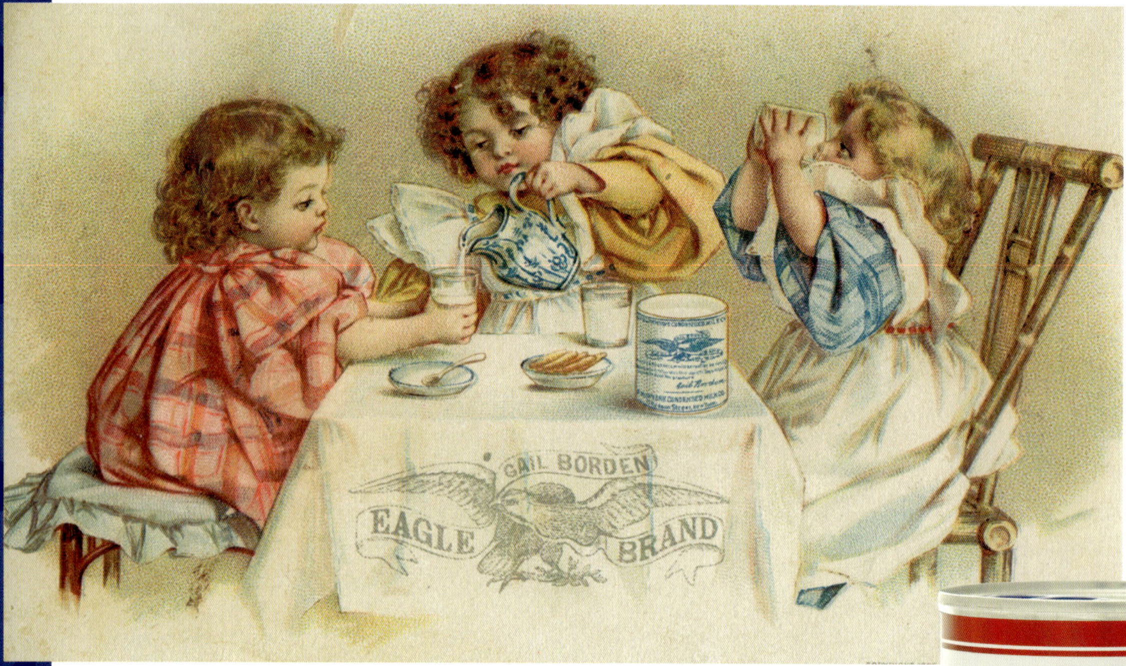

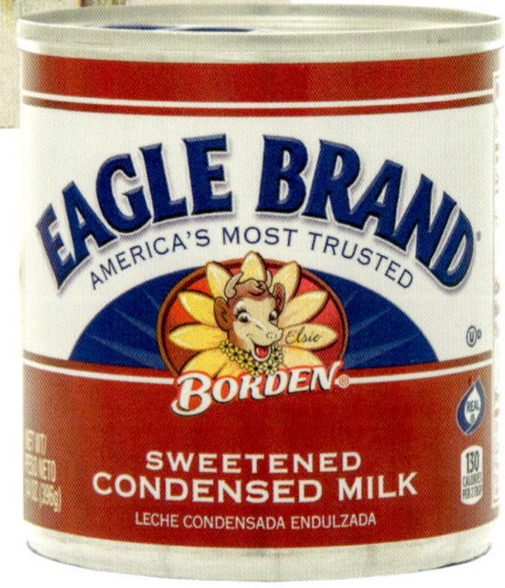

After witnessing children die on a long ship voyage due to contaminated milk, inventor Gail Borden Jr. (1801–1874) of Texas aimed to find a way to keep milk from spoiling. In 1856, he created a process for preserving milk without refrigeration by condensing it. Borden theorized that heating the milk at just the right temperature would kill any airborne contaminants and vaporize the water content without burning it, thus condensing the milk and allowing it to be stored securely and shipped over long distances. He used a vacuum pan he had seen used by the Shaker community and that the religious sect typically used to condense fruit juice. Borden eventually opened a factory that manufactured Eagle Brand condensed milk. Borden's product was very popular among Union Army soldiers during the Civil War, with the US government ordering so many units that Borden needed to open a separate factory just to keep up with demand. Soldiers returning from the war enjoyed Borden's condensed milk so much, they spread the word about it when they returned home, leading to more commercial success. The company's sweetened condensed milk has been credited with helping substantially reduce infant mortality rates in the country during the mid-to-late 1800s.

ABOVE LEFT: An early twentieth-century beverage trading card for Eagle Condensed Milk.
ABOVE RIGHT: The Eagle can featuring the now-iconic mascot Elsie the Cow, who made her debut in 1938, and was the star of Borden's exhibit at the 1939 World's Fair in New York.

1857 PASSENGER ELEVATOR

Elisha Otis (1811–1861) of Yonkers, New York, installed the first steam-powered passenger elevator in a department store in New York City on March 23, 1857. Years before, Otis had been toiling over his invention, working on a safety device attached to the elevator that would automatically prevent it from falling if the lifting chain or rope broke—a device he dubbed the "safety hoist." The elevator was in service for years hauling freight, but orders for the device were few. On September 20, 1853, Otis staged one of several public demonstrations at the Crystal Palace in New York City: He constructed an open-air elevator platform, rode it high in the air, then ordered someone to cut its lifting rope. Instead of plunging to the ground, the platform stopped after lowering just a few inches. After Otis's drama-filled demonstrations, orders for the elevator came flooding in, starting with the 1857 installation and soon followed by many more. Historians credit Otis's "safety elevator" with changing not just the skyline of America but the world, making it possible to build taller and taller buildings whose upper floors would not be easily accessible by stairs.

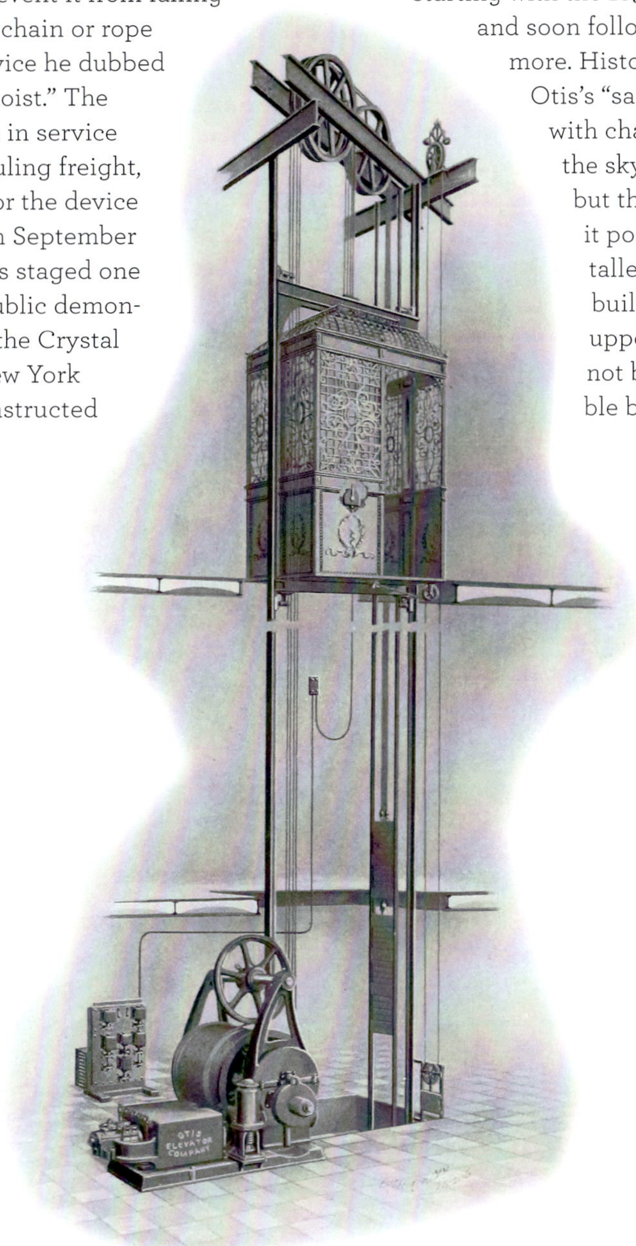

1858 BURGLAR ALARM

On February 21, 1858, businessman Edwin Holmes (1820–1901) installed the first burglar alarm in Boston. But Holmes did not invent the burglar alarm—he bought the patent the year before from Albert Augustus Pope just before Pope's death. The system electrically wired together a house's or business's entryways and windows, activating an alarm if the connection was broken. But in order for a business based on theft to be lucrative, actual burglaries needed to occur. Believing that more crime happened in New York, Holmes moved to the nearby city to try to boost his business. Holmes's innovations came largely in recognizing the commercial opportunity and in his skill at marketing and advertising, which helped his fledgling operation succeed despite a distrusting customer base. The alarm, which relied on electricity to work, was difficult to sell to potential customers because, at that time in history, people were fearful of electricity, worried that it might be dangerous. The Holmes Electric Protective Company was known to be among the first to use testimonials from customers and famous figures in its advertising along with other innovative marketing techniques.

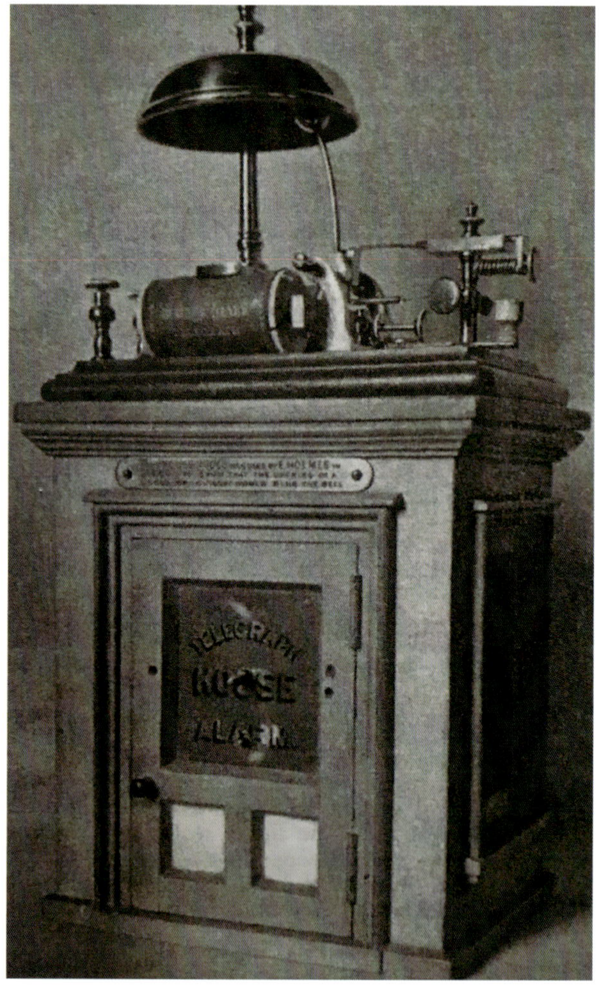

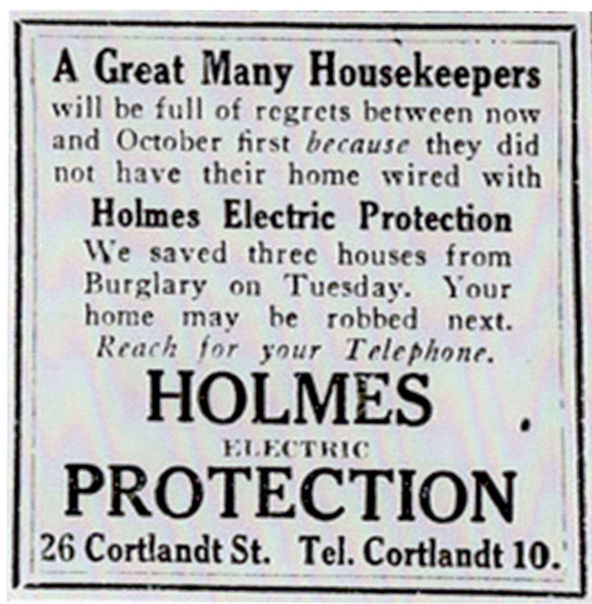

Holmes's original alarm, and the newspaper ad that helped sell them.

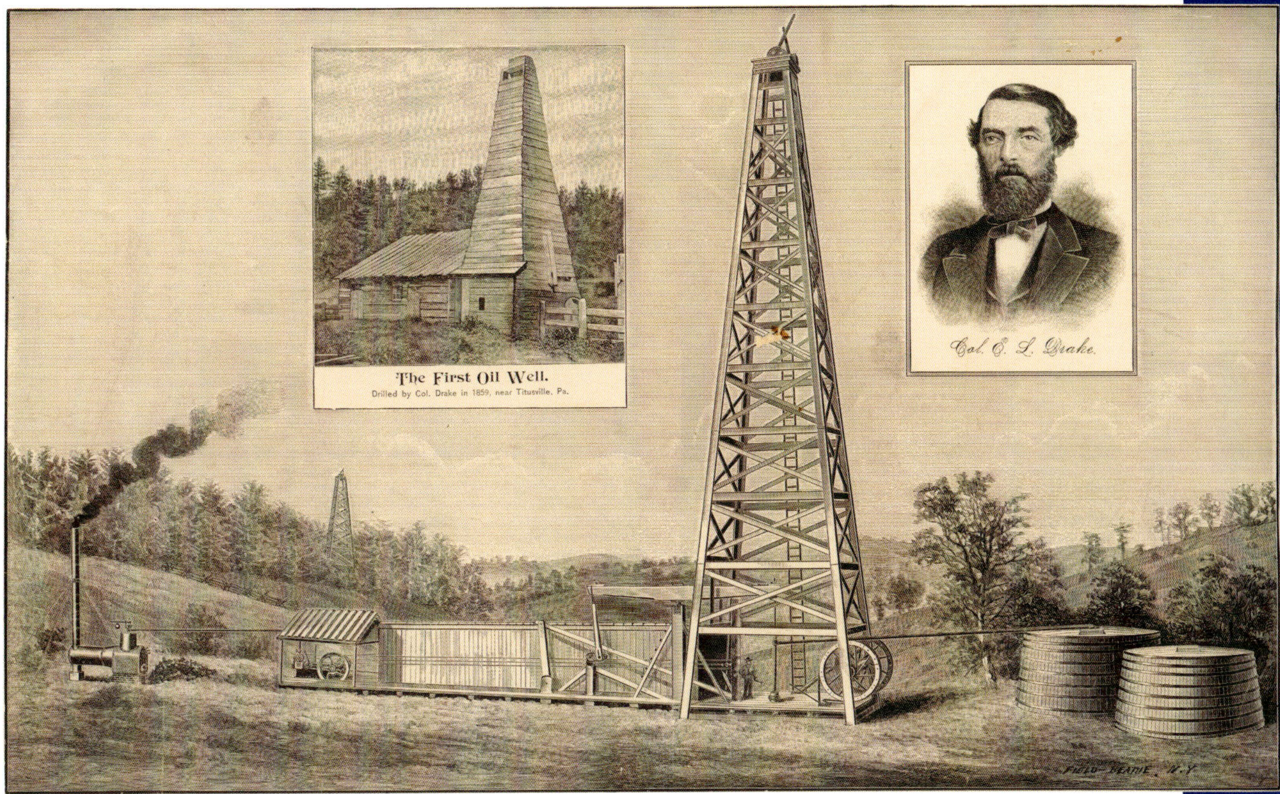

1859 OIL WELL

Edwin Drake (1819–1880) was a retired railroad conductor from New York. He was hired by the Seneca Oil Company to explore the area around Titusville, Pennsylvania, which was thought to have rock oil deposits—at that time, rock oil (later known as petroleum) was expected to be a viable substitute for the coal oil and whale oil that was used to produce lamp fuel. After experimenting with digging to reach the oil deposits and running out of company money, Drake persisted with his intuition that drilling was the answer to retrieving rock oil from the ground. Finally, at the end of August 1859, the wooden oil derrick Drake designed, and with the help of a team of men, found and extracted the oil that he had been long seeking. Sadly, inventing the world's first oil well would not immediately transform Drake's life—he did not seek a patent for the oil well and was soon fired by the company who hired him. Oil barons would step in to seek their fortunes with the Pennsylvania petroleum in what would be dubbed an "oil rush." Years later, the state of Pennsylvania awarded Drake an annuity to honor his innovation, though he died shortly after the honor.

1860 CARPET SWEEPER

The Industrial Revolution brought with it new luxuries, such as carpet, as well as new annoyances, such as increasing amounts of dirt. Enter inventor Daniel Hess (1831–1916) of West Union, Iowa, who received a patent for what he called a carpet sweeper, a sort of manual vacuum cleaner, in July of 1860. The device featured brushes on rollers that would agitate the fibers of the carpet to loosen the dirt embedded in them and a bellows system that created suction. Two water-filled chambers collected the dirt and dust that the bellows would direct into them. The idea behind the water chambers was to keep the debris in one place where it could be disposed of, rather than floating in clouds, as with traditional cleaning methods of the day. Hess's carpet sweeper was purportedly very difficult to use—it had to be constantly pumped by hand to work properly—and it is unclear whether the device was ever manufactured commercially. That said, one can imagine how Hess's innovation paved the way for more modern vacuum cleaners to come.

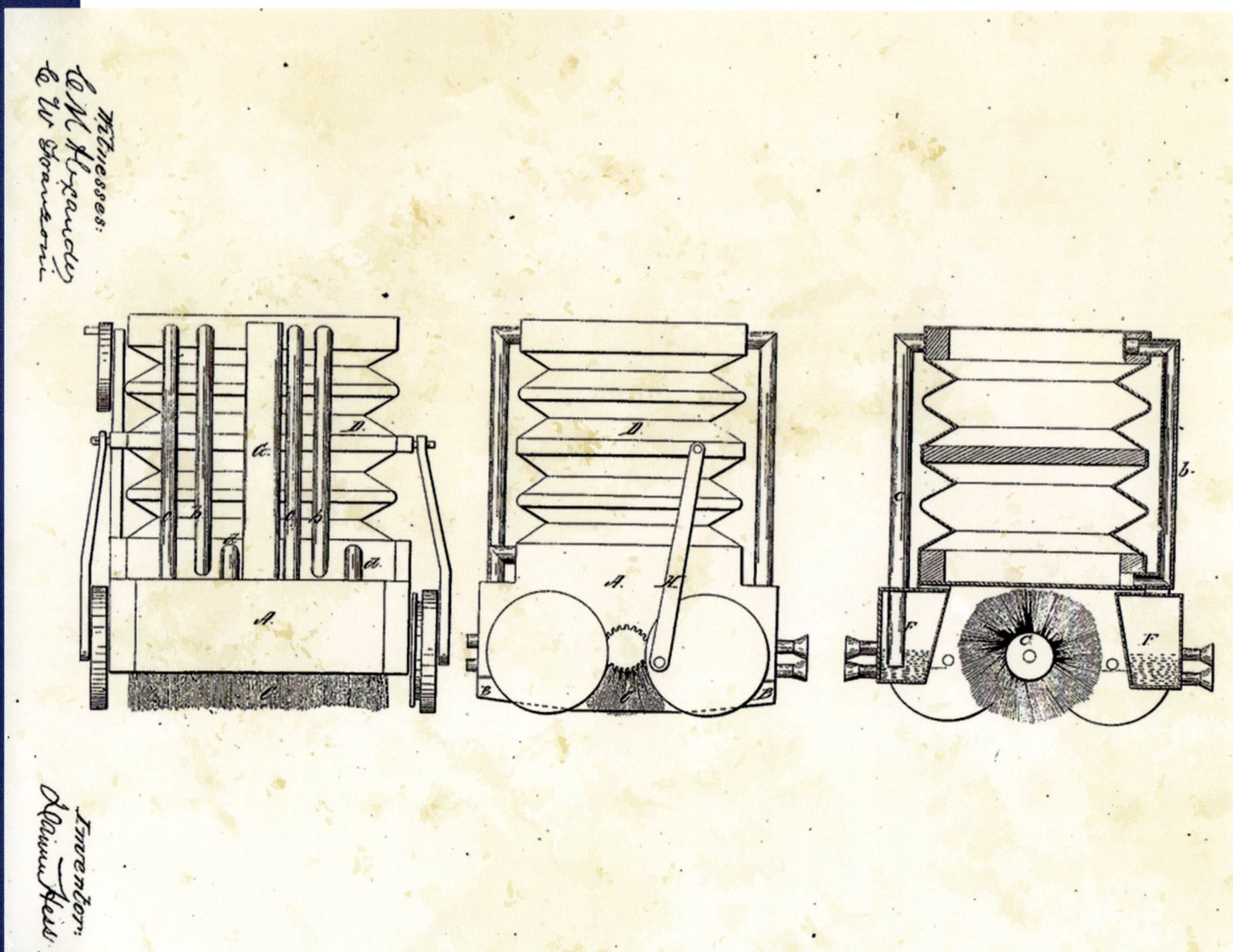

1861 TRANSCONTINENTAL TELEGRAPH SYSTEM

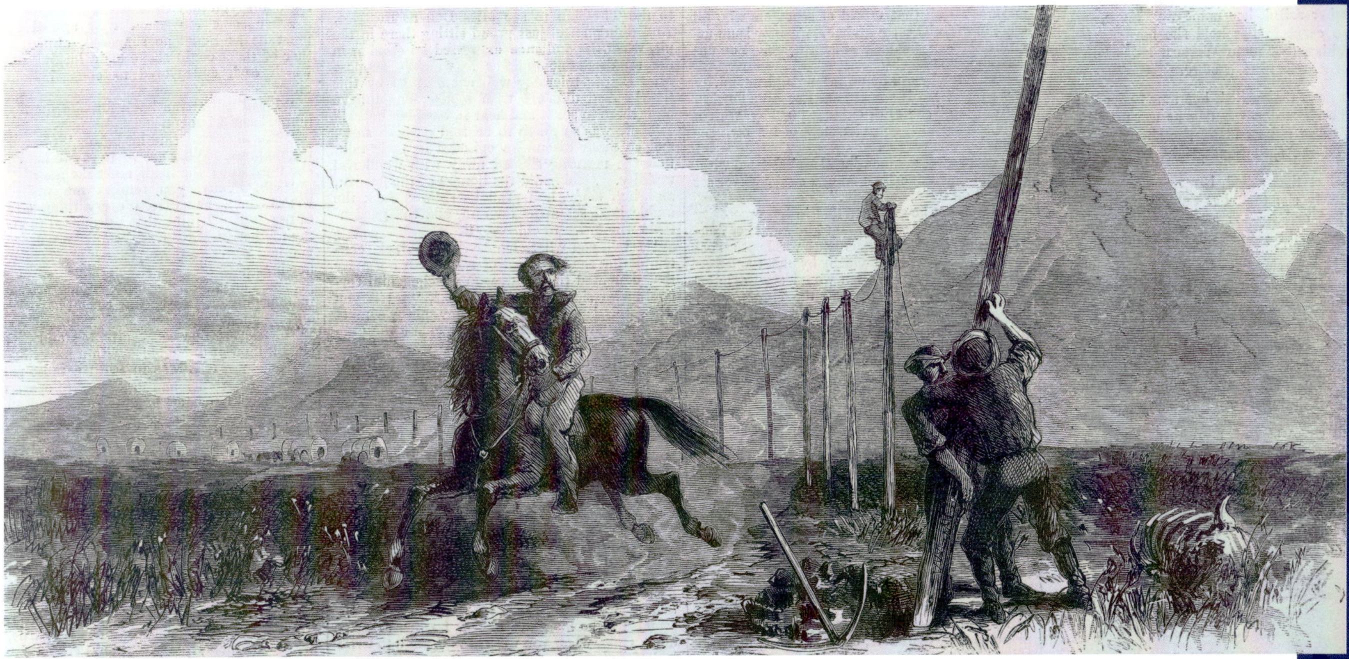

Thanks to Samuel Morse, the telegraph gained wide acceptance in 1844. But there remained more technological obstacles to overcome before it was widely used throughout the world. On October 24, 1861, a giant step was made toward that goal: The first transcontinental message was sent via telegraph from the chief justice of California, Stephen J. Field, from San Francisco to President Abraham Lincoln in Washington, DC. On that historic day, the Western Union company had just completed the connection between the western and eastern telegraph lines in Salt Lake City, Utah, enabling nearly instant communication across the country for the first time. The year prior, President James Buchanan had signed the Pacific Telegraph Act that would provide the funding for the construction and maintenance of the telegraph lines. From July to October of 1861, workers from the east and west raced to complete the telegraph lines that would allow communication in a fraction of the time it took to send a message across the country via the Pony Express, ensuring the demise of the horseback mail service. Communication would never be the same again.

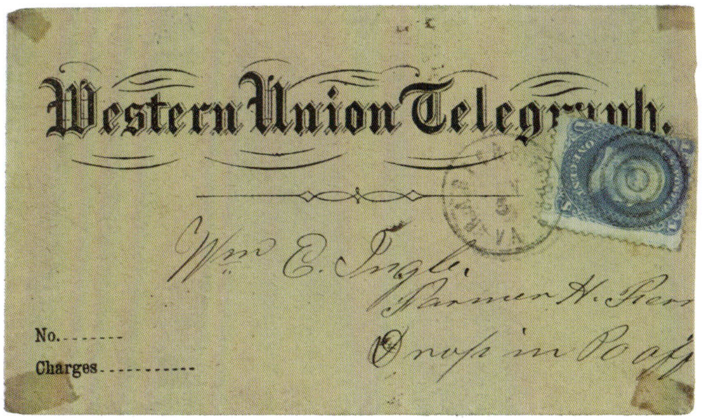

ABOVE: A Pony Express rider races past men putting up poles for the telegraph. **BELOW:** A Western Union Telegraph envelope, c. 1861.

1862 "BATTLE HYMN OF THE REPUBLIC"

Julia Ward Howe (1819–1910) was an American writer and abolitionist during the Civil War era. In 1861, she and her husband, Samuel Gridley Howe, a reformer and activist, were part of a group who were invited to Washington, DC to visit the Union troops after an important battle and assess their morale. Some believe that the Howes were entertained on that day by the Union army with popular patriotic songs of the era. These events are rumored to have inspired Julia to write her famous poem, "Battle Hymn of the Republic," which was published in *The Atlantic Monthly* in 1862. The poem was set to the tune of one of those songs she and her husband likely heard, "John Brown's Body," a popular Union marching song. Howe's song soon caught on among non-fighting residents and became one of the most famous American patriotic songs of all time—the song most closely identified with the Union army's cause in the Civil War. In 1908, Howe became the first woman elected to the American Academy of Arts and Letters.

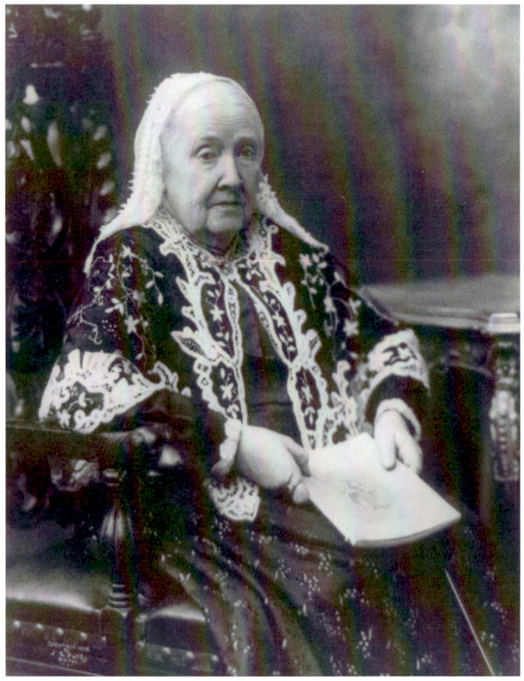

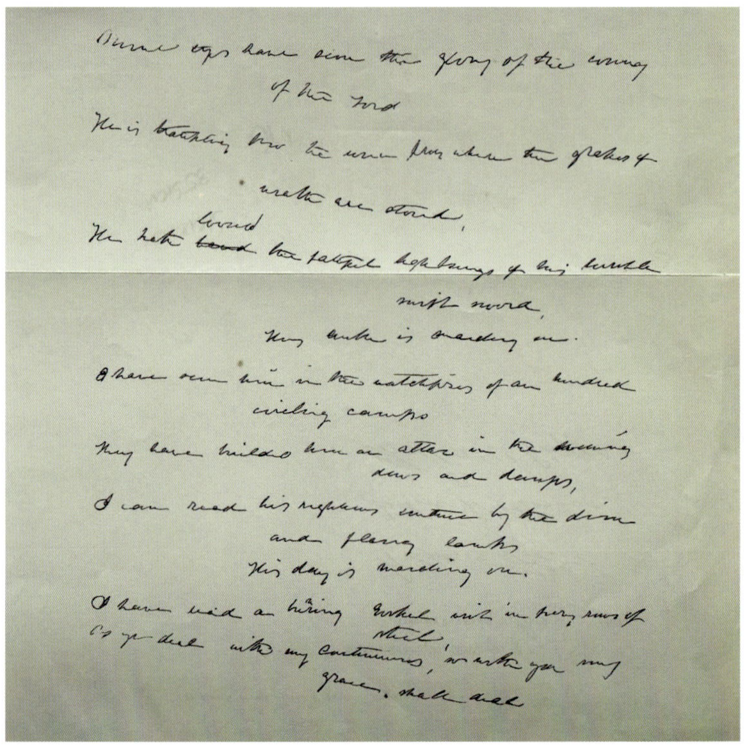

Mine eyes have seen the glory of the coming
of the Lord
He is trampling out the wine press where the grapes of
wrath are stored,
He hath ~~[illegible]~~ loosed the fateful lightnings of his terrible
swift sword,
His truth is marching on.
I have seen him in the watchfires of an hundred
circling camps
They have builded him an altar in the evening
dews and damps,
I can read his righteous sentence by the dim
and flaring lamps
His day is marching on.
I have read a fiery gospel writ in fiery rows of
steel,
As ye deal with my contemners, so with you my
grace. shall deal

TOP: Julia Ward Howe, c. 1902. **BOTTOM:** Julia's handwritten lyrics. **RIGHT:** A late nineteenth-century song sheet featuring the lyrics and interesting notes about who sang it and when.

Battle Hymn

OF THE REPUBLIC.

Sung by our prisoners in Libby Prison on hearing of the Brilliant Victory gained after the hard fought Battle of Gettysburg, Pa.

And as sung by Parson McKay, fresh from Libby Prison, before the United States Army Christian Commission, Washington, D. C.

Mine eyes have seen the glory of the coming of the Lord:
He is trampling out the vintage where the grapes of wrath are stored;
He hath loosed the fateful lightning of His terrible swift sword:
His truth is marching on.
Chorus.—Glory, glory Hallelujah,
Glory, glory Hallelujah,
Glory, glory Hallelujah,
His truth is marching on.

I have seen Him in the watch-fires of a hundred circling camps:
They have builded Him an altar in the evening dews and damps;
I can read His righteous sentence by the dim and flaring lamps:
His day is marching on.
Chorus.—Glory, glory Hallelujah, &c.

I have read a fiery gospel, writ in burnished rows of steel:
As ye deal with my contemners, so with you my grace shall deal;
Let the Hero, born of woman, crush the serpent with his heel.
Since God is marching on.
Chorus.—Glory, glory Hallelujah, &c.

He has sounded forth the trumpet that shall never call retreat;
He is sifting out the hearts of men before His judgment seat:
Oh! be swift, my soul, to answer Him! be jubilant my feet!
Our God is marching on.
Chorus.—Glory, glory Hallelujah, &c.

In the beauty of the lilies, Christ was born across the sea,
With a glory in his bosom that transfigures you and me:
As He died to make men holy, let *us* die to make men FREE,
While God is marching on.
Chorus.—Glory, glory Hallelujah, &c.

SOLD BY
A. ANDERSON, No. 420 South Tenth Street, Philadelphia. where all Ballads bearing his name may be had; also, all the most Popular Songs of the times, Wholesale and Retail.

1863 ROLLER SKATES

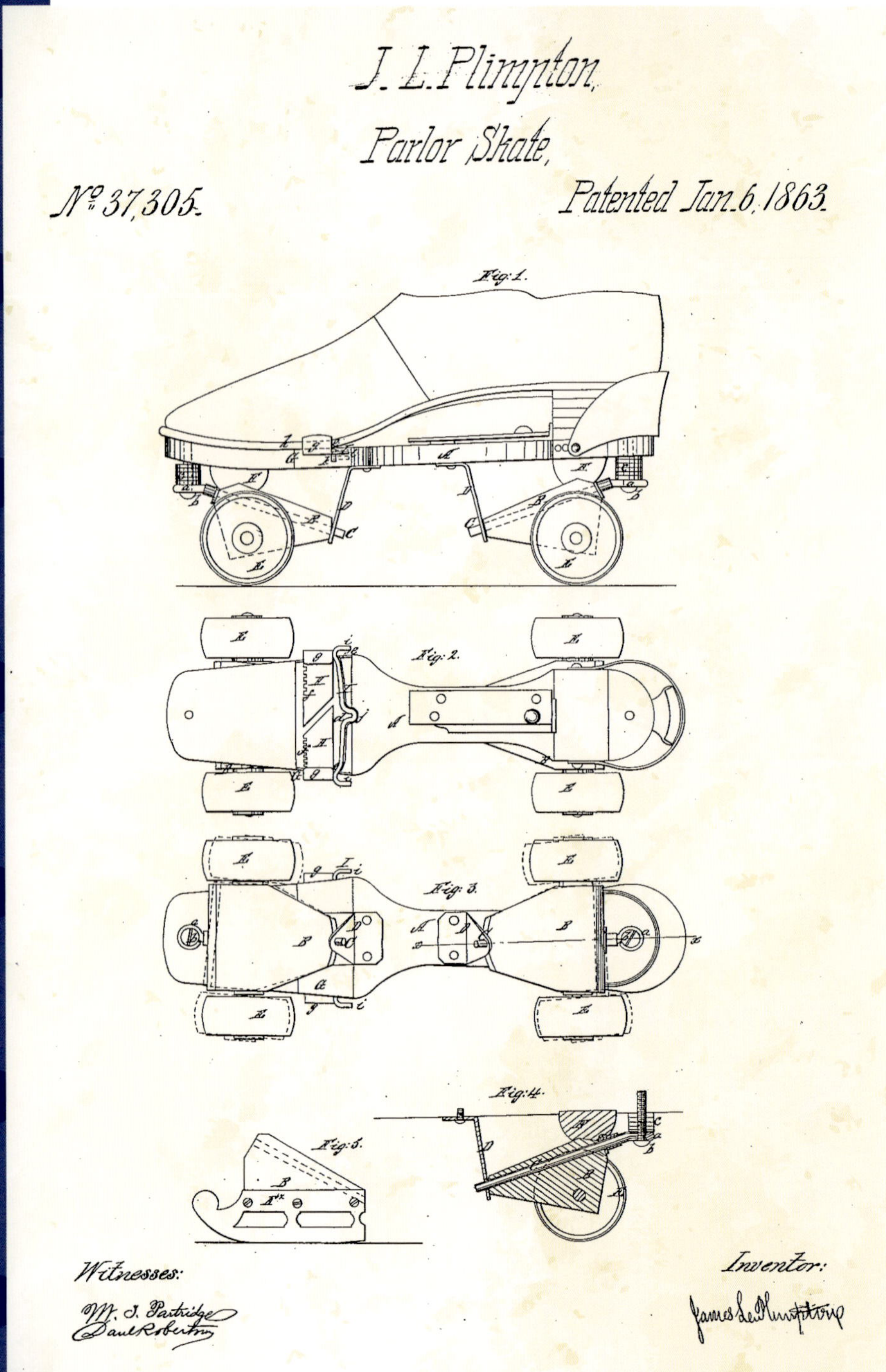

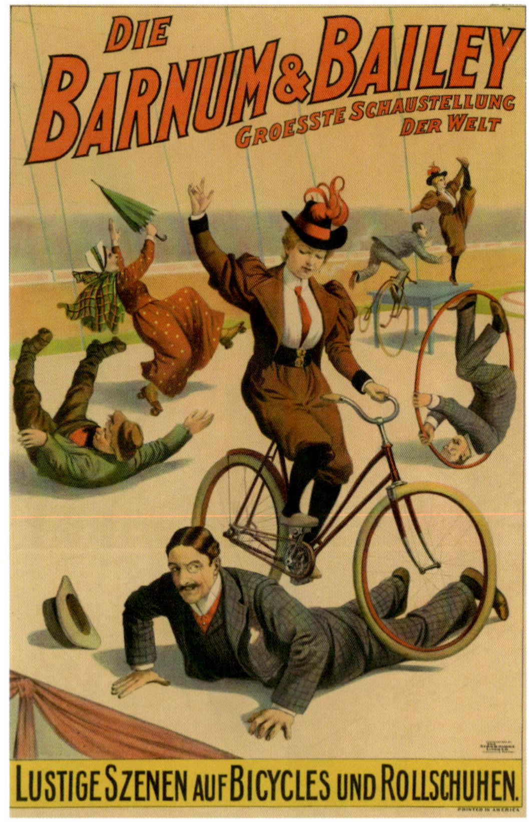

LEFT: Plimpton's patent drawing for his skates. RIGHT: A Barnum & Bailey poster shows the dangers of roller skating.

On January 6, 1863, Massachusetts-born machinist James Plimpton (1828–1911) received a patent for a device that would by 1871 launch a national pastime: roller skates (and the rinks in which they were enjoyed). Plimpton's skates featured four wheels each, attached to a boot via a springlike device. The skates could be maneuvered by leaning one's body weight in the direction they intended to go. The intention was to create a skate that could be enjoyed as easily on a smooth floor as it could be on ice—Plimpton had found ice skating to be a boon to his declining health a few years prior. Roller skating caught on with the elite in New York—the city where Plimpton and the clientele he courted lived—and would soon expand to other cities and to Western Europe. Plimpton's brilliance was not just in the skates he invented but in his business acumen. He was known for creating brilliant marketing campaigns, leasing the skates instead of selling them, and offering franchising rights for his roller rinks, a model that made him wealthy but that he needed to constantly defend.

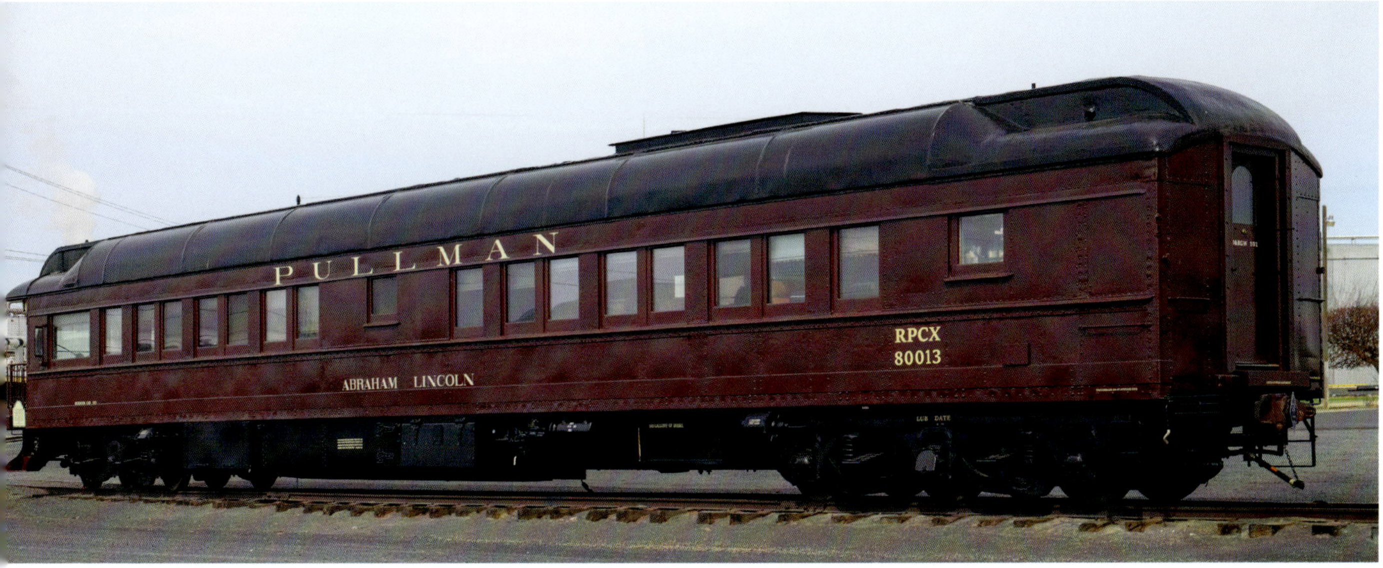

1864 PULLMAN SLEEPING CARS

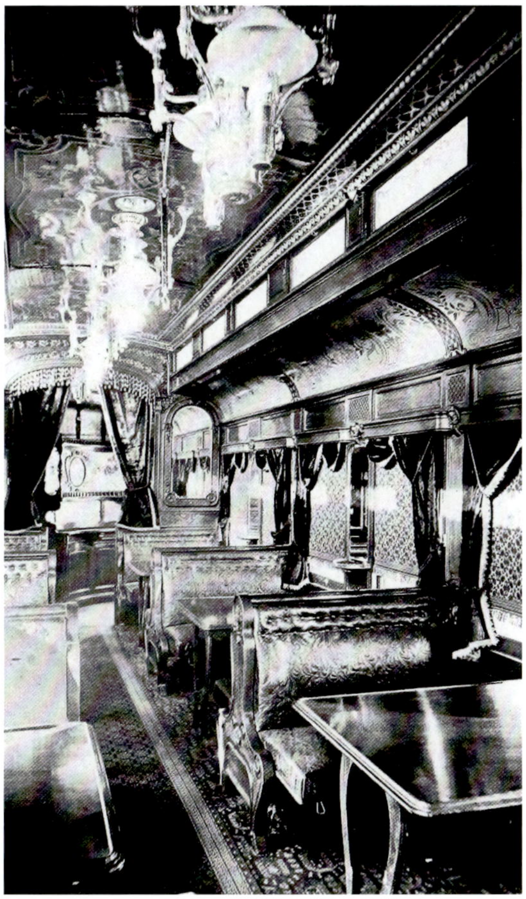

Chicago-based engineer and inventor George M. Pullman (1831–1897) designed his eponymous railroad cars for overnight travel. They were the first cars to offer clean and comfortable compartments for slumber, a far cry from the cramped, dirty, and stuffy sleeping cars of the day. Catering to the wealthy traveler seeking a luxurious experience, the Pullman sleepers became immediately popular. The cars included such comforts as electric lighting—including fancy chandeliers—furniture made of the finest materials, and four-star service from professional porters. They would accommodate regular passenger travel during the day in a comfortable, living room–like setting, then easily convert to beds for sleeping, with both lower and upper berths. After President Lincoln's assassination, a Pullman car was used for the last leg of the journey to his funeral. The publicity around the trip ensured that the Pullman sleeping cars became a household name.

The opulence of a Pullman car.

1865 STETSON HAT

When you think of the Stetson hat, it's probably notPhiladelphia that immediately comes to mind. Butthe iconic hat, with its wide brim and high crown—emblematic of a stereotypical Western cowboy—was invented by John B. Stetson (1830–1906), a Philadelphia hat manufacturer, in 1865. Stetson originally fashioned the hat for himself from a beaver pelt on a trip west to protect himself from the harsh weather. Legend has it that upon returning to Philadelphia, he made felt models of the hat, originally called "Boss of the Plains," and sent them to hat dealers in the west. With its natural insulating and waterproofing qualities, the hat was lightweight, versatile, and sturdy, perfect for cowboys and ranchers who made their living in the harsh western weather conditions. Fortunately for Stetson, the hat caught on. It became so popular that well-known entertainers such as "Buffalo Bill" Cody, Calamity Jane, Annie Oakley, and Will Rodgers all wore Stetsons in their acts. Thanks to John B. Stetson's ingenuity, by the early twentieth century, Stetson hats became synonymous with the American West.

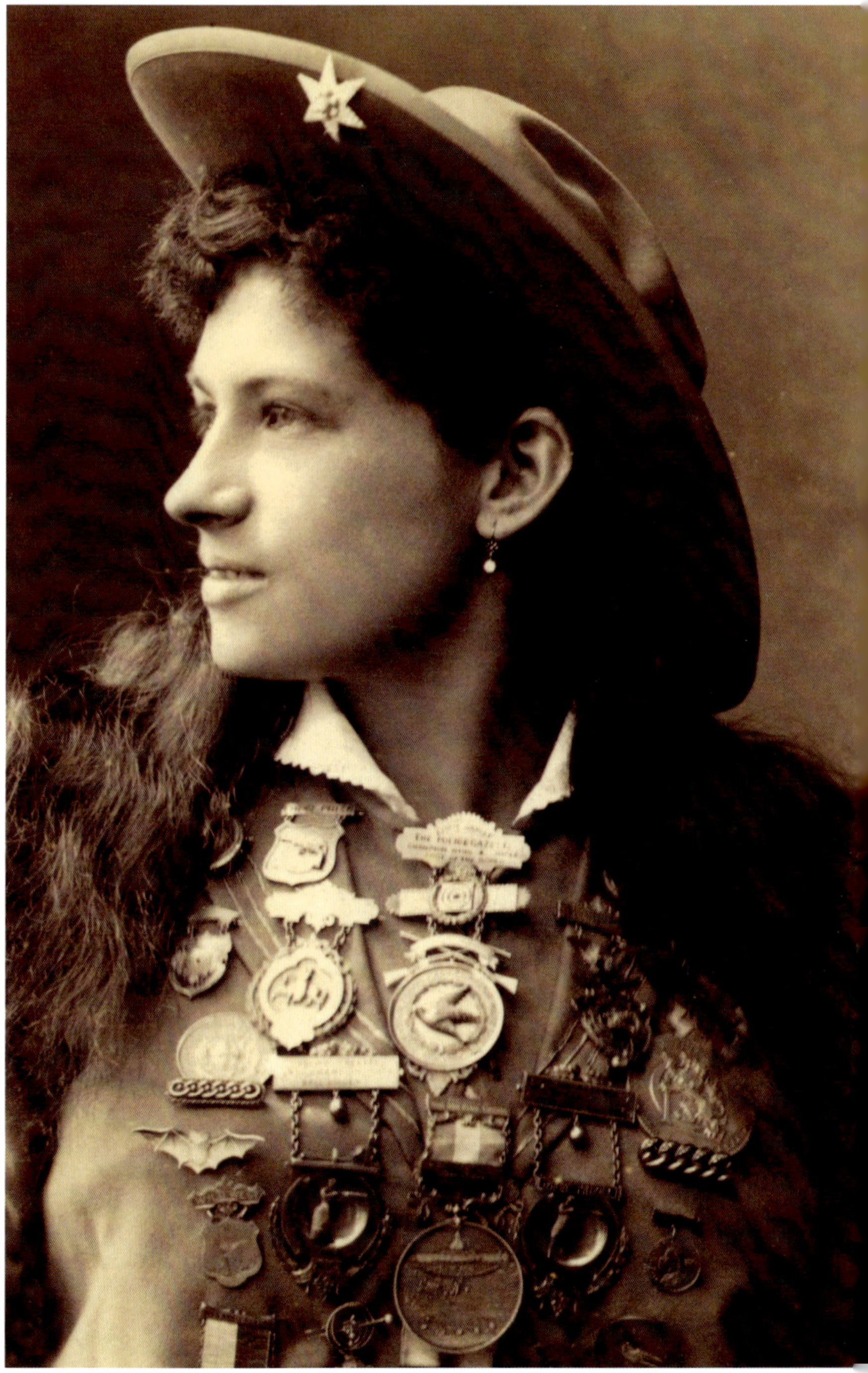

Legendary sharpshooter Annie Oakley in her custom-made Stetson.

1866 GINGER ALE

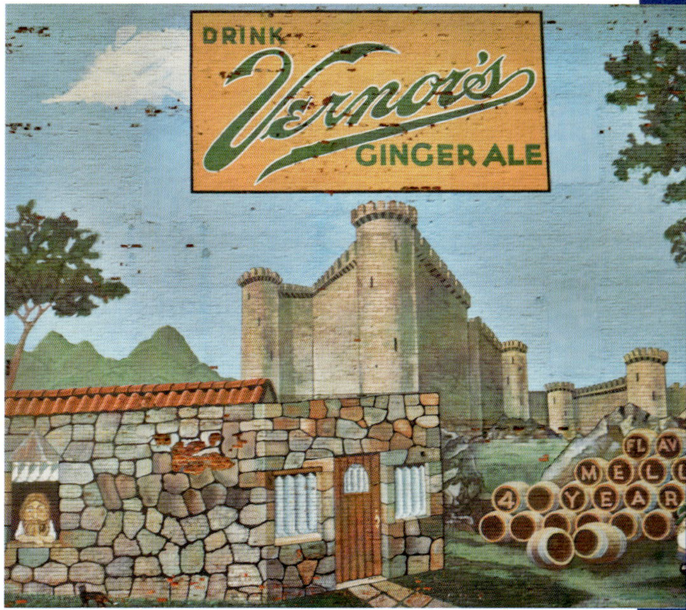

A 1932 mural advertising Vernor's in Flint, Michigan.

Created by Detroit pharmacist James Vernor (1843–1927), Vernor's Ginger Ale is one of the United States' oldest soft drinks. There are a few stories about the origin of the soda, one of the most popular being the following: Vernor worked as a junior clerk at a drug store and was experimenting with flavors, including vanilla and ginger. He left his untried mixture in an oak barrel before leaving to fight in the Civil War. When he returned from the war a few years later, he was delighted by the aged mixture he found waiting for him in the barrel and served it to eager customers. Though this story may be embellished, Vernor did open his own pharmacy in 1866 and served his special concoction there first, now with added carbonic acid for a powerful carbonated punch. Soon, he began to sell the soda to other soda fountains, with the caveat that they install special equipment in order to properly serve it. Still produced today, the highly carbonated, intensely ginger-flavored drink continues to have a cult following.

1867 PAPER CLIP

On April 23, 1867, Samuel B. Fay (birth and death date unknown) received a patent for what would become one of America's most ubiquitous innovations: the paper clip. Originally called a "Ticket Fastener," Fay's design was a simple loop of wire intended primarily for attaching tickets or tags to fabrics without causing damage. Even though his bent-wire construction was much simpler than today's familiar double-loop "gem" design, Fay's patent did recognize the device's potential for binding papers together. Fay's original clip wouldn't gain widespread commercial use until the 1890s, but his innovation laid the groundwork for a revolution in office organization. This humble invention would ultimately transform how Americans organize and manage paperwork, proving that revolutionary innovations often come in the smallest packages.

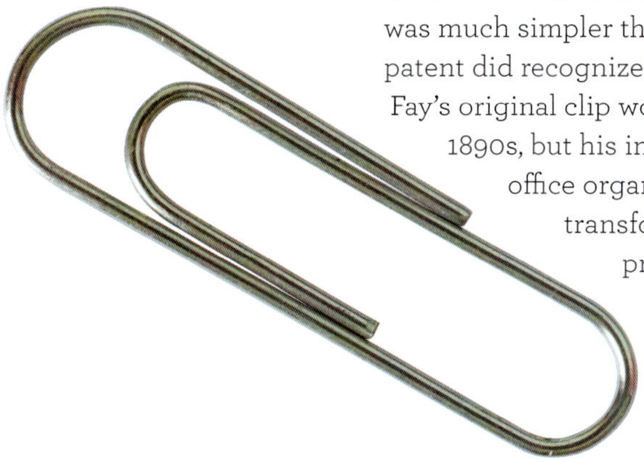

1868 *LITTLE WOMEN*

The children's book *Little Women,* first published by Massachusetts author Louisa May Alcott (1832–1888) on September 30, 1868, was revolutionary in its day for many reasons. First, few women of the era became writers. Indeed, Alcott wrote many of her early works under a pseudonym in order to make money—she wasn't necessarily proud of the writing she was churning out. Second, the story of the four March sisters—Meg, Jo, Amy, and Beth—was a groundbreaking and intimate portrayal of young women in the post–Civil War era, representing their individuality and distinct life paths. That each character could make a choice that suited them best was not a notion most people held. Finally, Alcott wrote an unusually feminist storyline for the era, giving the protagonist, Jo, in particular, notions of gender equality not normally presented in literature of the day. Some scholars purport that Alcott's book was so popular among women that it created one of the first waves of feminism. Transcending run-of-the-mill classic literature, Alcott's *Little Women*—which has never been out of print—manages to stay relevant to audiences to this day, with its evergreen messages and relatable character struggles.

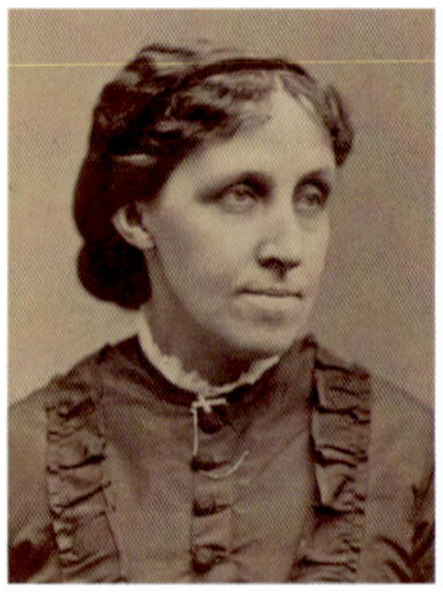
Louisa May Alcott, c. 1870.

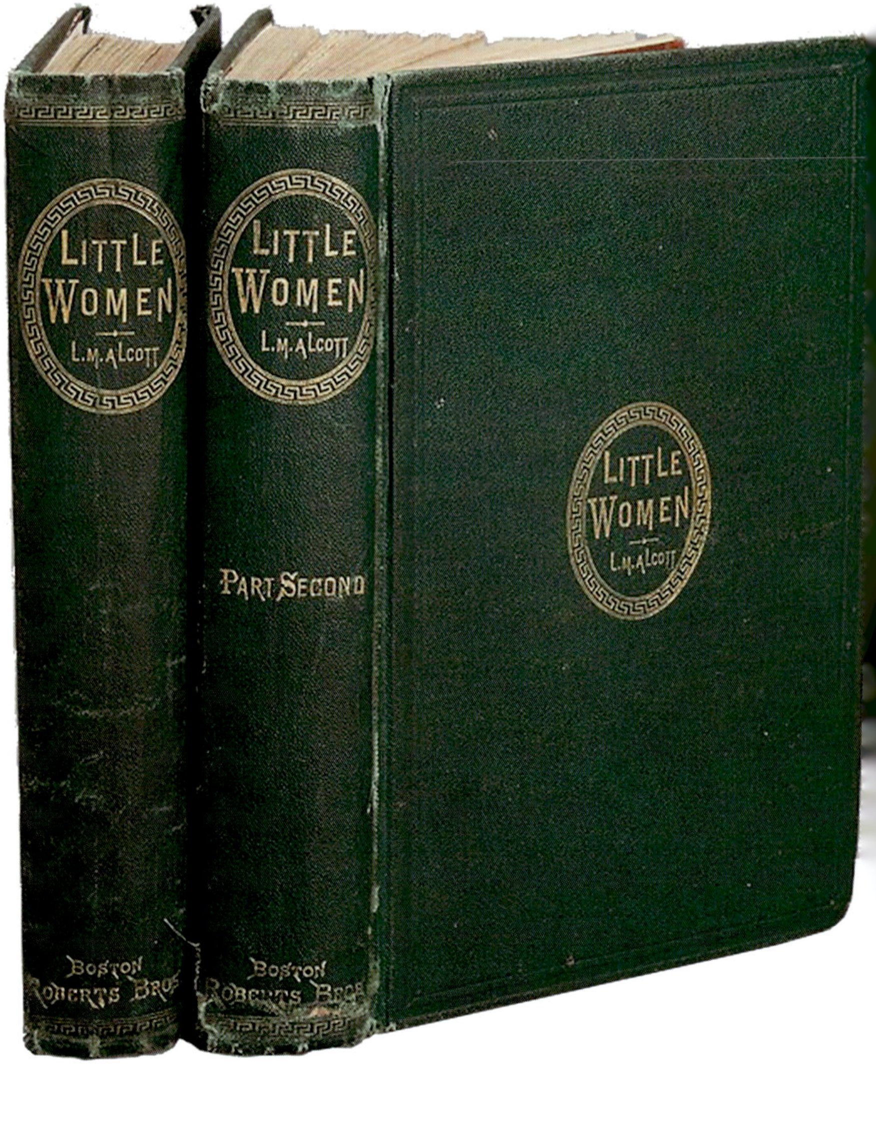

1869 TRANSCONTINENTAL RAILROAD

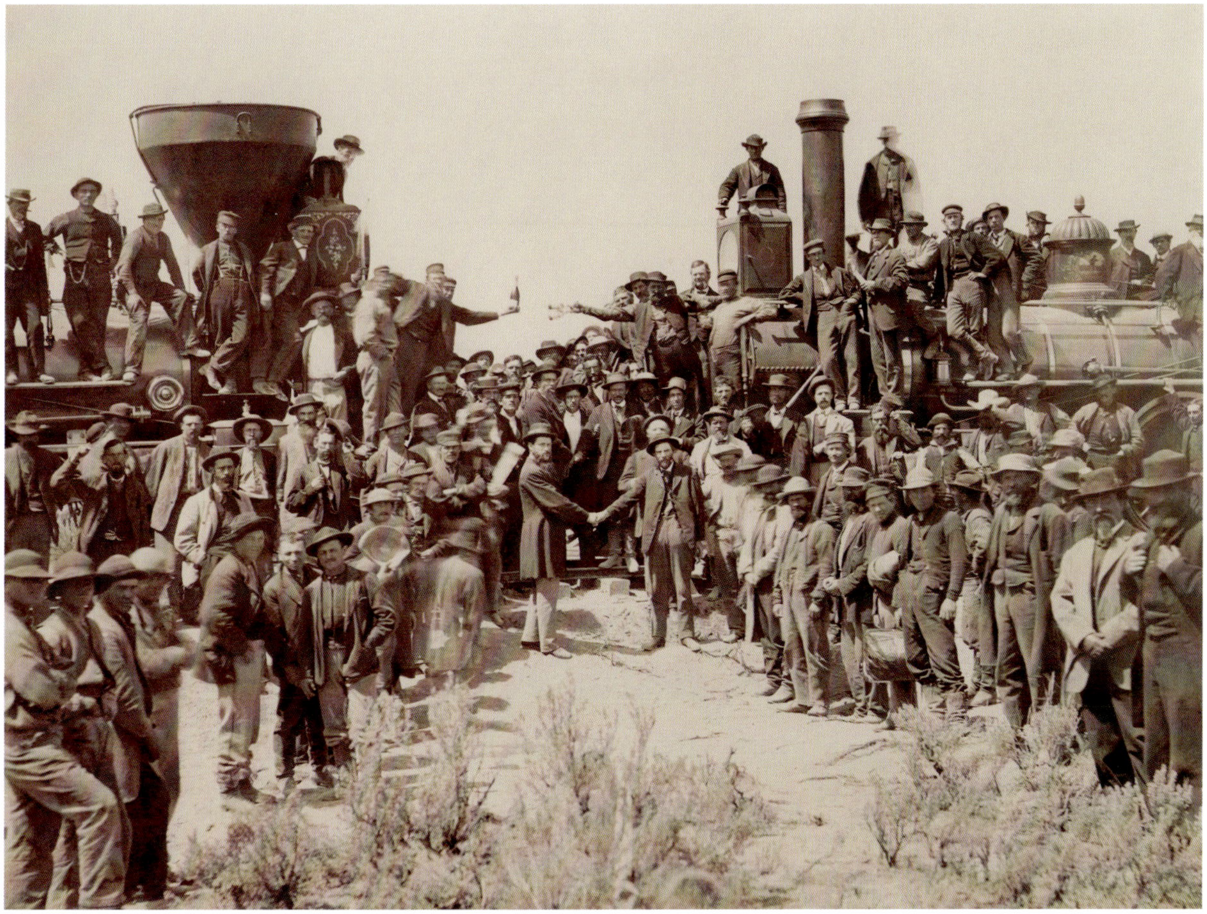

The celebration ceremony for the completion of the railroad, 1869.

On May 10, 1869, a ceremonial golden spike was driven in Promontory Summit, Utah, to signal the completion of the transcontinental railroad. For the first time in history, it would be possible for passengers and freight to obtain passage on one train from coast to coast. But reaching this milestone date was not an easy task. It would take Congress to appropriate funds, the passing of the Pacific Railroad Act in 1862, public land grants and loans, and the work of the Union Pacific and Central Pacific Railroad companies and their crews to lay the two thousand miles of track to complete the job. It should be mentioned that the building of the railroad relied heavily on immigrant muscle—especially Chinese workers for the Central Pacific section, who faced harsh working conditions and discrimination. In addition, the newly constructed railroad made the struggle of Native Americans ever more difficult with colonizers. Nevertheless, the section of track was completed early and under budget. The significance of the transcontinental railroad, also called the Overland Route, made passenger travel safer, encouraged American business growth, and increased intellectual pursuits and public discourse.

1870 CHEWING GUM

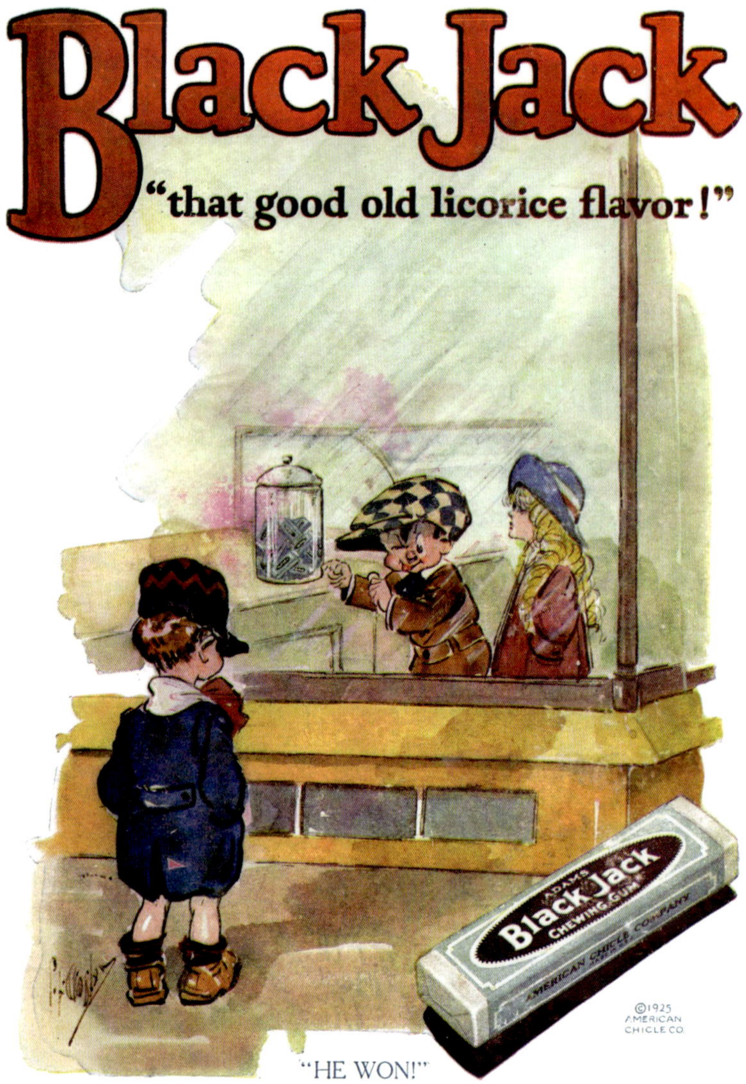

Chewing gum has been a known quantity in North America for centuries in the form of chicle, a resin obtained from a tree native to central America. In 1870, American scientist and inventor Thomas Adams (1818–1905) of New York City received a supply of chicle and experimented with it—boiling it and hand-rolling it to create pieces of chewing gum. Adams first sold his gum to local drug stores to wide appeal; it appeared he had a hit on his hands. He and his son tried to keep up with demand. In 1871, Adams created a flavored gum called Black Jack, which tasted like licorice. That same year, Adams patented a machine that could mass-produce his gum for commercial sale. In 1888, Adams sold his Black Jack and Tutti Frutti gum flavors from a vending machine in a New York subway station. Though Adams did not invent chewing gum, his efforts to mass-produce it and promote it arguably gave birth to the chewing gum industry in America.

ABOVE: A 1925 ad for Black Jack gum.
RIGHT: Adams's Tutti Frutti gum boasted not only a great flavor but relief from indigestion and dyspepsia.

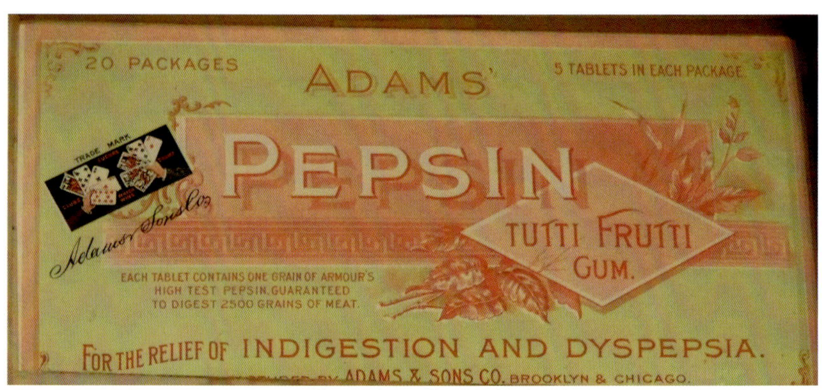

1871 LEVI'S JEANS

On May 20, 1873, Reno, Nevada, tailor Jacob Davis (1831–1908) and Levi Strauss (1829–1902), a German immigrant who had a dry goods store in nearby San Francisco, received a patent for pants reinforced with metal rivets. A couple years prior, Davis came up with the idea for the pants when he witnessed miners' workwear ripping and wearing out at the same stress points. To solve the problem, Davis affixed metal rivets to the pocket corners and at the base of the fly. The financial backing for the immensely popular garments came from Strauss, who was accustomed to selling clothing and other items to miners and settlers flooding through town during the Gold Rush. The garments would soon come to be known as "Levi's jeans." Though the original design didn't feature it, denim, a fabric that had been worn by enslaved and immigrant laborers for generations, would eventually become synonymous with the jeans. The first Levi's jeans, later dubbed the 501 model, quickly became a bestseller. Today, Levi's continues to influence fashion while maintaining its connection to American heritage. The brand's legacy proves that functional, well-designed clothing can transcend its original purpose to become a culture and style across generations and continents.

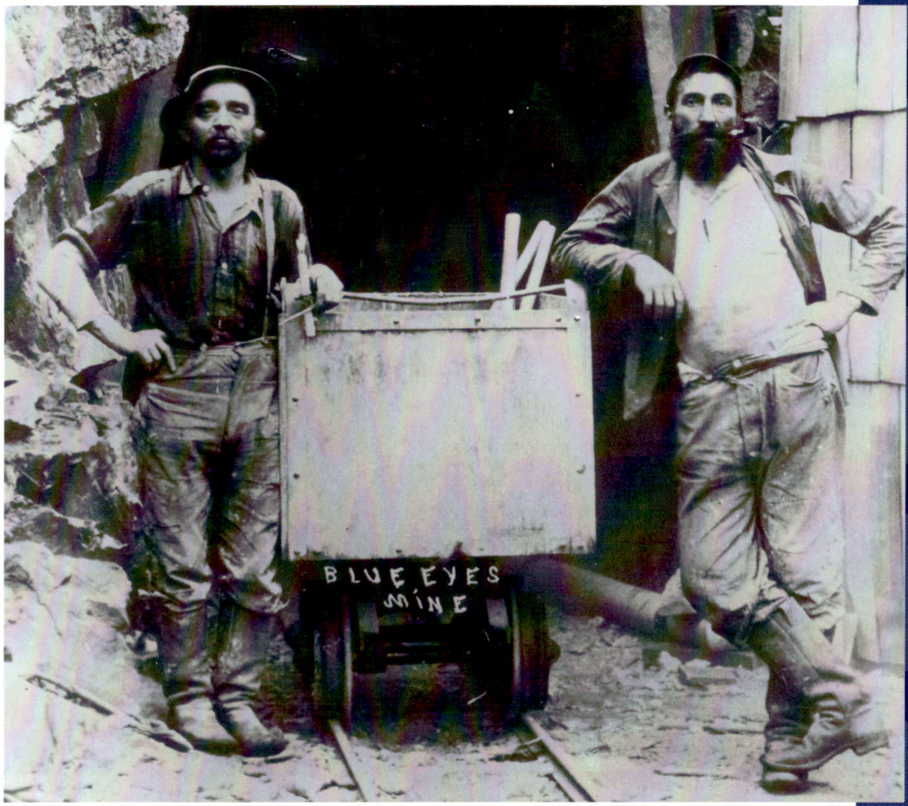

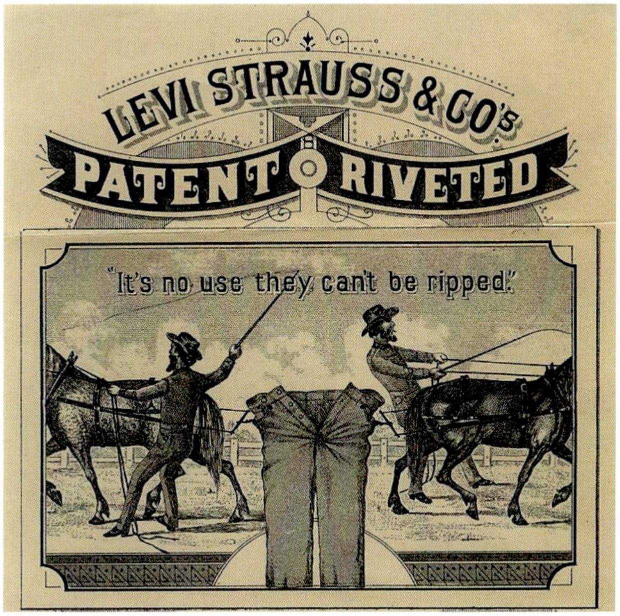

TOP: Gold miners in Levi's stand outside the Last Chance Mine Gold in California, c. 1882. **RIGHT:** An early ad for the jeans that "can't be ripped."

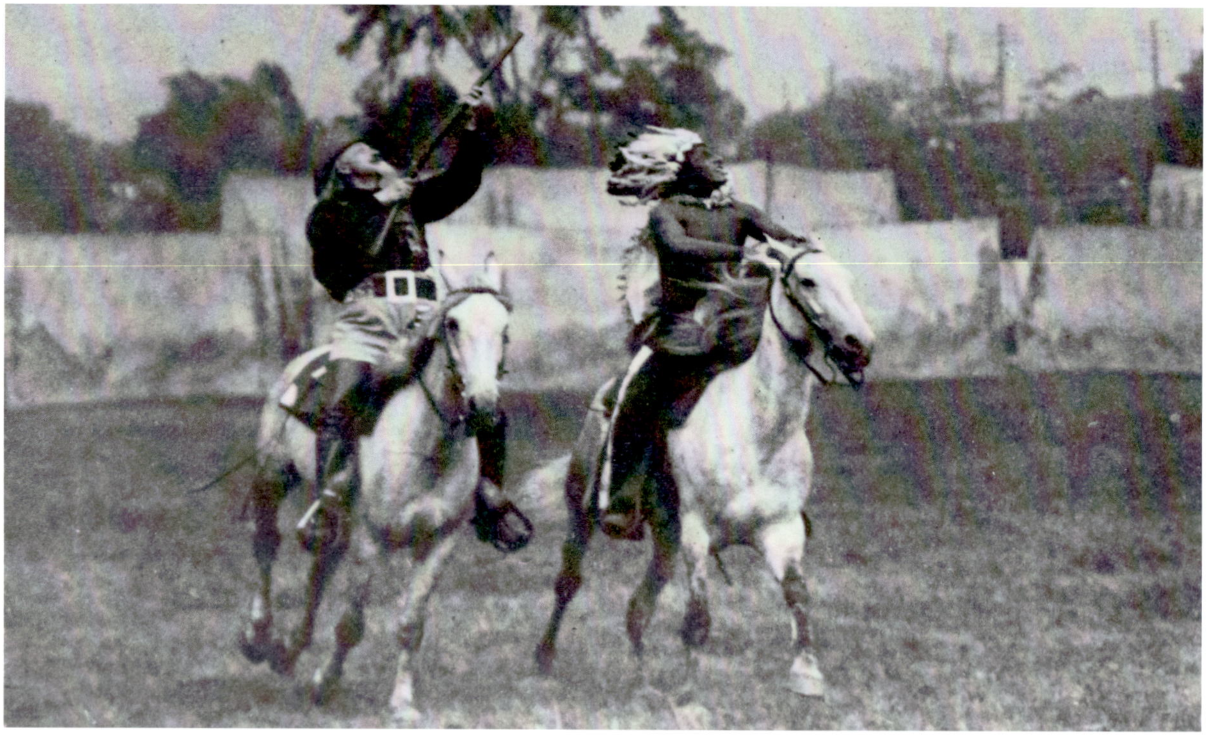

1872 THE RODEO

With the expansion of the railroad and the fencing of the land at the end of the open range system in the late nineteenth century, cowboys were in less and less demand. Nevertheless, there was still a lot of pride in the job. The sport of rodeo was created as a way of supporting the cattle industry in North America, honoring the skills typically used by a working cowboy. Ranchers offered bonuses for cowboys competing in the sport, participating in events focused on riding abilities, trick roping, and horse tricks. Rivalries between ranchers developed as rodeo's popularity increased. The first official rodeo was held in Cheyenne, Wyoming, in 1872. As rodeos popped up across the country, some also included Wild West shows with popular entertainers of the day, such as "Buffalo Bill" Cody. The popularity of rodeos soon grew to such a level that many western towns began to hold annual ones, including Cheyenne with its Frontier Days beginning in 1897 and Pendleton, Oregon's Round-Up in 1910. It is a tradition that endures today, helping preserve and romanticize cowboy culture long after the end of the open-range era and maintain strong connections to rural communities.

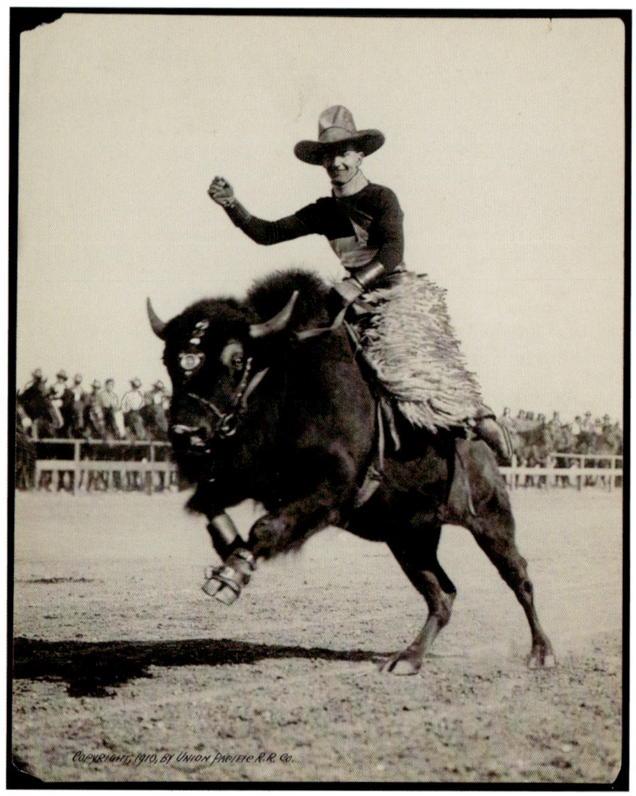

TOP: Buffalo Bill Cody doing his trick shooting during a Wild West show, c. 1907. **BOTTOM:** Rodeo star Verne Elliott on a bucking buffalo at the Cheyenne Frontier Days, 1910.

1873 THE POSTCARD

A card sent through the mail without an envelope, the postcard, has long been a convenient and popular way to send a quick note or to remind a loved one that you are thinking of them while traveling—and the tradition still endures today. The first postcard produced in America was created by the Morgan Envelope Company of Springfield, Massachusetts, in 1873. Just the year prior, the US Congress approved that such a card could be made available with postage preprinted on it, the "penny postcard." On these cards, one side was reserved for the message, and the other was meant for the address. The Morgan Envelope Company was awarded the contract to produce the government postcards by then postmaster general, John Creswell (1828–1891). Other types of postcards without the preprinted postage cost double the amount to mail, two cents. The tradition of sending souvenir postcards with pictures on the front is believed to have started during the Chicago World's Fair in 1893. The legacy of sending postcards is a fascinating blend of personal communication, cultural documentation, and social ritual. These small rectangular pieces of cardboard became one of the most democratic forms of correspondence, accessible to people across all social classes.

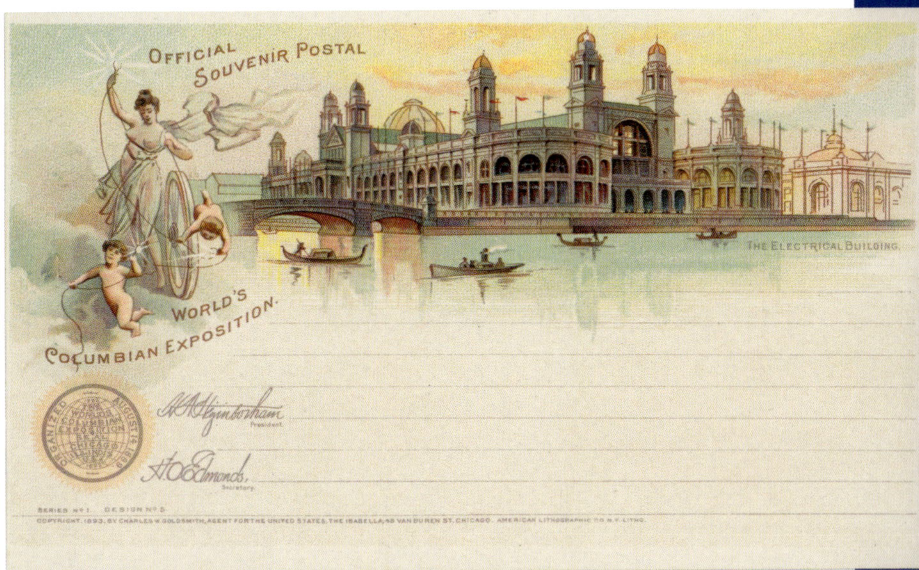

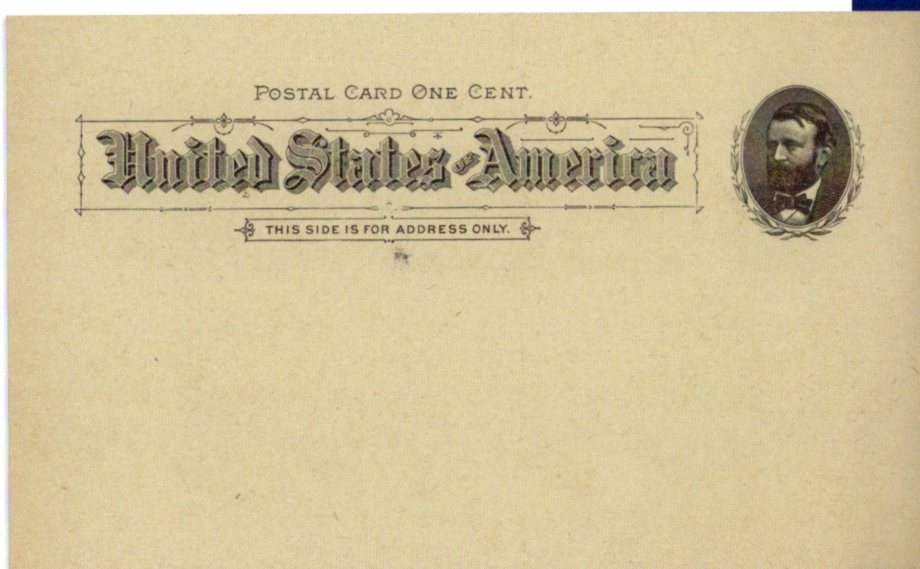

A souvenir postcard from the Chicago World's Fair, 1893.

1874 THE TYPEWRITER

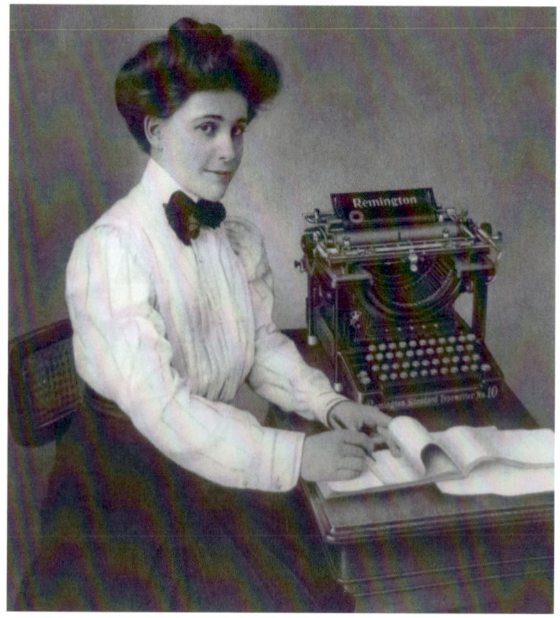

American inventor Christopher Latham Sholes (1819–1890) of Milwaukee, Wisconsin, is credited with creating the typewriter. Records show that Sholes created at least four practical models and received up to three patents for working typewriter machines, then sold the rights to E. Remington and Sons, a New York–based gun company. The "Remington" typewriter was the first model for commercial sale in 1874 and boasted the ability to create written messages at a speed far faster than that of pen and paper. Though the first typewriter models were primitive, Sholes kept innovating—later improvements included a shift bar, allowing lowercase and uppercase type; a space bar; and an innovation that still impacts written communications today, the QWERTY keyboard (his initial, alphabetically arranged keyboard was prone to frequent jamming). Remington typewriters were known at the time to be on the cutting edge of typewriter technology—the continual advances kept the brand a relevant and reliable one throughout their heyday. It has been said that the typewriter is the most significant business invention of the pre-computer age.

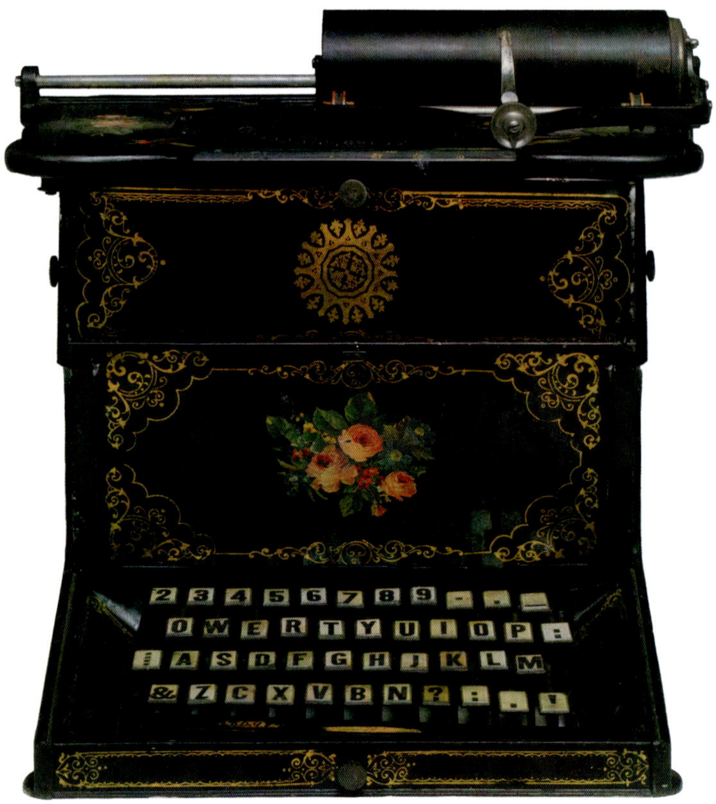

A Sholes and Glidden Type-Writer, c. 1874.

1875 ELECTRIC DENTAL DRILL

On January 26, 1875, George Green of Kalamazoo, Michigan (1832–1892) received a patent for the first electric-powered dental drill, which he had invented seven years before in 1868. The slow, arduous, and painful process for the dental patient was likely a motivator for Green's research. Green's new drill was powered via a bellows system and operated by a motor controlled by a foot pedal. His drill exponentially sped up the removal of decayed tissue and dramatically reduced the amount of suffering for his patients. Looking at Green's invention through a modern lens, less pain and better hygiene was only the beginning of the electric dental drill's contributions: it has led to less medical waste, shorter appointment times, the ability to see more patients in a given day, and more revenue for the dentist, among other offerings.

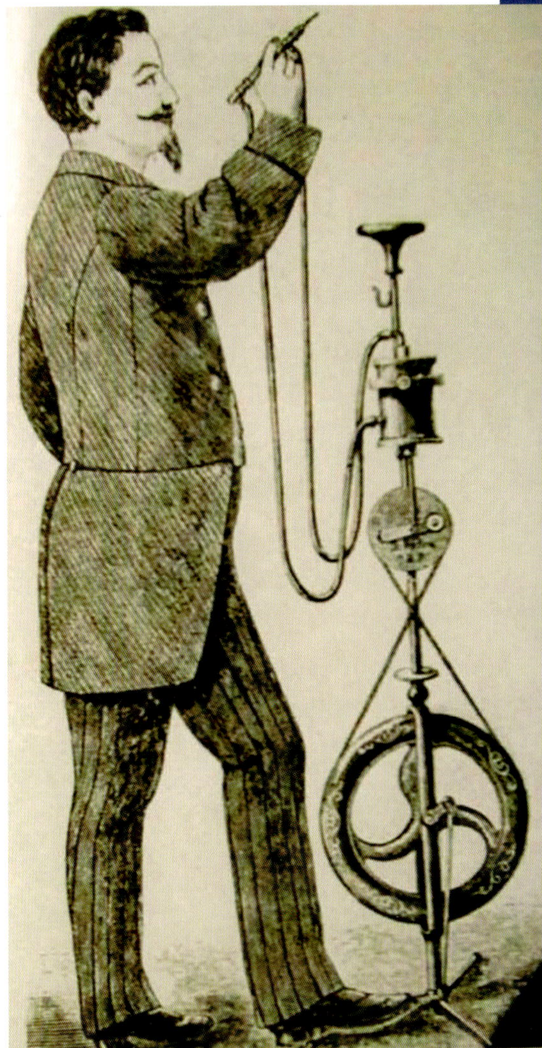

1876 THE TELEPHONE

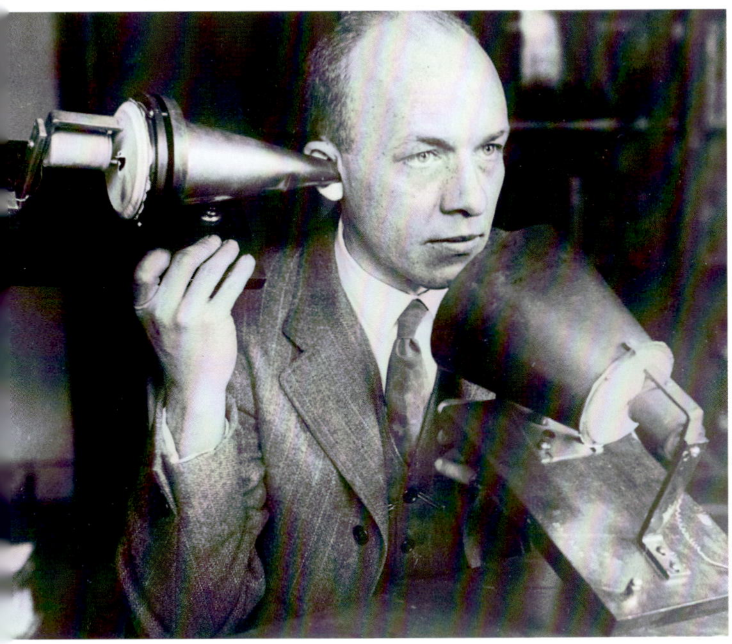

Although he was a prolific inventor in his lifetime, Alexander Graham Bell (1847–1922) is probably most famous for inventing the telephone. While experimenting with improving telegraph technology in Boston, Bell figured out how to transmit sound over a telegraph wire. He received a patent for the telephone on March 7, 1876—some argue it to be one of the most valuable awarded by the patent office in its history. Within a few days, with the help of his lab assistant, Thomas A. Watson, Bell developed the technology into an actual prototype—his "electrical speech machine." It is hard to understate what an impact the telephone has had on American society and beyond: It has fundamentally reshaped human communication, business, culture, infrastructure, and technology since its invention.

Bell's first telephone in use.

1877 THE PHONOGRAPH

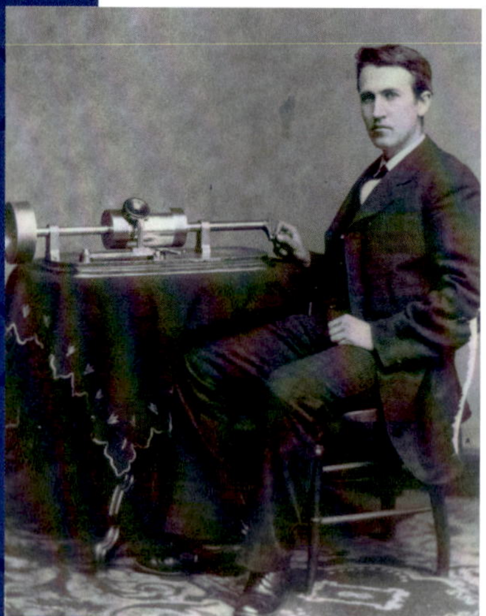

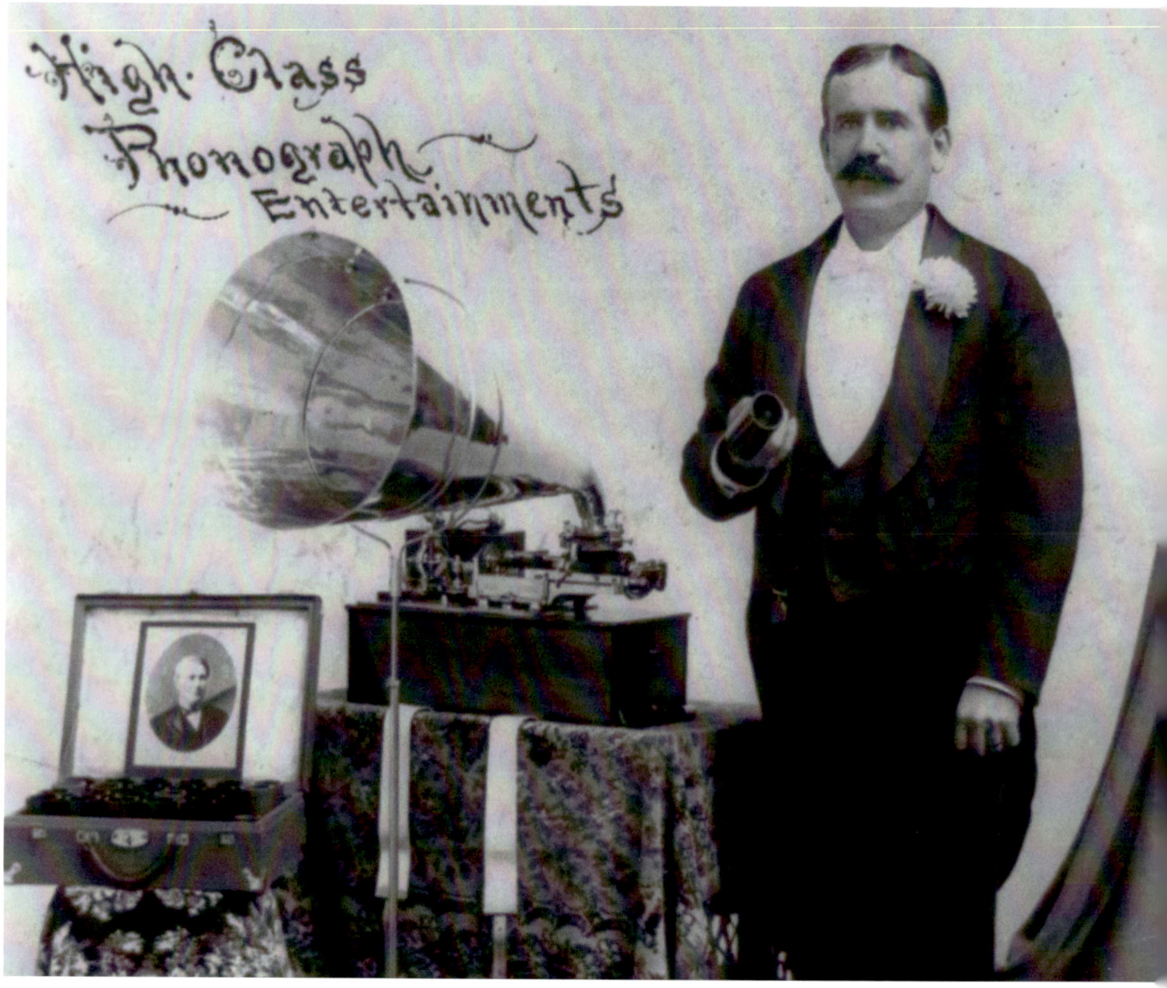

ABOVE: Edison demonstrating his "speaking phonograph" at the National Academy of Science, 1878. **RIGHT:** A man holding the cylinder for the new Edison machine, c. 1885.

Thomas Edison (1847–1931) at one time held a world record for having the largest number of American patents. Though he had lost his hearing when he was twelve, there are reports that the phonograph, invented in 1877 in Menlo Park, New Jersey, was Edison's favorite invention. The device revolutionized human communication by capturing and reproducing sound for the first time in history. Edison's initial design used a hand-cranked cylinder wrapped in tinfoil, onto which sound vibrations were mechanically etched as grooves. Edison initially envisioned the phonograph primarily for business dictation and preserving the voices of deceased loved ones. However, it quickly evolved into an entertainment medium, transforming how people experienced music and storytelling. The technology underwent significant improvements through the 1880s and 1890s, with better recording materials and enhanced sound quality. The phonograph laid the foundation for the music industry, establishing concepts of mass media in the twentieth century. The phonograph also preserved cultural heritage, capturing voices, languages, and musical traditions that might otherwise have been lost. Its influence flows through every subsequent audio technology, from radio broadcasting to modern digital streaming, making it one of history's most important inventions.

1878 BURPEE SEED CATALOG

The name Burpee has been synonymous with fruit, vegetable, and flower seeds for more than a century. Its fame dates back to 1878, when Philadelphia resident Washington Atlee Burpee (1858–1915) formed his eponymous seed company, W. Atlee Burpee & Company. Even at the beginning, Burpee was fixated on hybridizing new fruit, vegetable, and flower varieties for better flavor, appearance, and longevity, among them the Fordhook lima bean, Surehead cabbage, and iceberg lettuce—which could be shipped long distances more efficiently than any other lettuce variety. Another innovation of Burpee's was starting a mail-order business that facilitated the purchase of a large variety of seeds via the Burpee catalog. The catalog was filled with entertaining sales copy and charming illustrations that delighted customers. At the time of his death, when his son, David, took over the company, Burpee's firm was the largest seed company in the world. Equally passionate about produce and blooms, the younger Burpee encouraged the American public to grow Burpee seeds in backyard Victory Gardens during World War II. The innovations of the Burpee Seed Company continued, and it remains one of the most beloved and reliable seed companies to this day.

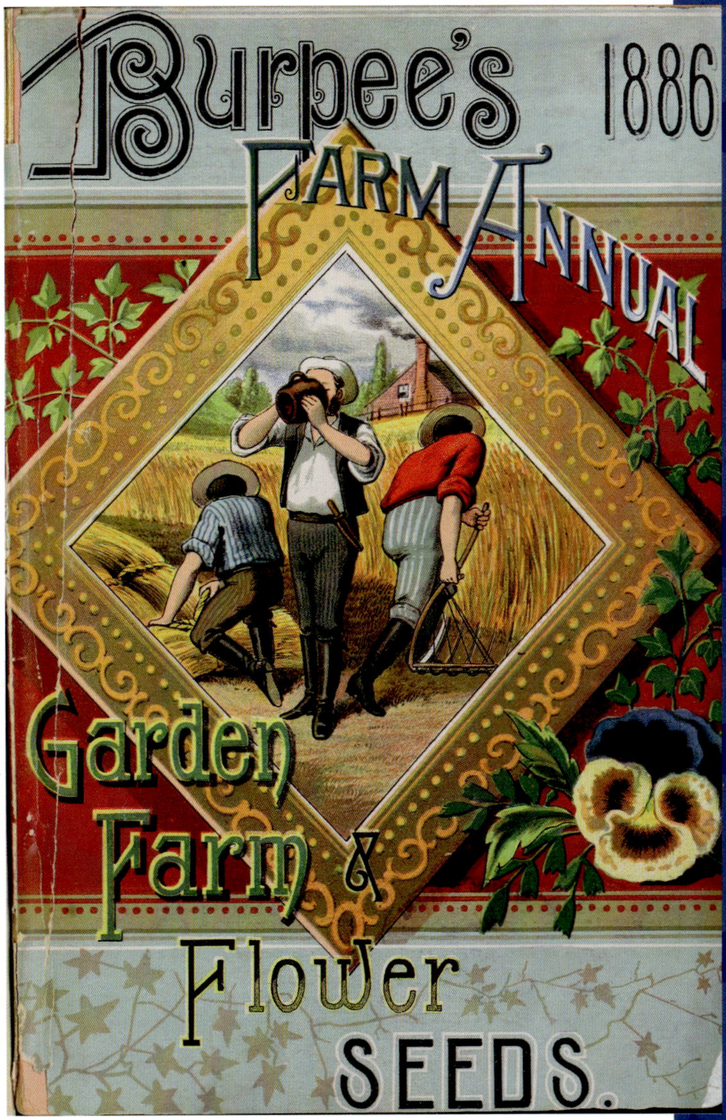

The 1886 Burpee's catalog.

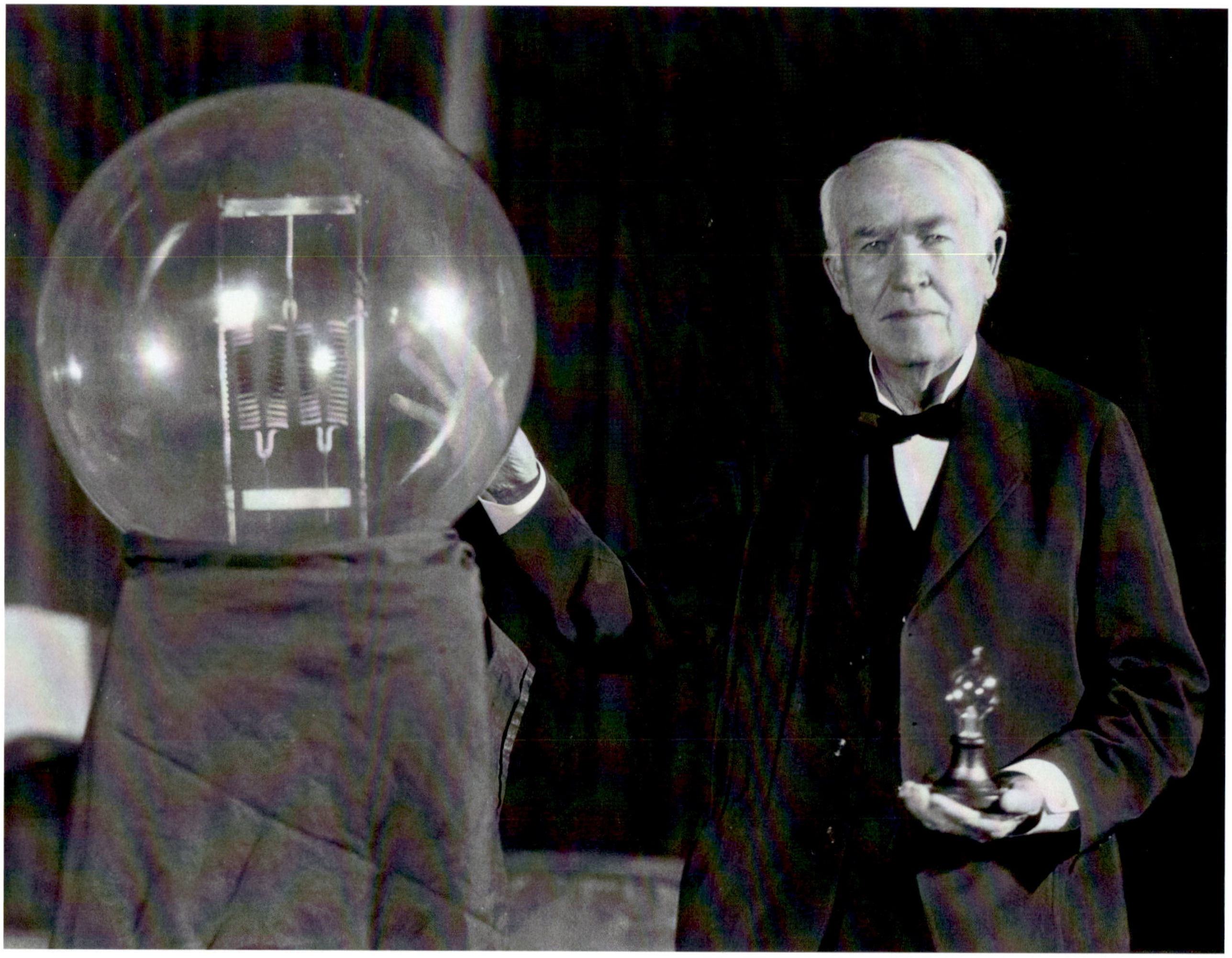

Edison holds a replica of his sixteen-candlepower incandescent bulb next to a modern-day fifty-thousand-watt version in 1929.

1879 INCANDESCENT LIGHT BULB

The incandescent light bulb, perfected by Thomas Edison (1847–1931) in 1879, revolutionized life in the late nineteenth century by extending productive hours beyond daylight. Though Edison didn't invent the first electric light, he created the first practical, long-lasting bulb using a carbonized bamboo filament that could burn for over twelve hundred hours. His innovation lay not just in the bulb itself but in developing an entire electrical distribution system to power it. The incandescent light bulb enabled the rise of modern cities, as streets and buildings could be safely illuminated at night. Factories could operate around the clock, dramatically increasing productivity. Social life expanded as people could gather after dark, altering daily rhythms. Beyond its practical applications, the incandescent bulb became a symbol of innovation and progress. It spurred the development of electrical infrastructure worldwide and established electricity as an essential utility. Though largely replaced by more efficient LED technology today, Edison's light bulb remains an icon of human ingenuity.

1880 MODERN LIFE RAFT

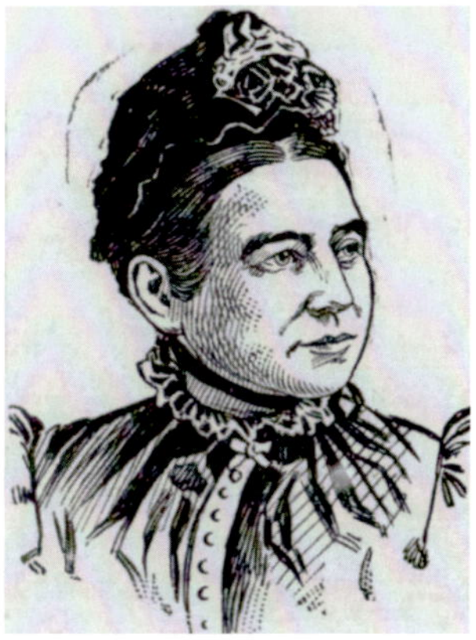

Maria E. Beasley (1836–1913) of Philadelphia was a prolific inventor, receiving patents for fifteen recorded items in the United States between 1878 and 1898. But her invention of the collapsible life raft, patented in 1880 and 1882, stands as one of the most consequential maritime safety innovations of the nineteenth century. Her design featured metal floats that enabled compact storage while allowing rapid deployment during emergencies, reversable functionality in the event of overturning, an adjustable guardrail, and more. Each raft could accommodate forty-seven people while occupying significantly less storage space than traditional wooden lifeboats. Because Beasley's life rafts allowed for more people to be saved, they expanded survival possibilities for all passengers regardless of social status. Beasley's collapsible life raft fundamentally transformed maritime safety protocols and continues influencing modern emergency equipment design today.

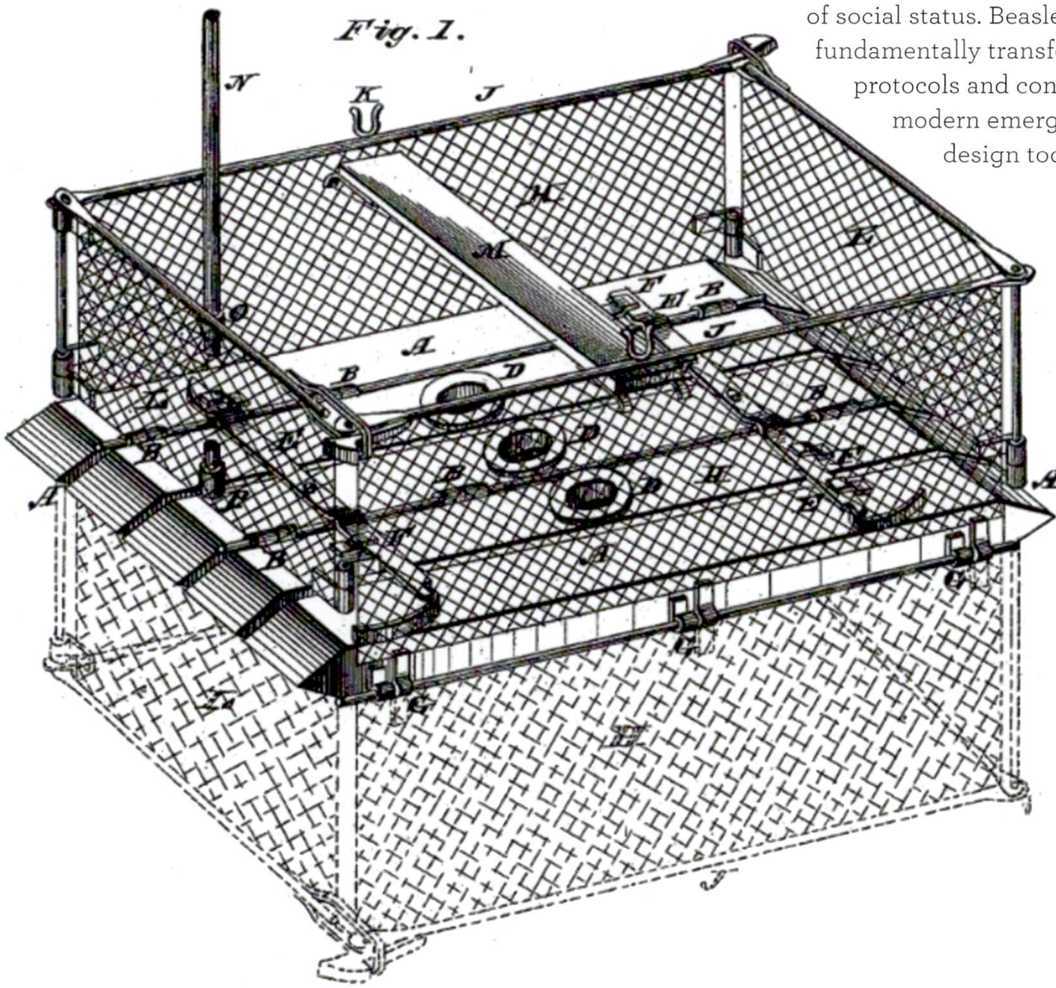

Drawings of Maria E. Beasley and her patent design.

1881 HERDIC HORSE-DRAWN CAB

The Herdic cab, invented in 1881 by Peter Herdic (1824–1888) in Williamsport, Pennsylvania, pioneered urban transportation in late nineteenth-century America. This horse-drawn vehicle featured a distinctive low-hung body with the entrance at the back, making it easier for passengers to board compared to traditional carriages. The driver sat on an elevated front seat, providing better visibility and control. The Herdic's design prioritized passenger comfort and accessibility, featuring large windows and spacious interiors. The cab's innovative, low center of gravity structure made it more stable and safer than conventional vehicles of the era. Initially operating in Williamsport, the Herdic cab quickly spread to major cities including New York, Philadelphia, and Washington, DC. Though they were eventually replaced by motorized vehicles in the early 1900s, the Herdic cab's emphasis on user-friendly features and safety solutions helped shape the modern taxi—they were also painted bright yellow—and public transit systems. Its brief but significant presence marked a transitional period between traditional carriages and the motorized transportation that would follow.

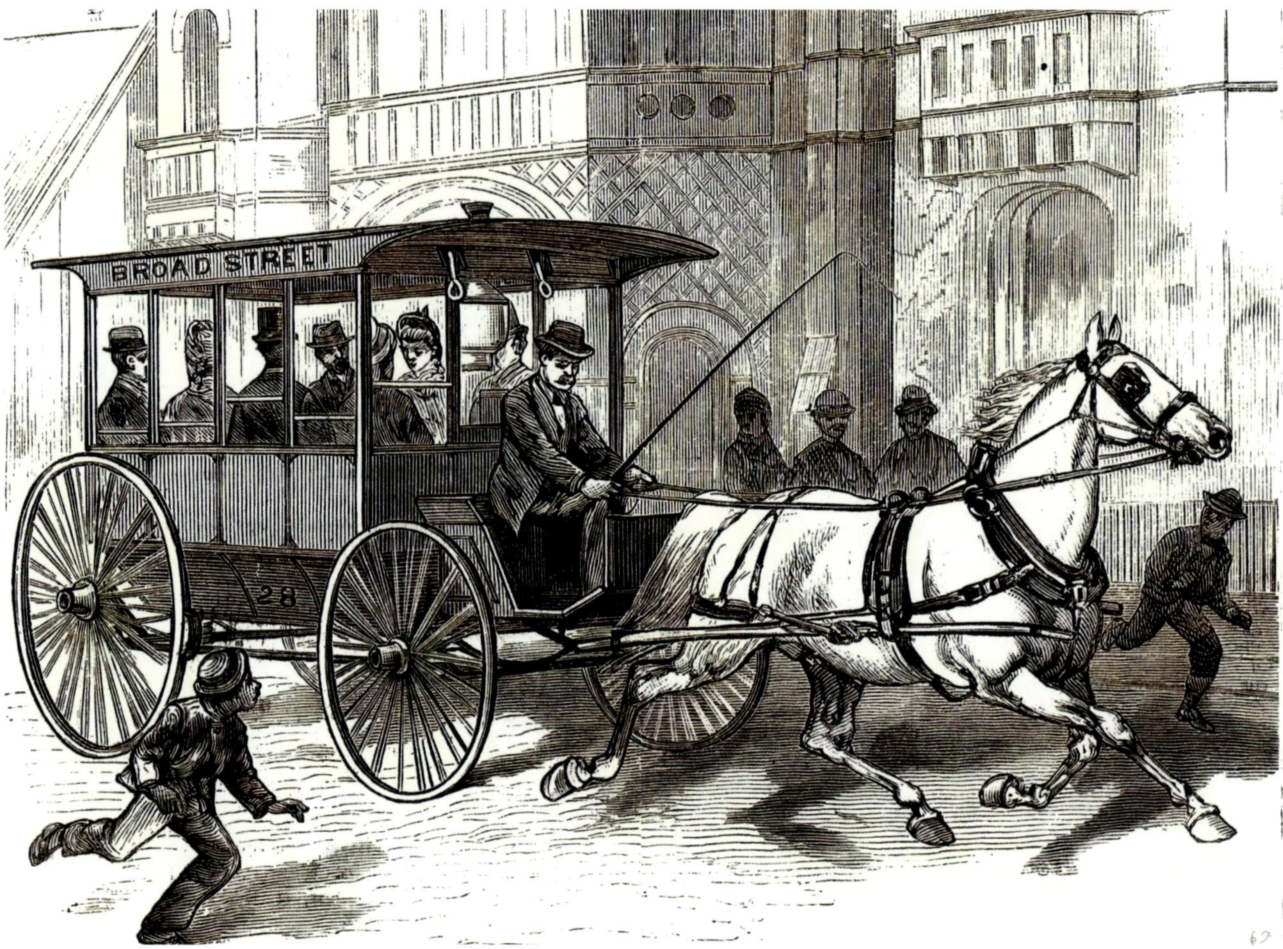

Herdic coaches arrive in Philadelphia, c. 1881.

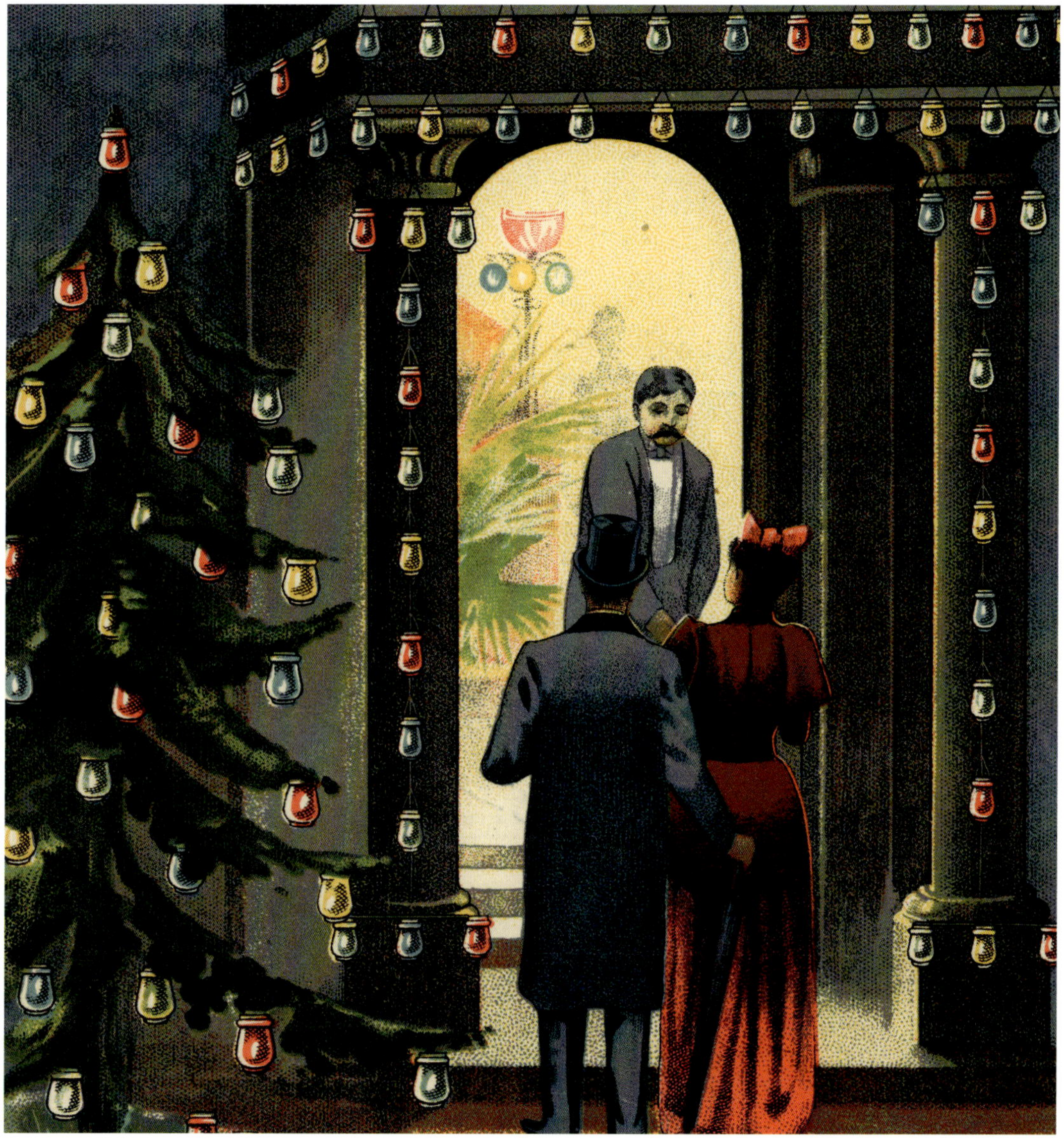

1882 CHRISTMAS LIGHTS

The first electric Christmas lights were invented in 1882 by Edward H. Johnson (1846–1917), an associate of Thomas Edison, who hand-wired eighty red, white, and blue light bulbs and strung them around an outdoor Christmas tree at his New York City home and illuminated them for passersby to view. He then called a reporter to boast about his achievement. This stunt—and Johnson's tradition of adding more and more lights to the tree each year—earned him a reputation as the Father of Electric Christmas Tree Lights. Electric lights replaced the dangerous candles that had traditionally illuminated Christmas trees, which frequently caused house fires. Initially, they were a luxury only the wealthy could afford. The first commercially produced sets appeared in the 1890s, but they were expensive and required professional installation. It wasn't until the 1920s that mass production made them accessible to middle-class families, and by the 1930s, they had become a popular holiday tradition. Johnson's invention transformed Christmas celebrations worldwide, making holiday decorating safer and more elaborate.

1883 BROOKLYN BRIDGE

The Brooklyn Bridge, completed in 1883, stands as one of America's greatest engineering triumphs and a symbol of human determination. Designed by John Augustus Roebling (1806–1869), the project was taken over by his son Washington Roebling (1837–1926) after John tragically died from tetanus following a ferry accident while surveying the bridge site. When Washington became incapacitated by decompression sickness from working in the underwater caissons, his wife Emily Warren Roebling (1843–1903) stepped in to oversee the construction, becoming one of the first female field engineers. The bridge was a marvel of nineteenth-century engineering, featuring innovative steel cables and Gothic Revival stone towers. It was the world's longest suspension bridge at the time. The aesthetic inspired countless artists, writers, and poets of the day, including Hart Crane's epic poem "The Bridge." The Brooklyn Bridge became a powerful symbol of progress, connecting the Manhattan and Brooklyn boroughs and facilitating the growth of New York City as a whole. Today, it remains an iconic landmark, attracting millions of visitors annually while continuing to serve as a vital transportation link.

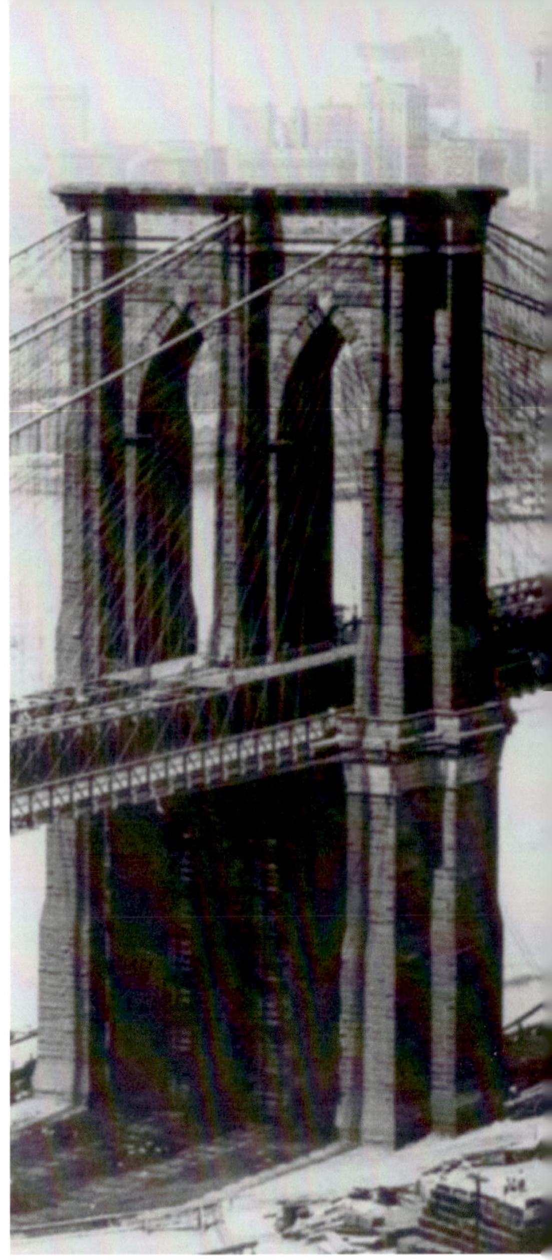

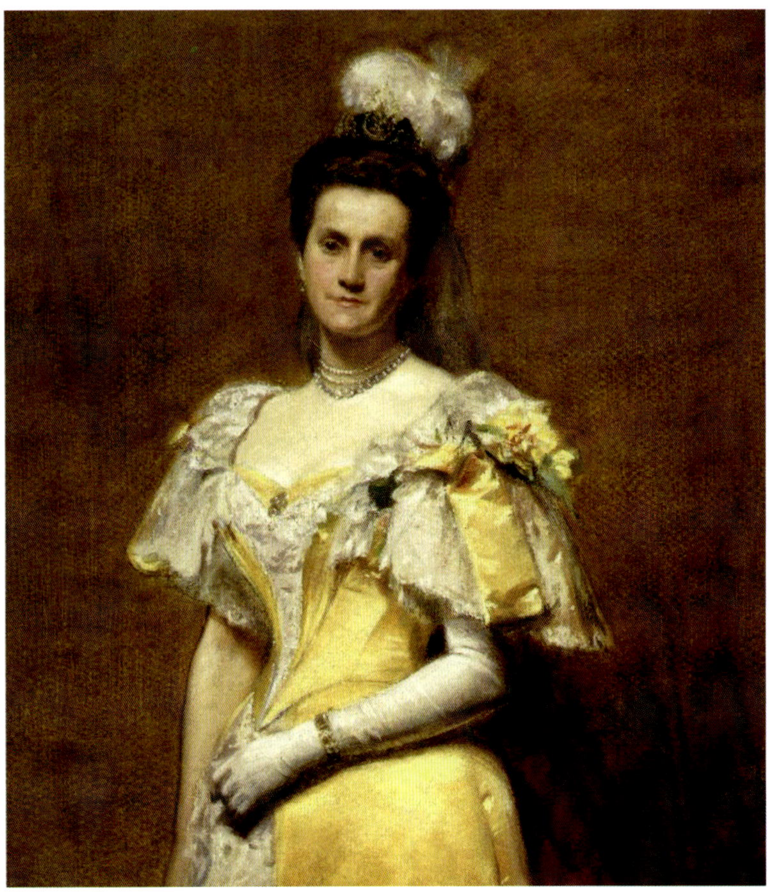

LEFT: Portrait of Emily Warren Roebling, who oversaw completion of the bridge.
ABOVE: View of the bridge, c. 1901.
RIGHT: People walking the promenade, c. 1897.

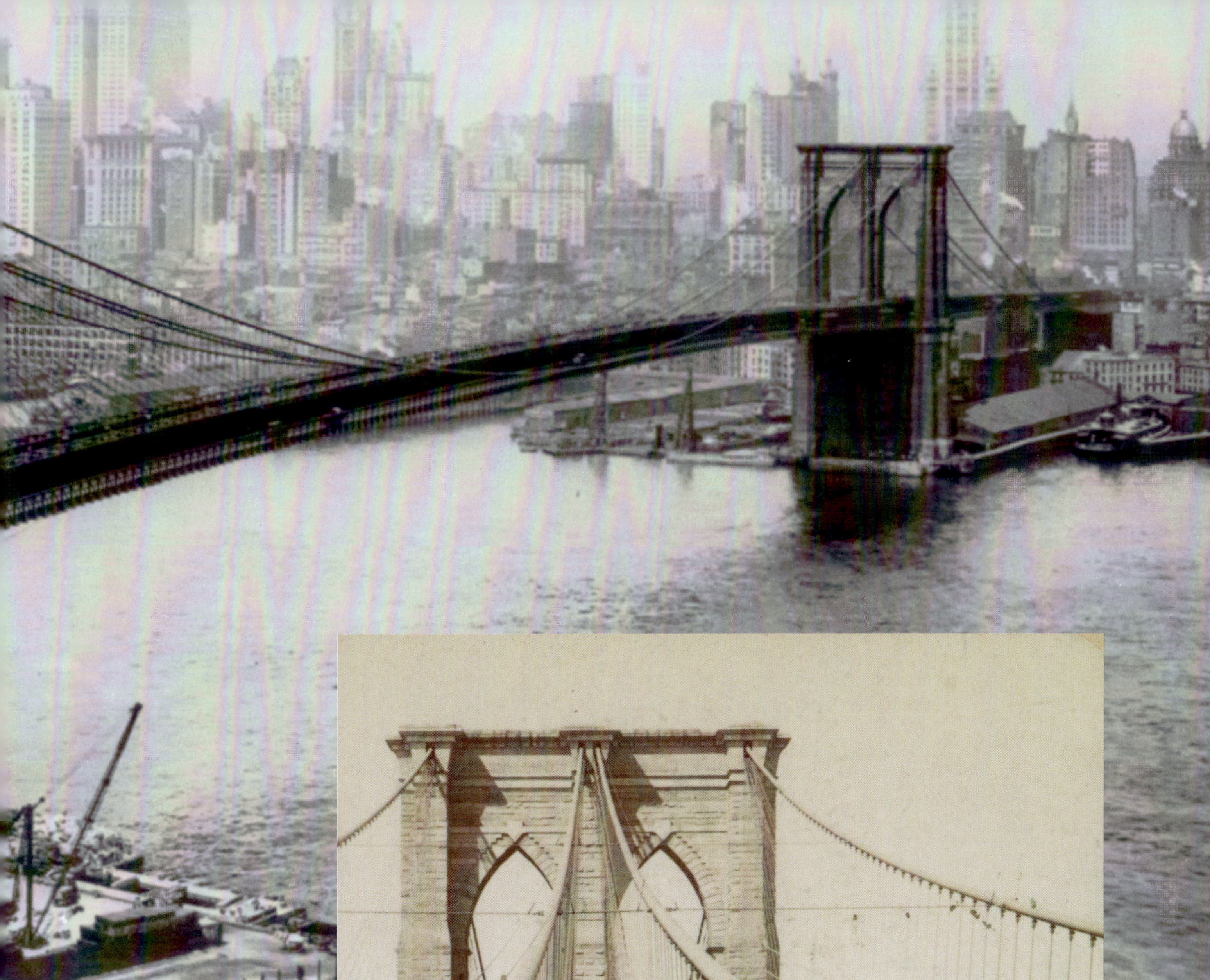

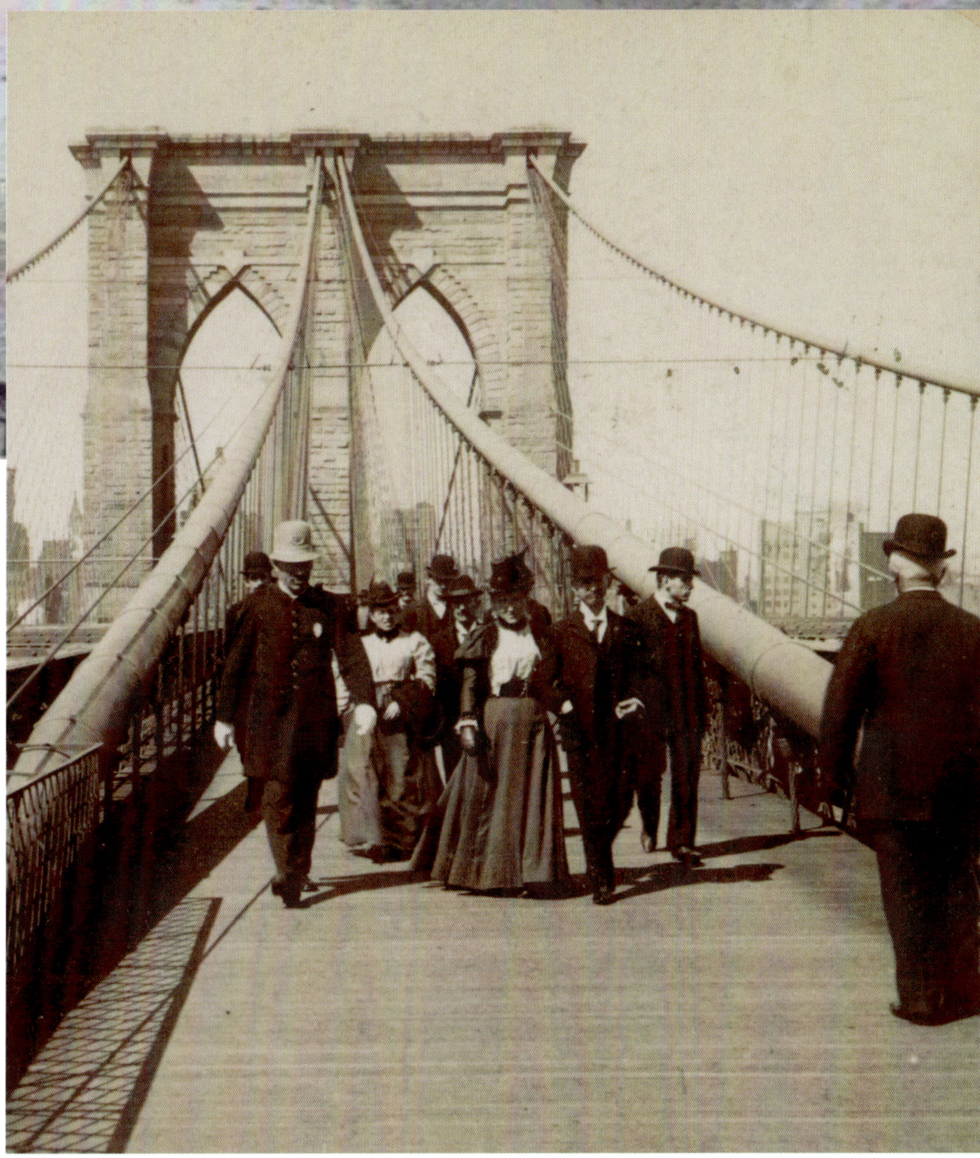

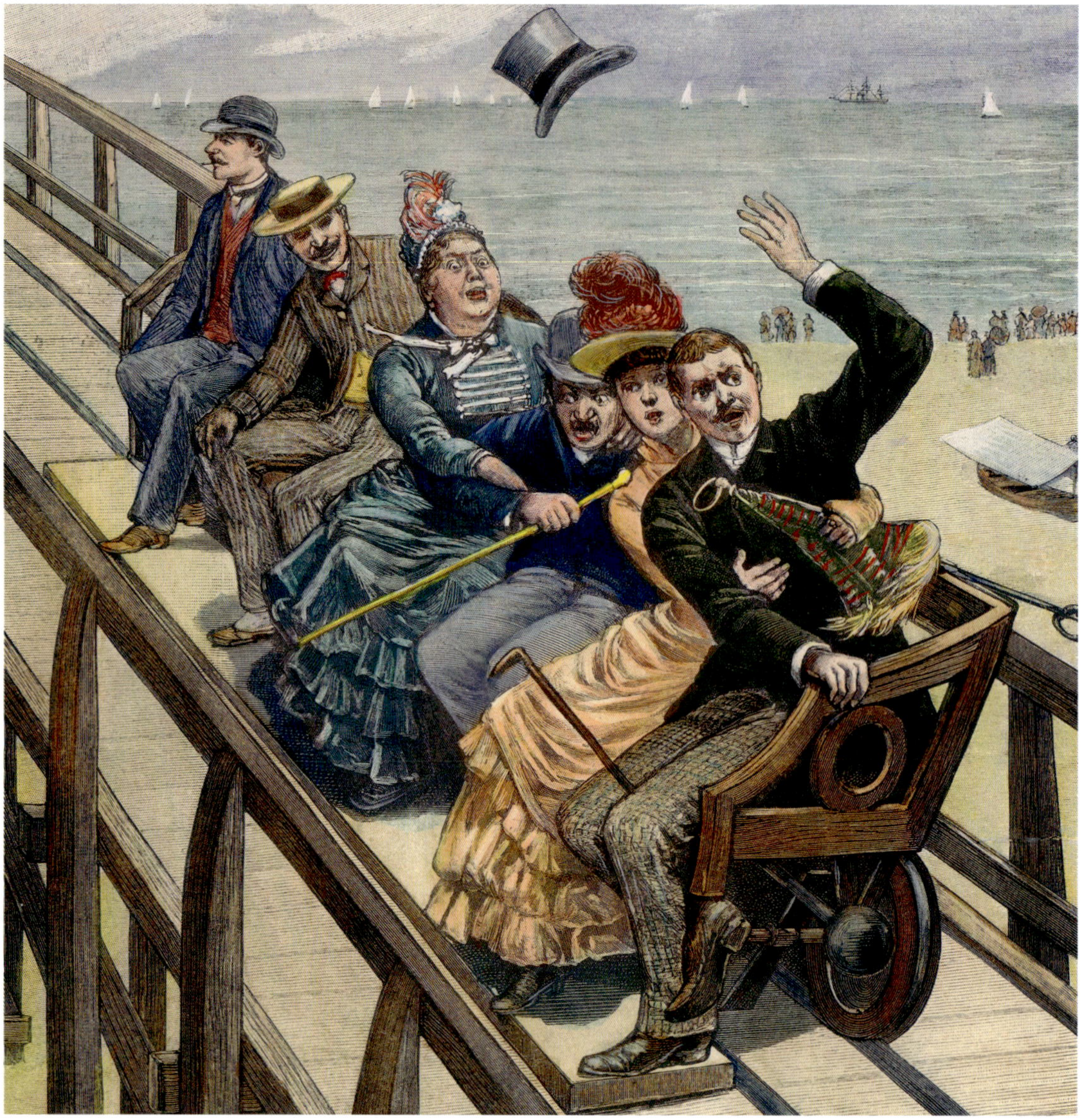

1886 engraving of the Switchback Railway.

1884 ROLLER COASTER

Coney Island in Brooklyn became the birthplace of America's first roller coaster when the Switchback Railway opened on June 16, 1884. Created by LaMarcus A. Thompson (1848–1919), this pioneering ride led to Thompson modifying an existing patent to receive his own for a gravity-powered "Roller Coasting Structure" the following year. The Switchback bore little resemblance to today's high-speed thrill rides. Passengers had to climb to the top of a tower to board, where the car was manually pushed from the loading platform. The ride crawled along at approximately six miles per hour. Unlike modern coasters with their front-to-back seating, Thompson's design featured side-by-side arrangements. Most notably, the journey was one-way only—riders had to climb a second tower at the end to catch a return trip to the starting point. Despite these limitations, this novel form of entertainment quickly captivated the public. Thompson continually refined his invention, introducing faster speeds, loops, steeper drops, and other innovations that enhanced the coaster experience. His ingenuity sparked widespread interest, and by the turn of the century, roller coasters created by Thompson and other inventors had become among the most popular attractions at amusement parks nationwide—a legacy that endures today.

1885 SKYSCRAPER

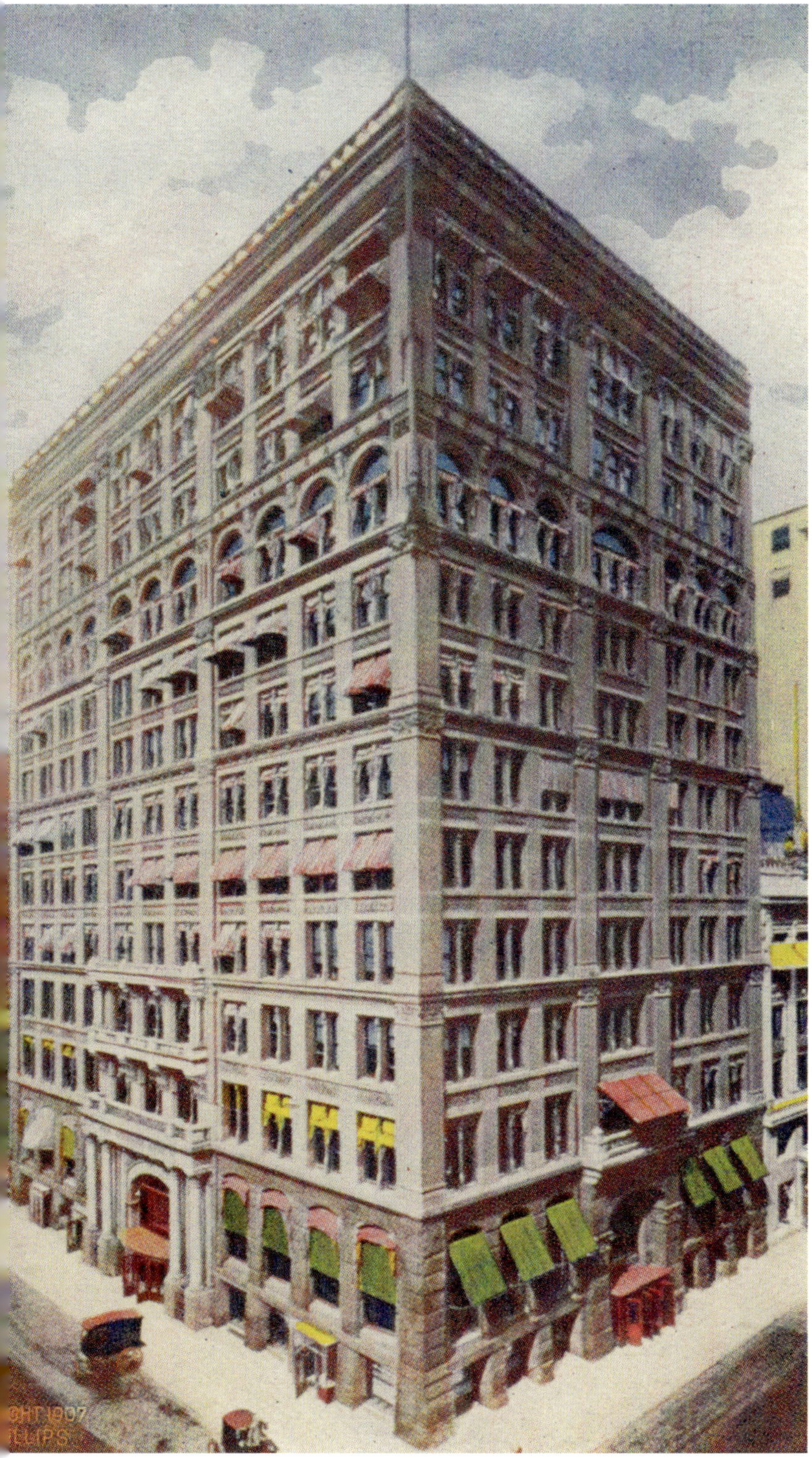

A ten-story building would hardly be considered a skyscraper today, but the Home Insurance Building in Chicago, built in 1885 by American architect and civil engineer William Le Baron Jenney (1832–1907), is considered the first one ever built. What made this building groundbreaking was its innovative steel-frame construction. Unlike traditional masonry buildings that relied on thick walls to support their weight, Jenney's design used a skeleton of steel beams to carry the building's load. This allowed for thinner walls, larger windows, and greater heights than previously possible. The building's facade featured extensive glass windows, maximizing natural light and creating the distinctive appearance that would define modern skyscrapers. The Home Insurance Building established the architectural principles that enabled the vertical expansion of cities worldwide. Though demolished in 1931, its influence persists in every modern skyscraper. The building proved that cities could grow upward rather than outward, fundamentally reshaping urban development and making possible the dense metropolitan centers we know today.

Postcard of the Home Insurance Building in Chicago, c. 1907.

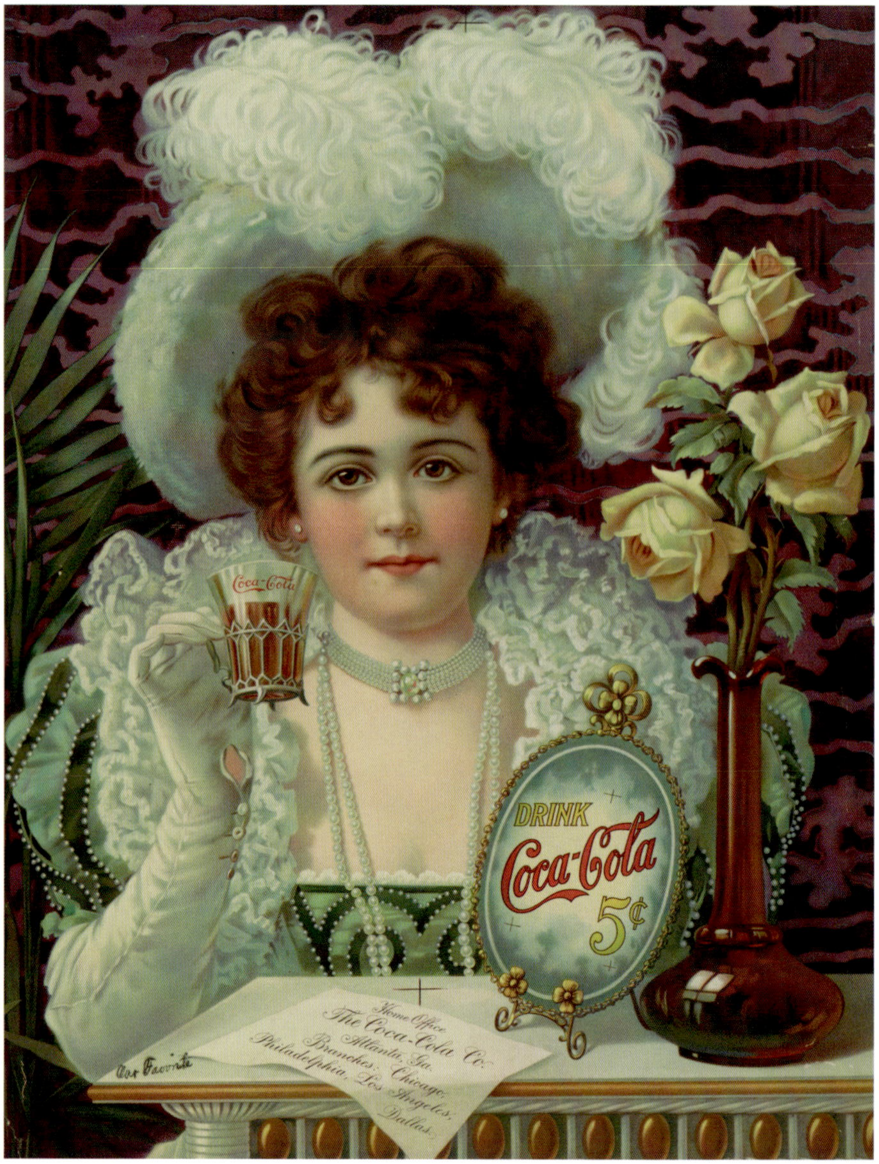

An ad for Coca-Cola, c. 1900.

1886 COCA-COLA

Dr. John Stith Pemberton (1831–1888) was an Atlanta-based pharmacist who, on May 8, 1886, introduced a product to the American public that would change the country forever. He served the world's first Coca-Cola at the counter of Jacob's Pharmacy in downtown Atlanta. According to legend, Pemberton created the recipe for the syrup from a combination of extracts from the coca leaf and kola nut as a health tonic. Dr. Pemberton's bookkeeper, Frank M. Robinson, is credited with naming the beverage "Coca-Cola" (changing the K in *kola nut* to a C) as well as designing the distinctive script logo still used today.

Though the beverage was initially well-received, Coca-Cola's popularity was slow to develop. Sadly, Dr. Pemberton did not live to see the full potential of his creation. Shortly before his death, he sold the rights to the formula to Atlanta businessman Asa Candler, who marketed it aggressively and was responsible for its phenomenal growth in the late 1800s. By the turn of the century, Coca-Cola was sold across the United States and Canada. Pemberton's legacy remains intact to this day: Coca-Cola Classic continues to be one of the world's most recognizable beverage brands.

1887 GRAMOPHONE RECORD

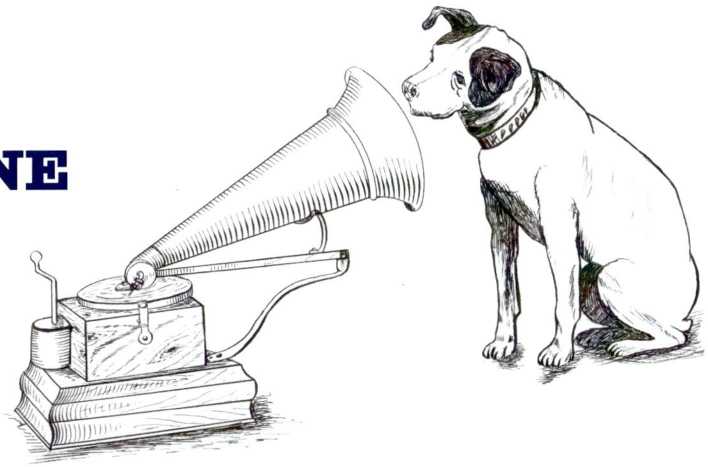

Emile Berliner (1851–1929), a German immigrant who settled in Washington, DC, patented the gramophone record on November 8, 1887, revolutionizing how humanity experienced and preserved music. Unlike Thomas Edison's earlier cylindrical phonograph, Berliner's flat disc design proved more practical for mass production and storage. His innovation used a lateral cutting technique that etched sound waves into grooves spiraling from the record's edge to its center. After experimenting with different materials— rubber, glass, and celluloid among them—Berliner discovered that a shellac compound made by the Duranoid Company worked best for his discs. The material held deep grooves that allowed the stylus to fit snugly as the disc spun on a turntable, while remaining durable enough to resist breakage. Berliner's discs offered significant advantages over Edison's cylinders: they required only a fraction of the storage space, could be mass-produced more easily, and were simpler to label.

In the 1890s, the gramophone expanded internationally as Berliner secured patents in other countries. The gramophone's legacy endures as the foundation of recorded music culture.

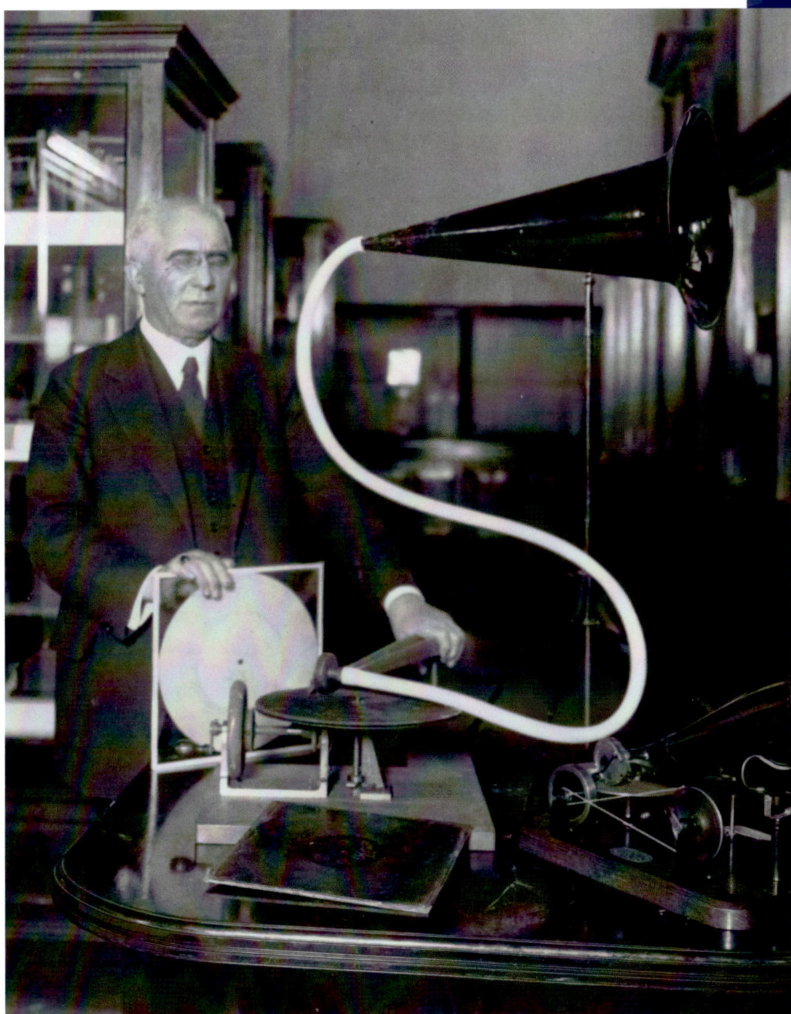

Emile Berliner with a model of his phonograph, and his trademark registration for "His Master's Voice" brand gramophones.

1888 KODAK CAMERA

George Eastman (1854–1932) founded the Eastman Kodak Company and revolutionized amateur photography in America. One of his first innovations was a portable, handheld box-style camera with a revolutionary feature: instead of using a glass-plate negative for each exposure, as had been the norm in cameras of the era, the Kodak camera came preloaded with a hundred-exposure roll of film that was sealed inside. After shooting the hundred images, the photographer would return the entire camera to the Kodak factory, where the film was developed and the camera was reloaded with film. The prints—small circular images—and camera were then mailed back to the customer for a fee. Another attractive feature of this early camera was its ease of use. The shutter was simple to set by pulling a string, and the image was captured by simply pressing a button. Before the Kodak, photography required extensive technical knowledge, expensive equipment, and darkroom skills. "You press the button, we do the rest" became the company's marketing slogan. It democratized photography by making it accessible to ordinary people for the first time.

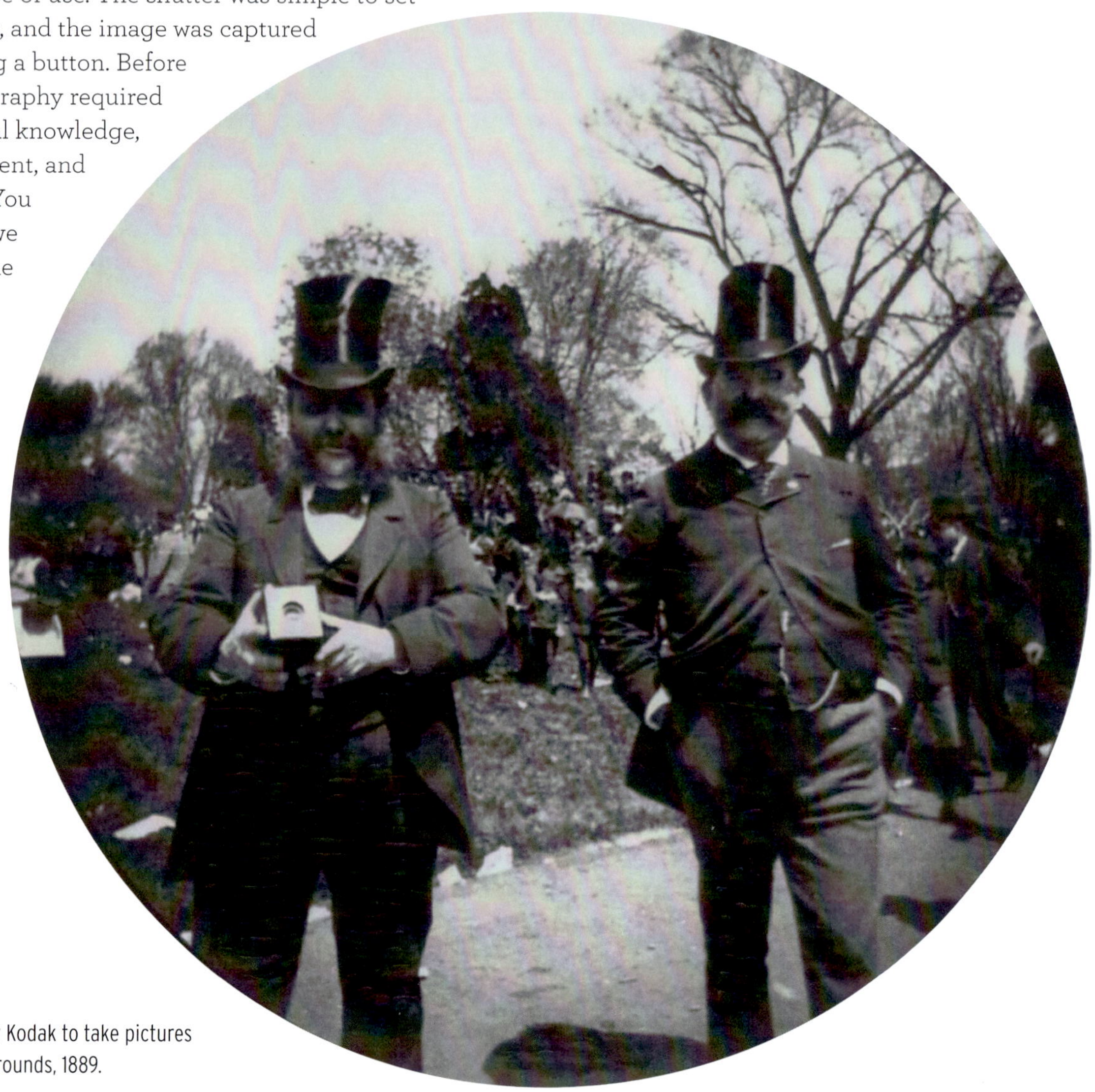

A tourist uses his new Kodak to take pictures on the White House grounds, 1889.

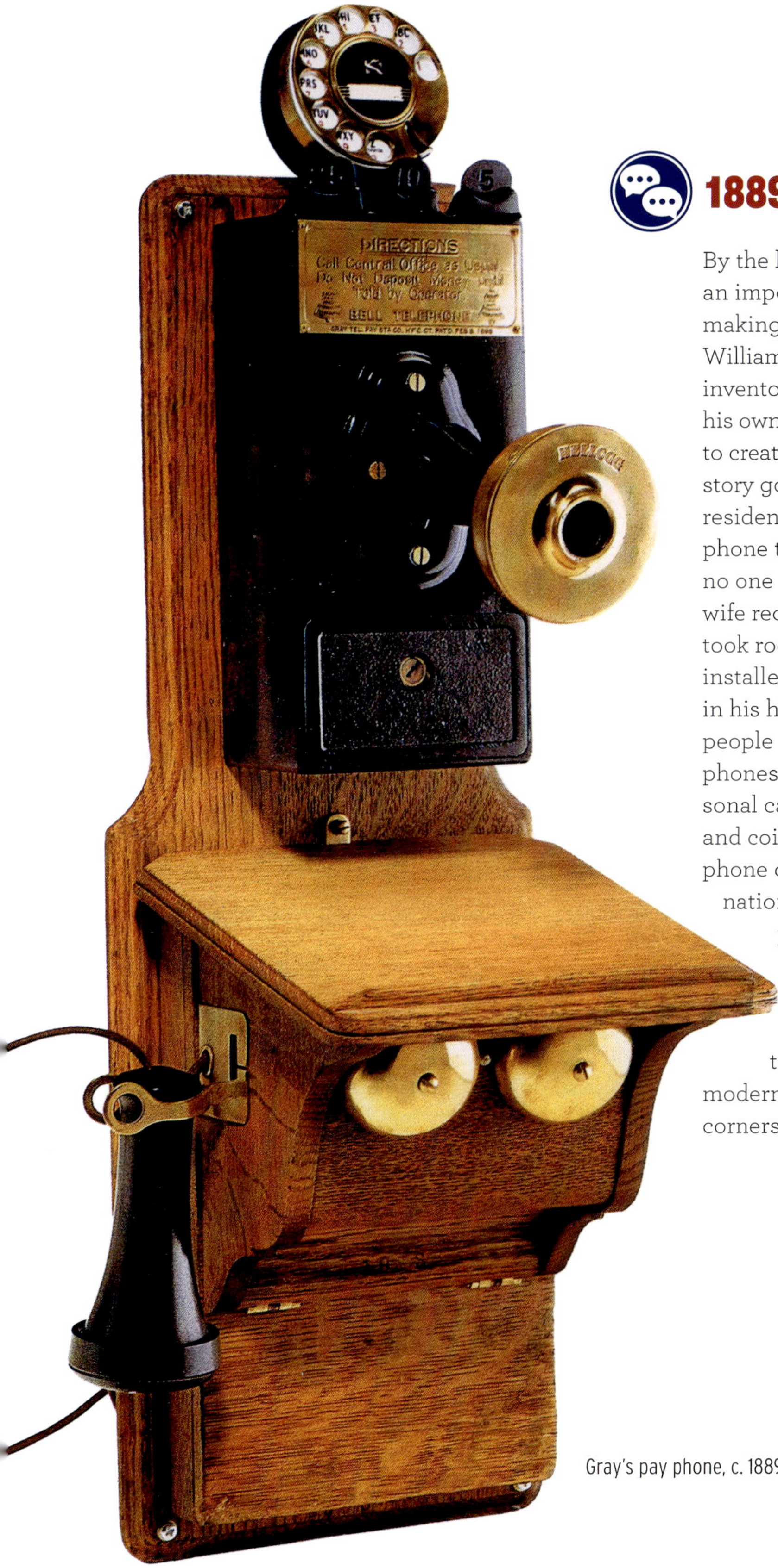

Gray's pay phone, c. 1889.

1889 PAY PHONE

By the late 1800s, the telephone had become an important part of American society, but making a call was not always straightforward. William Gray (1850–1903), credited as the inventor of the pay phone, was inspired by his own difficulty in finding a working phone to create his groundbreaking invention. The story goes that when the Hartford, Connecticut, resident was desperately trying to locate a phone to summon a doctor for his ailing wife, no one nearby would help him. After Gray's wife recovered, the idea for a public telephone took root. Following several prototypes, Gray installed the first pay phone outside a bank in his hometown in 1889. For the first time, people without the means to afford private phones could make both business and personal calls wherever they could find a phone and coins to operate it. Gray founded his pay phone company in 1891, expanding the service nationwide and installing the coin-operated phones in wooden cabinets or booths. Though pay phones are now difficult to find in our mobile phone era, Gray's innovation once made these devices a ubiquitous symbol of modern technology and public safety on street corners across America.

1890 HOLLERITH TABULATING MACHINE

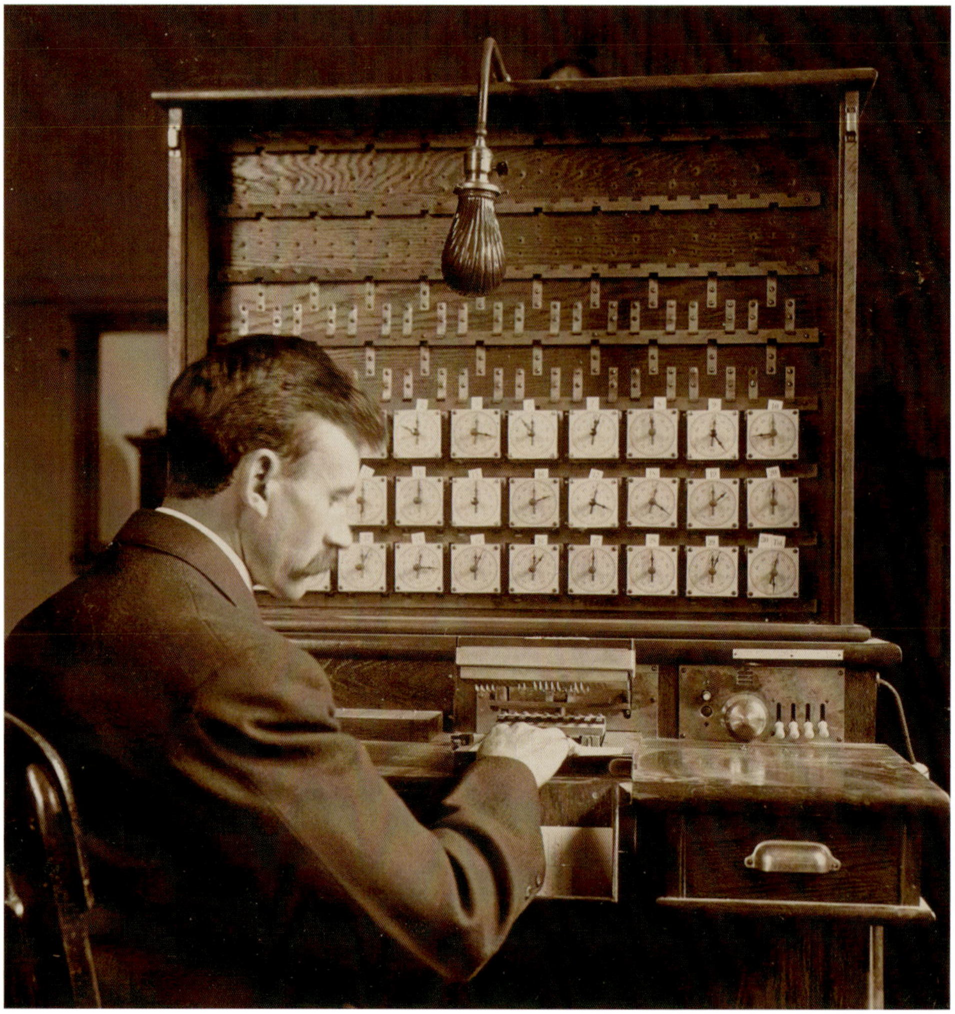

Herman Hollerith (1860–1929), a statistician, engineer, and inventor from Washington, DC, revolutionized data processing by creating the tabulating machine used in the 1890 US census. This groundbreaking invention dramatically accelerated data counting and became a crucial predecessor to the modern computer. Hollerith's comprehensive system consisted of four essential components: punch cards for data recording, a punching device for entering information onto the cards, collection boxes for organizing the processed cards, and a tabulator for reading the recorded data. The system's effectiveness was recognized internationally when it won a gold medal at the 1889 Paris World's Fair, one year before its implementation in the census. Despite facing significant competition, Hollerith's approach proved overwhelmingly superior in comparative testing. His success across both American and European markets enabled him to establish his own company, which through a series of strategic mergers eventually evolved into IBM—now one of history's most influential manufacturers of business and personal computers.

An employee of the US Census Bureau using a Hollerith Tabulating Machine, c. 1908.

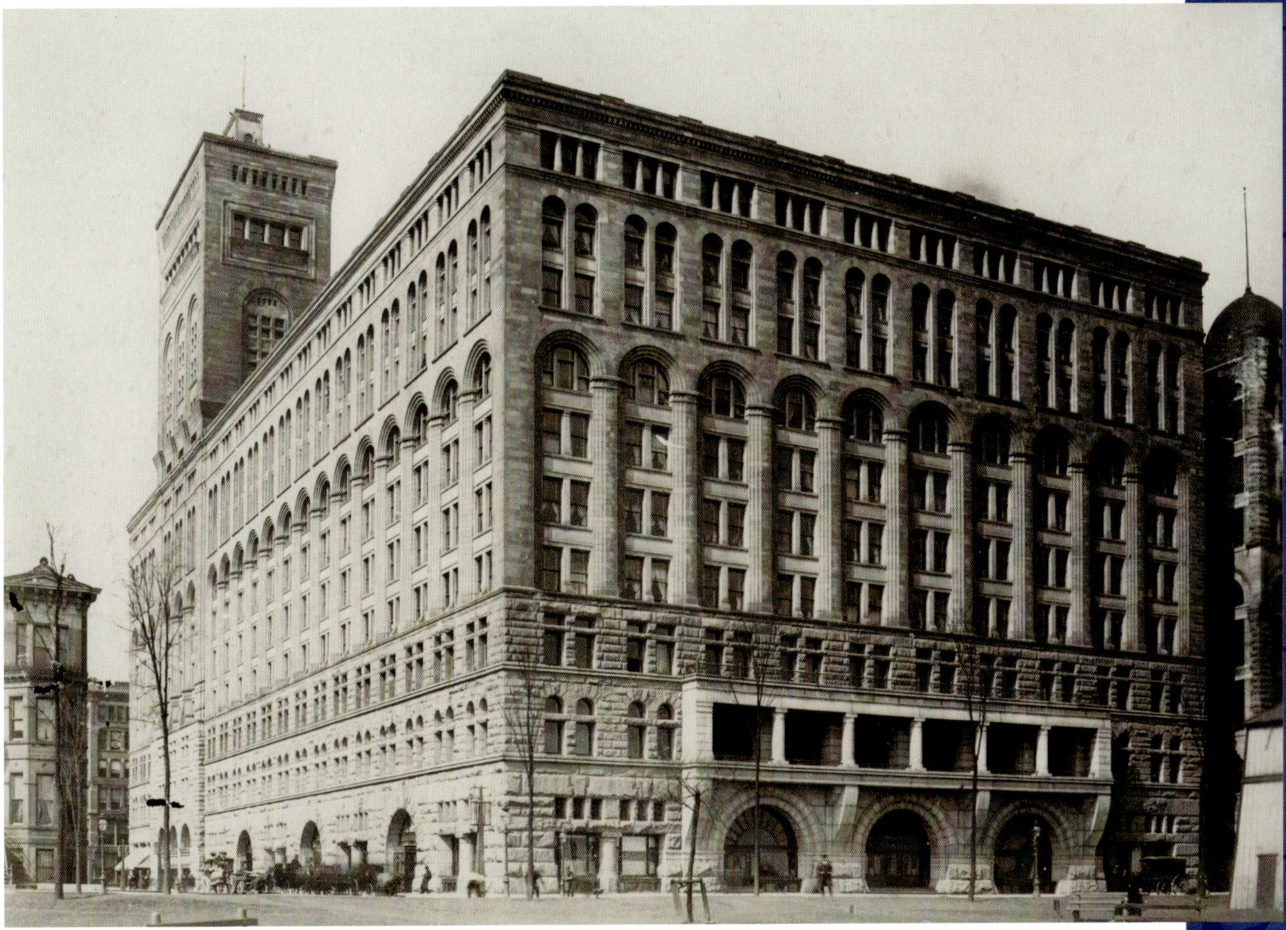

1891 CARNEGIE HALL

On May 5, 1891, Carnegie Hall, located at Seventh Avenue and 57th Street in New York City, held its first performance, launching a five-day classical music festival. Though Pyotr Ilyich Tchaikovsky participated as a guest conductor during the opening festivities, the inaugural concert was primarily conducted by Walter Damrosch. Since then, Carnegie Hall has become one of the most treasured concert halls and performance venues in the world and served as the home of the New York Philharmonic Orchestra until the orchestra moved to Lincoln Center in 1962. The historic hall was commissioned and completely financed by famed industrialist Andrew Carnegie (1835–1919) and designed by architect William B. Tuthill (1855–1929) in an Italian Renaissance Revival style. Though it was slated for demolition in the 1950s, Carnegie Hall is widely regarded as the most famous performance venue in the world. It was designated a National Historic Landmark in 1962 and continues to attract the world's top talent to its revered stage. Virtually every significant American musician, and many international artists, spanning classical to pop music and everything in between, has performed at Carnegie Hall.

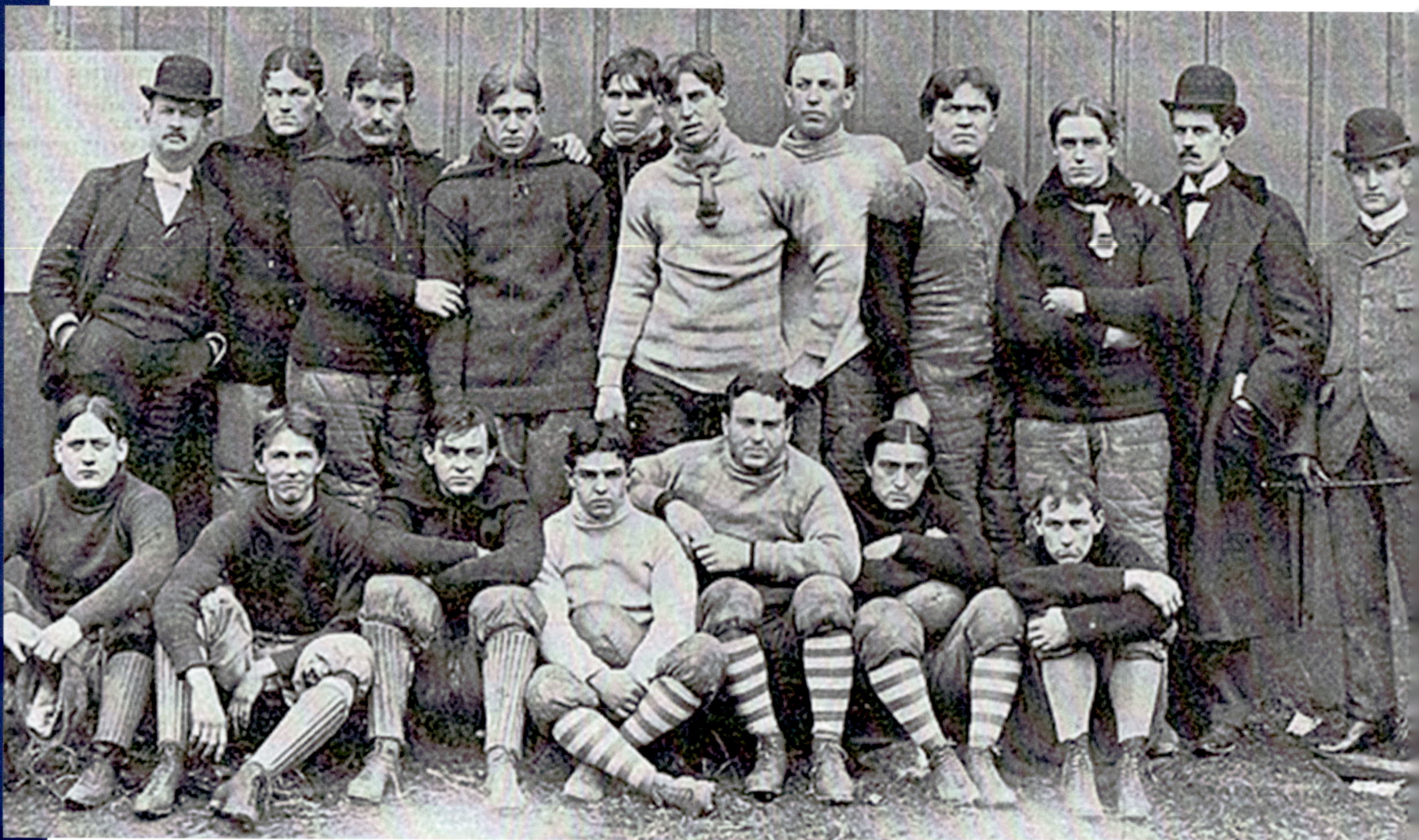

1892 PRO FOOTBALL

The first person to be paid to play football was William "Pudge" Heffelfinger (1867–1954), in a game played on November 12, 1892. The amount paid to him was $500—equal to more than $16,000 in today's dollars. Football was extremely popular at the time, attracting intense competition among athletic clubs. Heffelfinger had been a star player during his college years at Yale. He was recruited from Chicago for the occasion and played for the Allegheny Athletic Association (AAA), a Pittsburgh, Pennsylvania, area club, against their staunch rival, the Pittsburgh Athletic Club (PAC). The game was played at Recreation Park, on Pittsburgh's north shore. The AAA team won the game when Heffelfinger scored the only touchdown after recovering a fumble—it seems the investment in his athletic prowess was worth it. Heffelfinger would later become a football coach and writer. He was inducted into the College Football Hall of Fame in 1951 and the Pittsburgh Pro Football Hall of Fame in 2015. Heffelfinger's decision to accept payment helped legitimize football as a spectator sport worthy of financial investment, paving the way for the massive industry professional football has become today.

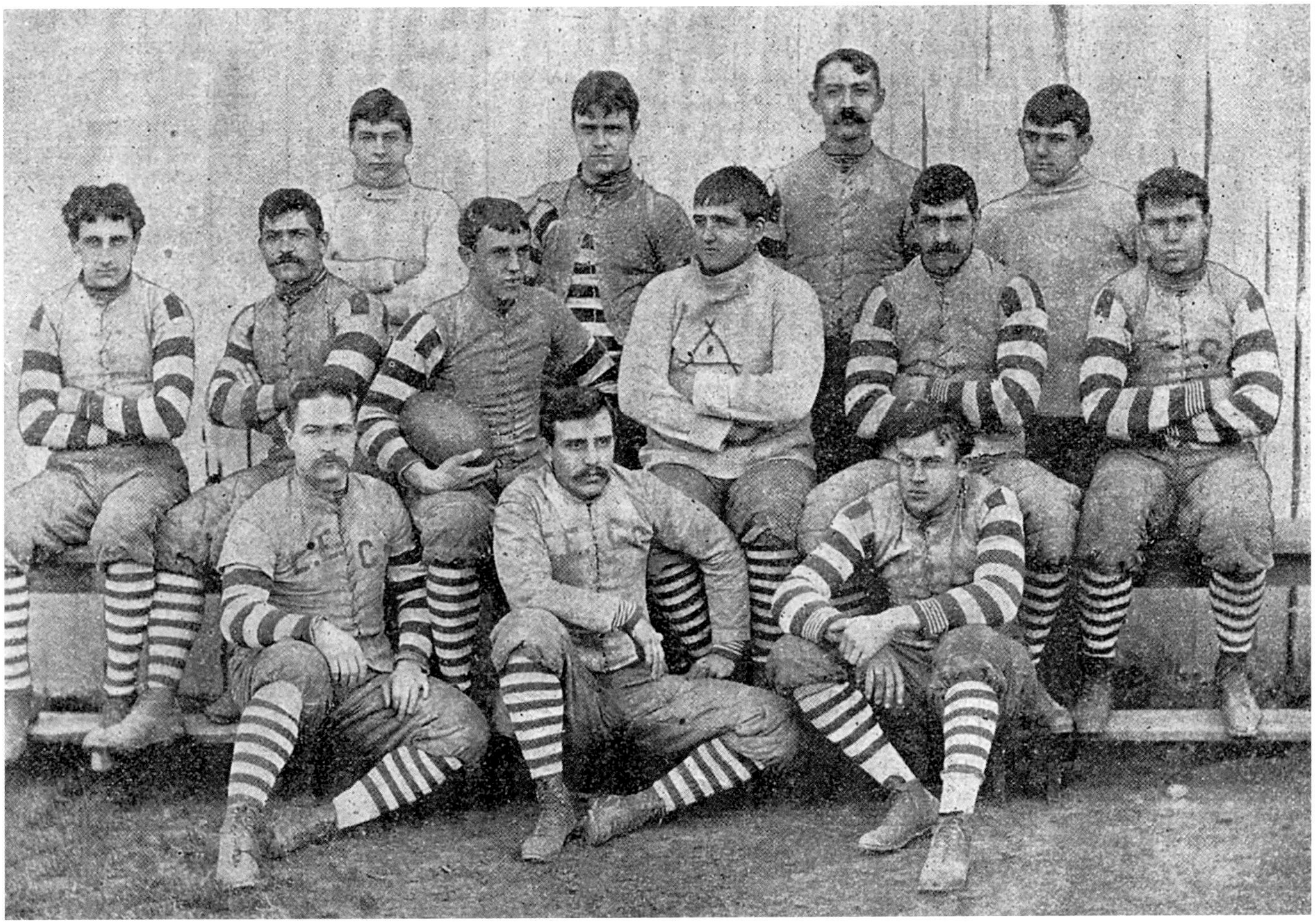

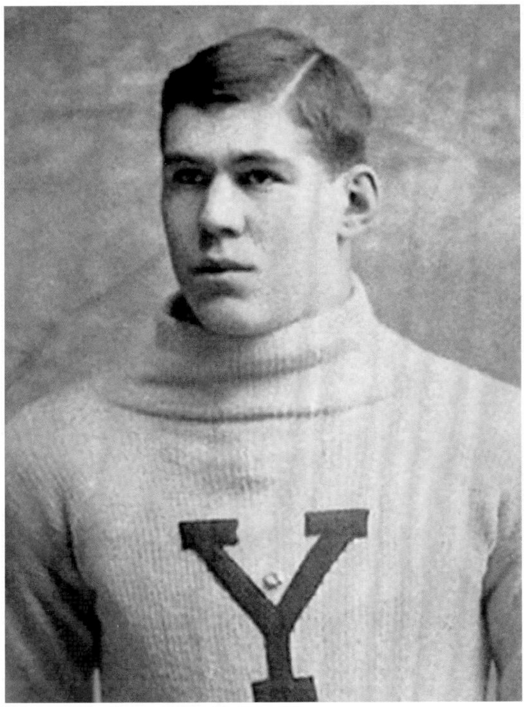

TOP LEFT AND RIGHT: The 1892 teams that played the first game, the Allegheny Athletic Association and the Pittsburgh Athletic Club.
LEFT: William "Pudge" Heffelfinger during his Yale days.

1893 THE DISHWASHER

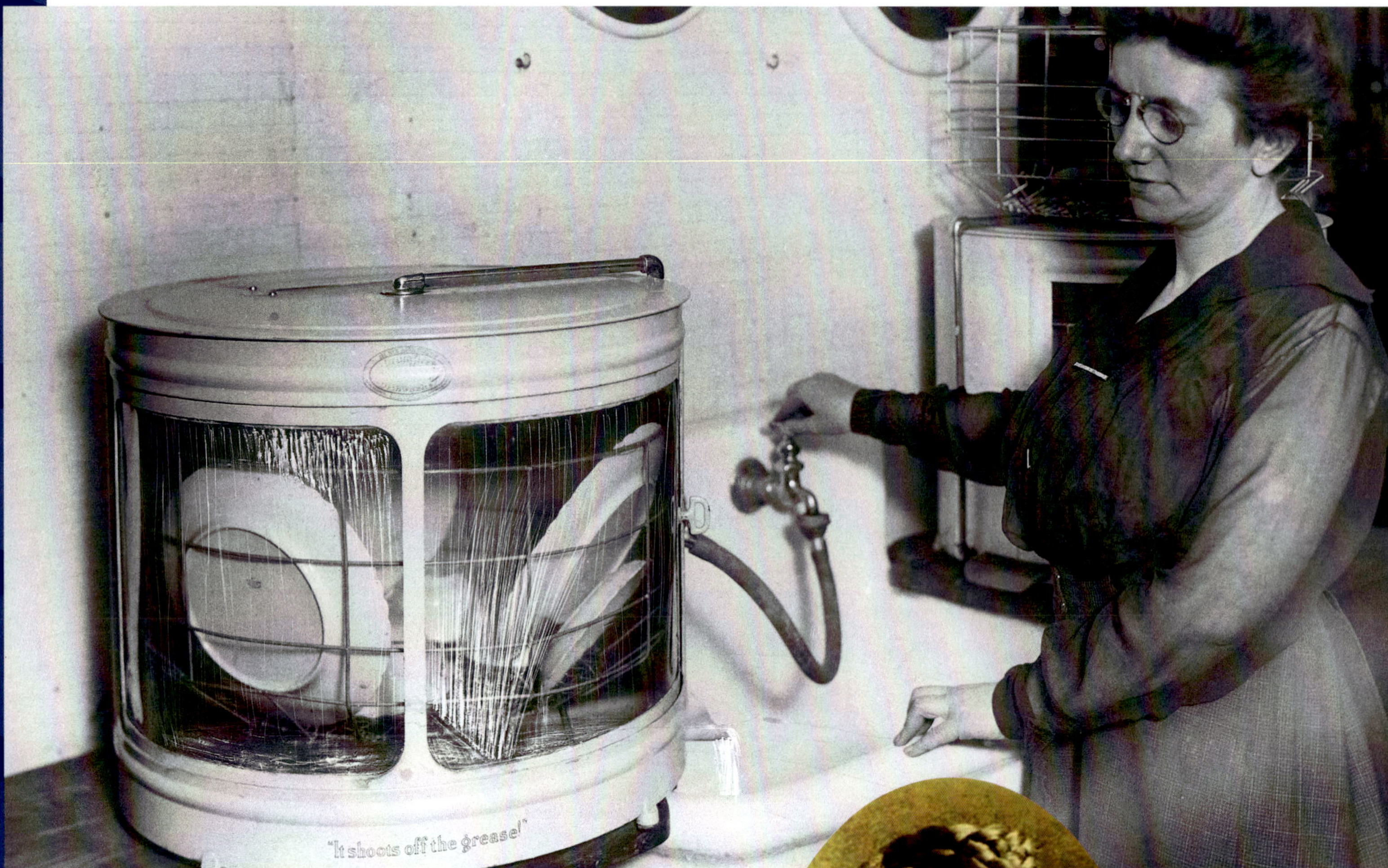

Chicago socialite Josephine Cochrane (1839–1913) invented the first practical dishwasher and showcased it at the 1893 World's Fair. Her motivation was threefold: protecting her heirloom tableware from chipping by servants, avoiding handwashing china herself, and making money as a recently widowed woman facing mounting debts. Her Garis-Cochrane Manufacturing Company's dishwashing machine impressed judges so much they awarded it top prize—the only device there invented by a woman. Unlike previous dishwashing devices, Cochrane's used water pressure instead of scrubbers and featured special racks to hold dishes in place. She received her patent in December 1886. Success proved elusive for a woman entrepreneur in that era, but the World's Fair provided crucial exposure. Her machine could clean and dry more than two hundred dishes in about two minutes, allowing her to demonstrate its capabilities, sell to businesses, open a factory, and expand operations. Cochrane's dream was finally realized in the mid-twentieth century when dishwashers became common household items. She was inducted into the National Inventors Hall of Fame in 2006.

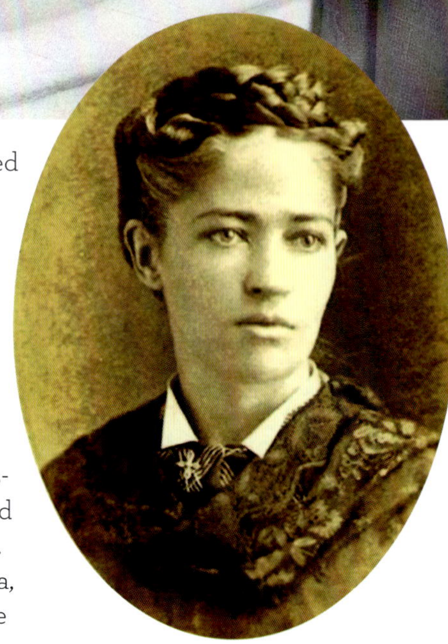

TOP: Cochrane's invention led to dishwashers becoming commonplace in homes, such as this one, in 1920. **BOTTOM:** Josephine Cochrane, c. 1893.

1894 KINETOSCOPE

Thomas Edison (1847–1931) is credited with more than a thousand inventions, but one of the more interesting ones involves a man sneezing. The kinetoscope, a precursor to the motion picture projector, came out of Edison's desire to "do for the eye what the phonograph did for the ear," as he allegedly wrote in a note to the US Patent Office in 1888. On January 7, 1894, Edison presented what came to be known as "Fred Ott's Sneeze" on the viewing device—depicting his employee sneezing dramatically for the camera—at a parlor in New York City. The invention, co-created with his assistant William Dickson (1860–1935), worked by quickly passing a series of still images (taken on a camera that employed celluloid film, called a kinetograph) in front of a light source (perhaps Edison's incandescent light bulb?) as the spectator watched through a peephole. Viewers could watch short kinetoscope films one after the other on kinetoscope viewers in their own preferred order (similar to the way modern viewers enjoy digital content today). Though Edison didn't remain long in the motion picture business, it's easy to trace the innovation of the kinetoscope to the massive popularity of movies and video content today.

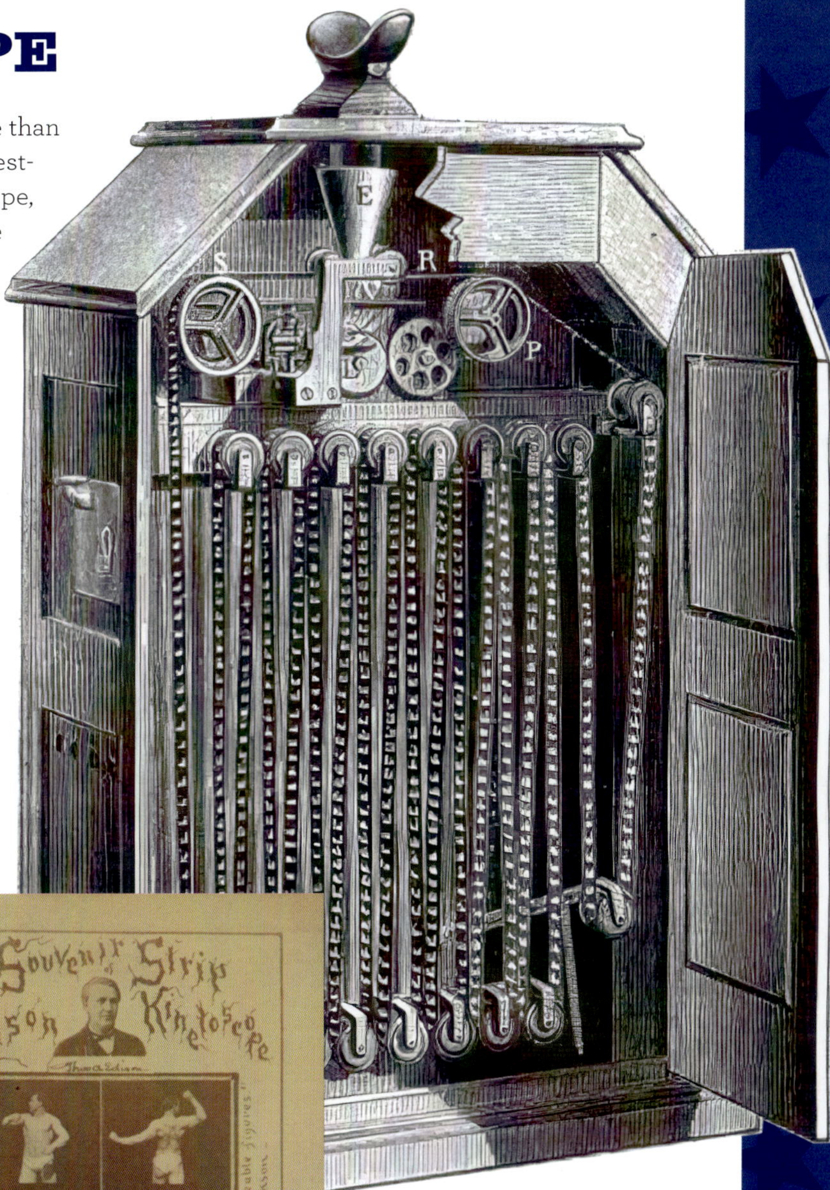

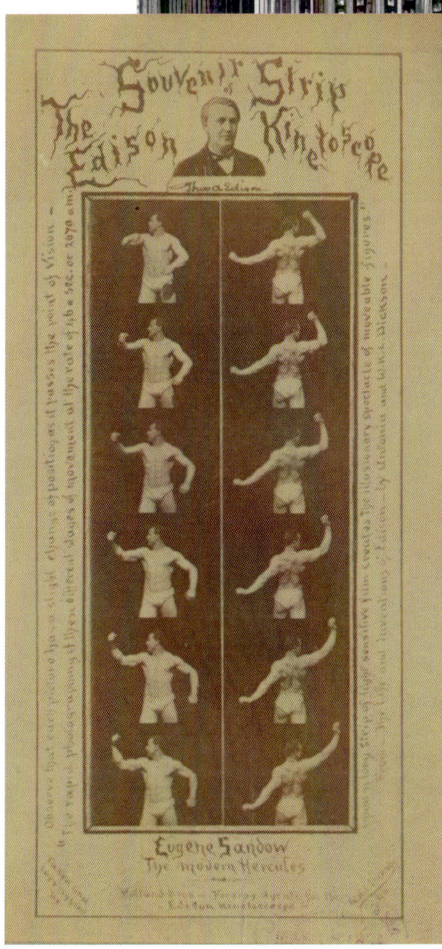

TOP: Edison's Kinetoscope. **RIGHT:** An 1894 souvenir strip of *The Modern Hercules*, Eugen Sandow.

1895 PEANUT BUTTER

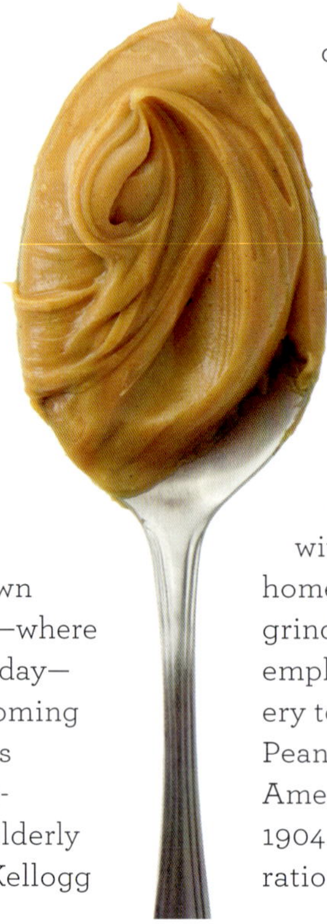

The earliest known mention of peanut butter comes from the ancient civilizations of the Aztecs and Incas. But how did it become one of the most ubiquitous items in American pantries? Although Marcellus Gilmore Edson patented a peanut paste in 1884, his was only for the use of it in candy. But the distinction of creating peanut butter as we know it today belongs to Dr. John Harvey Kellogg (1852–1943), the physician and nutritionist of cereal fame. At his well-known sanitarium in Battle Creek, Michigan—where the Kellogg Company is still based today—Dr. Kellogg created a health spa welcoming patients who wished to cure numerous health conditions. As part of his plant-based diet protocol, and to serve his elderly patients who had difficulty chewing, Kellogg developed a food compound to help his patients: an easily digestible paste made from boiled, ground almonds or other less expensive nuts. He applied for a patent for the nut butter in 1895, but because he didn't specify the type of nut used, the patent didn't hold. However, A-list visitors to the spa, such as Henry Ford and Sojourner Truth, helped establish Kellogg's treat as a delicacy. The peanut butter craze began, with magazines of the time encouraging homemakers to follow Kellogg's lead and grind peanuts themselves. Around 1900, an employee of Kellogg's spa invented machinery to make peanut butter on a large scale. Peanut butter was formally introduced to Americans at the St. Louis World's Fair in 1904 and it became part of US Armed Forces rations during both World Wars.

1896 FIRST MOVIE THEATER

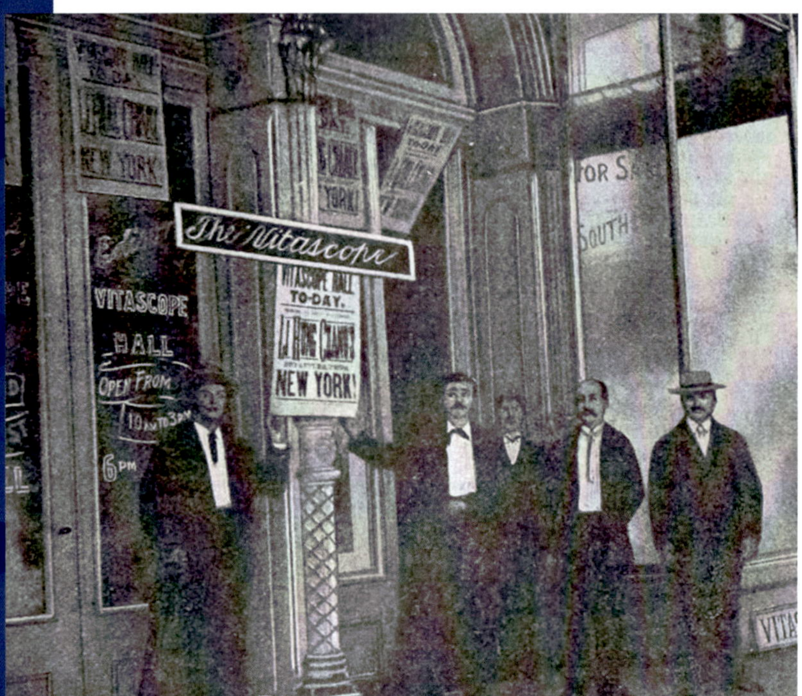

The Vitascope Hall in New Orleans, c. 1896.

On July 26, 1896, the first indoor commercial movie theater with fixed seats opened on Canal Street in New Orleans, Louisiana. The theater was called Vitascope Hall, named after the Vitascope projector on which the films were shown—a device patented by Thomas Armat the prior year. Promoters Walter J. Wainwright (birth and death dates unknown) and William "Pop" Rock (1853–1916) established the theater in an unadorned storefront that accommodated four hundred people, showing two performances daily for ten cents per show. The programs were short, stringing together brief skits of often humorous scenes. For an additional charge, patrons could glimpse the camera after the show. Though the theater didn't remain in business for long, possibly due to the limited supply of films at the time, it foreshadowed the modern movie theater. Within a decade, moving picture theaters would become a staple of entertainment in American neighborhoods for generations to come.

1897 PLAYER PIANO

The player piano (or pianola), invented in 1895 and patented in 1897 by Edwin S. Votey (1856–1931), revolutionized home entertainment by enabling pianos to play automatically using perforated paper rolls. This self-playing piano used a pneumatic system where foot pedals or electric motors powered bellows that created suction, drawing air through holes in the paper rolls to activate the piano keys. The instrument reached peak popularity in the early 1900s, with companies like Aeolian producing thousands of units for middle-class households. The pianola allowed families without musical training to experience complex classical pieces and popular songs in their homes. Famous composers of the day, such as Gershwin, Debussy, and Stravinsky, recorded songs for the pianola, which provide invaluable insights into early twentieth-century performance practices that would otherwise have been lost. Though the pianola's popularity waned with the rise of radio and phonographs in the 1920s, the concept of automated musical reproduction laid the groundwork for today's programmed musical instruments and sampling technology.

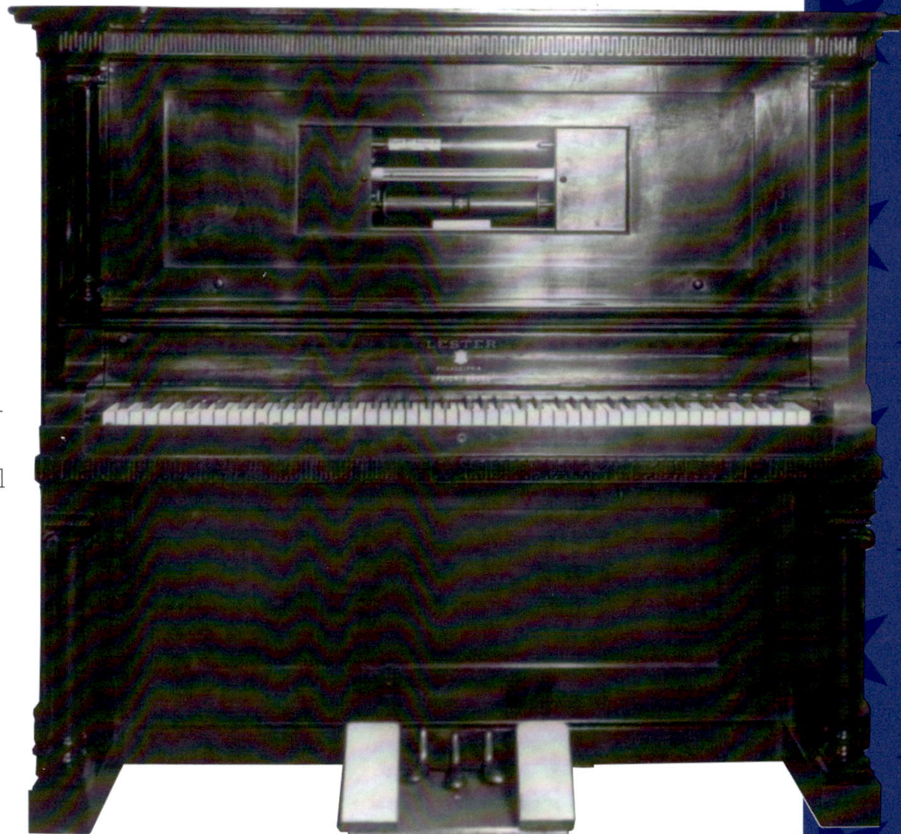

ABOVE: An upright player, c. 1895. **BELOW:** A paper roll installed and ready to play.

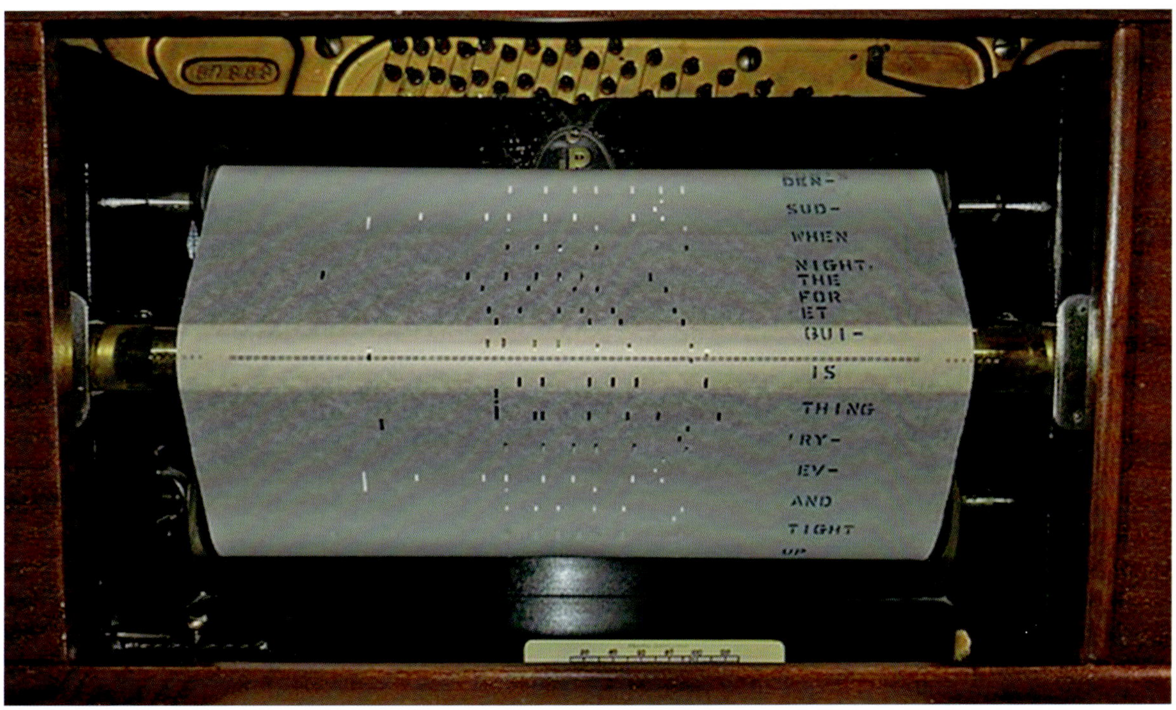

1898 BRUSH FOR AFRICAN AMERICAN HAIR

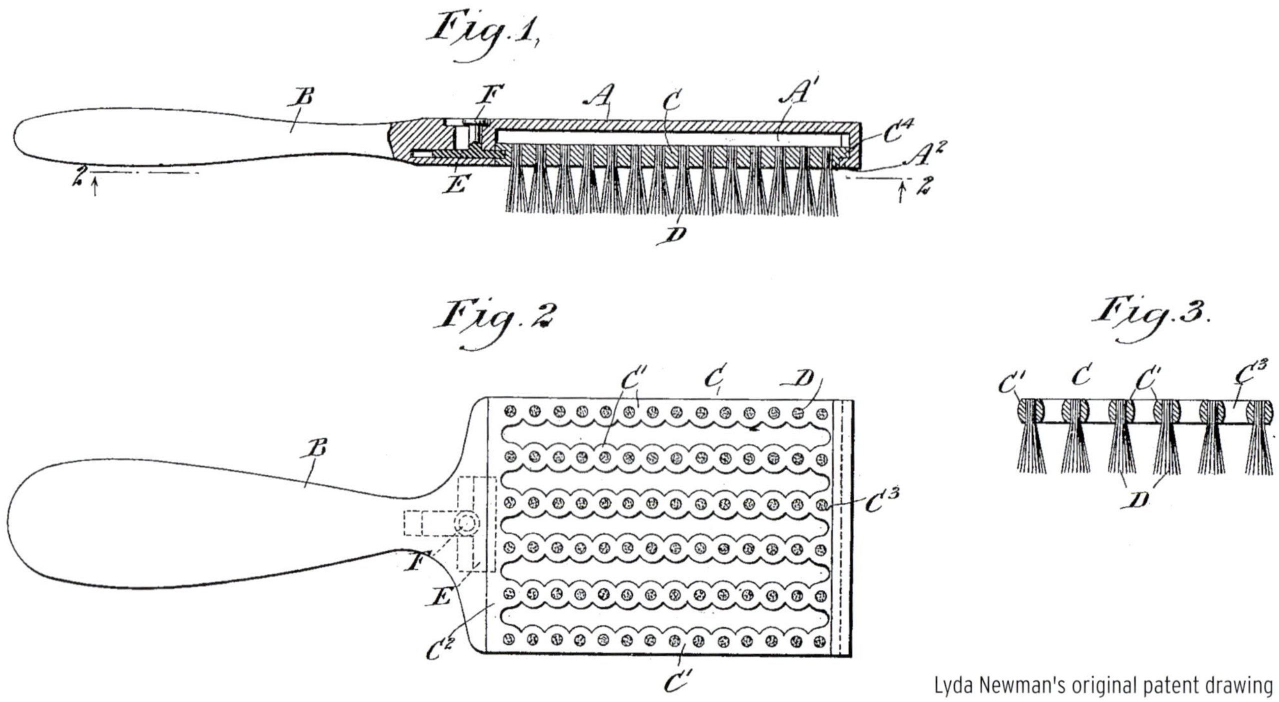

Lyda Newman's original patent drawing for her revolutionary hairbrush.

Women's rights advocate Lyda D. Newman (1885–Unknown), who is remembered for fighting for all women to gain the right to vote, is also known for a specially designed hairbrush that transformed beauty care for African American women. Newman received a patent for her hairbrush on November 15, 1898— making her one of the first Black women to receive a US patent— when she was living in New York City and working as a hairdresser. Most hairbrushes at the time used animal hair, but it was too delicate to work effectively on the coarse texture of African American hair and was also difficult to clean. Newman's innovative brush featured firm, durable synthetic bristles instead of soft, easily breakable animal hair; a recessed chamber into which debris could settle during styling; detachable bristles that could be easily cleaned and whose position could be customized for improved effectiveness; and an air chamber that helped wet hair dry efficiently. The brush's relatively affordable price made it accessible to a wider range of income levels than luxury hairbrushes. Newman is said to have paved the way for future Black women haircare pioneers, including Madam C. J. Walker.

1899 "MAPLE LEAF RAG"

Ragtime music, a highly syncopated genre of instrumental piano music, is considered uniquely American. It originated from honky-tonk bars in the southern states, influenced by the popular minstrel show songs of the day. Scott Joplin (1867 or 1868–1917) is considered the father of the music style and wrote dozens of ragtime pieces. In 1899, his song "Maple Leaf Rag," a delightful, spirited piece, helped introduce the American public to the genre. Legend has it that when the sheet music was issued, it sold more than seventy-five thousand copies in six months and went on to sell millions after that. The success of the song spurred a nationwide craze, inspiring more and more songs composed in the ragtime style. Although Joplin's music was mostly forgotten after his death, proof of his long-lasting legacy came posthumously: in 1976, the country's bicentennial year, he was awarded a Pulitzer Prize for his contributions to American music, and in 1983, his face adorned a US postage stamp. Joplin's music was adapted for an Oscar-winning score for the 1973 movie *The Sting*, which reintroduced the man and his music to pop culture nearly sixty years after his death.

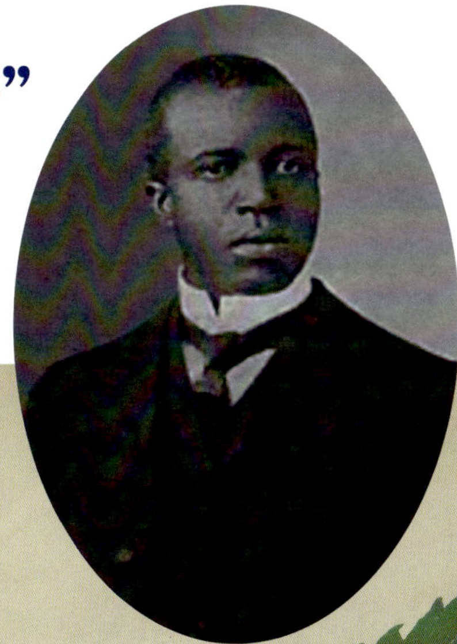

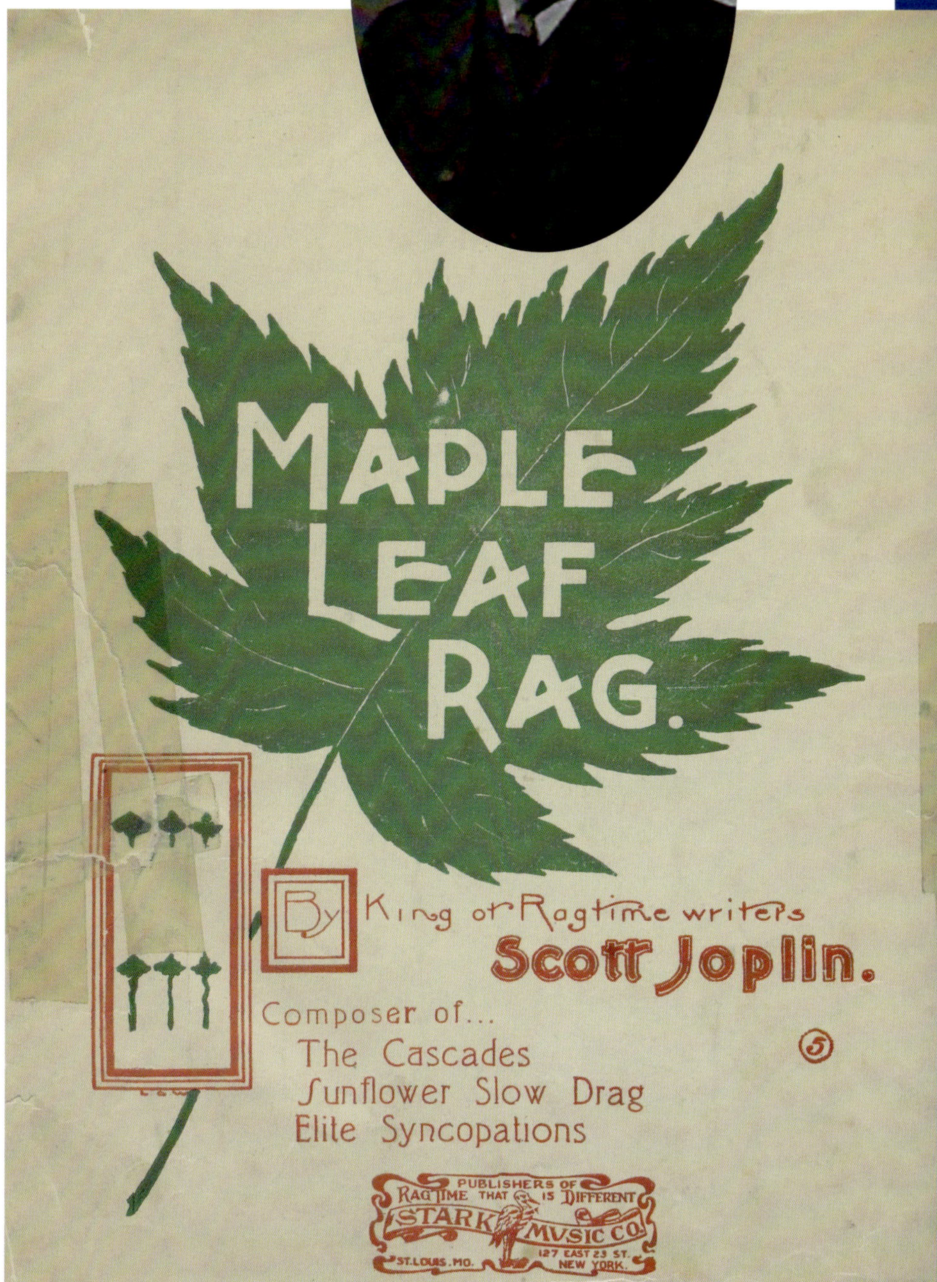

TOP: Scott Joplin, c. 1900. **BOTTOM:** The sheet music for "Maple Leaf Rag."

1900 ESCALATOR

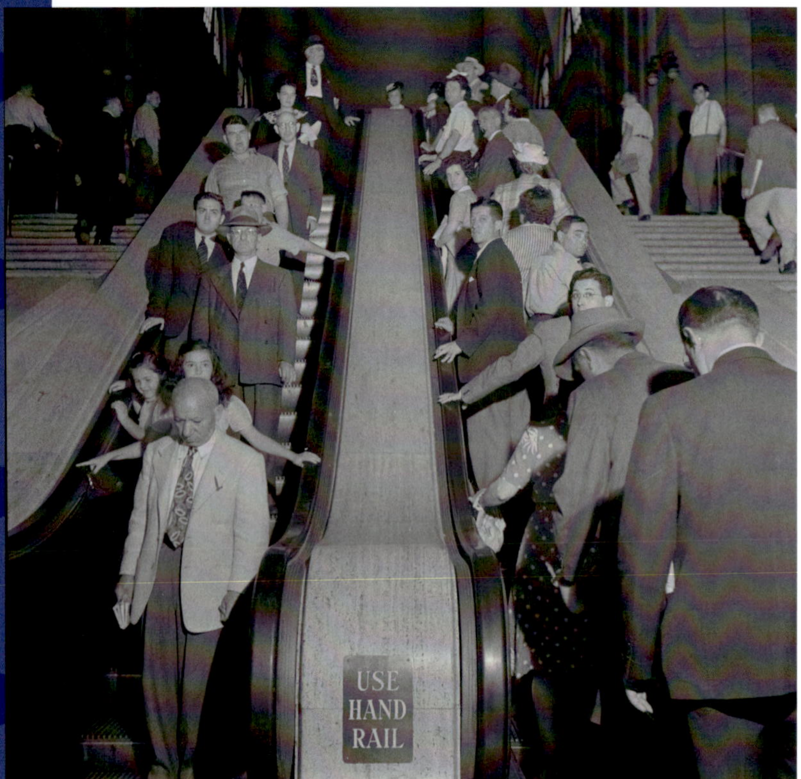

Inventor Charles D. Seeberger (1857–1931), an employee of the Otis Elevator Company, created what we know today as the escalator—mechanical steps that move along a continuous track. Seeberger is also credited with conceiving the word *escalator* itself. The invention was showcased at the Paris Exposition Universelle in 1900, where it won the highest honors. In 1910, the Otis company bought the rights to the escalator from Seeberger and continued to manufacture it commercially. The significance of the escalator in American and global life cannot be overstated. In retail, moving people between floors with minimal effort increased selling potential exponentially—stores could expand vertically at a fraction of the cost of horizontal expansion. In public transportation, escalators efficiently moved people in and out of underground railways in cities like New York and London, helping expand their use.

The escalators at New York City's Pennsylvania Station, c. 1942.

1901 SAFETY RAZOR

The safety razor revolutionized personal grooming when King Camp Gillette (1855–1932) patented his design in 1901, though it didn't achieve commercial success until 1903. Gillette's innovation replaced the dangerous straight razor with a protective guard between the blade and skin, making shaving far safer and more accessible to the average person. But Gillette's genius lay not just in the safety razor's design but also in its business model. Gillette sold the handles cheaply, then sold the disposable blade cartridges for a profit. This approach made quality shaving affordable while generating recurring revenue. The safety razor allowed men to shave safely at home without barber visits. During World War I, the US military issued Gillette razors to soldiers, creating millions of new customers and establishing clean-shaven faces as the American standard.

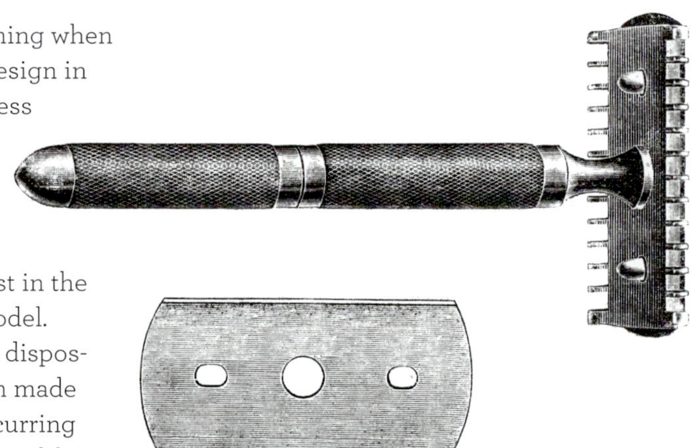

1902 THE TEDDY BEAR

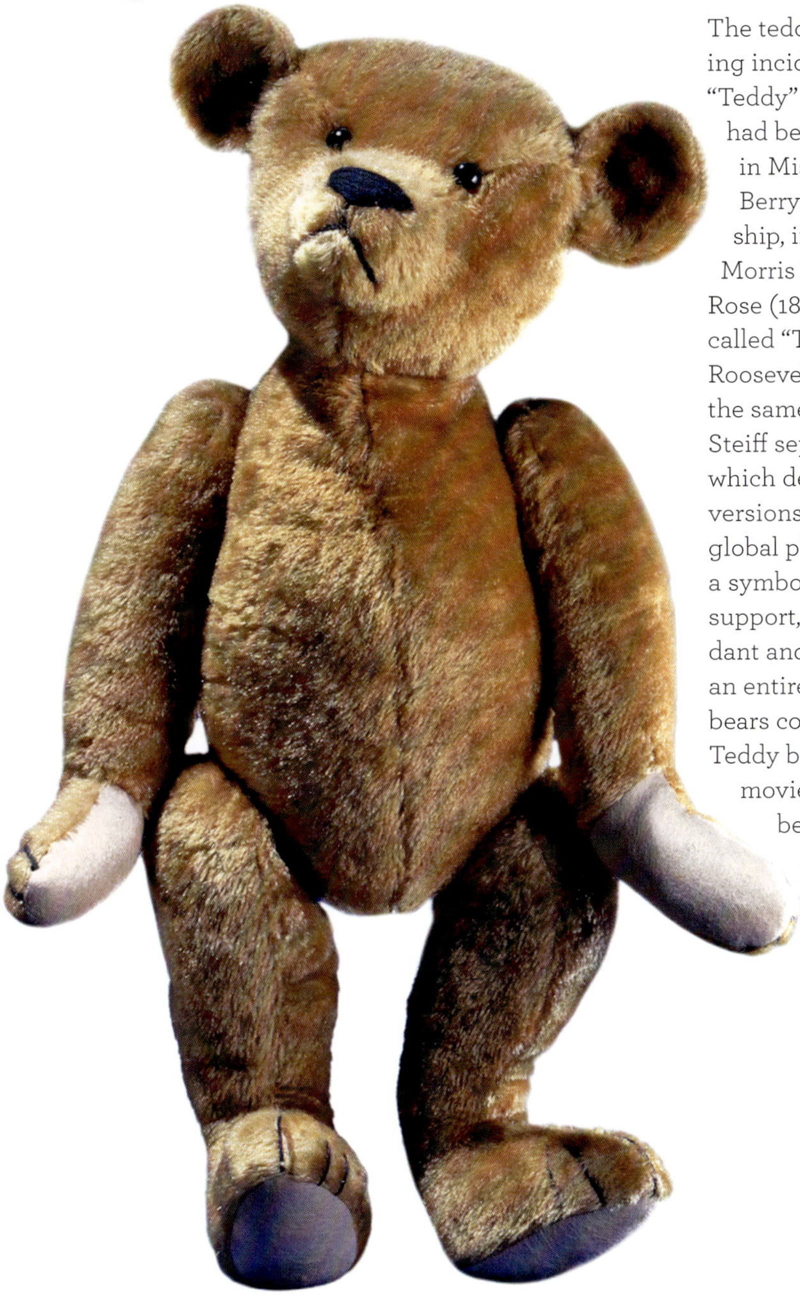

The teddy bear originated from a famous hunting incident in 1902 when President Theodore "Teddy" Roosevelt refused to shoot a bear that had been tied to a tree during a hunting trip in Mississippi. Political cartoonist Clifford Berryman illustrated this act of sportsman ship, inspiring Brooklyn candy-shop owners Morris Michtom (1869–1938) and his wife Rose (1867–1937) to create a stuffed bear they called "Teddy's bear." They sought and received Roosevelt's permission to use his name. (Around the same time, German toymaker Margarete Steiff separately developed her own stuffed bear, which debuted at a 1903 Leipzig toy fair.) Both versions became wildly popular, launching a global phenomenon. The teddy bear became a symbol of innocence, security, and emotional support, often serving as children's first confidant and transitional object. The toy sparked an entire industry of collectibles, with vintage bears commanding high prices at auctions. Teddy bears have appeared in countless books, movies, and cultural references, such as the best-selling children's book *Corduroy* by American author Don Freeman. Today, they remain one of the most beloved and recognizable toys worldwide.

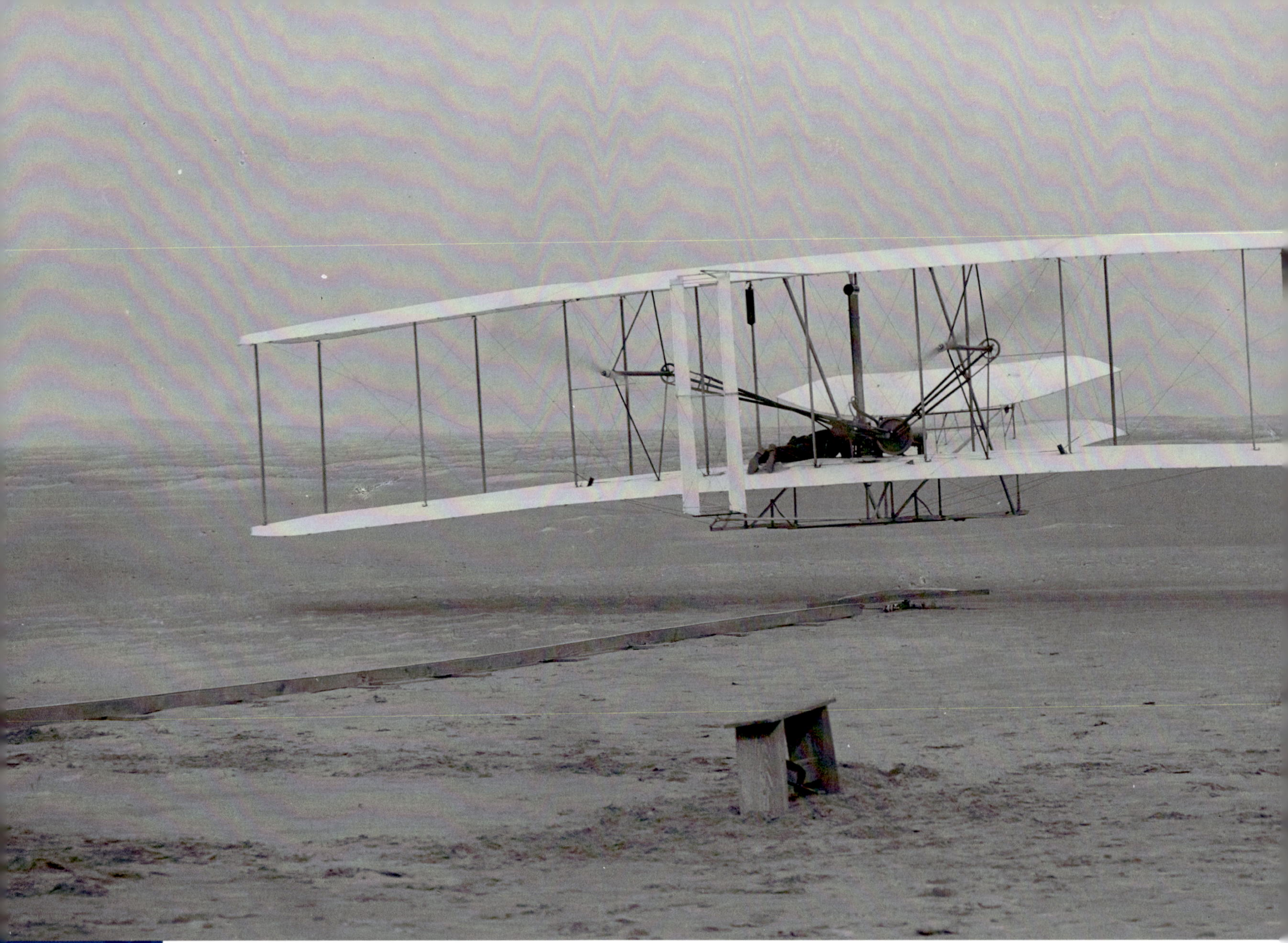

1903 THE AIRPLANE

The airplane was invented by siblings Orville Wright (1871–1948) and Wilbur Wright (1867–1912), who achieved the first powered, sustained, and controlled "heavier-than-air" human flight on December 17, 1903, near Kitty Hawk, North Carolina. Their Wright Flyer flew for twelve seconds and covered 120 feet, marking a pivotal moment in human history. The brothers' success came from their systematic approach to solving flight problems, including their development of wind-tunnel testing and their focus on control systems. It's hard to overstate the airplane's impact on humanity. It revolutionized transportation, shrinking global distances and making international travel accessible to millions. Aviation became essential for commerce, enabling rapid cargo delivery and creating entire industries around air transport. Militarily, aircraft changed warfare tactics and strategy, from World War I reconnaissance to modern air-power projection. Beyond transportation, aviation fostered cultural exchange, connected isolated communities, and enabled humanitarian missions during disasters. The technology also spurred innovations in materials science, aerodynamics, and manufacturing processes that benefited other industries. Today's commercial aviation industry employs millions worldwide and continues the Wright brothers' legacy of pushing the boundaries of human mobility and technological achievement.

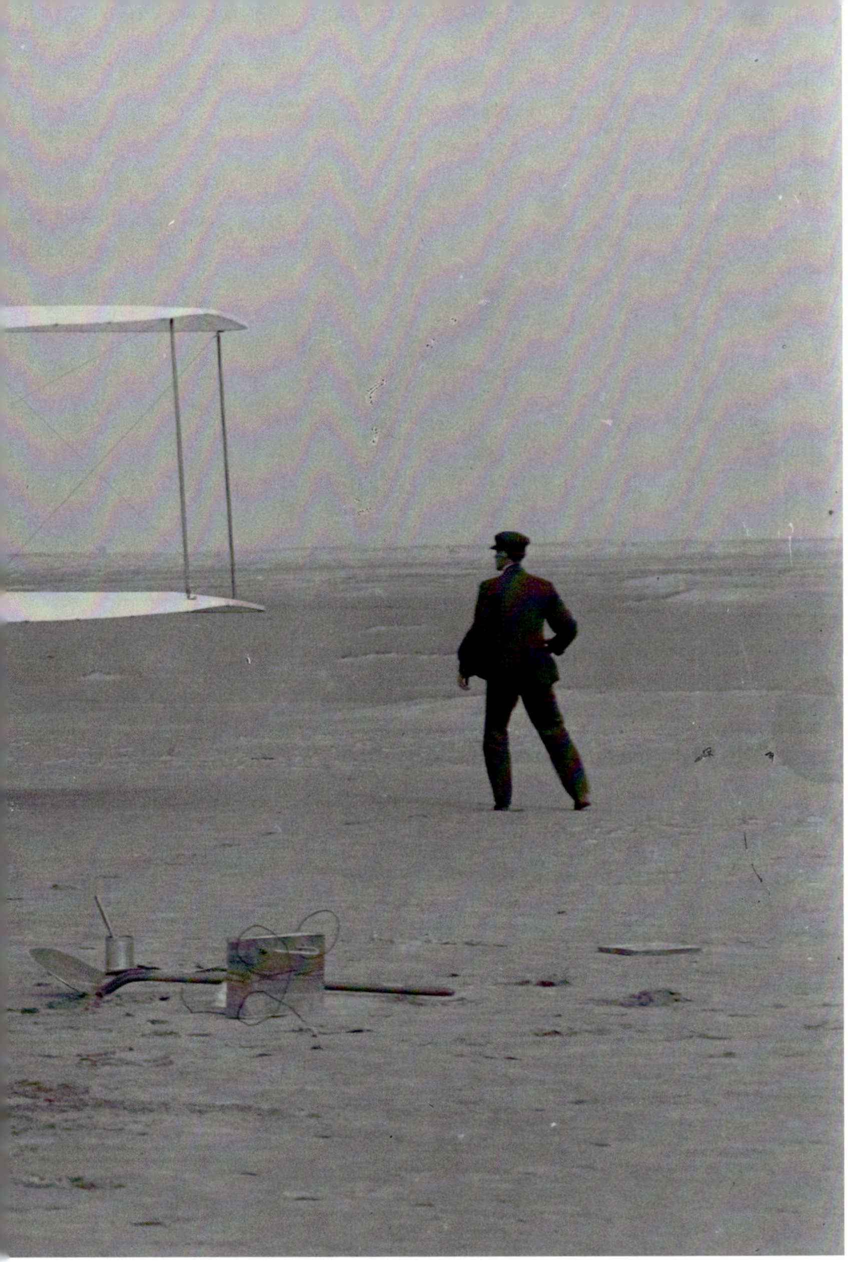

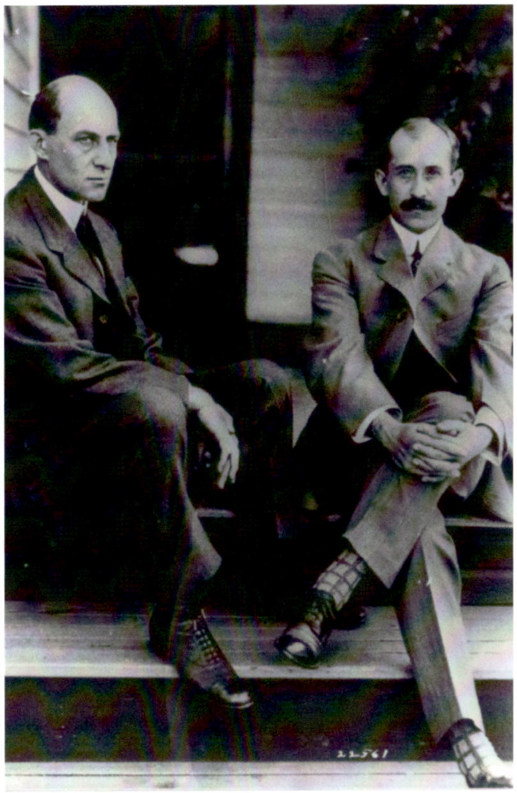

LEFT: The first flight at Kitty Hawk, December 17, 1903. **ABOVE:** The brothers Wright—Wilber and Orville, 1909.

1903 Crayola Crayons

Crayola made the first widely successful wax crayons in 1903. The company was started by cousins Edwin Binney (1866-1934) and C. Harold Smith (1860-1931) under their company name Binney & Smith in Pennsylvania. The name "Crayola" was coined by Binney's wife, Alice, as a combination of the French word for chalk (*craie*) and "ola" from "oleaginous," meaning "oily"—a reference to the paraffin wax used to make the crayons. The original assortment came in eight colors: black, blue, brown, green, orange, purple, red, and yellow. The same variety is still included in the brand's eight-crayon box today. Crayola has since dominated the American crayon market and remains a must-have on school-supply lists every year.

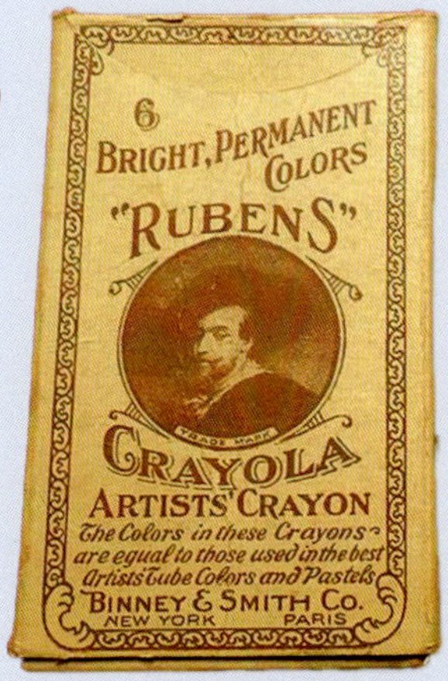

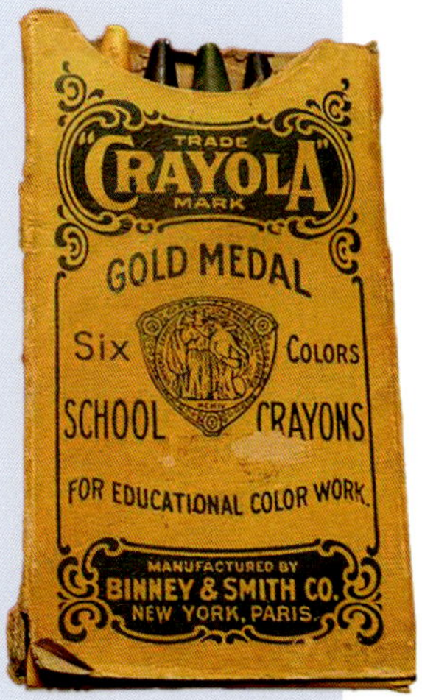

1904 CONTINUOUS TRACK PROPULSION

Prior to 1904, heavy tractors relied on conventional wheels that would sink into muddy soil when working the land during wet seasons, limiting their effectiveness. Benjamin Holt (1849–1920), an inventor from Stockton, California, sought to solve this problem by perfecting a tractor equipped with continuous, self-laying metal tracks designed to help farmers increase their productivity. In 1904, Holt successfully tested the first practical track-type tractor. His improved track system distributed the weight of heavy equipment evenly across a larger surface area, preventing the machine from sinking into soft ground. Holt named his invention the "Caterpillar," and it quickly demonstrated potential for applications beyond farming, including construction, infrastructure development, and military operations. The company eventually established manufacturing operations in Peoria, Illinois. The original Caterpillar became the foundation for modern bulldozers and military tanks. During World War I, Holt licensed his track design for use in armored vehicles. After World War II, Holt's company thrived by supporting global reconstruction efforts, and the Caterpillar brand has maintained its position as one of the world's most respected names in construction and mining equipment.

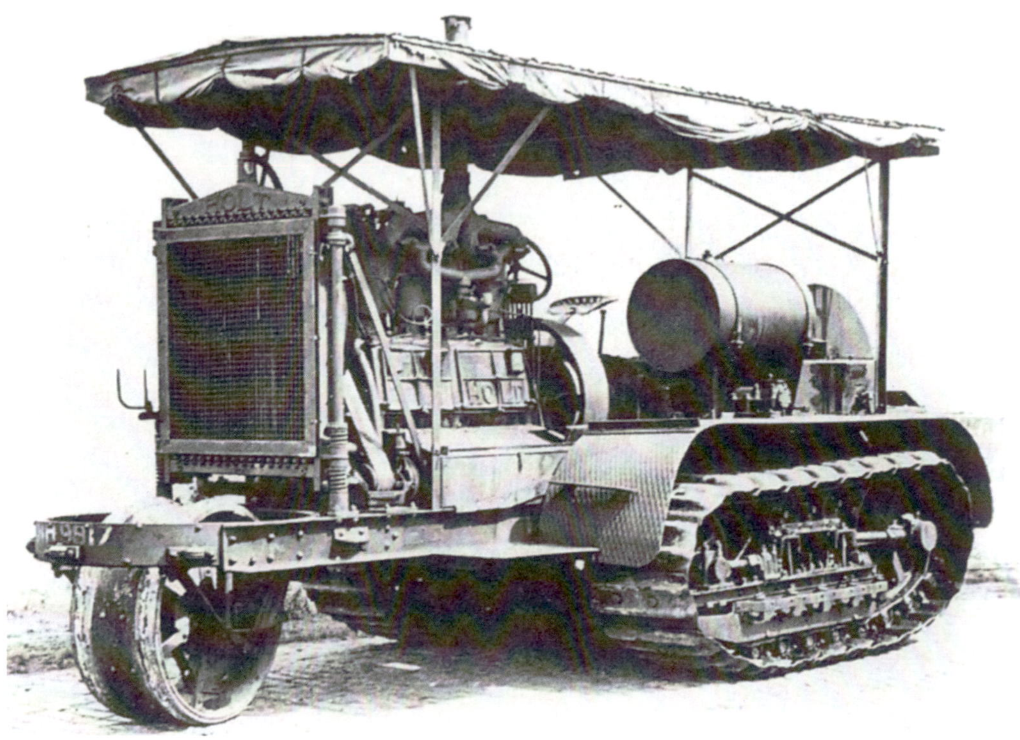

The Holt tractor, c. 1914.

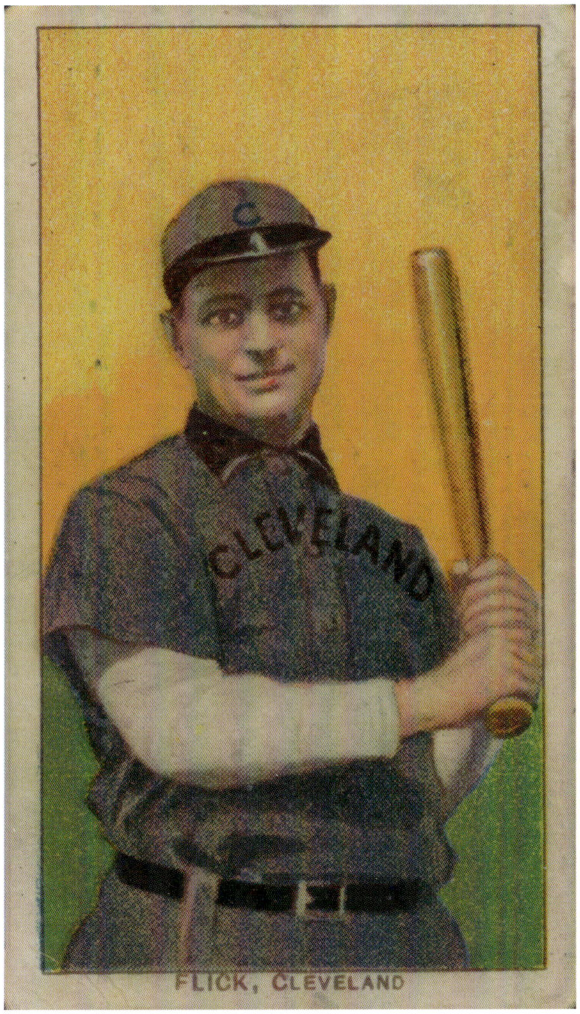

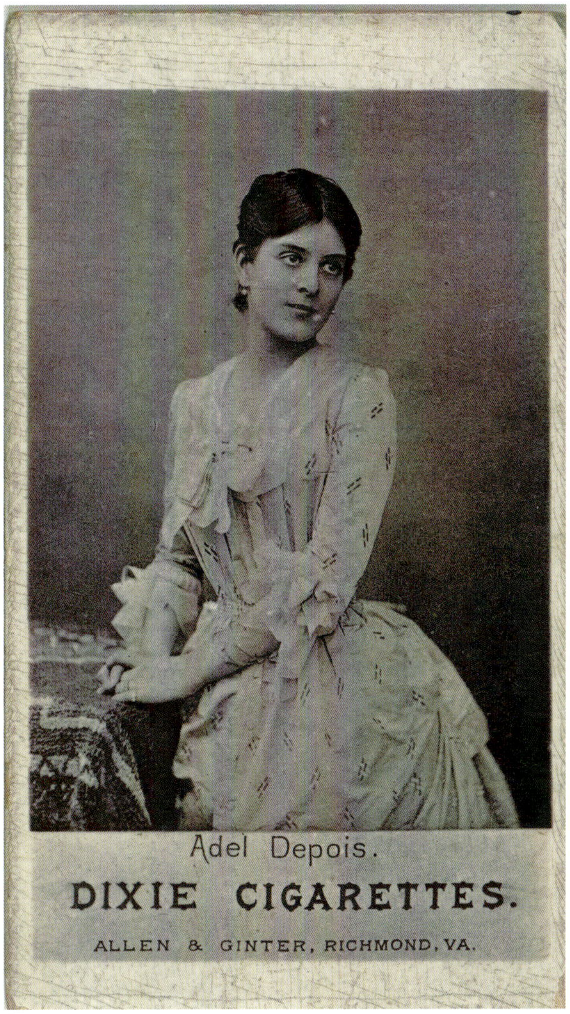

1905 COLLECTIBLE TRADING CARDS

Cigarette cards were well established and very popular by 1905. The first cigarette cards were introduced around 1875 to stiffen cigarette packaging and advertise different brands. By the early 1900s, major cigarette manufacturers produced elaborate card series featuring everything from baseball players and actresses to educational topics like geography, natural history, and military uniforms, and cigarette cards had established an almost fanatical following. The cards served dual purposes: encouraging brand loyalty while educating the public in an era before widespread television or radio programming. Families would eagerly collect complete sets, with children enjoying the fruits of their parents' smoking habits. The artistic quality of the cards was notable, featuring detailed lithographic illustrations and later high-quality photography. Cigarette cards became serious collectibles, laying the foundation for modern trading card culture. Cigarette card production was discontinued in order to save paper during the World Wars and was never fully reintroduced. Their influence endures in modern trading card culture, and they remain highly valued by collectors today. Museums currently recognize cigarette cards as important historical documents that offer unique glimpses into early twentieth-century popular culture, advertising techniques, and social values.

LEFT: A 1909 Sovereign Cigarettes card featuring Elmer Flick of the Cleveland Naps. **RIGHT:** Adel Depois from the Actresses series of cards promoting Dixie Cigarettes, c. 1888.

1906 CORN FLAKES

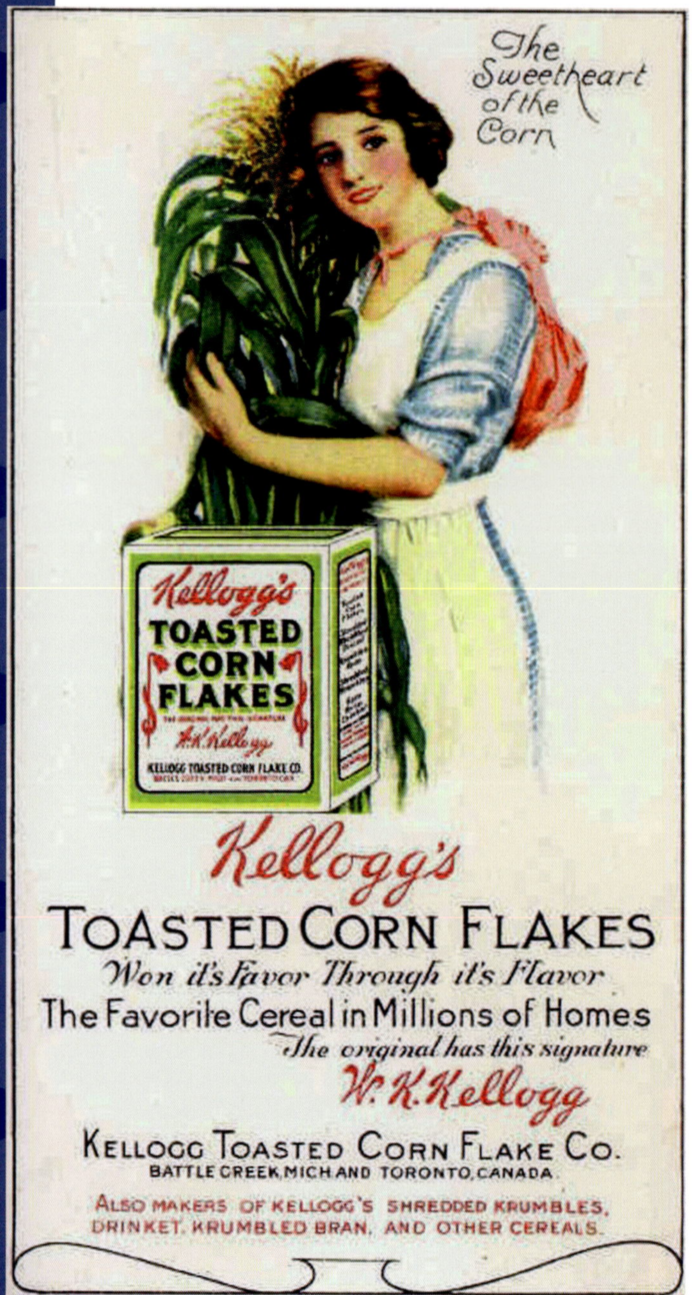

Dr. John Harvey Kellogg (1852–1943) and his brother William Keith Kellogg (1860–1951) came up with Corn Flakes, the first dry, flaked breakfast cereal, at a Battle Creek, Michigan, sanitarium and health spa where John was the medical director and popular wellness guru of the day. Cereals were commonly served at the institution as a health remedy. One story purports that the brothers were experimenting with a wheat dough to make cereal when they accidentally left a batch of it out overnight. The next day, they rolled out the slightly fermented dough into thin sheets and then baked it, forming thin, crisp flakes. William continued to develop the recipe and discovered a corn-based dough produced a superior texture. After years of alleged mistreatment by his brother, William bought the rights to the corn flake recipe, left the sanitarium, and founded the Battle Creek Toasted Corn Flake Company in 1906. He tweaked the recipe to appeal to the general public and began manufacturing it and advertising the company. Will Kellogg's innovation remains relevant even today: Corn Flakes still ranks among America's top ten favorite cereals, demonstrating the lasting appeal of this accidental discovery that revolutionized breakfast.

An ad for "The Favorite Cereal," c. 1910.

1907 HAIR PRODUCTS FOR BLACK WOMEN

Born Sarah Breedlove to formerly enslaved parents, Madam C. J. Walker (1867–1919) was a pioneer in African American hair products for women. After working as a sales representative for a product that promised to grow hair—she herself experienced hair loss—Walker launched her own line of hair products for Black women. In 1907, Walker and her husband traveled widely to promote her products, called the "Walker System," which included her own hair loss formula, lotions, pomade, and styling tools customized for African American hair. Walker became known as the first self-made female millionaire in America, which she accomplished through her direct-sales approach to Black women and by employing an army of sales associates she dubbed "beauty culturists." At the height of her business, Walker had several thousand Black saleswomen throughout the country selling such products as "Wonderful Hair Grower," "Glossine," and "Tetter Salve," for growing and styling African American hair.

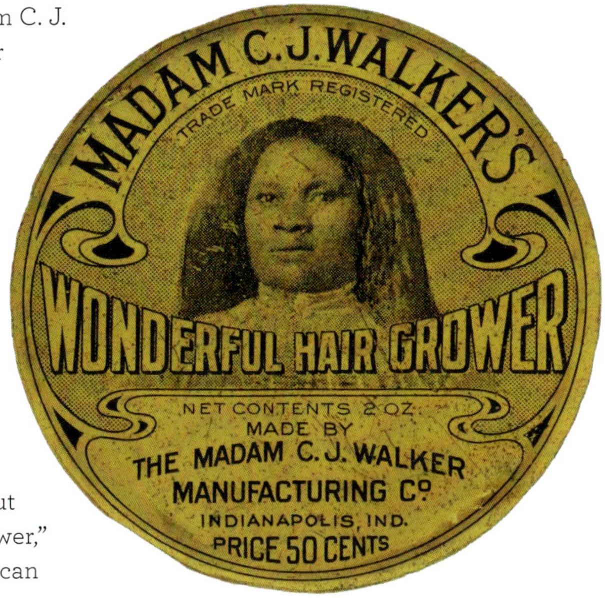

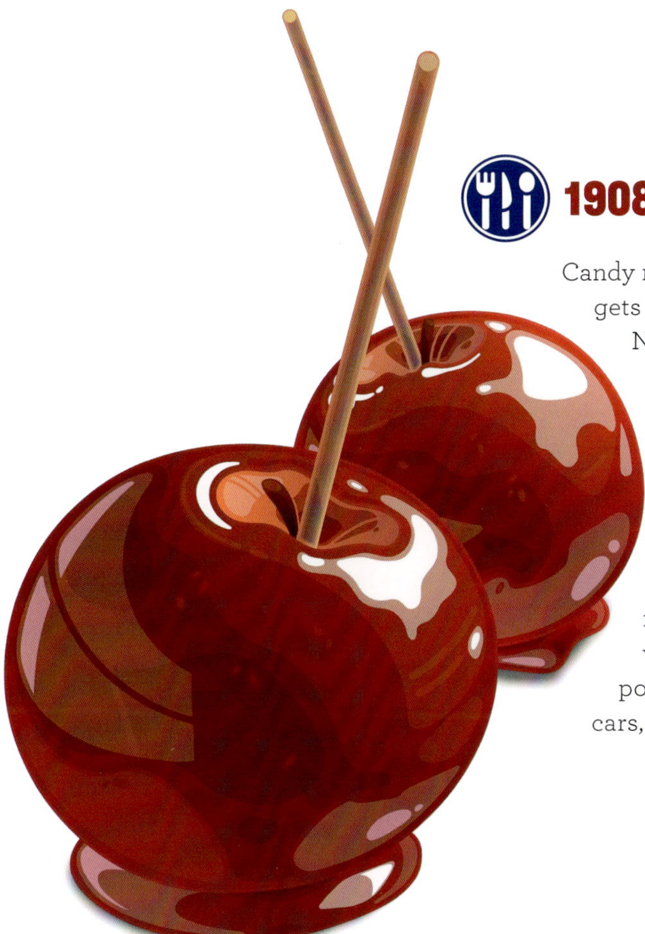

1908 CANDY APPLE

Candy maker William W. Kolb (birth and death dates unknown) gets the credit for inventing the candy apple in Newark, New Jersey, in 1908. Some legends claim that Kolb didn't intend for the candy apples to be eaten—rather, he was simply looking for a way to display the bright red cinnamon candies he had on hand, so he melted them down, dipped the apples into them, and put them on display in the shop window. But customers coveted the apples, and Kolb sold them for five cents apiece. As in Kolb's day, the most authentic candy apples are red in color and cinnamon in flavor, though there can be variations. The color of "candy-apple red" lives on in popular culture—appearing in the glossy finish of fancy cars, stiletto heels, lipstick, and nail polish.

1909 NAACP

The National Association for the Advancement of Colored People (NAACP) was founded on February 12, 1909, in New York City by a diverse group of civil rights activists, including W. E. B. Du Bois, Ida B. Wells-Barnett, and white allies like Mary White Ovington and Moorfield Storey. The organization emerged in response to white violence against the Springfield, Illinois, Black community in 1908 and the deteriorating conditions faced by African Americans in the post-Reconstruction era. The NAACP's founding mission centered on securing political, educational, social, and economic equality for all people and eliminating racial hatred and discrimination. From its beginning, the organization adopted a strategic approach focusing on legal challenges, lobbying, and public education campaigns. The NAACP's legacy is monumental in American civil rights history. Its legal strategy culminated in landmark Supreme Court victories, most notably *Brown v. Board of Education* (1954), which declared school segregation unconstitutional. The organization also played crucial roles in dismantling Jim Crow laws, advancing voting rights, and supporting the broader civil rights movement of the 1950s and 1960s. Today, the NAACP continues to advocate for civil rights, addressing issues like criminal justice reform, economic inequality, and voting rights protection, maintaining its position as America's oldest and largest civil rights organization.

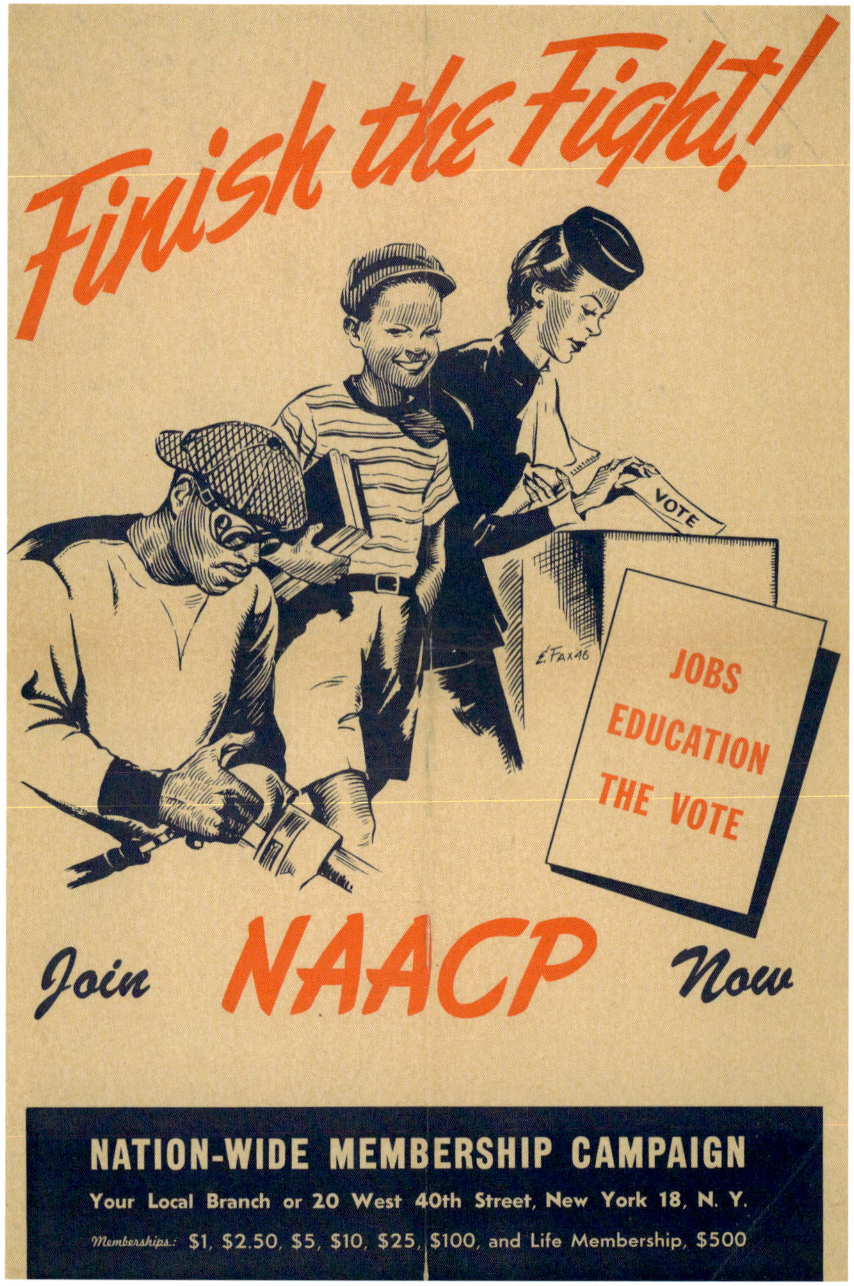

A promotional poster for the NAACP, 1946.

1910 DUCTILE TUNGSTEN FILAMENT

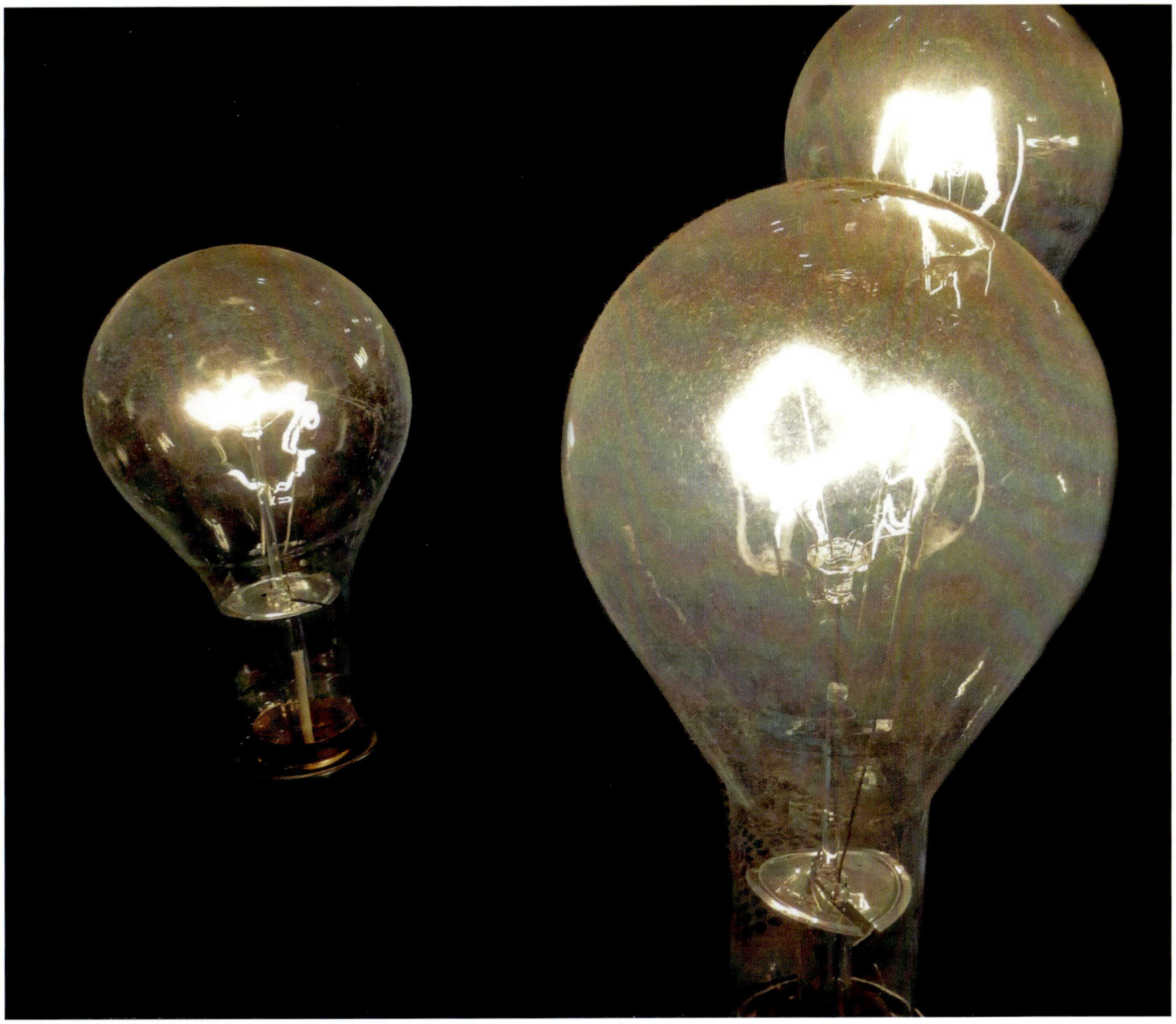

The ductile tungsten filament revolutionized electric lighting when William David Coolidge (1873–1975) of General Electric developed it in 1910. Before his discovery, incandescent bulbs used fragile carbon filaments or brittle tungsten wires that frequently broke during manufacturing and use. Coolidge's innovation involved creating a process to make tungsten ductile—malleable enough to be formed into fine, flexible wires without cracking. The key lay in Coolidge's technique of heating tungsten powder and mechanically working it through a series of special processes that aligned the metal's crystalline structure. The impact was transformative. Ductile tungsten filaments burned brighter, lasted significantly longer, and operated at higher temperatures than their carbon predecessors. This development made electric lighting more practical and economical, accelerating the transition from gas lighting to electricity in homes and businesses worldwide. Today, tungsten filaments remain the standard in incandescent bulbs, though LED technology has largely superseded them. Coolidge's work with tungsten also influenced other high-temperature applications, from vacuum tubes to aerospace components, establishing the metal as a critical industrial material.

1911 ELECTRICAL IGNITION SYSTEM

In 1911, Charles F. Kettering (1876–1958), an engineer from Dayton, Ohio, made a pivotal contribution to automotive development through his company Delco: the electric self-starting ignition system. Prior to this innovation, drivers had to use a metal hand crank to start their engines—a process that was not only physically demanding but also dangerous. Kettering drew on his experience from working at a cash-register company, where he had invented an electric motor that opened cash drawers at the touch of a button. He adapted this technology to create a key-operated starting system for automobiles. The first electric starter was installed in a Cadillac on February 17, 1911, and quickly spread to other General Motors models. Kettering received his patent for the device on August 17, 1915. General Motors eventually acquired Delco, and the electric starter became standard equipment in American automobiles.

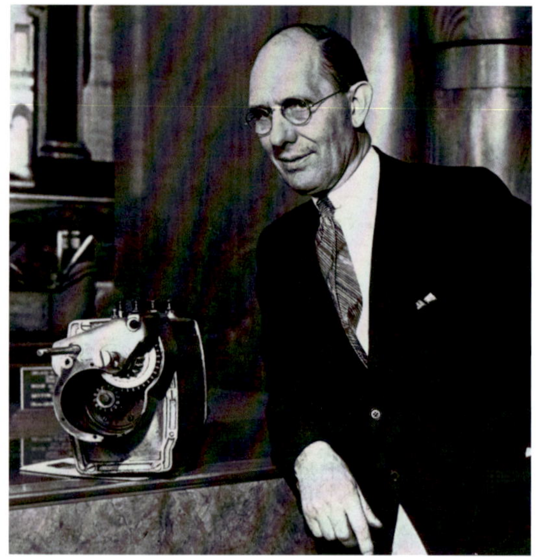

Kettering with a model of his electric self-starter.

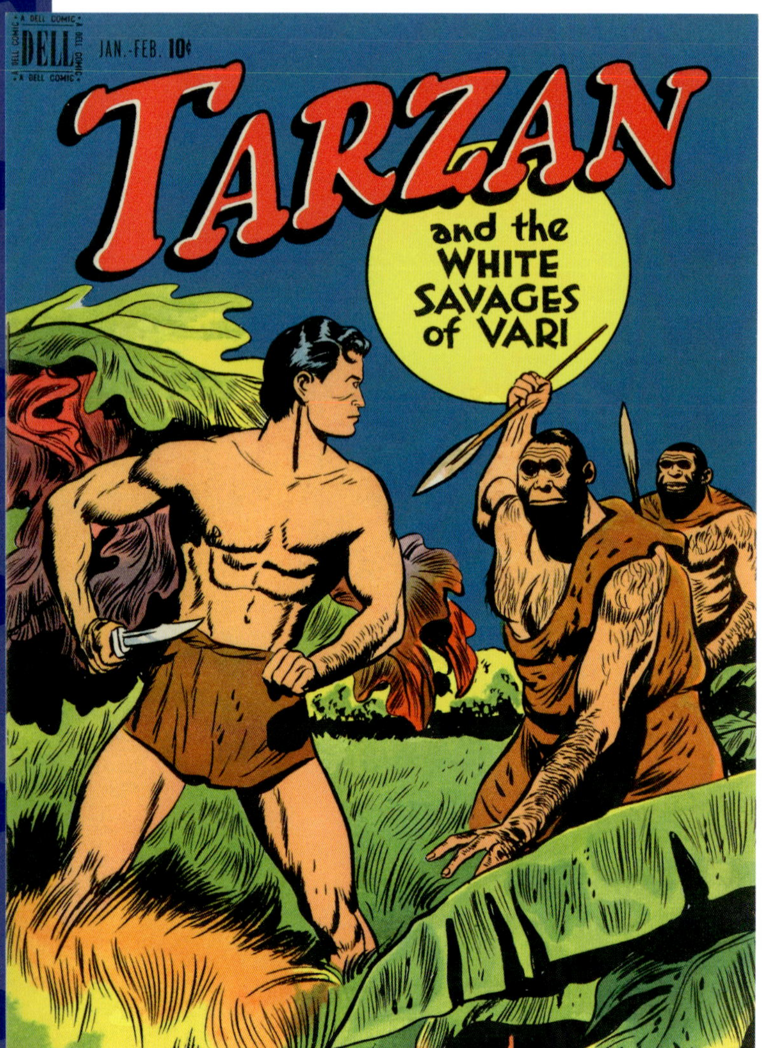

1912 TARZAN

Transmedia—the practice of telling a story across multiple platforms—may seem like a recent phenomenon, but it has existed for centuries. One of the most enduring examples is Edgar Rice Burroughs's Tarzan, among the most beloved pulp characters of all time. Burroughs (1875–1950) first introduced his orphaned British nobleman raised by apes in the pulp magazine *The All-Story* with "Tarzan of the Apes." The story's immense popularity spawned a magazine serial, a novel of the same title, and multiple sequels that sold more than twenty-five million copies worldwide. Within a few years, Hollywood adapted the story for film, launching a series of remakes that continued for decades. Tarzan became a fixture in comic books, radio programs, Broadway productions, and eventually television.

Dell Comics Issue 1 of *Tarzan*, January 1948.

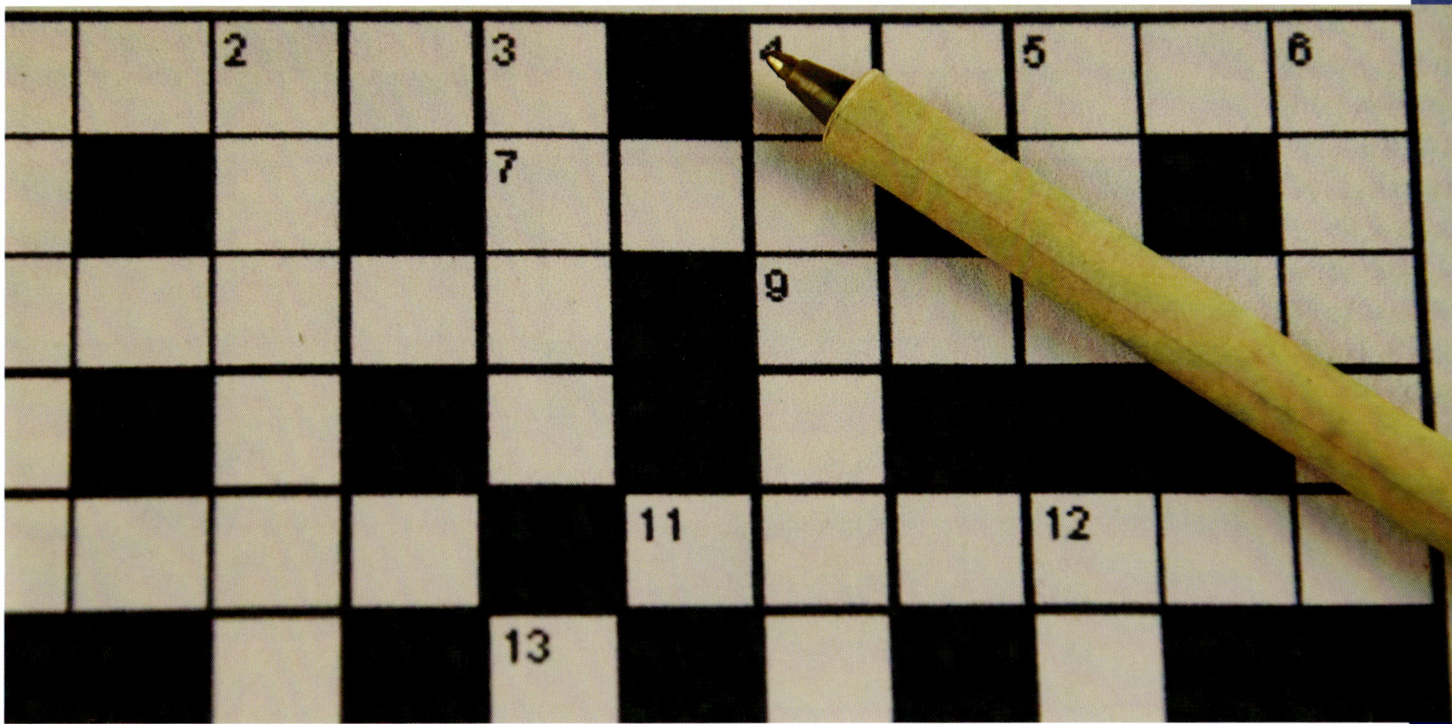

1913 THE MODERN CROSSWORD PUZZLE

Arthur Wynne (1871–1945) created the first American-style crossword puzzle, which was published on December 21, 1913, in the *New York World* newspaper, in the paper's "Fun" section. Wynne's original puzzle was a riff on a game he knew from childhood, called Magic Squares, which he modified for adults. The puzzle differed from what we know today, as it was diamond-shaped and lacked internal black squares, but the basic form of interlocking blank squares and a list of clues to words that would fill in the squares was the same. *New York World* owned the rights to the puzzle (they discouraged Wynne from copyrighting his invention), but they didn't think the game was of significance, so they didn't do anything to secure the rights. Other newspapers soon offered crossword puzzles, and by the 1920s, solving crosswords had become a massive craze. History shows us that Wynne's ingenuity was not a passing fad—crosswords are credited with helping distract citizens during World War II and a new word, *cruciverbalists*, was even created to describe crossword puzzles' devoted fans.

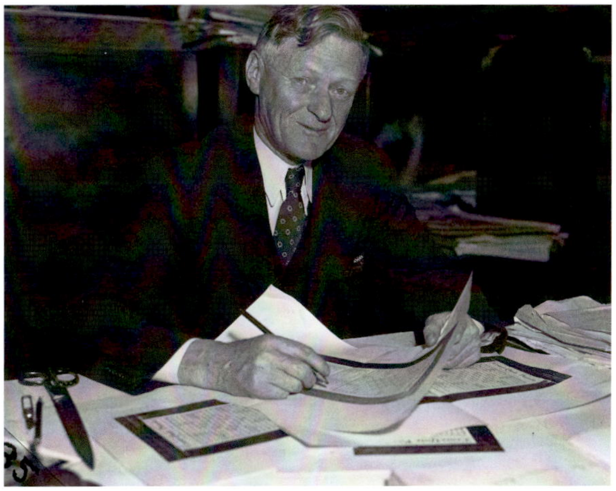

Editor and crossword inventor Arthur Wynne working on his next puzzle.

1914 MOTHER'S DAY CARD

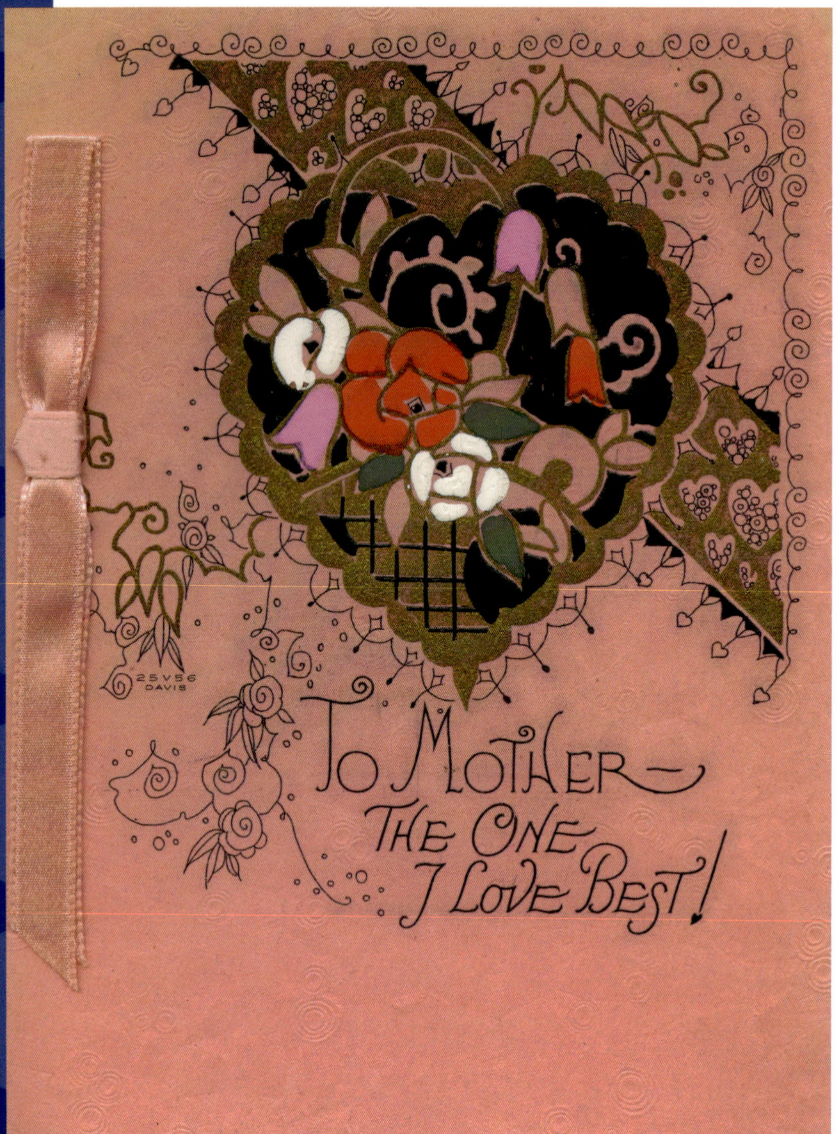

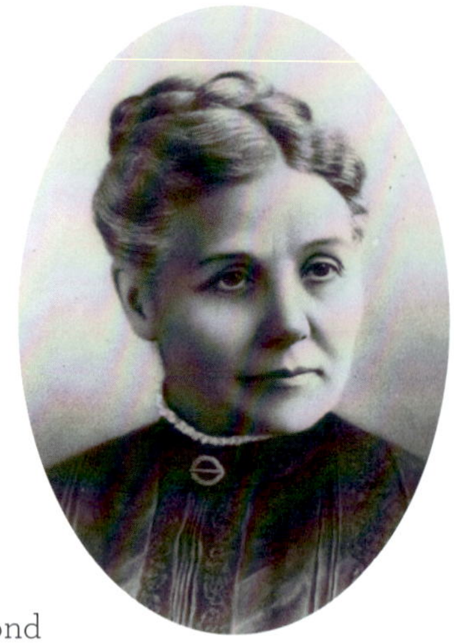

President Woodrow Wilson officially established Mother's Day as a national holiday in America in 1914. Anna Jarvis (1864–1948) created the holiday to honor her mother, who had organized "Mother's Day Work Clubs" during her lifetime to teach women proper childcare. The holiday is celebrated on the second Sunday of May each year, chosen to commemorate the anniversary of Jarvis's mother's death. Jarvis originally envisioned the holiday as an opportunity for people to send personal, heartfelt letters of gratitude to their mothers. However, greeting card companies quickly commercialized this concept, manufacturing preprinted cards that allowed consumers to bypass the personal effort Jarvis had intended. This commercialization deeply troubled Jarvis. By 1943, her disappointment had grown so intense that she launched a petition campaign to have Mother's Day repealed entirely. Despite her efforts to reverse the holiday's commercial trajectory, Mother's Day has continued to thrive as one of the highest-revenue-generating holidays in the United States. Today, Americans purchase more than one hundred million Mother's Day cards annually, a testament to both the holiday's enduring popularity and the very commercialization that its founder so vehemently opposed.

RIGHT: Anna Jarvis's mother, for whom she created the holiday.

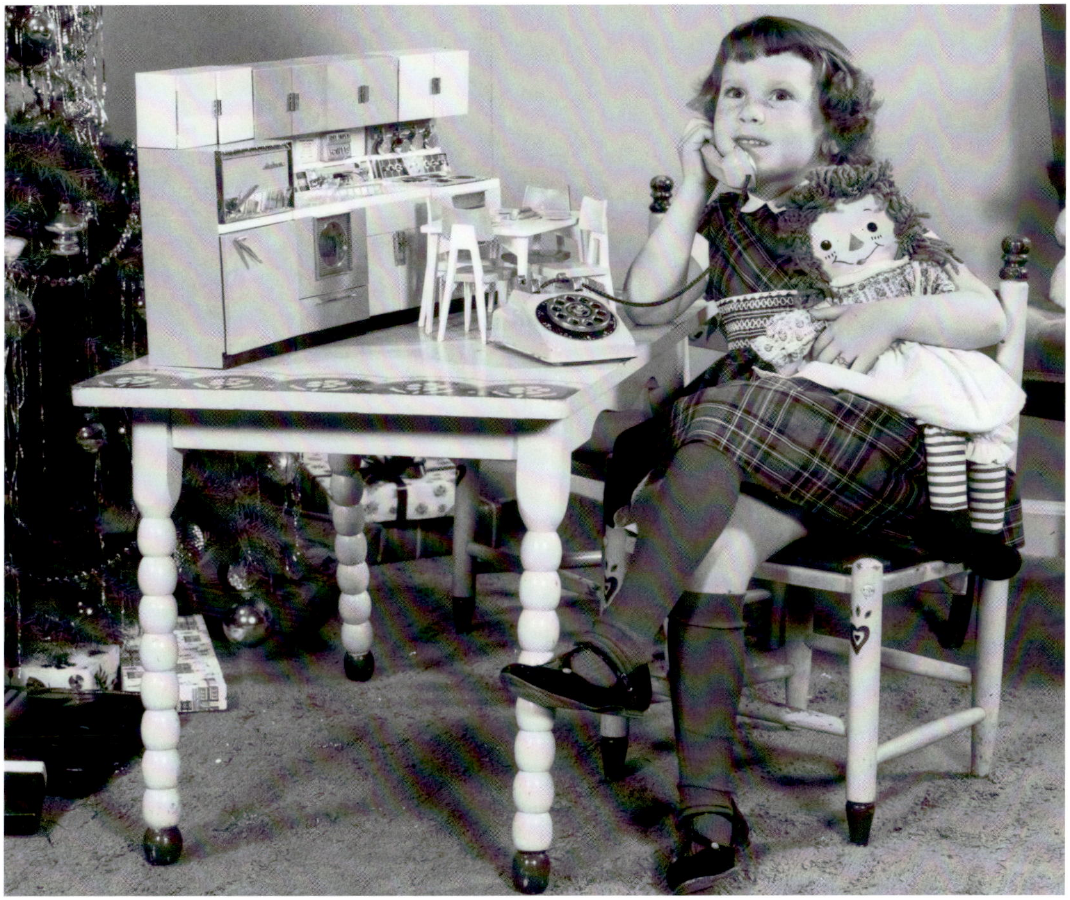

A child clutches her Raggedy Ann doll, c. 1950.

1915 RAGGEDY ANN DOLL

American author and illustrator Johnny Gruelle (1880–1938) is perhaps best known as the creator of the Raggedy Ann doll and her companion, Raggedy Andy. Raggedy Ann came to life in 1915 when, according to legend, Gruelle's daughter Marcella brought him an old rag doll. He drew a face on it with a distinctive triangle nose and named her Raggedy Ann. Perhaps Gruelle knew he would have a hit on his hands someday, because he both filed a design patent for the doll and registered a trademark on the name "Raggedy Ann" the same year. Three years later, Gruelle published a book of Raggedy Ann stories, soon joined by her brother, Raggedy Andy, which the publisher marketed alongside the dolls. The merchandising strategy proved a hit. Gruelle wrote one new book per year for twenty years to fuel children's imaginations with Raggedy Ann and Andy adventures. The dolls are still for sale today, both in new form and as prized vintage collectibles. The Raggedy Ann doll was inducted into the National Toy Hall of Fame in 2002, and Raggedy Andy followed in 2007.

1916 SELF-SERVICE GROCERY STORE

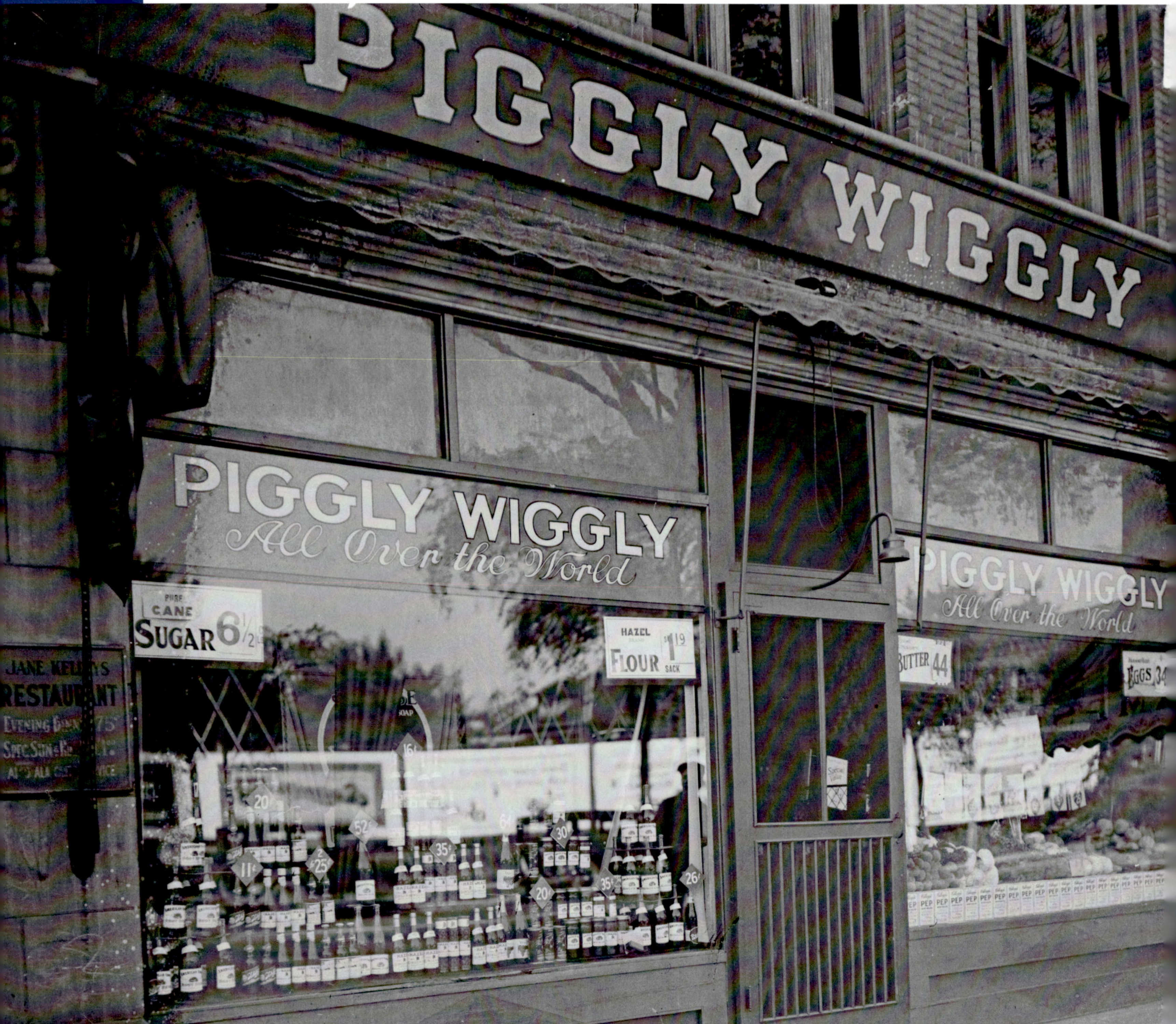

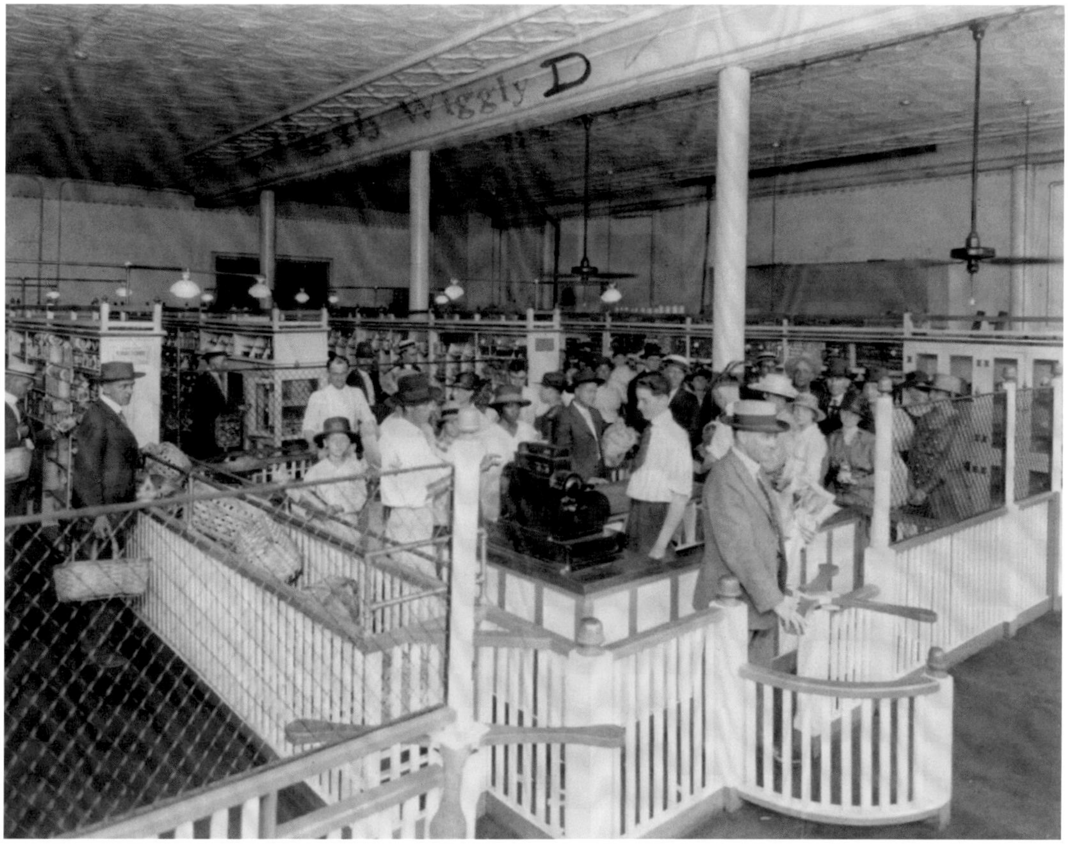

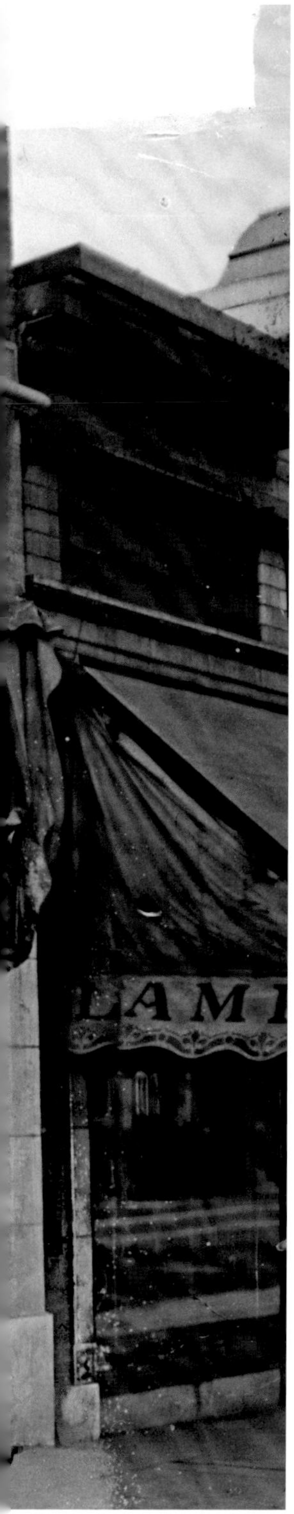

The first true self-service grocery store, Piggly Wiggly, opened in Memphis, Tennessee, on September 6, 1916. It was founded by grocer Clarence Saunders (1881–1953). Prior to the Piggly Wiggly, the shopping experience was vastly different: shoppers gave a list of desired goods to a clerk behind a counter who would gather their order from the store's shelves. Saunders brainstormed the time-saving solution of having shoppers gather the items themselves. There was much skepticism about the concept at the time, and many thought it would fail. But the goods in the store were less expensive than in the general stores of the previous era, in part because there were more items to choose from and in part because customers could self-select their own goods—the self-service model caught on like wildfire. Saunders had succeeded in proving his critics wrong and soon after franchised the Piggly Wiggly concept to hundreds of other retailers. This made him a wealthy man, and the concept of self-service grocery stores became the dominant model in the twentieth century and beyond.

LEFT: A Piggly Wiggly storefront in Austin, Texas, c. 1926. **ABOVE:** The interior of a Piggly Wiggly near Memphis, Tennessee, c. 1918.

1917 FIRST JAZZ RECORDING

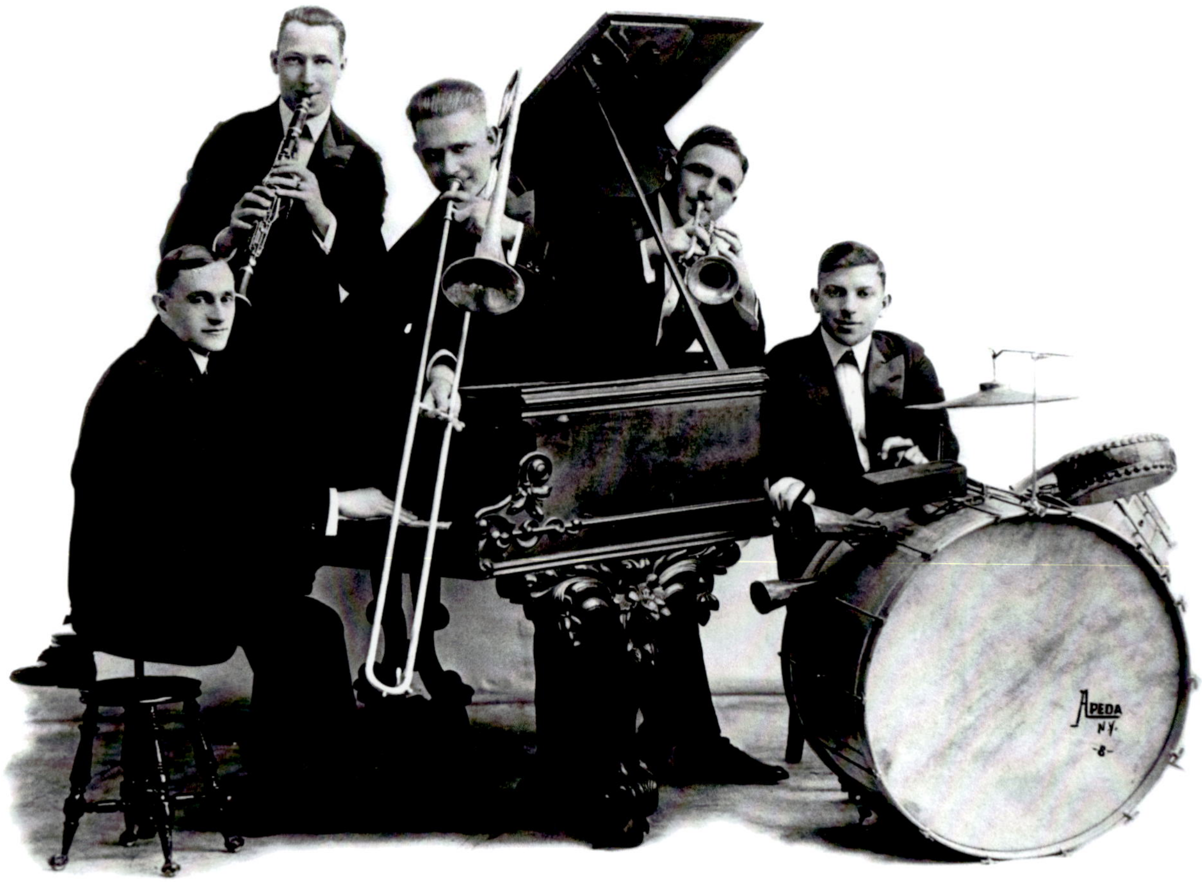

The Original Dixieland Jazz Band, c. 1917.

On February 26, 1917, in New York City, a group of New Orleans–based musicians called the Original Dixieland Jazz Band (ODJB) recorded "Livery Stable Blues." Though much debated, many consider this the first jazz recording in history. The irony is striking: Whereas jazz music is deeply associated with the Black experience, this pioneering recorded performance came from an all-white group. The recording featured a syncopated melody, high energy, entertaining "barnyard" animal sounds, and a playful approach. Music scholars trace its influences to New Orleans dance music and African American blues traditions. The Victor Company, known for conservative musical tastes, took a chance on including the ODJB in their catalog— a gamble that paid off handsomely when "Livery Stable Blues" sold nearly a million copies.
For the ODJB, jazz represented more than commercial success; it was a way to challenge established musical conventions. Though they achieved mainstream popularity, they maintained an outsider identity that would later echo in the punk rock movement of the mid-1970s.

1918 BLOOD BANKS

During World War I, US Army doctor Oswald Hope Robertson (1886–1966) was instrumental in developing methods to store blood for transfusions. He established the first blood bank—essentially a refrigerated storage system using glass bottles to preserve blood. Working with the British Army in France, Robertson pioneered a system that collected type O negative blood, the universal donor type, and treated it with anticoagulants to prevent clotting. This preserved blood could be stored for up to twenty-six days and transported to battlefield locations where transfusions were desperately needed. The British government recognized Robertson's life-saving contributions by awarding him the Distinguished Service Order. After World War I, blood banks gradually gained acceptance in medical practice, though their expansion to civilian healthcare took considerable time. By World War II, the American Red Cross had become the primary administrator of the United States' national blood collection program, gathering more than thirteen million pints of blood for the war effort. This national blood-banking system established during wartime continues to operate today, forming the foundation of America's blood supply infrastructure.

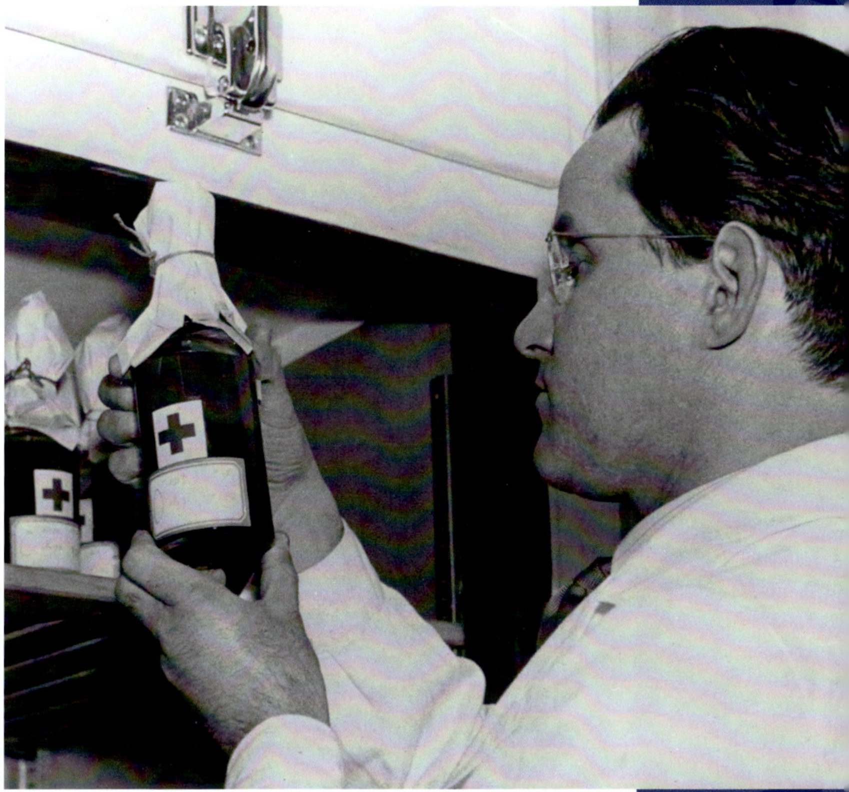

A doctor inspects the supply at a blood bank in Sweden, c. 1930.

1918 Color-Coordinated Makeup

Max Factor Sr. (1877–1938) is considered by many to be the inventor of modern makeup. In 1918 he introduced the Color Harmony product line, using a philosophy that the makeup a woman wore should complement her natural skin tone and hair color. Because he was a Hollywood makeup artist, the products were initially available only to actresses, but they proved so popular that he soon developed a line of face powder for the general public. Factor's innovations in makeup can be considered forerunners to today's wide assortment of foundations, tinted creams, and tinted powders.

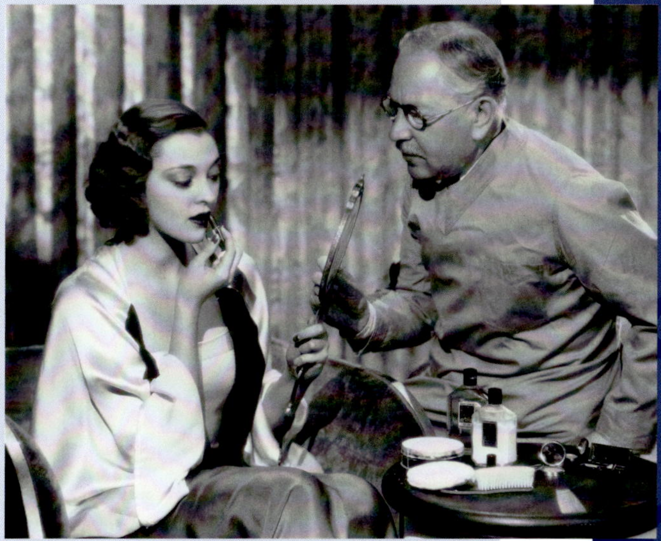

1919 KITCHENAID FOOD PREPARER

The KitchenAid Food Preparer was introduced in 1919 by the Hobart Manufacturing Company of Troy, Ohio. The company was already manufacturing high-power, eighty-quart mixers for commercial bakeries, restaurants, and military kitchens, where their powerful motors and bowls that could be moved up and down revolutionized the efficiency of food preparation. With its sights now on the retail market, the Hobart company announced a scaled-down, ten-quart version for home use under the KitchenAid brand. Legend has it that the model's name emerged when a company executive took the new mixer home to test it and returned with feedback, calling it "the best kitchen aid I've ever had." Though the new mixers were expensive, they eventually gained favor for their quality, reliability, and ability to tackle jobs previously deemed tough, tedious, and time-consuming for the home baker. The machine, now commonly referred to as a stand mixer, went through cosmetic redesigns throughout the decades but largely functions as it did when it was first introduced in 1919. To this day, KitchenAid holds a dominant market share in the stand mixer category year after year, solidifying its legacy.

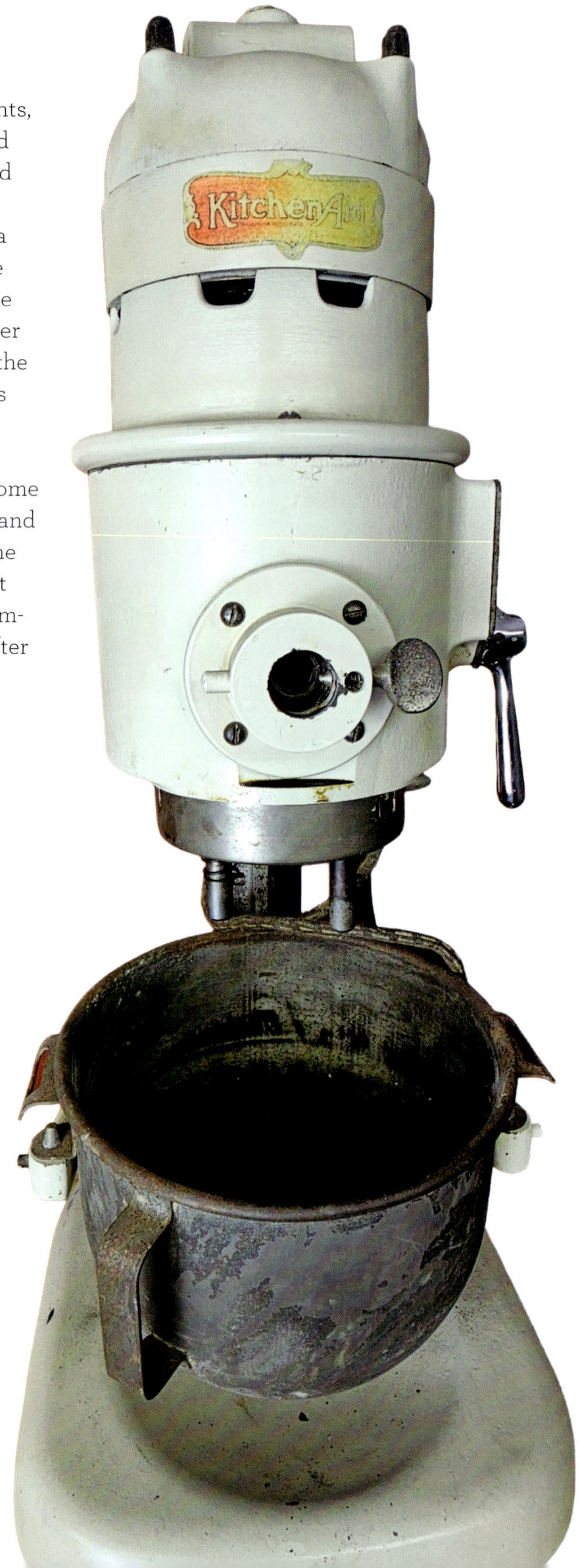

The Kitchenaid H-5 mixer, introduced in 1922.

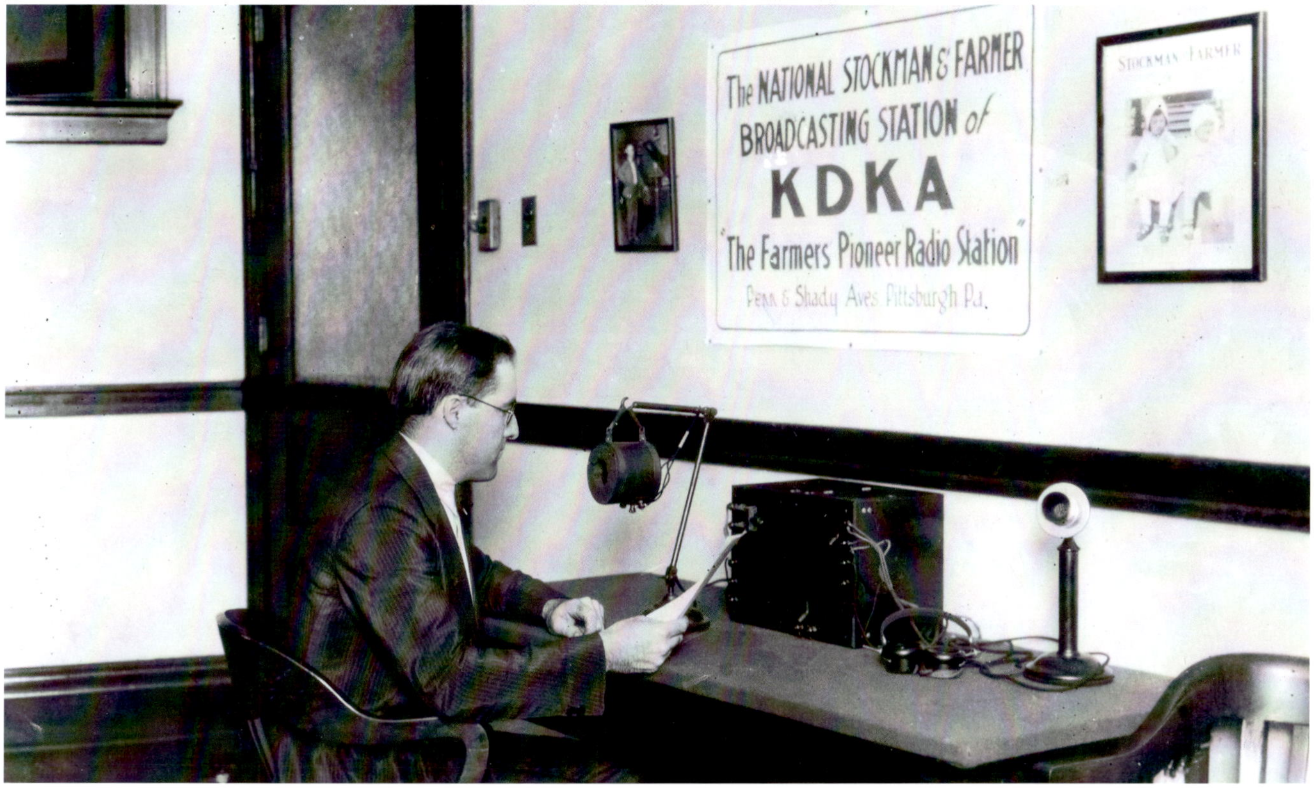

Broadcaster Frank Mullen reads his copy on the air at KDKA, c. 1922.

1920 FIRST COMMERCIAL RADIO STATION

In 1920, Westinghouse, a leading manufacturer of radios, conceived a revolutionary strategy for expanding its market reach: The company would produce programming to broadcast over radio waves. On November 2 of that year, this vision became reality when station KDKA went on the air. The timing proved fortuitous—election-day coverage allowed listeners to hear results of the Harding–Cox presidential race hours before morning newspapers could report them. KDKA's local success inspired other companies to enter radio broadcasting, marking the beginning of a media revolution. Radio fundamentally transformed American society by dismantling the geographical barriers that had confined news to local newspapers and community gossip. Radio created America's first truly shared national experience, enabling simultaneous participation in events across vast distances and forging a collective consciousness that crossed regional boundaries. The medium established advertising as a dominant economic force, granting businesses unprecedented direct access to consumers in their homes. Politicians, entertainers, and social activists could now communicate directly with audiences without intermediaries. This inaugural broadcast represented more than a technological milestone—it catalyzed the emergence of mass media, transformed journalism into its modern form, and helped give birth to popular culture as we know it today.

1921 BETTY CROCKER

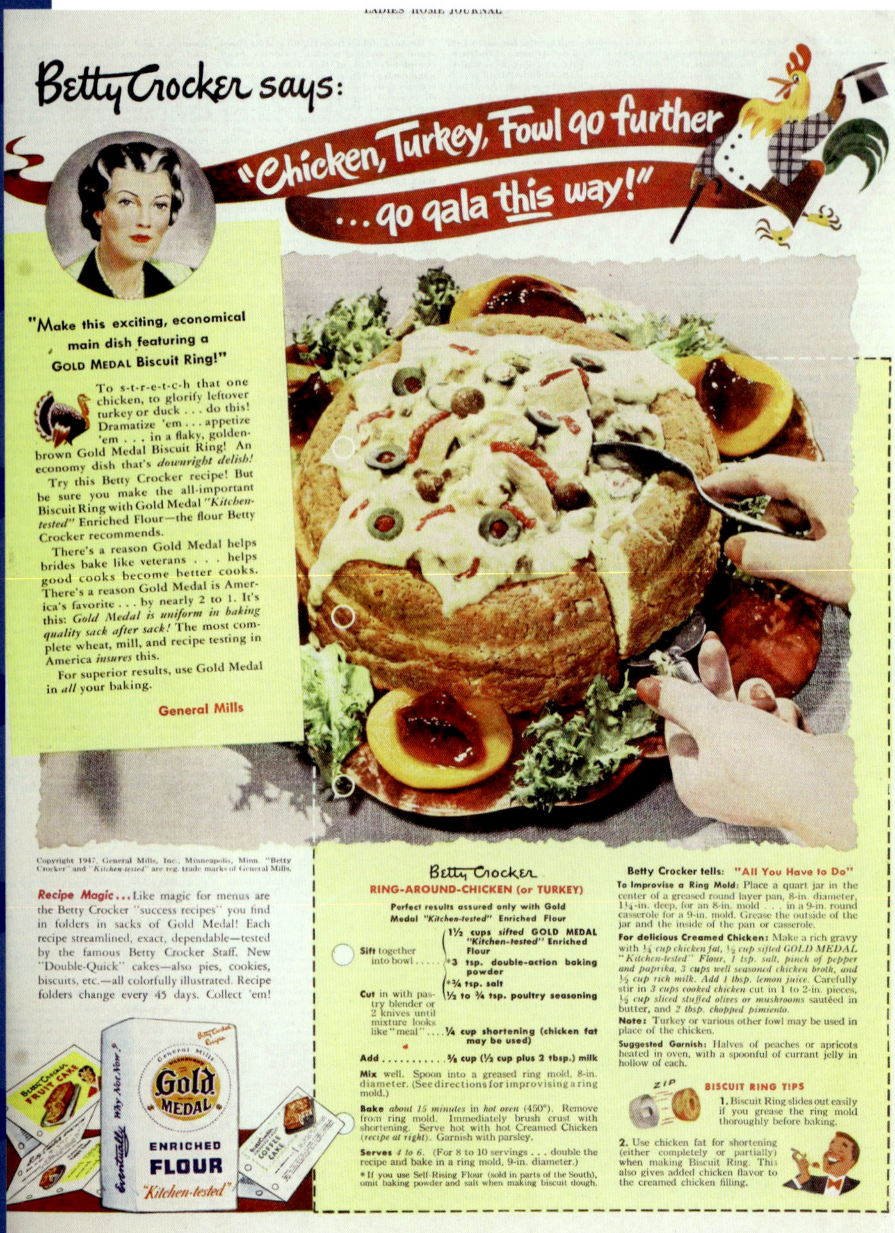

A Betty Crocker recipe ad, chock full with her signature tips, from a 1948 *Ladies Home Journal*.

Created by the company that would become General Mills in 1921 as a fictional spokesperson, Betty Crocker became one of America's most trusted household names. The persona was invented when the company found itself overwhelmed with questions from home cooks seeking baking advice after a promotional contest generated an unexpected volume of responses. Popular lore states that "Betty" was chosen for its friendly appeal, whereas "Crocker" honored a former company executive. During the Great Depression and World War II, Betty provided guidance to American homemakers facing rationing and food shortages. Her radio programs and recipe booklets helped families stretch budgets while maintaining nutritional standards, establishing her as an essential wartime resource for countless households. Betty Crocker's cultural impact extended far beyond cooking instructions. She embodied evolving ideals of American womanhood, initially representing traditional domesticity but gradually adapting to reflect changing attitudes toward women's roles in society. Her image was periodically updated to mirror contemporary demographics and social expectations, ensuring her relevance across decades. Today, Betty Crocker's face continues to appear on cake mixes, cookbooks, websites, and countless other products. Her enduring presence demonstrates how this fictional character transformed American eating habits and family dynamics for generations—and continues to do so in the modern era.

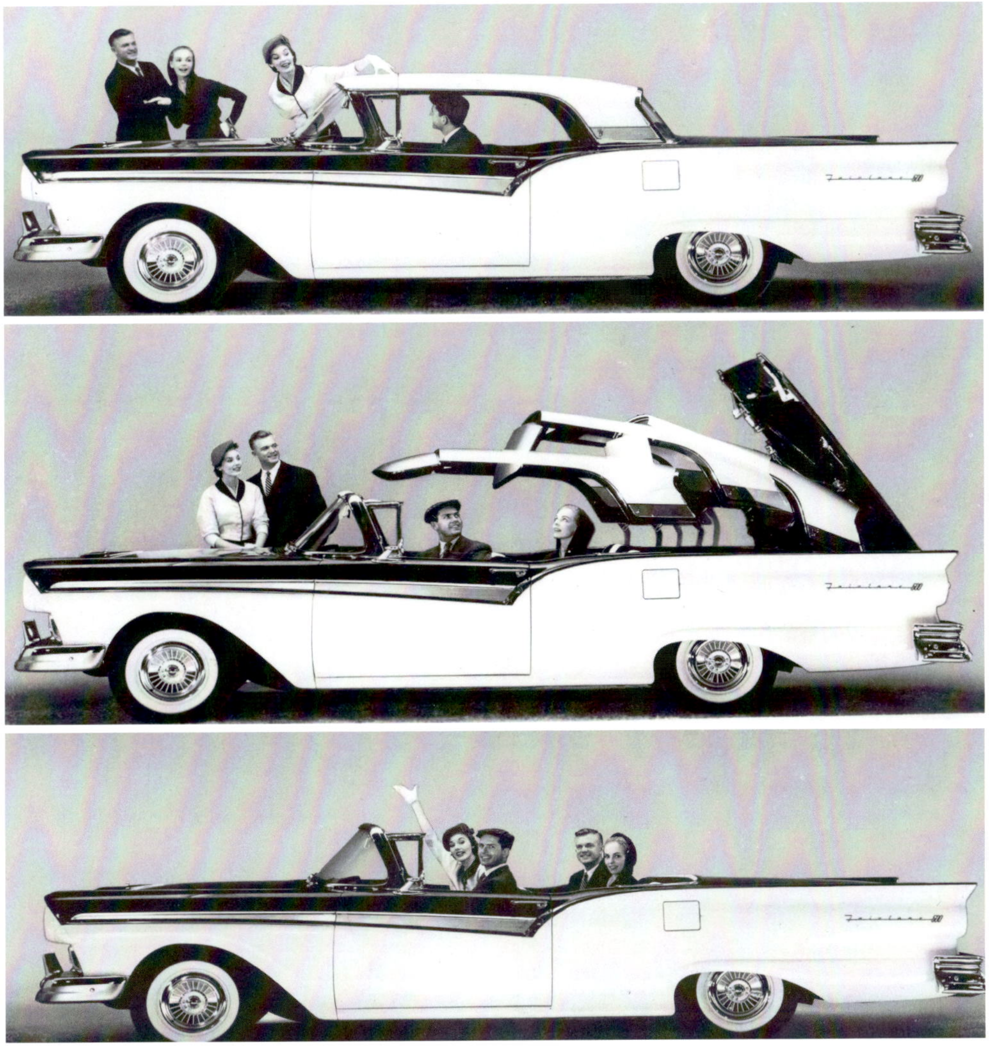

Hardtop convertibles hit their zenith with the 1957 Ford Fairlane 500 Skyliner.

1922 THE CONVERTIBLE

In 1922, Ben P. Ellerbeck (birth and death dates unknown) modified a 1919 Hudson coupe to feature a manually retractable convertible hardtop for the Automobile Body Builders Exhibition in New York City. Though he was credited with inventing the first convertible hardtop and received the first patent for such a design, his innovation was never put into production. Within a few years, however, hardtop convertibles were being manufactured by other American automakers including Buick, Cadillac, Chrysler, and Lincoln. Convertible cars quickly became powerful symbols of freedom, prosperity, and America's love affair with automotive innovation. By the 1930s, convertibles had gained widespread popularity and represented luxury and leisure in an increasingly mobile society. During the economic boom following World War II, they evolved into iconic symbols of American prosperity and optimism, embodying the carefree spirit that defined the 1950s and 1960s. Hollywood films and advertising campaigns celebrated these vehicles as emblems of youth culture and the emerging suburban lifestyle. Though today's convertibles are primarily associated with warm-weather locales and special occasions—appearing at homecoming processions and holiday parades—Ellerbeck's original invention remains a lasting symbol of American automotive ingenuity and the innovative spirit that has driven the industry forward.

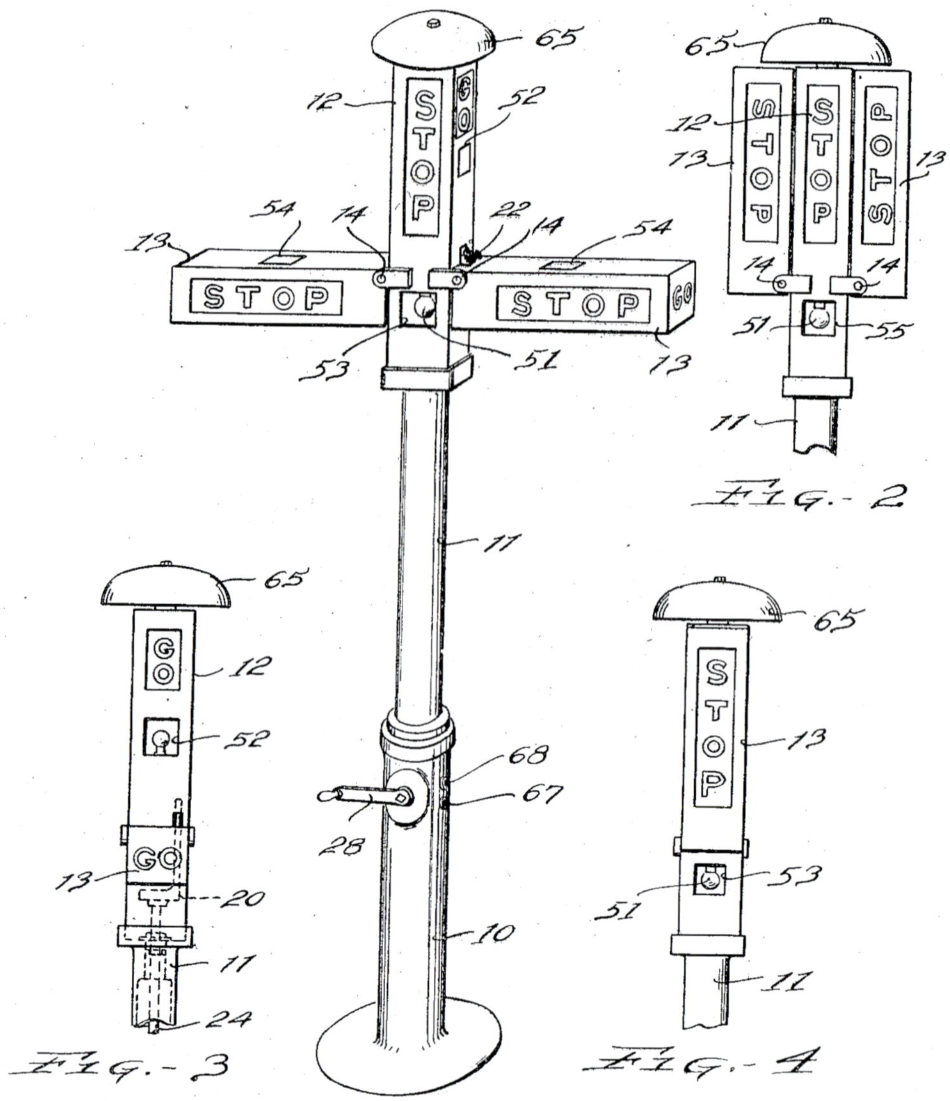

1923 THE TRAFFIC SIGNAL

The first electric three-way traffic signal was patented in 1923. It was invented by machinist, tailor, and entrepreneur Garrett A. Morgan (1877–1963) after he witnessed a collision at a Cleveland intersection. Morgan set about creating an improved traffic signal (at the time they consisted of just two signals—go and stop) to better control traffic on the congested city streets. The resulting T-shaped light added a "caution" light (the equivalent of the yellow light on modern traffic signals) that served to insert a buffer zone of time to help prevent incidents—vehicles already in the intersection were permitted to pass before new cars or carriages would proceed. Morgan ultimately sold his patent rights to General Electric for $40,000 (around $750,000 in today's dollars). The significance of Morgan's traffic signal goes beyond mere safety—it introduced Americans to automated regulation of daily behavior. In addition, the simple go-caution-stop system became a universal language, teaching citizens to accept technological authority and systematic rules around civic life.

The patent drawing for the traffic signal.

1924 KLEENEX

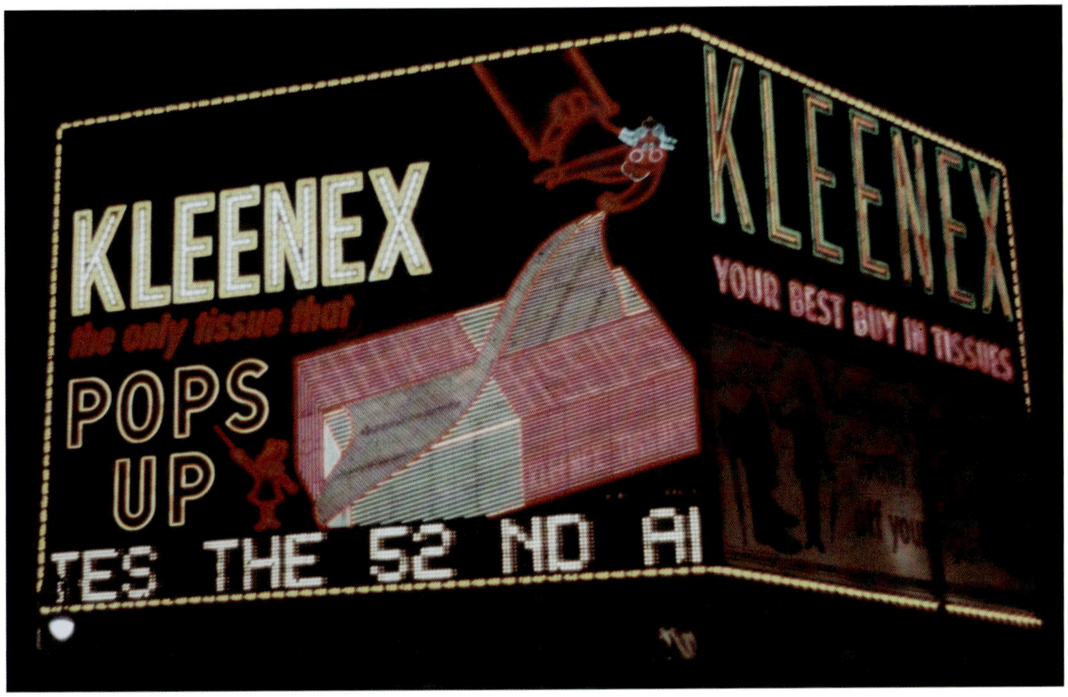

Kleenex "pops up" on a neon sign in Times Square, c. 1957.

In 1924, Kleenex, manufactured by the Kimberly-Clark Corporation, invented the facial tissue category, launching what would become one of the most successful brands in consumer history. Prior to its facial-tissue notoriety, however, a predecessor of the Kleenex brand had been used during World War I as a crepe paper filter for gas masks. Initially, Kleenex gained market share as a means to remove cold cream. In 1925, the company featured Kleenex tissues in advertisements with Hollywood actresses removing their makeup, cleverly positioning the product as a beauty accessory. The company then pivoted to the product's now-familiar role as a disposable handkerchief, demonstrating their success in reaching new markets. In subsequent decades, Kleenex continued to innovate, primarily through packaging improvements and by adding softening agents or antibacterial properties to the tissues. Most significantly, perhaps, Kleenex achieved the rare distinction of becoming a genericized trademark—where the brand name becomes synonymous with the entire product category.

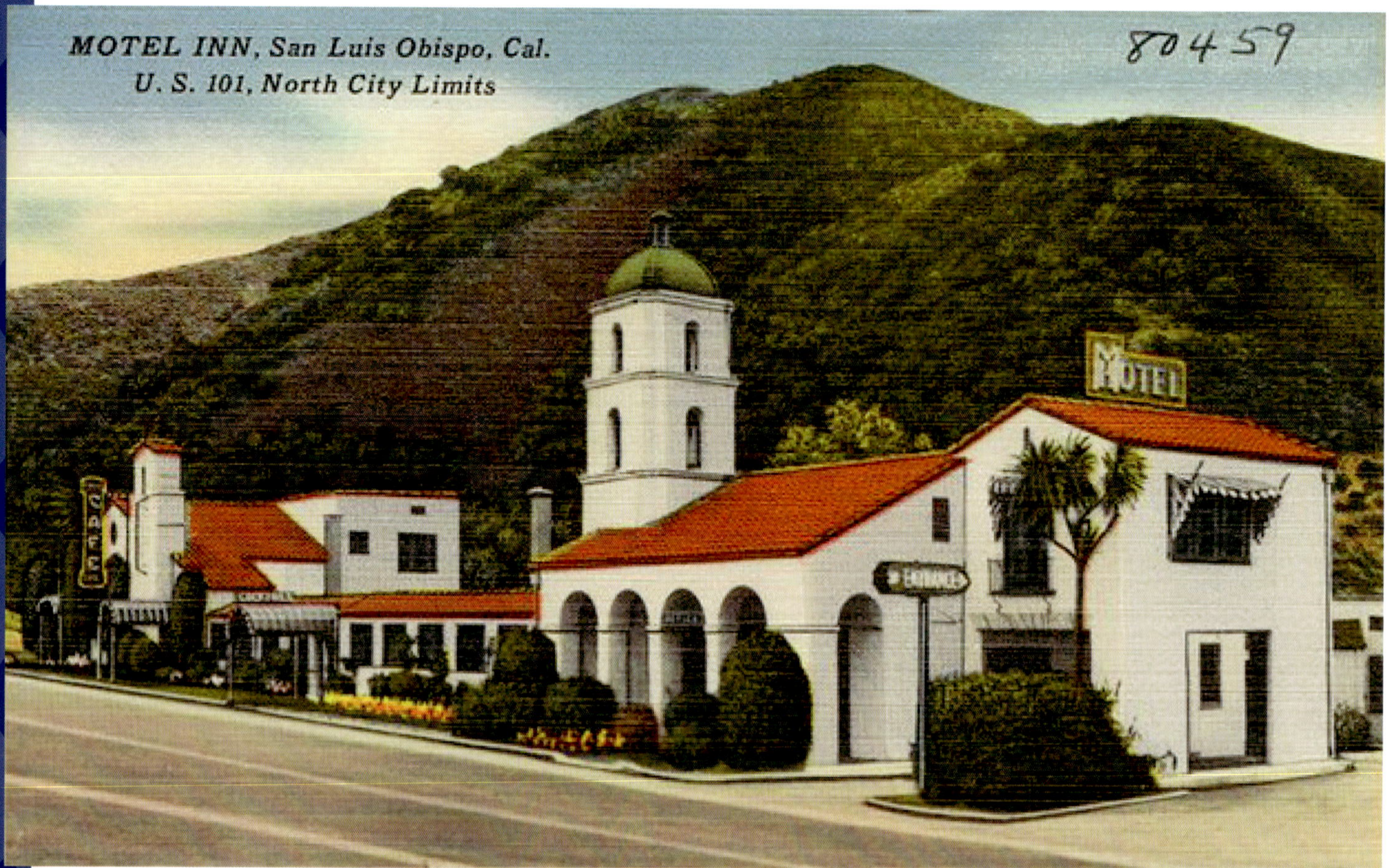

A postcard from the Milestone, c. 1925.

1925 THE MOTEL

The Milestone Mo-Tel, which opened on highway 101 in San Luis Obispo, California, in 1925, revolutionized travel in America by creating an entirely new lodging category. Designed by Pasadena, California, architect Arthur S. Heineman (1878–1972), the complex, which was built in the Spanish Revival style, combined the convenience of being able to drive right up to the door of the room with hotel amenities (the name was derived from combining the word "motor" with "hotel"). The impact of Heineman's motel was far-reaching. It offered affordable, convenient lodging for the growing number of Americans embracing automobile culture; unlike traditional hotels concentrated in city centers, the motels popped up along highways, which would change the landscape of American roadways. Within decades, motels dotted the country, becoming synonymous with American road culture, enabling family vacations and cross-country road trips, including the iconic Route 66 experience. Though the Milestone motel has long since closed, the industry it spurred grew to accommodate millions of travelers annually, fundamentally changing how Americans explored their country.

1926 LIQUID-FUELED ROCKET

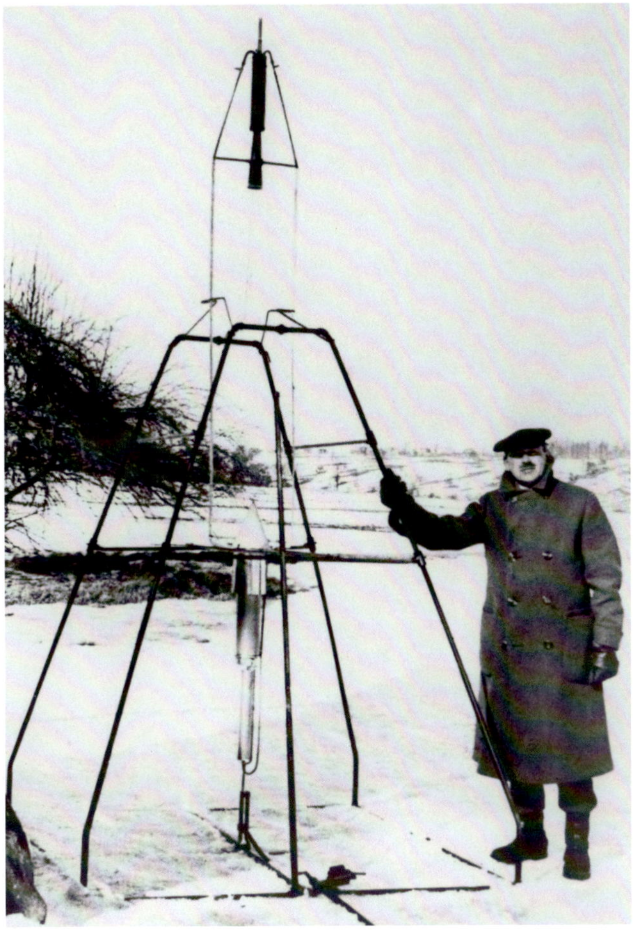

Goddard posing with his rocket, 1926.

On March 16, 1926, Robert H. Goddard (1882–1945) launched the world's first liquid-fueled rocket from a farm in Auburn, Massachusetts. Though the rocket reached just forty-one feet in altitude and flew for only two and a half seconds, the historic flight successfully demonstrated that liquid propellants could power rockets more efficiently than solid fuels. Like the Wright brothers' first airplane flight at Kitty Hawk, Goddard's achievement was a brief, seemingly modest moment that would ultimately change everything. His innovation proved that space travel was theoretically possible, transforming it from science fiction into scientific reality. The experiment showed that rockets could function in the vacuum of space, contradicting the widespread skepticism of scientists and the public who doubted such possibilities. With few witnesses present, Goddard kept his groundbreaking test secret for nearly a decade, concerned that others might steal his revolutionary technology. However, his liquid-fuel concept eventually became the foundation for future space exploration and sparked a technological revolution that continues to propel humanity toward the stars. Today, many regard Goddard as the father of modern rocketry for his pioneering contributions to space flight.

1926 Burma-Shave Road Signs

Road signs advertising Burma-Shave shaving cream first appeared on US Highway 65 near Lakeville, Minnesota, in 1926. Created by Allan Odell (1904–1994) to market the brushless shaving cream made by the Burma-Vita company, the advertisements were written across clusters of six signs spaced one hundred feet apart on stretches of rural roadways. They entertained weary motorists with clever rhyming wordplay that concluded with "Burma-Shave." The brilliant campaign transformed the struggling family business into a national success. The tradition, lasting until the early 1960s, revolutionized American roadside advertising and became an iconic part of the Great American road trip.

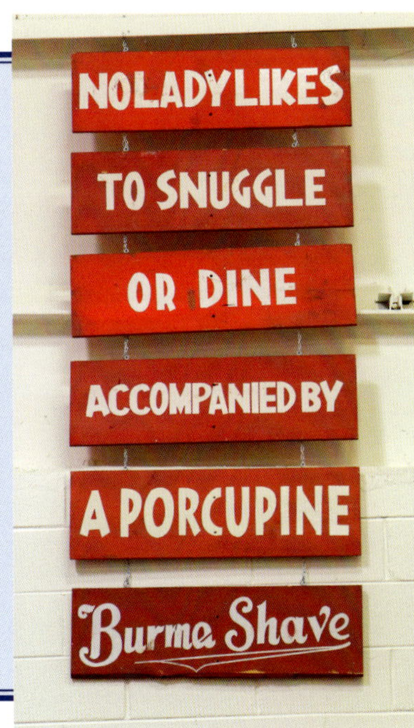

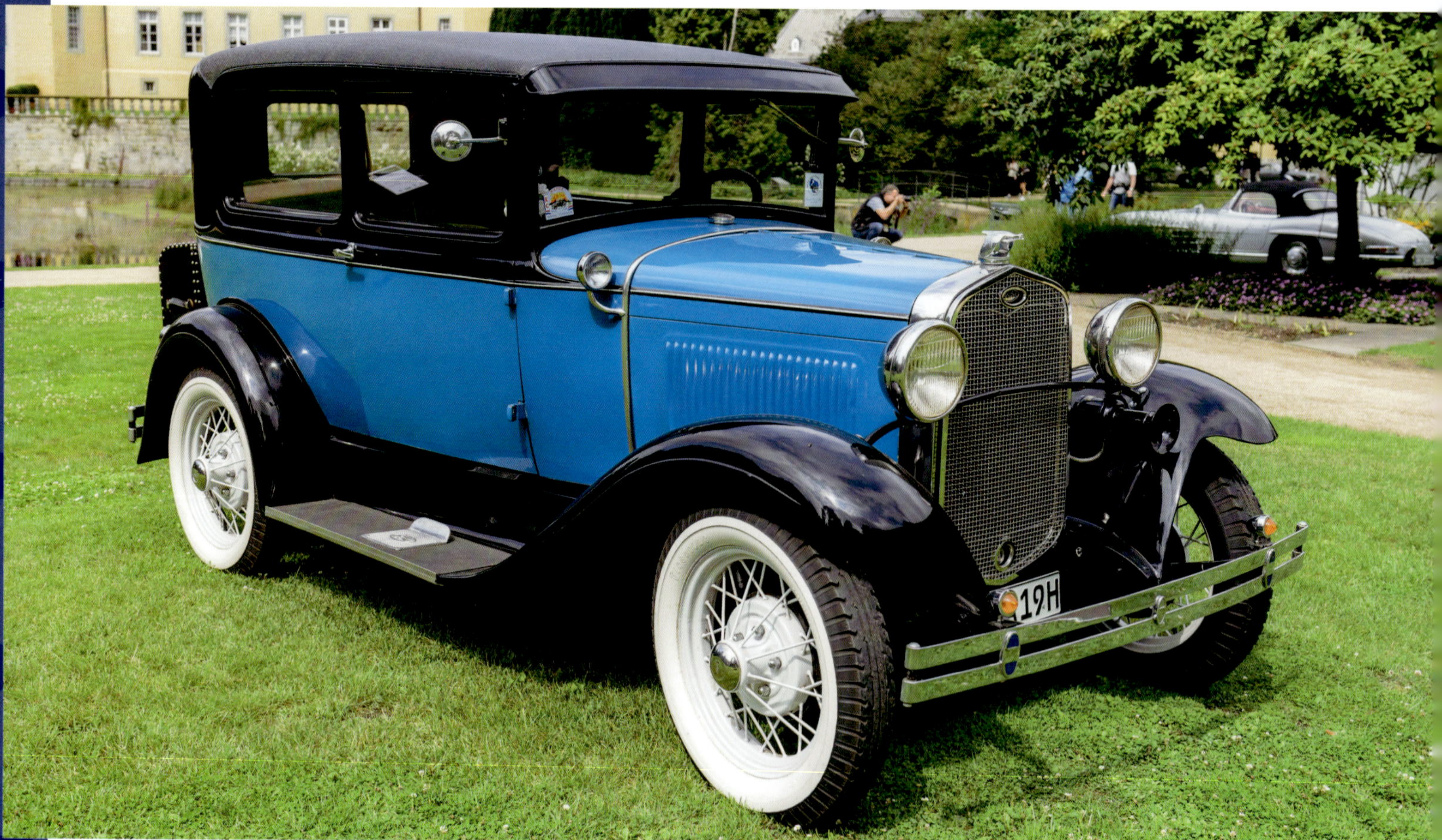

During its production, there were nine Model A body styles available; pictured above is the Tudor sedan body style.

1927 FORD MODEL A

American industrialist Henry Ford (1863–1947) is known for many things—inventing the assembly line, making cars more affordable, and the Model T automobile—but there is another Ford car that deserves recognition. The Ford Model A, launched in December 1927, represented a significant evolution in automotive design and manufacturing. At that time, the Model T's nineteen-year production run was ending, and the Ford company faced declining sales as consumers demanded more sophisticated features and styling. Other car manufacturers, especially General Motors, were taking a growing market share.

The Model A added modern amenities, including a conventional three-pedal system, a more powerful engine, hydraulic shock absorbers, and windshields made with safety glass. It also offered more choices in body style and color—no longer was black the only option. Even with all the options, the Model A was still competitively priced. Between 1927 and 1931, the Ford company produced more than four million Model A automobiles. The Model A is a symbol of Ford's ability to listen to consumer needs and maintain leadership during a crucial transition period in automotive history.

1928 BRASSIERE WITH CUP SIZES

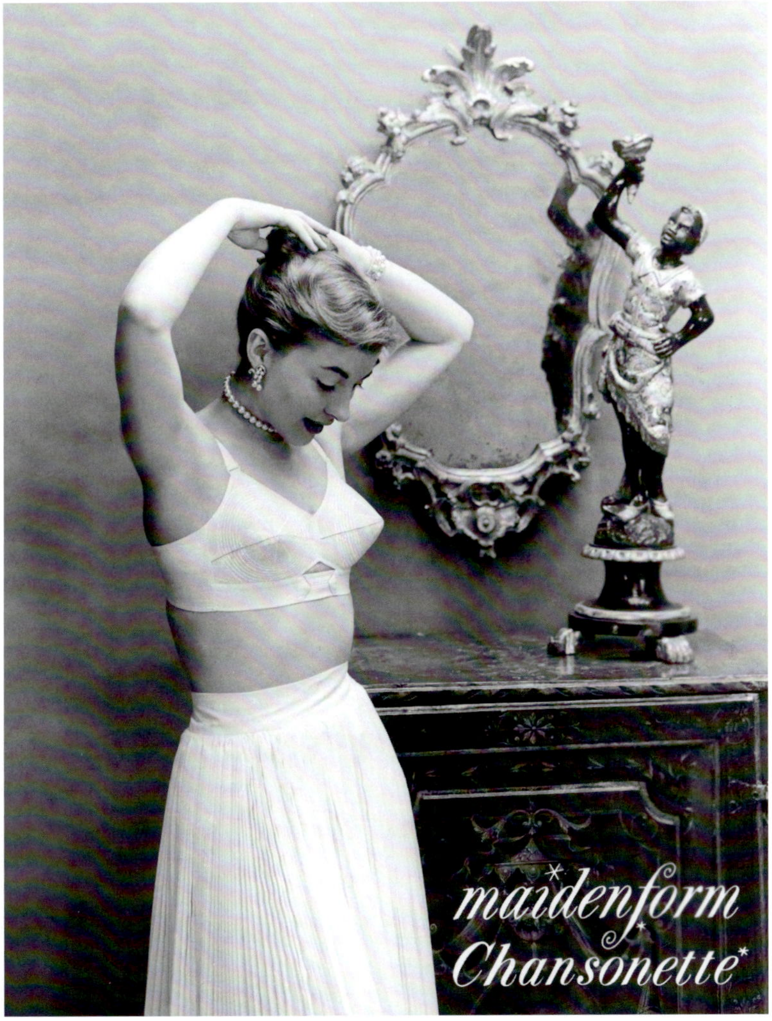

An ad for one of Maidenform's most popular brassieres, the Chansonette, c. 1949.

In the 1920s, women's brassieres had a simple, band-like construction with hooks in the back; they weren't particularly flattering to women's figures. Enter Ida Rosenthal (1886–1973), a Russian immigrant, and her business partner Enid Bisset, who together founded the Maidenform bra company in 1922 in Bayonne, New Jersey. The company was initially an offshoot of Enid's dressmaking company called "Enid Frocks." Rosenthal and Bissett began incorporating softly shaped bras into the dresses they were making to achieve a better fit. As these bras gained popularity, they transitioned from giving them away with each dress to selling them separately, eventually abandoning dressmaking entirely to focus on full-scale brassiere manufacturing in 1928. Maidenform pioneered the development of the modern bra and was the first to offer a truly fitted bra cup, revolutionizing women's undergarments. Ida Rosenthal is considered the creator of the modern bra, and her company became the most successful bra manufacturer in the world, producing and selling more bras than any other company for many years.

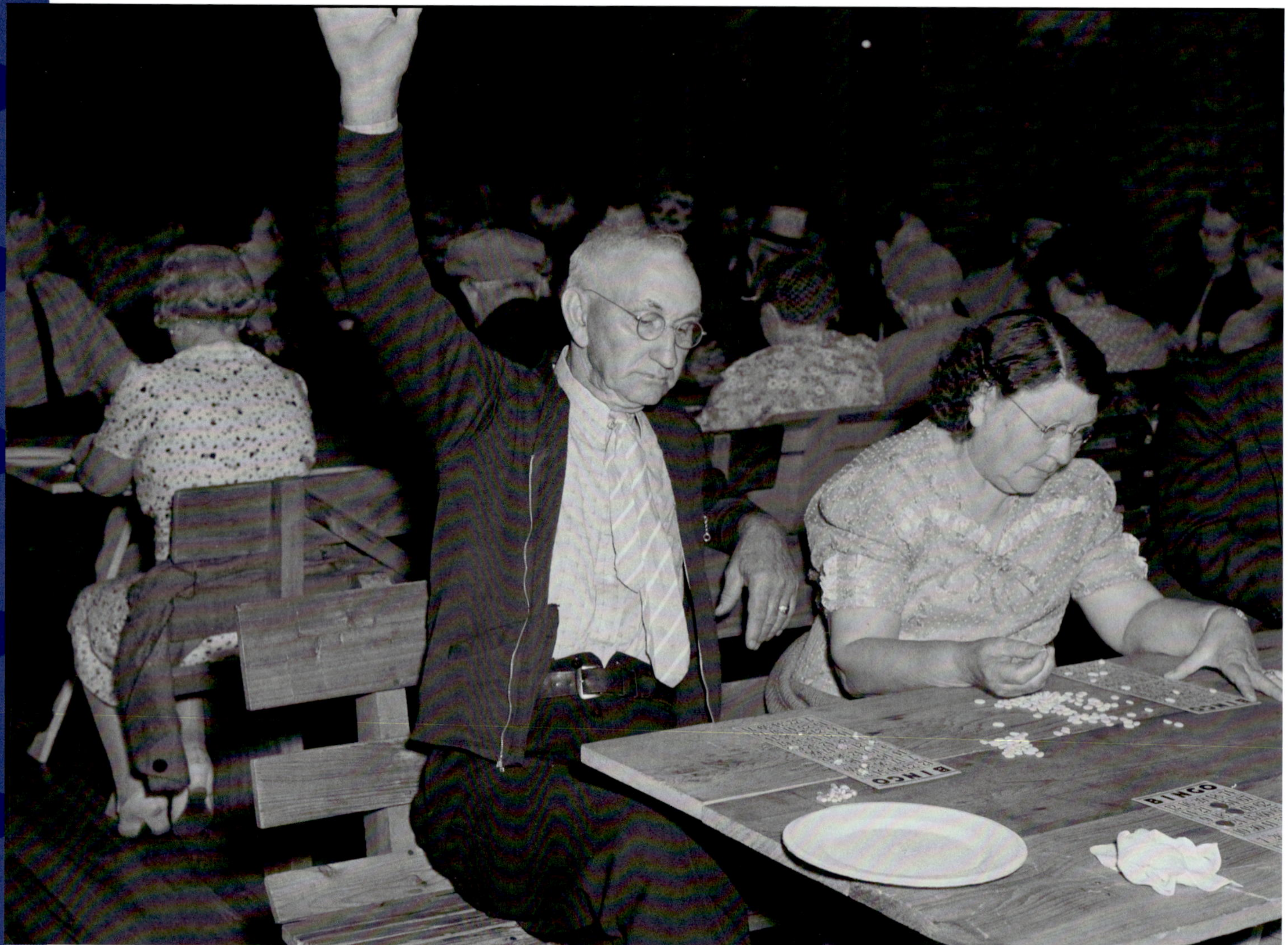

A man signals "bingo!" at a game in a Sarasota, Florida, trailer park, 1941.

1929 BINGO

The game of bingo evolved from a sixteenth-century Italian lottery game. The modern American version was created and popularized by New York toy salesman Edwin S. Lowe (1910–1986), who discovered the game at a carnival near Atlanta in December 1929. Players marked their cards for the game, originally called "Beano," with dried beans on cardboard sheets. Lowe brought the game back to New York and began hosting Beano parties. The name "bingo" allegedly originated when one of Lowe's friends accidentally shouted "Bingo!" instead of "Beano" upon winning, and Lowe thought the new name had a better ring to it. Due to his friends' enthusiastic response, Lowe began mass-producing the game. By the 1940s, bingo had spread throughout the United States, becoming a favorite offering for community fund-raising—the game fostered social interaction across diverse communities and age groups, creating a shared cultural experience. Bingo's simple rules and inclusive nature made it accessible to people regardless of education or social status, and its association with charitable causes gave it social legitimacy and purpose in American society.

1930 BIRDS EYE FROZEN FOODS

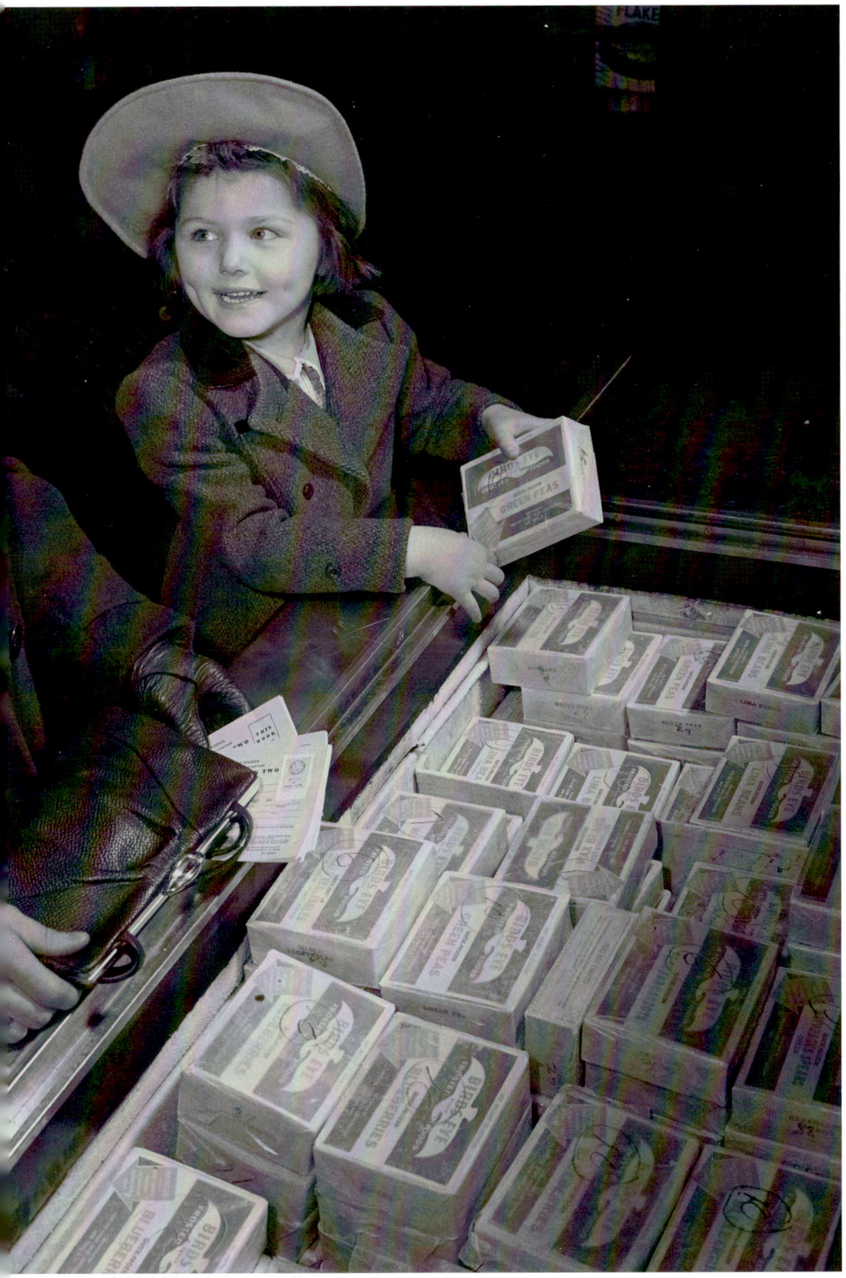

The first line of Birds Eye Frozen Foods was introduced at a market in Springfield, Massachusetts, on March 6, 1930, but the journey to market took several years of development. Clarence Birdseye (1886–1956), observing Inuit preservation techniques during his time in Labrador, noticed that rapidly frozen fish and caribou maintained better taste and texture than slowly frozen foods. He went on to invent a "quick-freezing" process that froze foods at extremely low temperatures between metal plates, preserving their cellular structure and nutritional value. This innovation solved the problem of ice crystals forming in slowly frozen foods, which damaged texture and flavor. Birdseye founded his company in 1922 and proceeded to refine his techniques. In 1929, he sold his company to General Foods for $22 million, a massive sum at the time. Though the products were slow to catch on initially, Birds Eye Frozen Foods transformed American food systems in multiple ways: they allowed farmers to preserve seasonal produce year-round, they changed retail with the introduction of frozen food sections in grocery stores, they altered family dynamics by reducing meal preparation time, and they enabled global food distribution. In short, Birdseye's innovation transformed the way Americans would eat for generations to come.

A mother during wartime holds her ration book as her daughter picks up some Birds Eye frozen vegetables, 1943.

1931 AIRSTREAM TRAILER

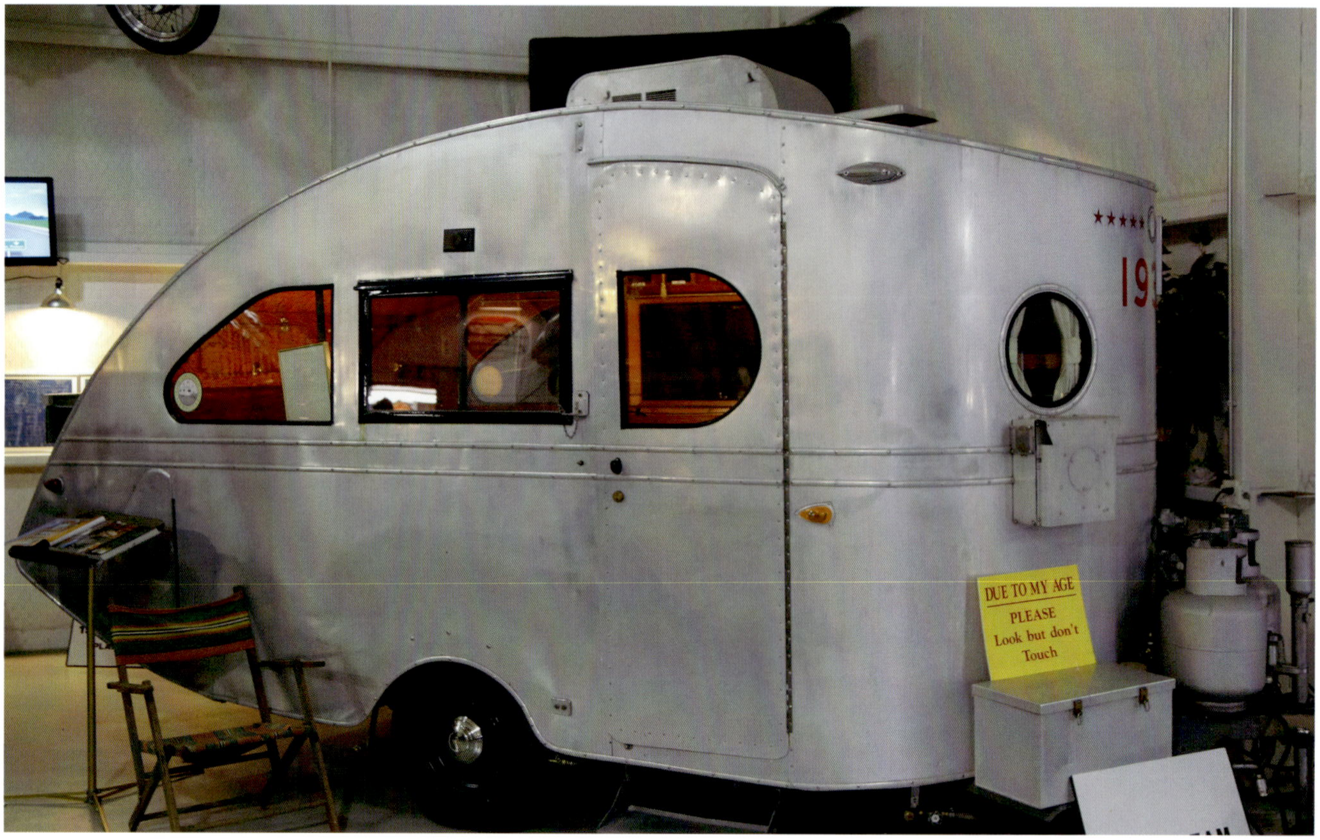

The Torpedo Cruiser was in production well into the 1940s; above is the 1935 version of the trailer.

In 1931, advertiser and publisher Wally Byam (1896–1962) came out with a design for the first Airstream trailer, "The Torpedo Car Cruiser." The inspiration came from frustration with existing travel trailers after a disastrous camping trip with his wife and experimentation with a tent and the chassis of a Model T. Byam's later designs featured the distinctive riveted aluminum construction and aerodynamic "silver bullet" shape that would become Airstream's signature aesthetic. During the Great Depression, Byam recognized that Americans still craved adventure despite economic hardship. The aluminum construction wasn't just stylish—it was also lighter and more durable compared to its wood-frame competitors. Byam's Torpedo offered consumers an affordable way to explore the country while maintaining homelike comfort. Within a year of its launch, more than a thousand Torpedos were on the road. Byam organized caravans that took owners on adventures worldwide, fostering a unique community culture that still persists today. The design has remained mostly unchanged since its inception, a testament to Byam's original vision. Today, the Airstream represents both nostalgia and luxury, with vintage models commanding premium prices while new ones continue to be manufactured. Airstream has managed to do what few other brands have done—maintain its status as an American icon.

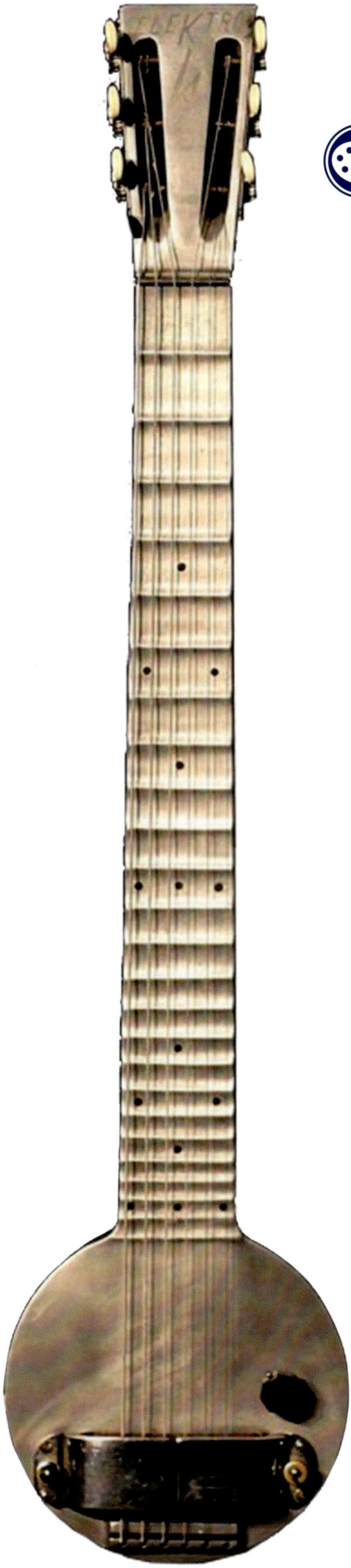

1932 THE ELECTRIC GUITAR

In the 1930s, loud jazz and big band music dominated popular entertainment, and musicians desperately needed to amplify their instruments for increasingly boisterous venues. Enter George Beauchamp (1899–1941) and Adolph Rickenbacker (1887–1976), who created the first commercially viable electric guitar: the legendary "Frying Pan." This revolutionary lap steel guitar earned its memorable nickname from its distinctive silhouette—a six-inch circular body attached to a long, straight neck that unmistakably resembled a cooking pan. Constructed from cast aluminum, the heavy, gleaming instrument looked and felt radically different from the traditional wooden guitars of the era. The body housed a groundbreaking electromagnetic pickup system that would define electric guitar technology for decades to come. This horseshoe-shaped pickup worked by encircling the strings and capturing their vibrations through electromagnetic induction. Players would rest the guitar horizontally across their lap, using a steel bar or slide to fret the strings while picking with their other hand—a technique that gave "steel guitar" its name. First available for purchase in 1932, the Frying Pan launched a musical revolution. The electric guitar's significance extends far beyond its technical innovation: it fundamentally transformed popular music, youth culture, and the music industry itself, cementing its place as one of the most culturally influential inventions of the twentieth century.

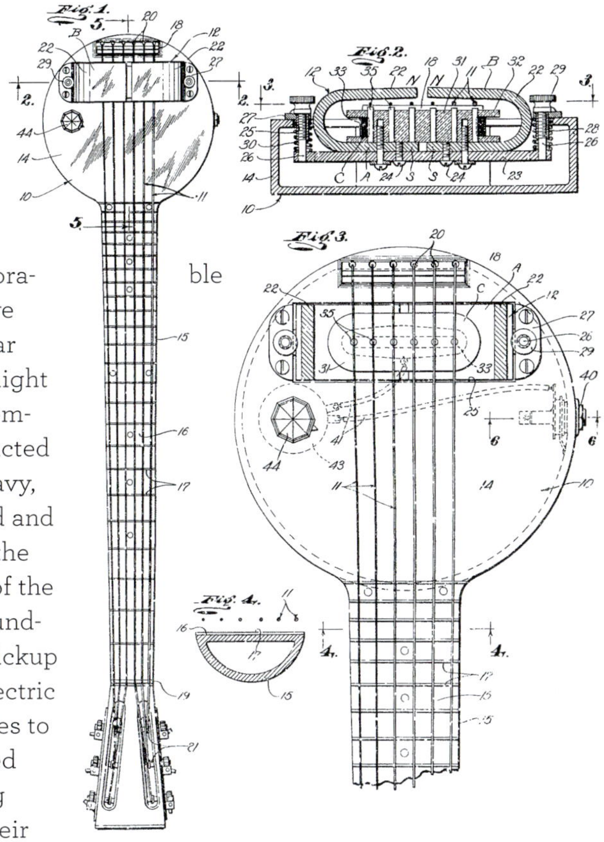

The Electro "Frying Pan."

1933 DRIVE-IN MOVIE THEATER

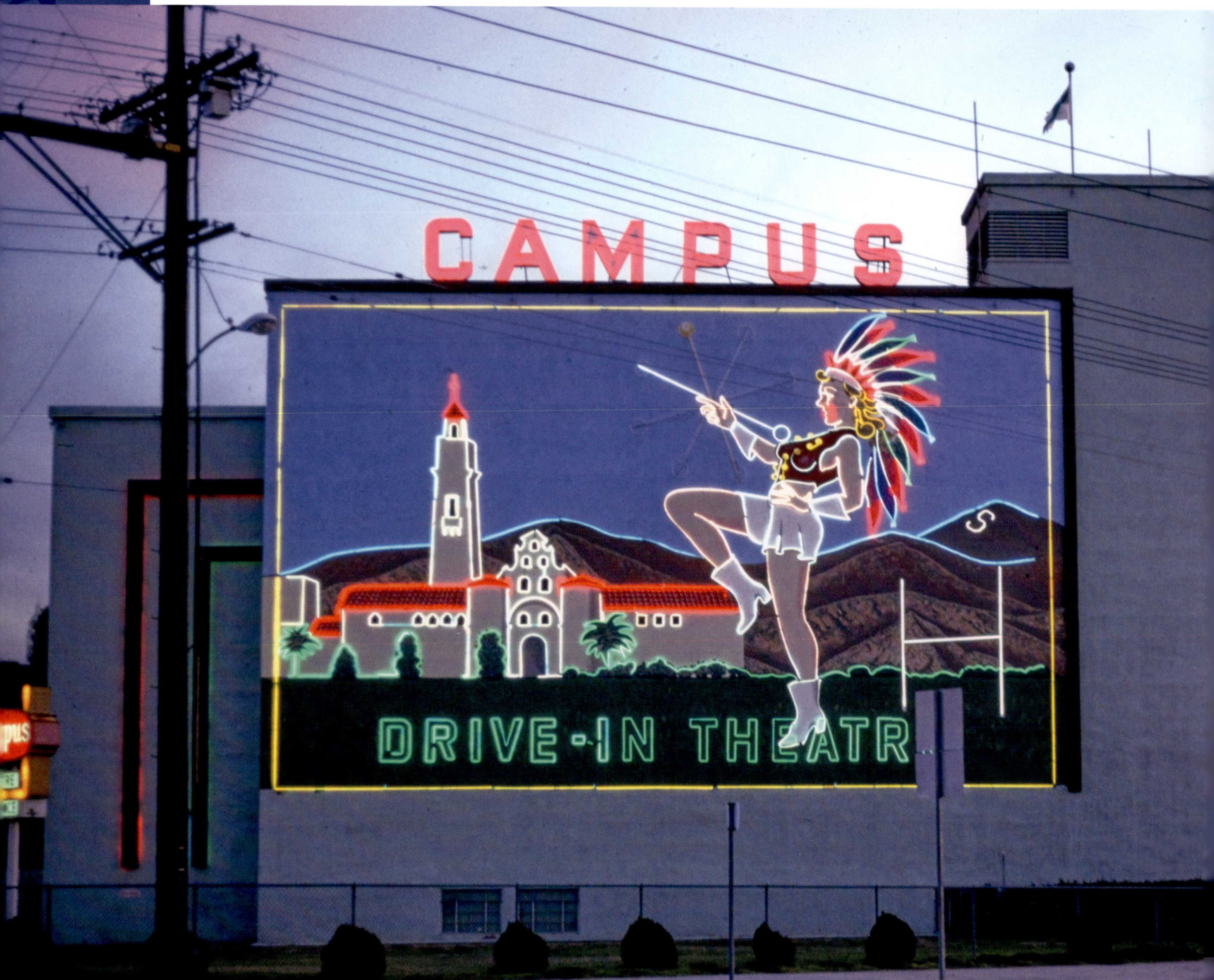

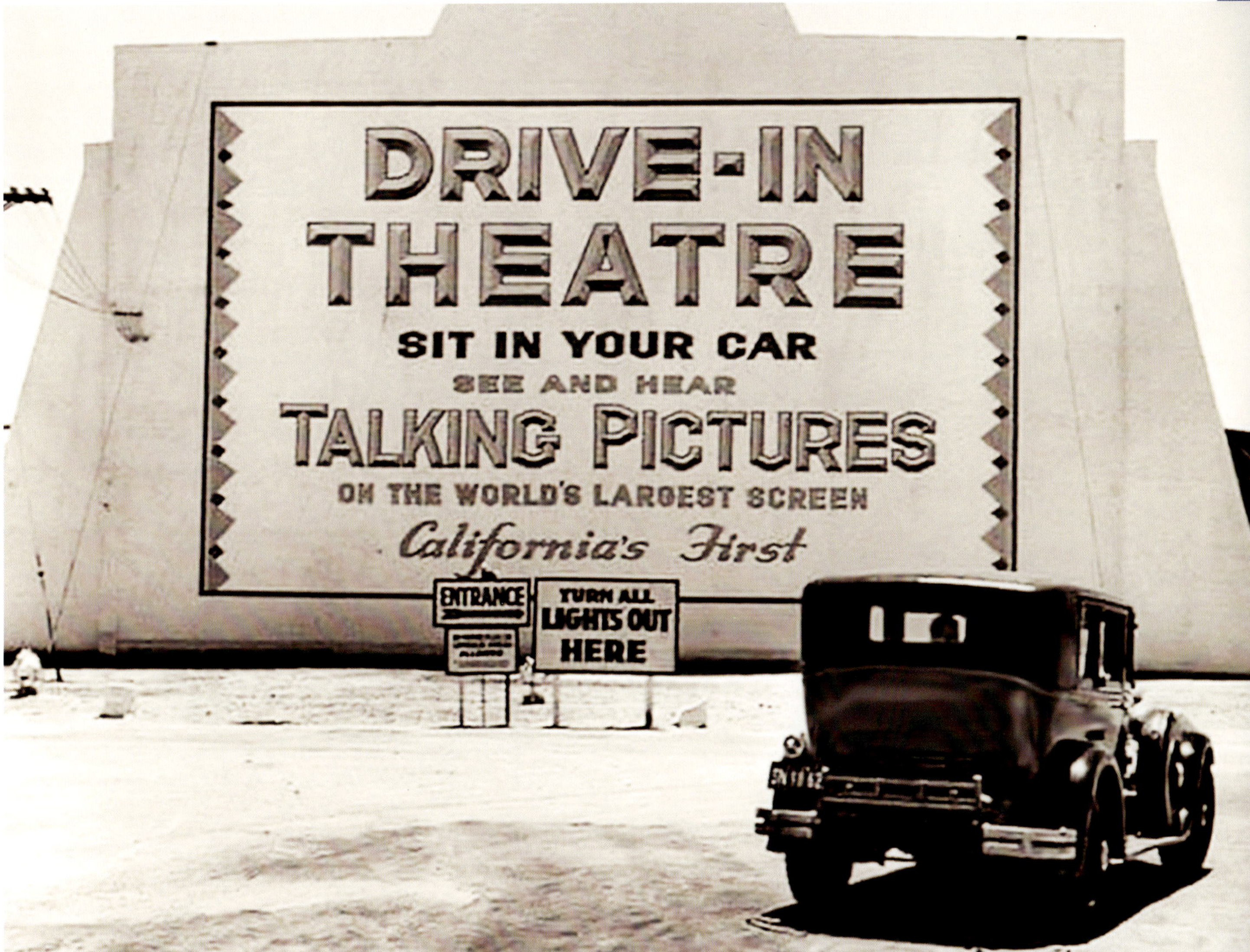

The first drive-in movie theater opened on June 6, 1933, in Camden, New Jersey. The idea for the theater came from entrepreneur Richard Hollingshead Jr. (1900–1975), who was frustrated by his mother's discomfort when visiting traditional theaters. To test the concept, Hollingshead experimented in his driveway with a projector, bedsheet screen, and his car radio for sound. He also tested various parking angles to ensure all cars would have clear views before opening his "Automobile Movie Theatre." The drive-in concept revolutionized entertainment by offering families an affordable, casual movie experience. Parents could bring their children in their pajamas, couples could enjoy privacy, and audiences could talk, smoke, or eat without disturbing others. Drive-ins exploded in popularity after World War II, peaking in the late 1950s with more than four thousand locations nationwide. Drive-in movie theaters became cultural icons representing American car culture and suburban expansion. Though the number of drive-ins eventually declined due to urban development, rising land costs, and new ways of enjoying movies, drive-in theaters remain nostalgic symbols of twentieth-century America.

LEFT: The Campus Drive-In in San Diego, built in 1948. **ABOVE:** California's first drive-in, the Pico, in Los Angeles, c. 1934.

1934 *FLASH GORDON* COMIC STRIP

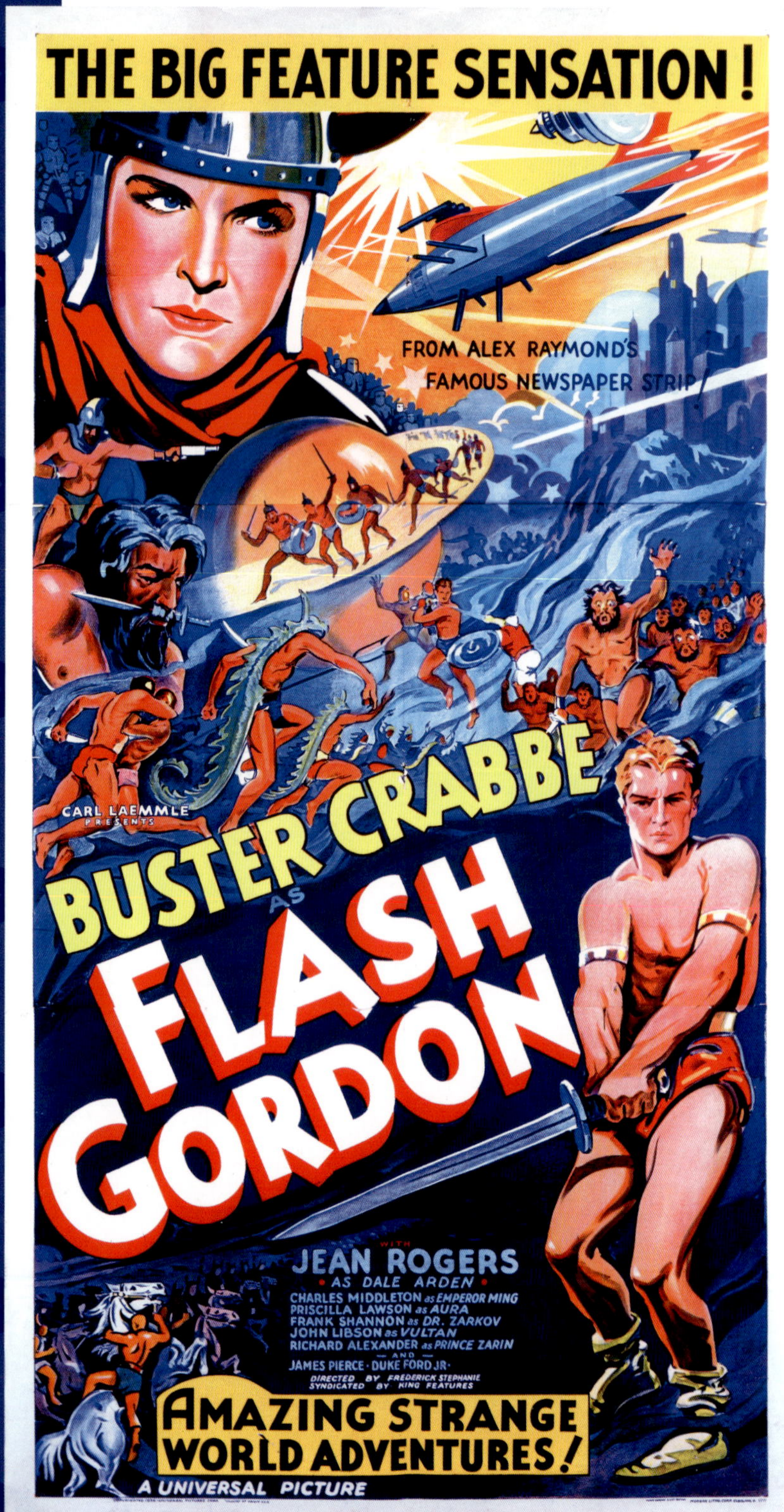

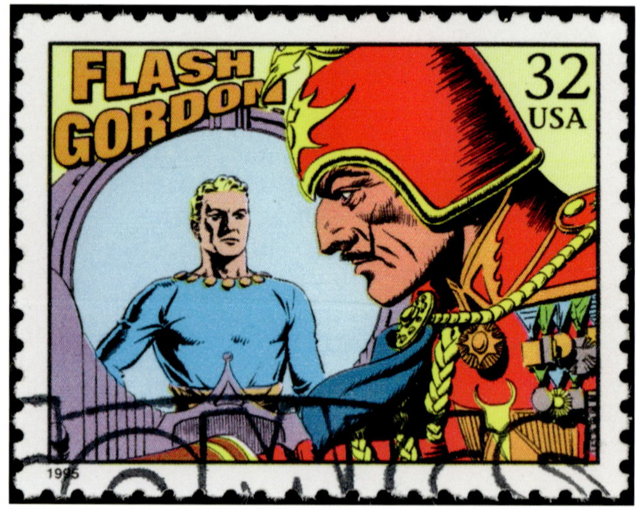

The *Flash Gordon* comic strip, created by Alex Raymond (1909–1956), debuted on January 7, 1934, in King Features Syndicate newspapers. Raymond was one of the finest comic strip illustrators of his era, and his superior artwork—combined with ghostwriter Don Moore's storylines—meant Flash Gordon quickly outshone its rival, the more established comic strip Buck Rogers. Raymond's detailed, cinematic art style brought new dimension and sophistication to newspaper comics. He offered lush alien landscapes, dramatic space battles, and exotic characters that captured readers' imaginations during the Great Depression. Raymond drew Flash Gordon and other characters such as Jungle Jim and Secret Agent X-9 during his prolific years, but in 1944 he gave upillustrating to enlist in the Marines during World War II. Raymond's Flash Gordon character had a long-lasting impact: It spawned radio shows, movies, and eventually television adaptations, making Raymond's creation one of the most enduring science fiction characters in popular culture.

LEFT: Flash first appeared on film in this 1936 feature. **ABOVE:** In 1995, Flash was honored with his own US stamp.

1935 MONOPOLY

Monopoly, one of the world's most popular board games, has a complex and ironic origin story. The game traces its roots to "The Landlord's Game," created by Lizzie Magie (1866–1948) to demonstrate the negative aspects of land monopolism. During the Great Depression, Charles Darrow (1889–1967) came across a modified version, refined it, and sold the rights to Parker Brothers in 1935. What makes the game's history particularly fascinating is its transformation of purpose: a game originally intended as a critique of capitalism evolved into a celebration of wealth accumulation. Since its commercial launch, Monopoly has become a global phenomenon, spawning localized versions featuring properties from cities worldwide—from London's Mayfair to Tokyo's Ginza districts. Monopoly serves as many people's first introduction to real estate investment, banking operations, and strategic economic thinking. The game's enduring popularity across generations demonstrates how games can both reflect and shape societal values. Despite the digital revolution, Monopoly has successfully adapted while maintaining its core identity. Today, Monopoly remains a cultural touchstone, spawning themed editions—from Star Wars to Disney—while preserving its appeal as family entertainment and accessible capitalism simulator. The game's journey from social commentary to mainstream success perfectly encapsulates how cultural artifacts can transcend their original intentions.

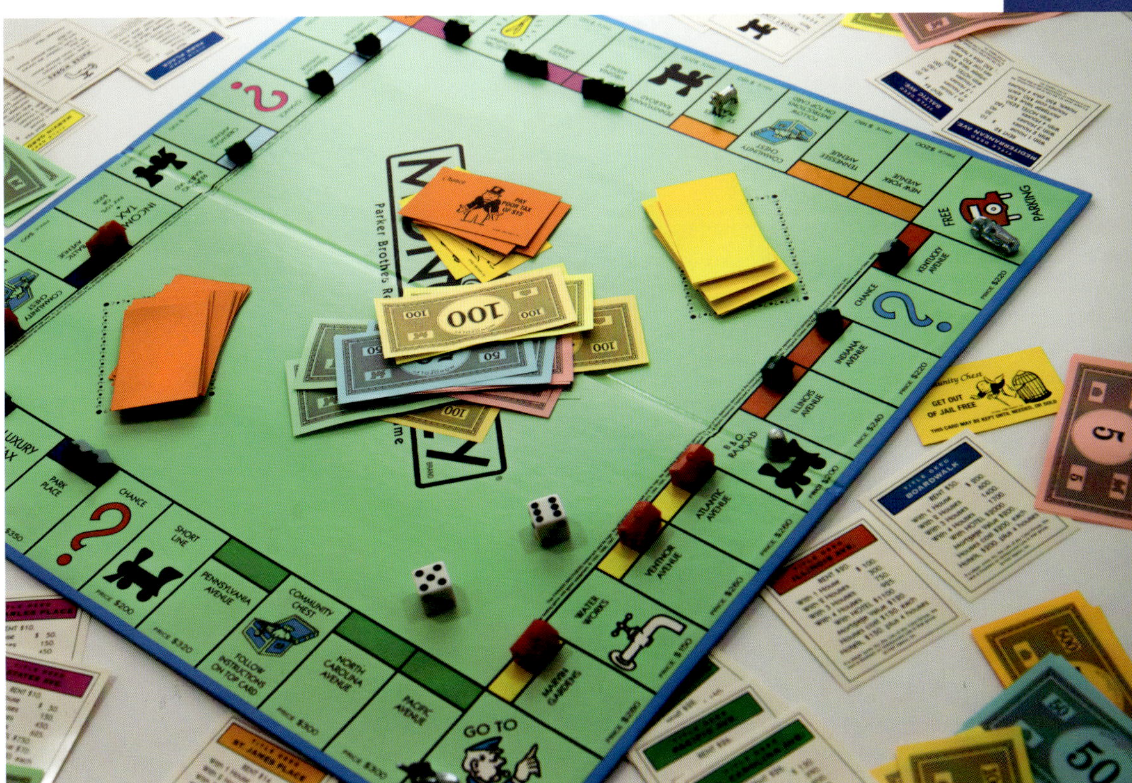

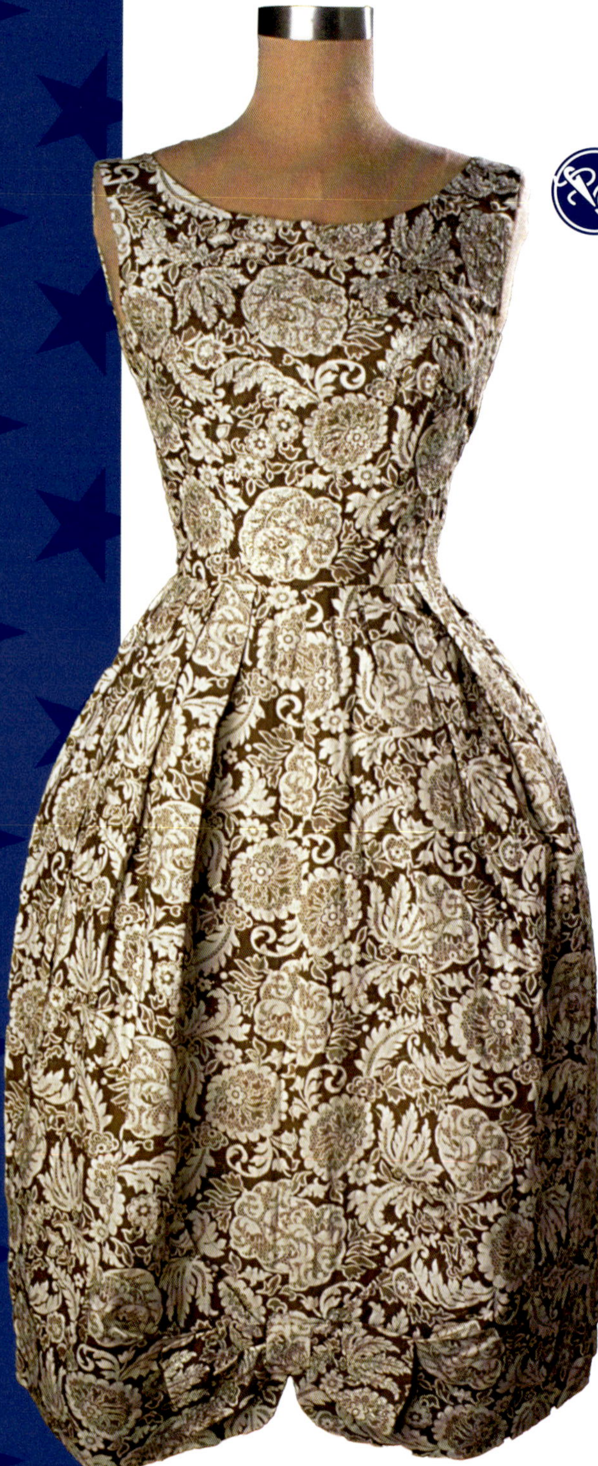

1936 FLOUR SACK DRESS

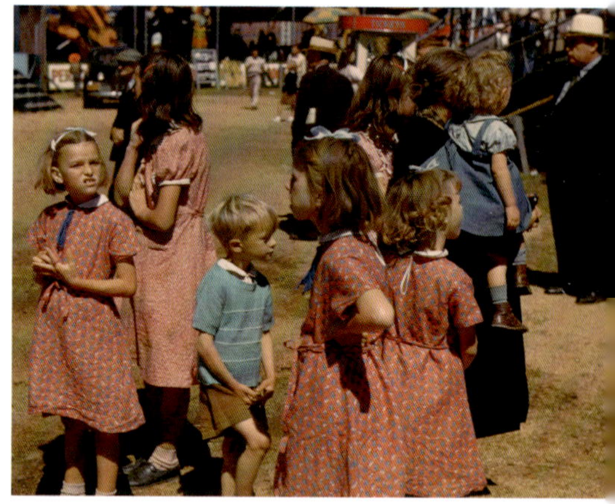

During the 1930s, flour, sugar, and feed were commonly sold in large cotton sacks. Frugal families discovered these sturdy vessels could be repurposed into clothing, with dresses made from flour sacks becoming particularly significant in rural communities where money was scarce but resourcefulness flourished. Women developed an art form around this necessity: They would carefully select sacks with complementary patterns; painstakingly bleach out company logos; and transform the fabric into everyday dresses, children's clothing, and household items such as curtains and towels. By 1936, choosing attractive flour sack patterns had become an integral part of the grocery shopping experience. This practice embodied both economic necessity and creative adaptation during one of America's most challenging periods. Manufacturers quickly recognized the trend and began producing sacks with attractive floral designs and vibrant patterns, essentially creating a secondary market for their packaging materials. The flour sack dress transcended mere practicality to symbolize the "make do and mend" philosophy that defined Depression-era America. These garments represented dignity in adversity and the transformative power of creativity when resources were limited. Today, flour sack dresses are celebrated as early examples of sustainable fashion and cherished artifacts of American folk culture, reminding us how necessity can spark innovation and beauty can emerge from unlikely sources.

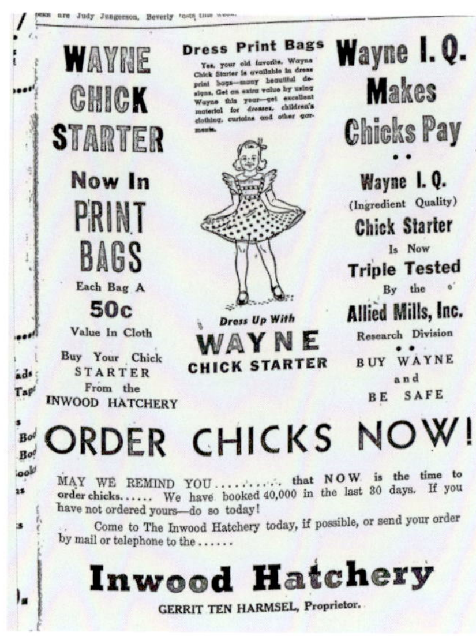

WAYNE CHICK STARTER

Now In PRINT BAGS

Each Bag A 50c Value In Cloth

Buy Your Chick STARTER From the INWOOD HATCHERY

Dress Print Bags

Yes, your old favorite, Wayne Chick Starter is available in dress print bags—many beautiful designs. Get an extra value by using Wayne this year—get excellent material for dresses, children's clothing, curtains and other garments.

Dress Up With WAYNE CHICK STARTER

Wayne I. Q. Makes Chicks Pay

Wayne I. Q. (Ingredient Quality) Chick Starter Is Now Triple Tested By the Allied Mills, Inc. Research Division

BUY WAYNE and BE SAFE

ORDER CHICKS NOW!

MAY WE REMIND YOU .... that NOW is the time to order chicks...... We have booked 40,000 in the last 30 days. If you have not ordered yours—do so today!

Come to The Inwood Hatchery today, if possible, or send your order by mail or telephone to the ......

Inwood Hatchery

GERRIT TEN HARMSEL, Proprietor.

CLOCKWISE FROM LEFT: A chicken feed company advertising their printed dress bags in the local newspaper, 1945; a feedsack dress made by Mrs. G. R. Overall of Caldwell, Kansas, that took second place in the National Cotton Council's Cotton Bag Sewing Contest, 1959; sisters wear their matching flour sack dresses to the Vermont state fair, 1941.

1937 CHAIR LIFT

The chairlift was invented by James Curran (1903–1968), a Nebraska-born railroad engineer. First installed at Sun Valley Resort in Idaho, the lift was introduced for the 1936–1937 season. Before chairlifts, skiers faced the arduous task of hiking uphill or relying on rope tows, which were both slow and physically demanding. Drawing inspiration from banana-loading systems used in tropical ports, Curran created an innovative transportation method that carried people in chairs suspended from an overhead cable twenty feet above the ground. The new system could transport four hundred skiers up the mountain per hour. The chairlift revolutionized skiing by making it accessible to a much broader audience and sparked the growth of ski resorts worldwide.

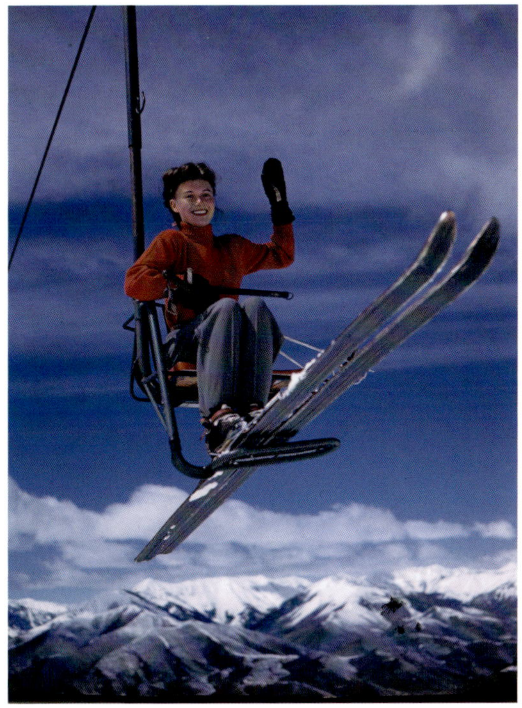

Actress Suzy Harris riding a chair lift, c. 1940.

1938 TOLL HOUSE COOKIES

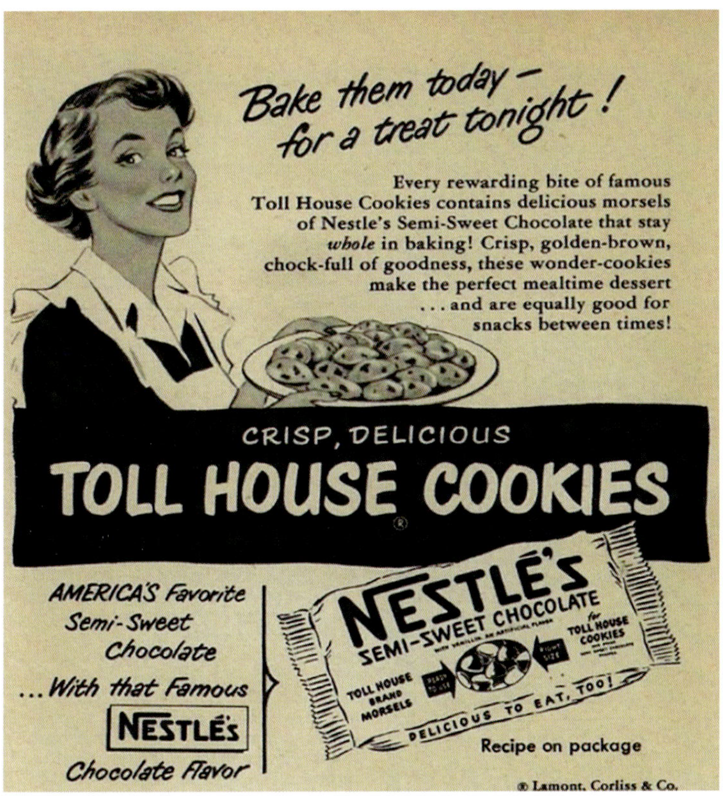

The chocolate chip cookie originated at the Toll House Inn, a roadside restaurant in Whitman, Massachusetts, owned by American chef Ruth Graves Wakefield (1903–1977). According to legend, Wakefield accidentally created the cookies when she added chopped chocolate to cookie dough, expecting it to melt throughout the mixture. When the chocolate retained its shape, the chocolate chip cookie was born. The story goes that Wakefield later sold her recipe to Nestlé for just one dollar and a lifetime supply of chocolate, allowing the company to print it on their semisweet chocolate chip packages. Wakefield's contributions were recognized posthumously when *The New York Times* featured her in their "Overlooked: Stories of Remarkable People" series.

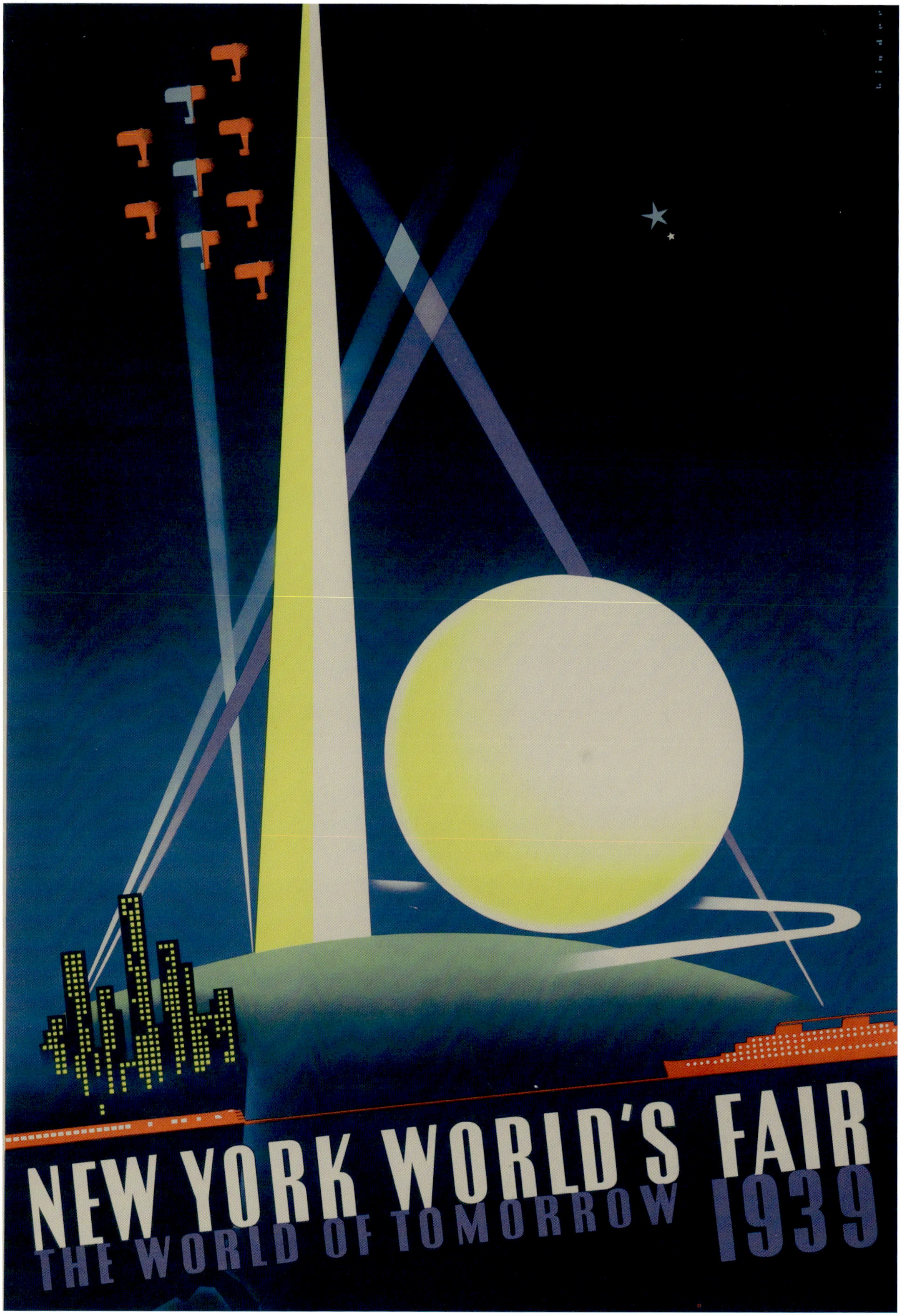
NEW YORK WORLD'S FAIR
THE WORLD OF TOMORROW
1939

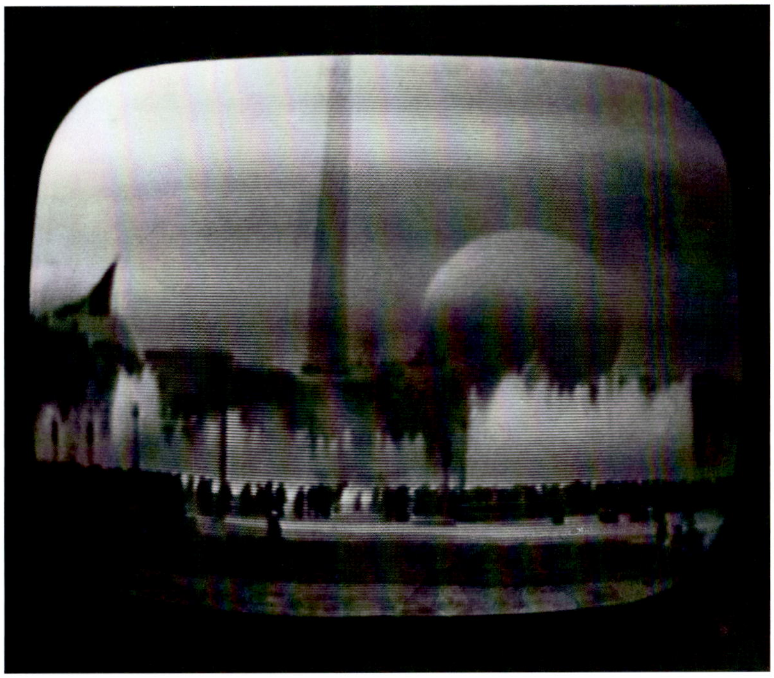

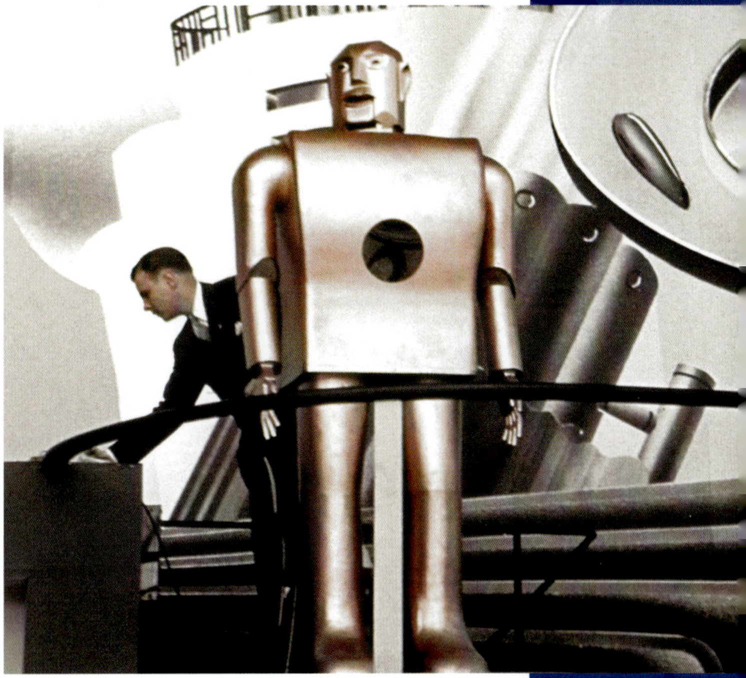

1939 WORLD OF TOMORROW

The 1939 New York World's Fair left a profound legacy despite its financial struggles. Themed "the World of Tomorrow," the fair transformed a former swamp and ash dump in Queens into what would become Flushing Meadows–Corona Park, providing New York with a valuable public space that continues to serve millions of visitors annually. The fair's greatest impact was its role as a showcase for emerging technologies that would reshape American life. New consumer products such as television, air-conditioning, nylon stockings, fluorescent lighting, color film, and even a crude fax machine were introduced to the public at the fair. General Motors' "Futurama" exhibit, which depicted the world of 1960, particularly captured visitors' imaginations and helped shape postwar suburban development patterns. Though declared an economic failure, the fair profoundly influenced consumer consciousness and left a lasting impact on American society, particularly in the role of mass-produced goods in daily life. The fair's vision of technological progress and modern living became a blueprint for the American dream in the decades that followed. With more than forty-five million visitors, it successfully planted seeds of optimism about the future just as World War II was beginning, making it a pivotal cultural moment in American history.

LEFT: The official poster for the fair, 1939. **ABOVE:** Two of the most impressive demonstrations at the fair were the very first television transmission (of the Trylon and Perisphere), and Elektro, the robot built by Westinghouse.

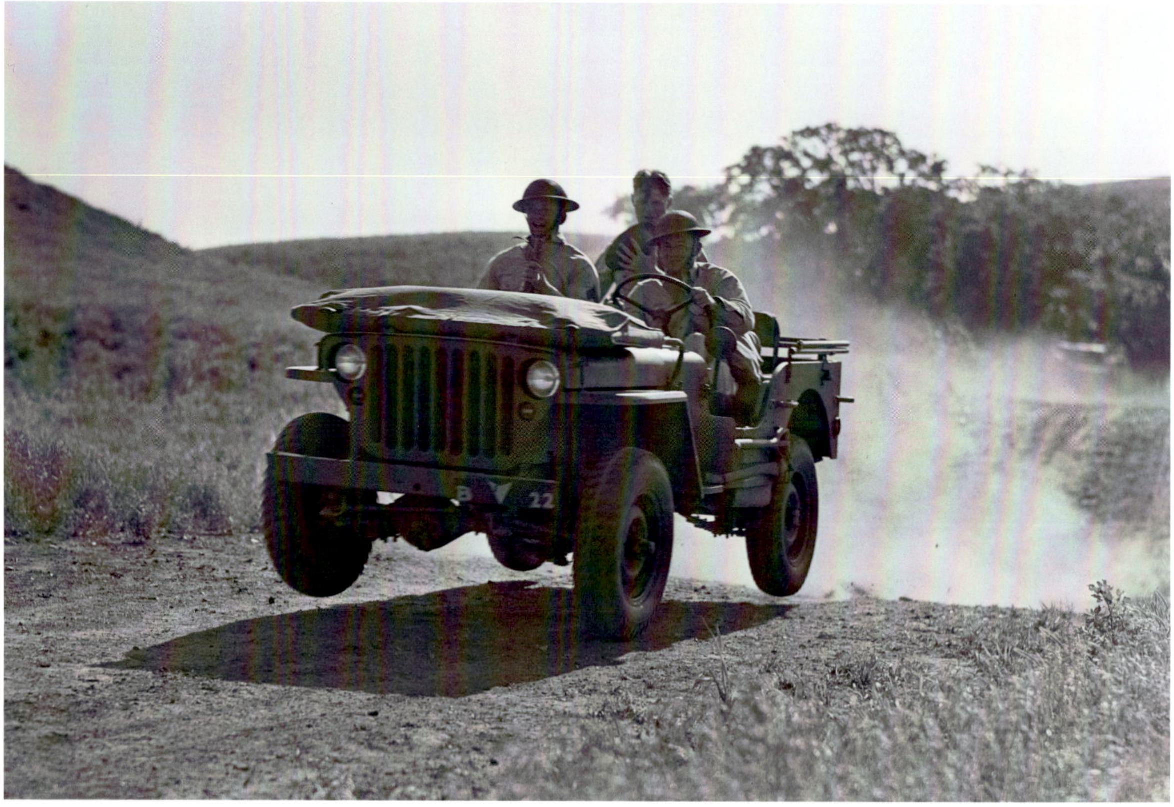

1940 THE JEEP

The Jeep was developed in 1940 through a collaboration between Willys-Overland and Ford Motor Company. The original Willys MB was engineered to fulfill the US Army's requirement for a lightweight, versatile reconnaissance vehicle capable of operating in diverse combat conditions. The vehicle's design emphasized practicality and durability, featuring four-wheel drive for enhanced traction, generous ground clearance, and a reinforced frame. The "Jeep" name likely stems from the military designation "GP" (General Purpose vehicle). During World War II, the Jeep proved indispensable across all theaters of operation. More than six hundred thousand units were produced, serving roles from reconnaissance and transportation to communication and medical evacuation. Its reliability and adaptability made it crucial for Allied forces. After the war, Willys recognized the civilian potential and adapted the design into the CJ (civilian Jeep) series, effectively creating the sport utility vehicle market decades before the SUV designation existed. Despite changing ownership multiple times, Jeep vehicles have maintained their reputation as symbols of American ingenuity, independence, and adventure. The brand continues to embody the rugged capability and go-anywhere attitude that made the original wartime vehicle legendary.

The First Cavalry Division of the US Army drives jeeps toward Manila to release American and Allied nationals held in Santo Tomas Internment Camp, March 1945.

1941 WONDER WOMAN

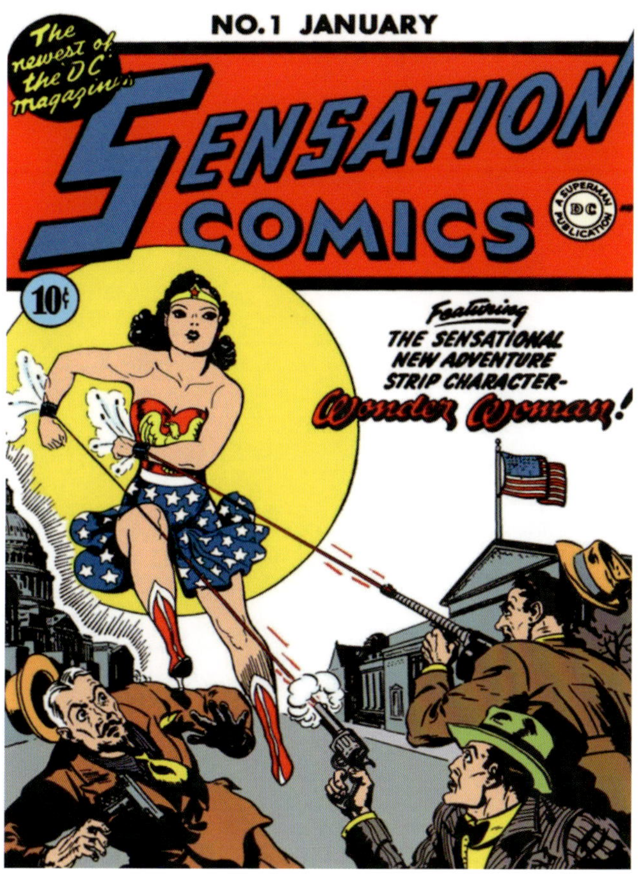

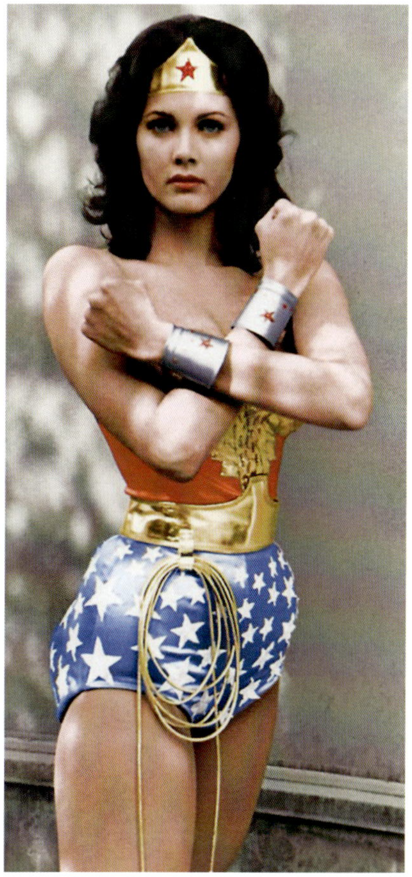

Wonder Woman was created by psychologist William Moulton Marston (1893–1947), who believed that women were inherently more loving and honest than men. Working with artist Harry G. Peter (1880–1958), Marston designed character Diana Prince's alter ego, Wonder Woman, as a powerful Amazon warrior who would demonstrate that women could be both strong and compassionate. The character debuted in *All Star Comics #8* in December 1941, during World War II when women were entering the workforce in unprecedented numbers. Marston's creation was revolutionary not only because Wonder Woman was a female superhero who was physically powerful but also because she was morally principled and emotionally intelligent—unlike other comic-book women of the era, who were typically portrayed as victims or love interests. Thanks to Marston, Wonder Woman became the hero of her own story, wielding her Lasso of Truth and indestructible bracelets to fight injustice. Wonder Woman's legacy as a feminist icon and cultural touchstone endures. The character has inspired generations of women and girls through comics, TV shows, and movies, proving that female characters can anchor successful franchises.

ABOVE LEFT: Wonder Woman's first solo appearance in a comic, 1942. **ABOVE RIGHT:** Lynda Carter as Wonder Woman in the 1970s TV show.

1942 DUCT TAPE

Duct tape was invented by the Johnson & Johnson company during World War II for the US military in order to provide a waterproof sealing solution for ammunition cases and equipment. The tape consisted of a fabric mesh backing coated with a rubber-based adhesive and topped with a polyethylene layer, making it both strong and water-resistant. The tape was originally made using cotton duck cloth and called "duck tape." The name shifted to "duct tape" after the war, when the construction industry adopted it for sealing heating and air-conditioning ducts (though ironically, it's not ideal for actual ductwork due to temperature extremes and adhesive failure over time). Duct tape gained cultural significance during the space program, where NASA used it for various repairs and modifications. The tape's versatility made it a household staple, spawning countless creative uses from fashion accessories to furniture repairs. Today, duct tape represents resourcefulness and temporary solutions, embodying the American philosophy that almost anything can be fixed with determination—and the right adhesive.

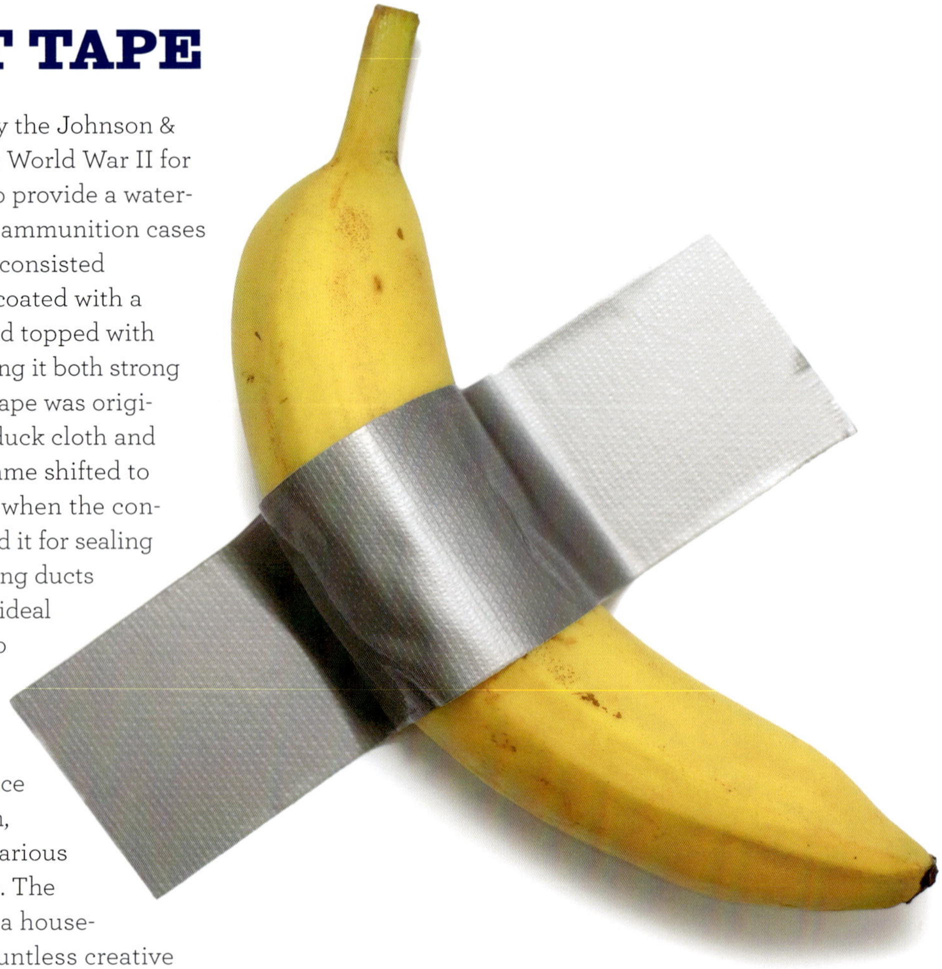

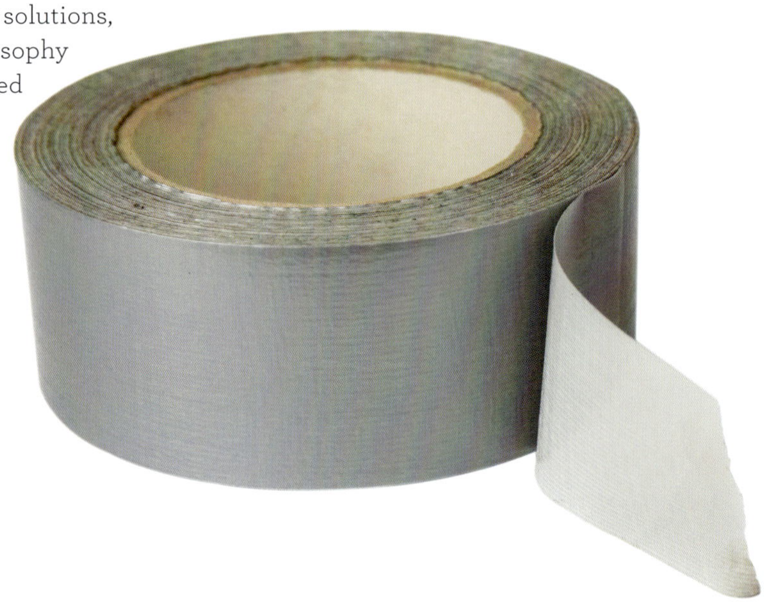

1943 ROSIE THE RIVETER

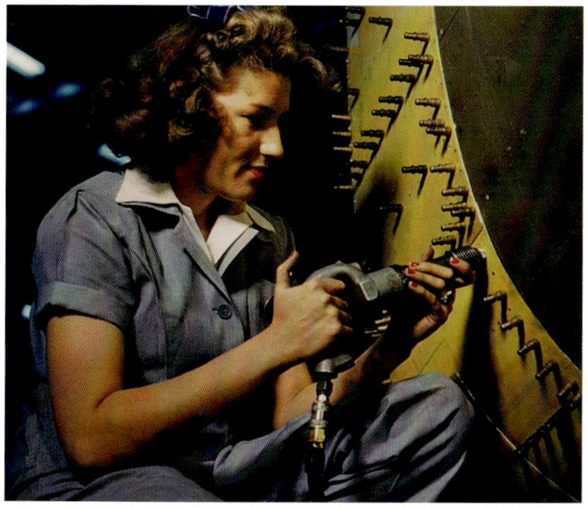

Norman Rockwell's (1894–1978) famous painting of Rosie the Riveter, appearing on the cover of *The Saturday Evening Post* on May 29, 1943, is one of the most powerful symbols of the determination of the American people—especially women—during World War II. The painting depicts a muscular woman in work clothes, eating a sandwich during her lunch break, with a rivet gun resting on her lap and an American flag behind her. The artwork captured the unprecedented entry of women into industrial jobs traditionally held by men, as millions of American women were called to join the workforce to support the war effort. Rockwell's Rosie embodied strength, patriotism, and competence, which challenged conventional gender roles of the era. Rockwell's painting is an enduring symbol of female empowerment and economic independence, inspiring subsequent generations of women entering the workforce. Today, Rosie the Riveter represents both historical achievement and ongoing struggles for workplace equality.

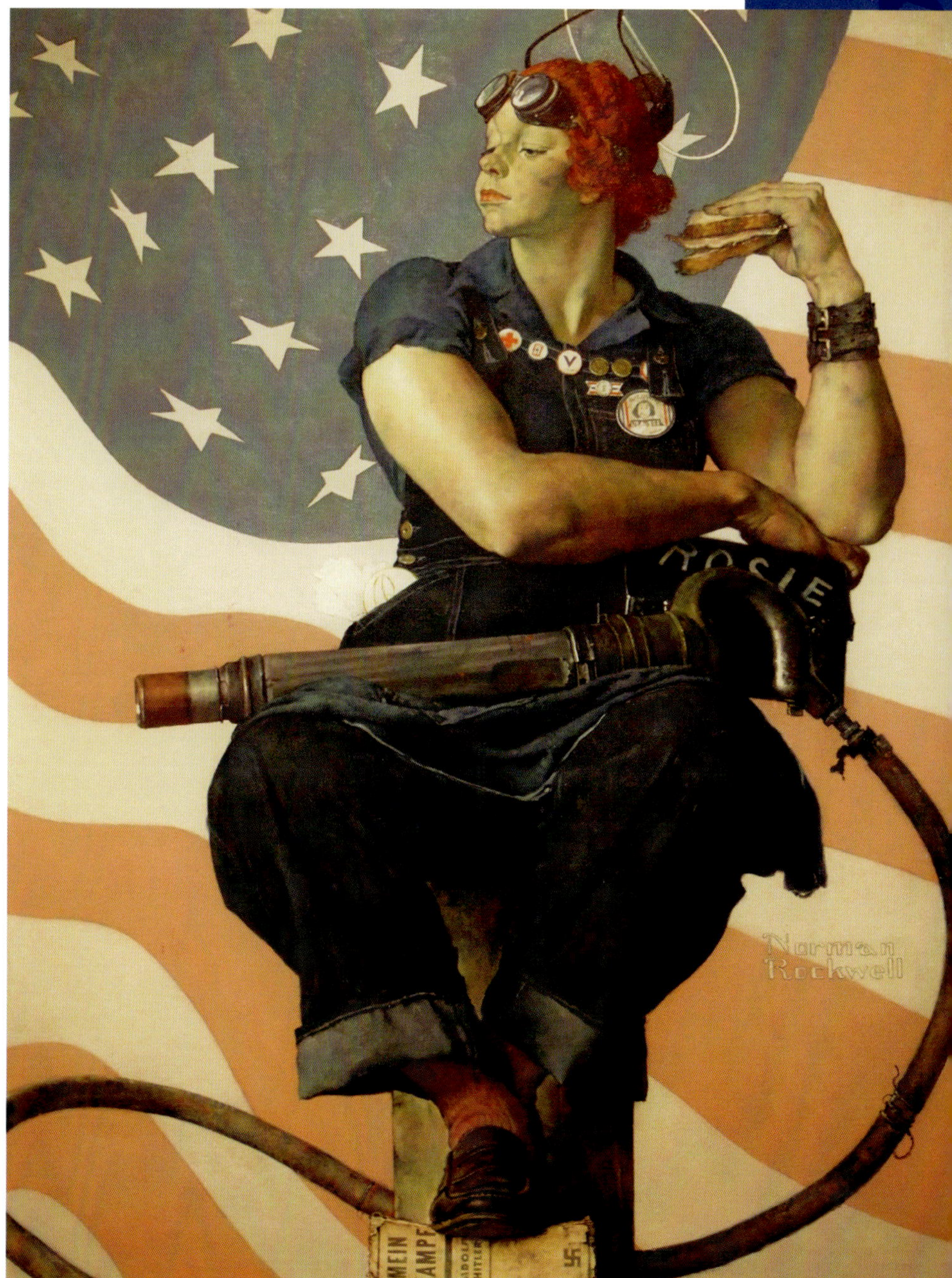

Rockwell's iconic painting, and the inspiration, a riveter working on a bomber at Consolidated Aircraft Corp in Fort Worth, Texas, 1942.

1944 SMOKEY BEAR

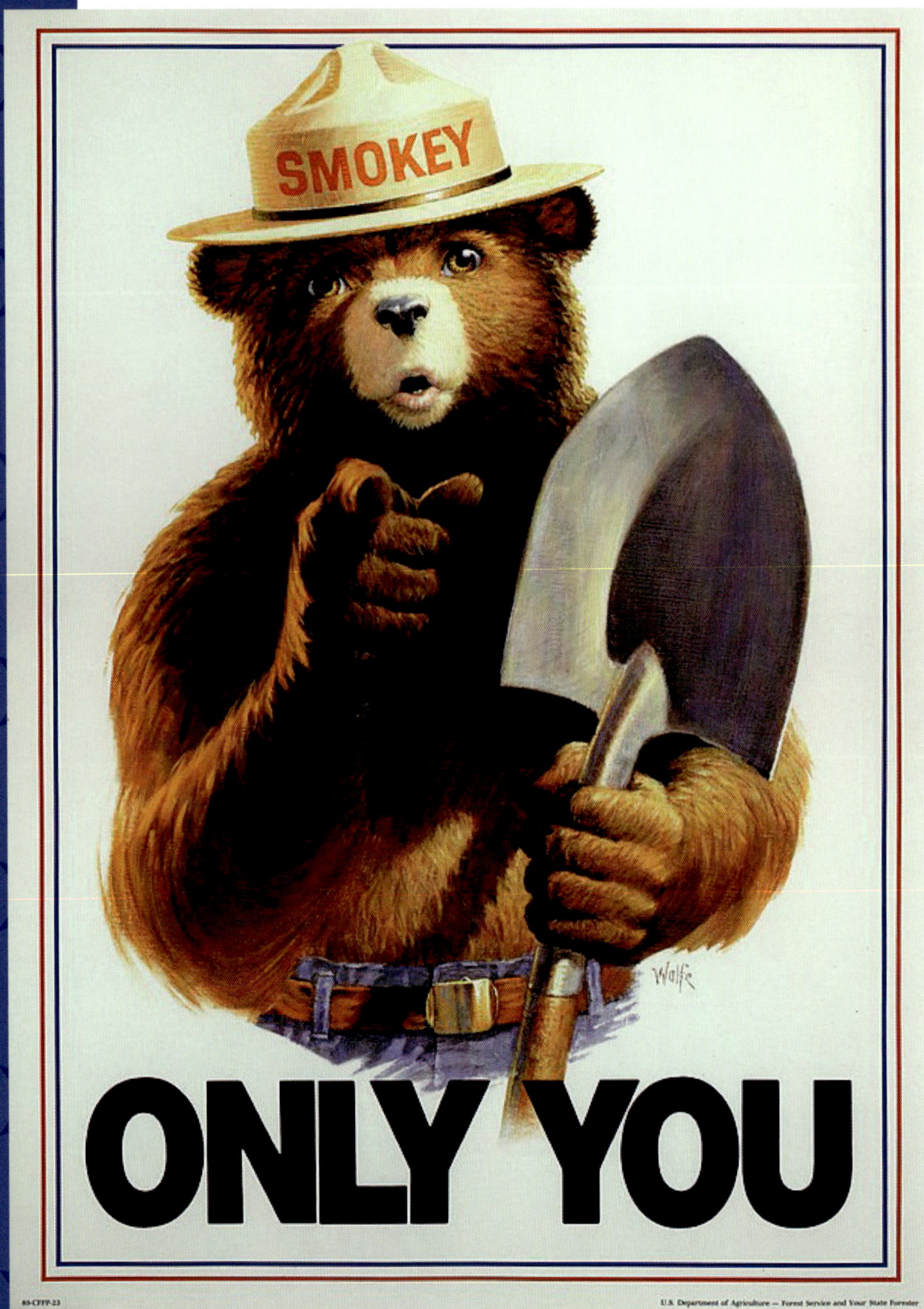

Smokey Bear has been America's beloved forest-fire-prevention mascot since 1944. Created during World War II when forest fires posed a serious threat to the timber resources needed for the war effort, Smokey emerged from a collaboration between the USDA Forest Service, the National Association of State Foresters, and the Ad Council. The character's iconic slogan, "Only You Can Prevent Forest Fires" (later changed to "Only You Can Prevent Wildfires"), became one of the most recognizable public service messages in American history. Smokey's gentle yet authoritative presence, complete with his iconic ranger hat and shovel, made fire prevention understandable to generations of Americans. The Smokey campaign has been credited with significantly reducing human-caused wildfires, and his influence helped shape public attitudes toward forest conservation and environmental responsibility. The character has adapted with changing times in multiple ways: visually, becoming more friendly-looking over the decades; in messaging, when "forest fires" became "wildfires" in 2001; and by incorporating modern fire science, as today's Smokey acknowledges that not all fires are harmful. Smokey remains a cultural icon today, representing not just fire prevention but America's broader commitment to protecting natural resources for future generations.

In a 1989 poster, Smokey Bear's iconic phrase has been shortened to "Only You."

1945 TEFLON

Teflon, the revolutionary nonstick coating, was invented accidentally in 1938 by Roy Plunkett (1910–1994), a twenty-seven-year-old chemist working for DuPont. While experimenting with refrigerant gases, Plunkett discovered that tetrafluoroethylene had polymerized into a white, waxy substance with remarkable properties—it was chemically inert, heat-resistant, and incredibly slippery. The material was classified during World War II, where it was used in clandestine operations such as the Manhattan Project, where its chemical resistance proved invaluable. In 1945, DuPont trademarked the name "Teflon" and registered it as a brand. The substance's classified status ended at this time, though its consumer breakthrough didn't come until the 1960s, when French engineer Marc Grégoire developed the first nonstick cookware. Teflon's legacy extends far beyond kitchen convenience. Its unique properties revolutionized industries, including aerospace, automotive, chemistry, electronics, and medicine. Despite environmental concerns about its production, Teflon's impact on modern life remains profound, demonstrating how a laboratory accident can transform into one of the most versatile materials ever created.

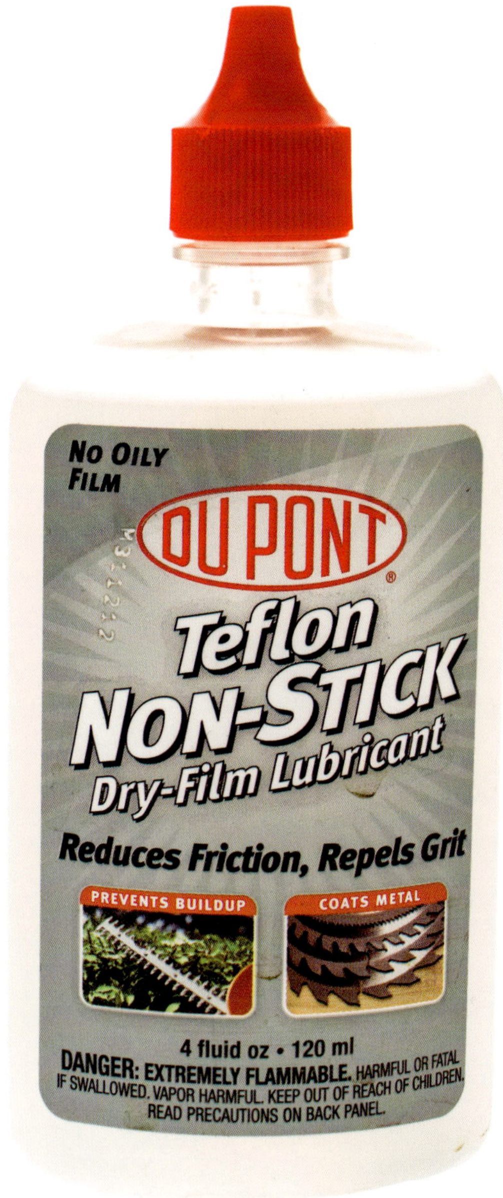

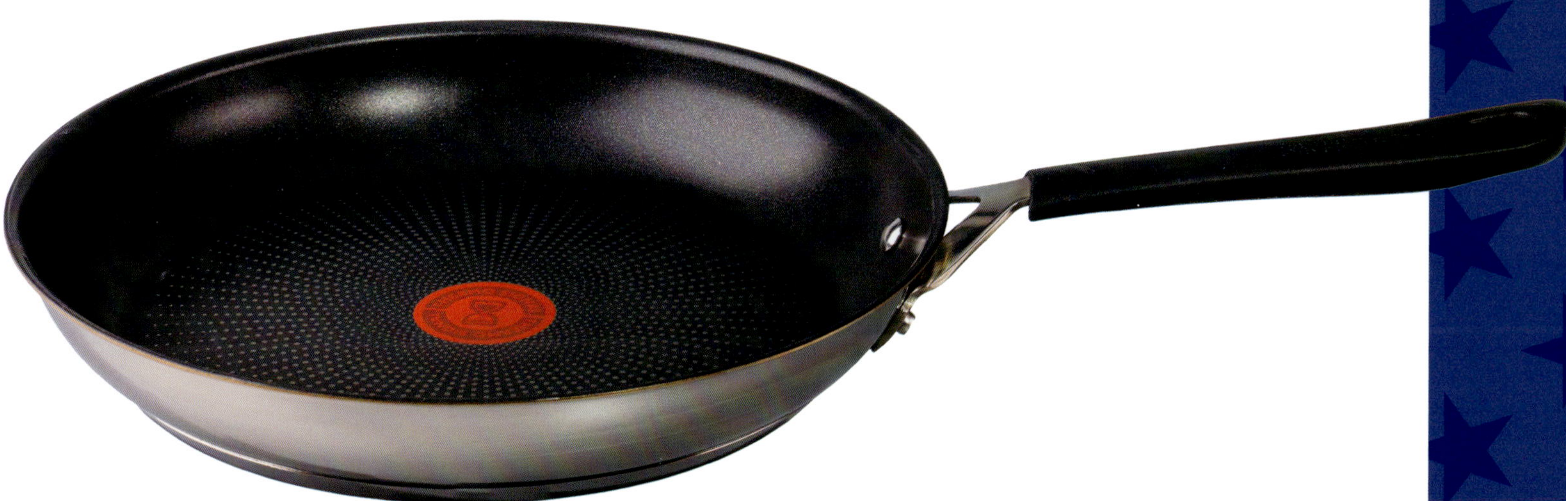

1946 TUPPERWARE

Earl Tupper (1907–1983) revolutionized food storage in 1946 when he invented Tupperware, using polyethylene plastic to create airtight containers with his patented "burping seal." This breakthrough technology kept food fresh longer than traditional storage methods, addressing a genuine household need in postwar America. However, the true cultural impact of Tupperware came from a woman named Brownie Wise (1913–1992). She pioneered the "Tupperware party" concept in the 1950s, a direct-sales model that brought demonstrations into homes and empowered thousands of women to become entrepreneurs. These gatherings became social events that strengthened community bonds while providing women with income opportunities during an era when career options were limited. The Tupperware brand became synonymous with preserving food and reducing waste. The party model influenced countless direct-sales companies, whereas the products themselves became cultural icons. Today, Tupperware containers remain ubiquitous in households worldwide, and the phrase "Tupperware party" endures social commerce—a testament to how a simple plastic container reshaped both kitchens and communities.

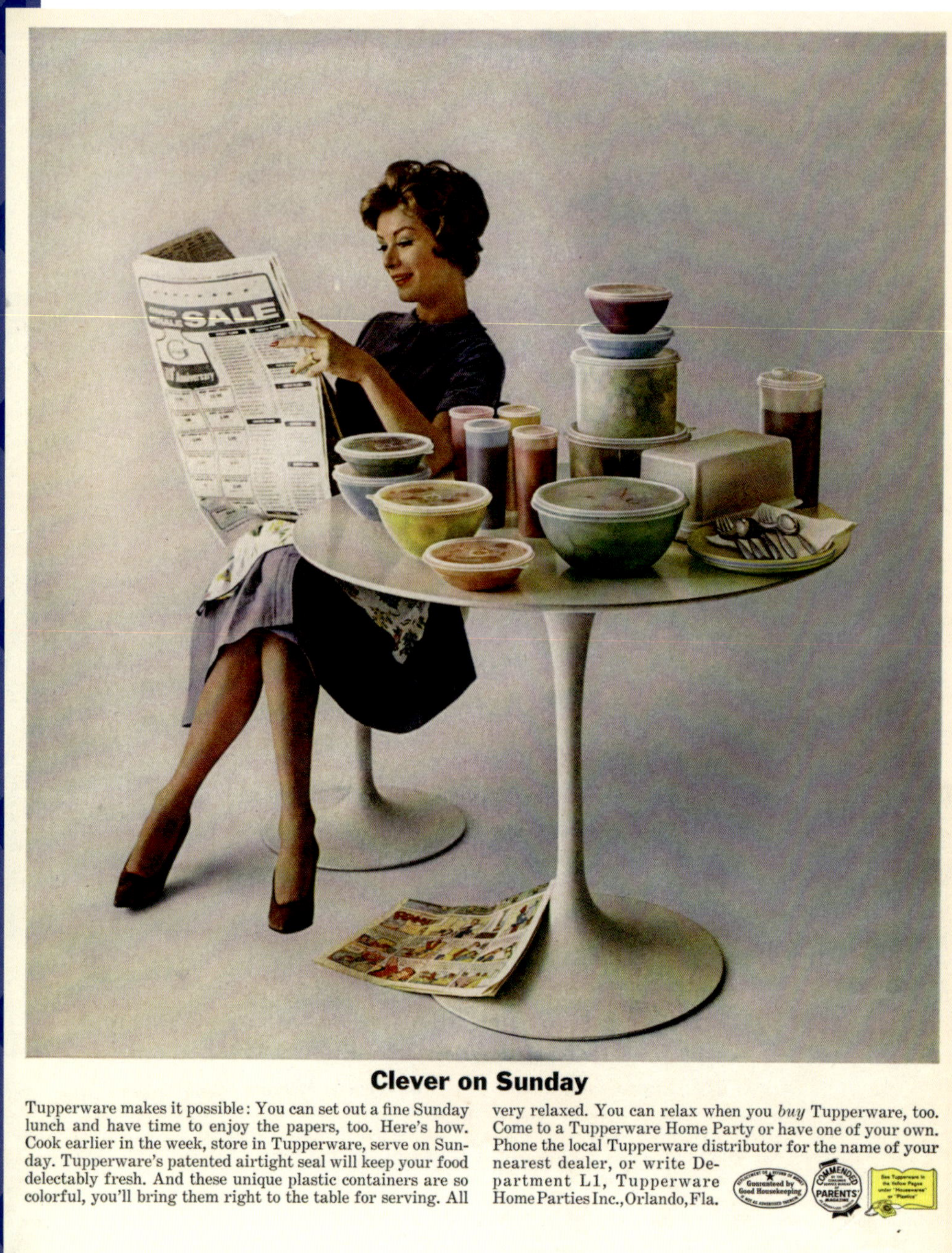

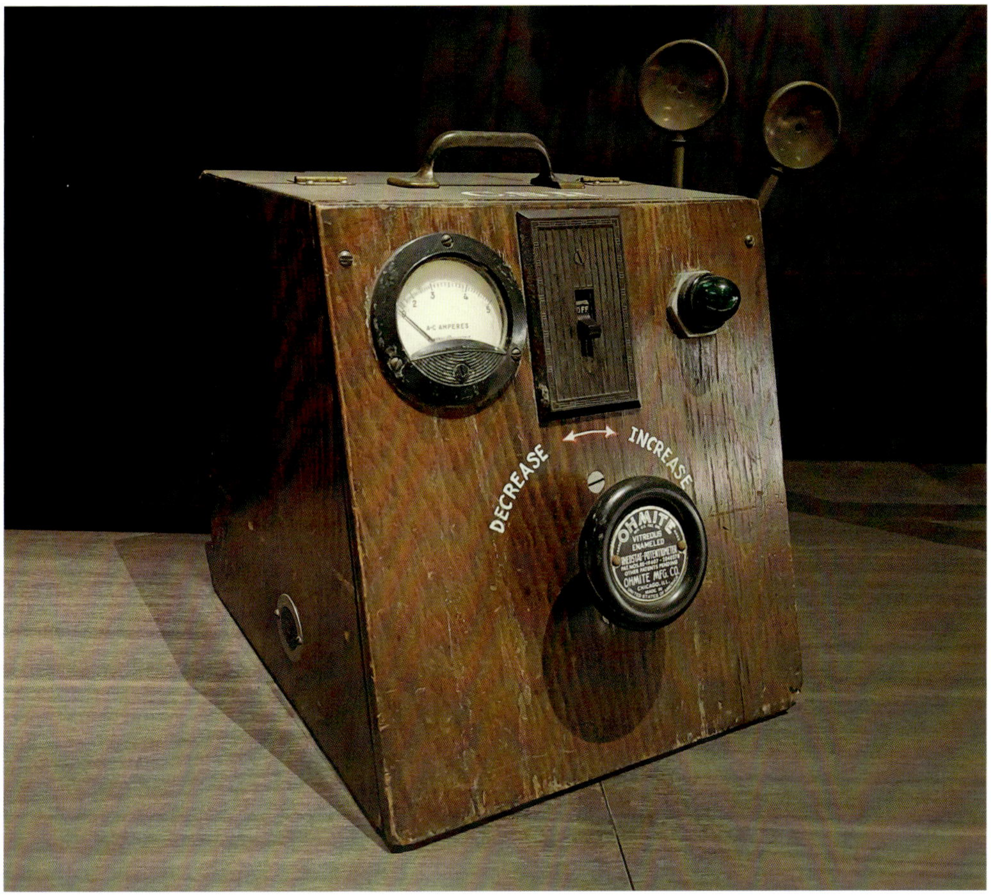

Beck's prototype of the defibrillator, 1947.

1947 DEFIBRILLATOR

The defibrillator was invented by cardiac surgeon Claude Beck (1894–1971) in 1947 at Case Western Reserve University in Cleveland, Ohio. During open-heart surgery, Case used the groundbreaking device to apply electrical shock to restore a normal heart rhythm in a human patient for the first time in history. But Beck didn't come up with the invention on his own—his work built upon earlier research by others who had studied electrical defibrillation in laboratory settings. Defibrillation technology evolved rapidly after that. In the 1960s, Bernard Lown developed the direct current defibrillator, which became the standard for hospitals. The 1970s saw the introduction of portable units for ambulances, whereas the 1980s brought implantable cardioverter-defibrillator (ICDs) for high-risk patients. Perhaps most significantly, automated external defibrillators (AEDs) were developed in the 1990s, making life-saving technology accessible to the general public. Today, defibrillators have saved millions of lives worldwide. They are standard equipment in hospitals, ambulances, and public spaces like airports and schools. Beck's invention represents one of modern medicine's greatest achievements: transforming sudden cardiac arrest from an almost certain death sentence into a treatable emergency.

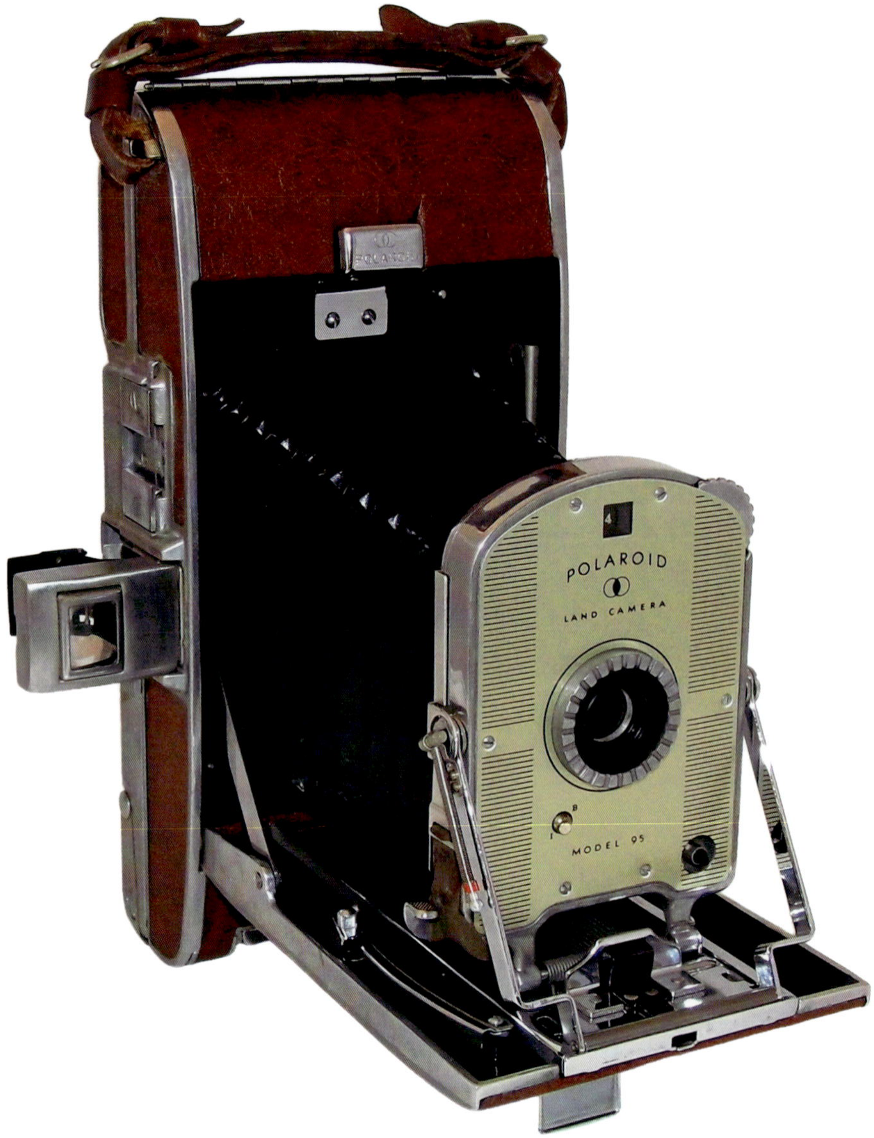

1948 POLAROID INSTANT CAMERA

Edwin Land's (1909–1991) invention of the instant camera transformed photography by delivering developed images within minutes. According to popular accounts, the breakthrough came when Land's three-year-old daughter wondered why she couldn't see a photograph immediately after it was taken—a question that sparked his development of a revolutionary self-contained film system in 1947. The first commercial Polaroid camera, the Model 95, debuted in 1948 to immediate acclaim. At its heart lay an ingenious chemical process: Specialized film packs contained multiple layers of chemicals that would spread and react as the photograph emerged from the camera, creating the iconic square format and that fascinating gradual reveal of the image. Beyond personal snapshots, Polaroid cameras proved invaluable in professional contexts. Medical professionals documented procedures, police captured crime scenes, and real estate agents provided instant property records—all situations where immediate visual confirmation held tremendous value. Though digital photography ultimately displaced instant cameras in mainstream use, Polaroid's influence endures. The distinctive aesthetic—those characteristic white borders, the slightly unpredictable color rendering, the tangible quality of holding a physical photograph—continues to shape modern photo applications and captivates artists who cherish the authentic, unrepeatable nature of analog instant photography.

1949 WALL-O-MATIC

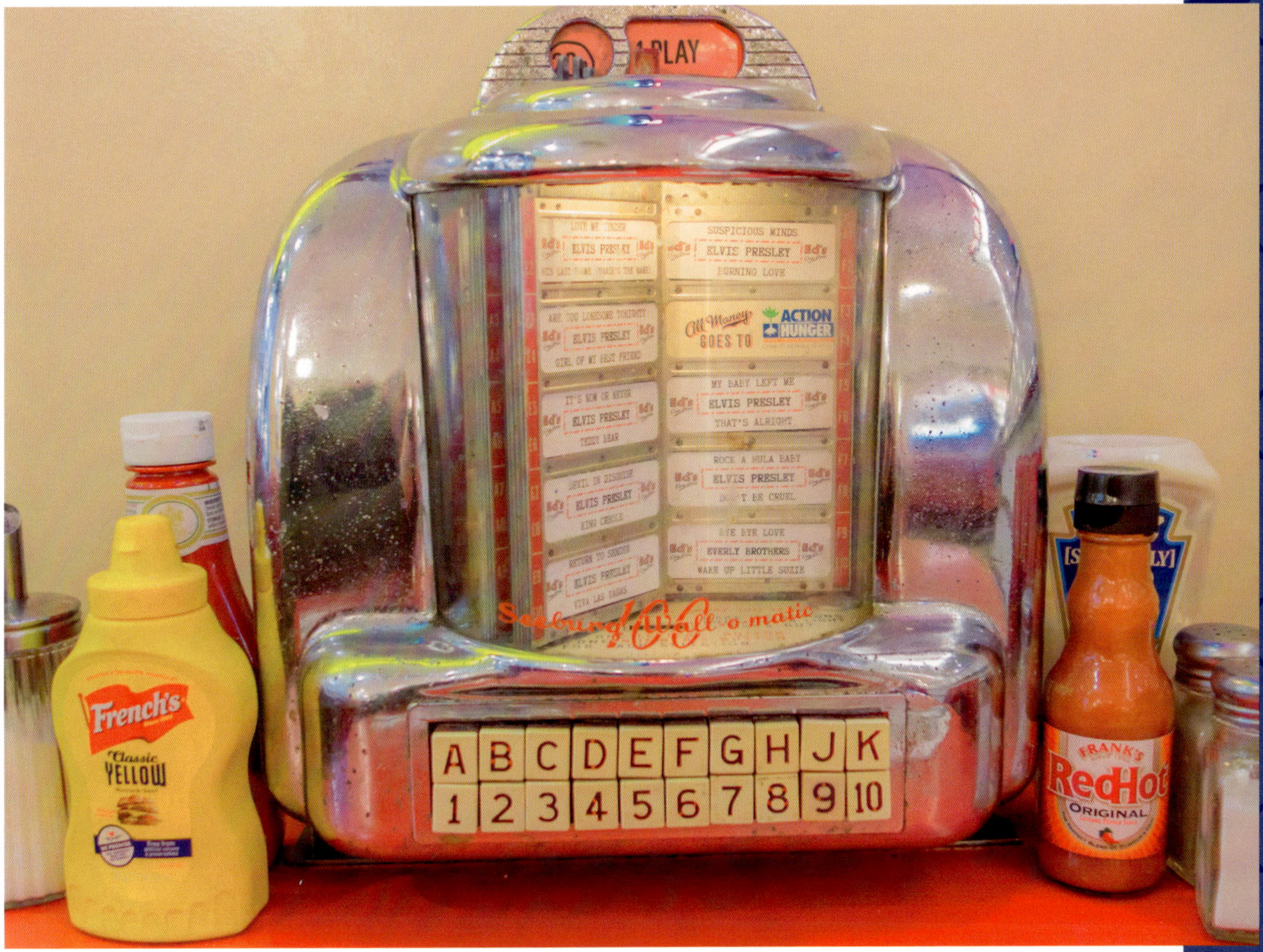

A Wall-O-Matic jukebox still in use at Ed's Diner in Swindon, England.

Wall-box type jukeboxes emerged in the 1940s and 1950s as an innovative extension of the traditional standalone jukebox. The Seeburg 3W1 Wall-O-Matic, introduced in early 1949 in Chicago, Illinois, stands as one of the most iconic wall-box models. It was designed to work with Seeburg's M100A jukebox, the first one-hundred-selection jukebox (all others at the time offered only twenty-four or forty selections). The wall-box revolutionized the dining experience by bringing music selection directly to customers' tables, creating an intimate, personalized atmosphere that became synonymous with American diner culture. Many Americans remember the wall-box experience fondly, recalling the tactile pleasure of dropping coins into the slot and carefully selecting their favorite songs while sharing meals with family and friends. The wall-box's widespread deployment established the remote music selection concept that would influence entertainment venues for decades to come. Their legacy endures today in nostalgic restaurant designs and modern Bluetooth reproductions that capture the retro aesthetic while incorporating contemporary technology. One could argue that the wall-box created a template for how we enjoy music today—through curated personal playlists, streaming services, and collaborative listening apps that allow us to customize what we listen to with unprecedented convenience and choice.

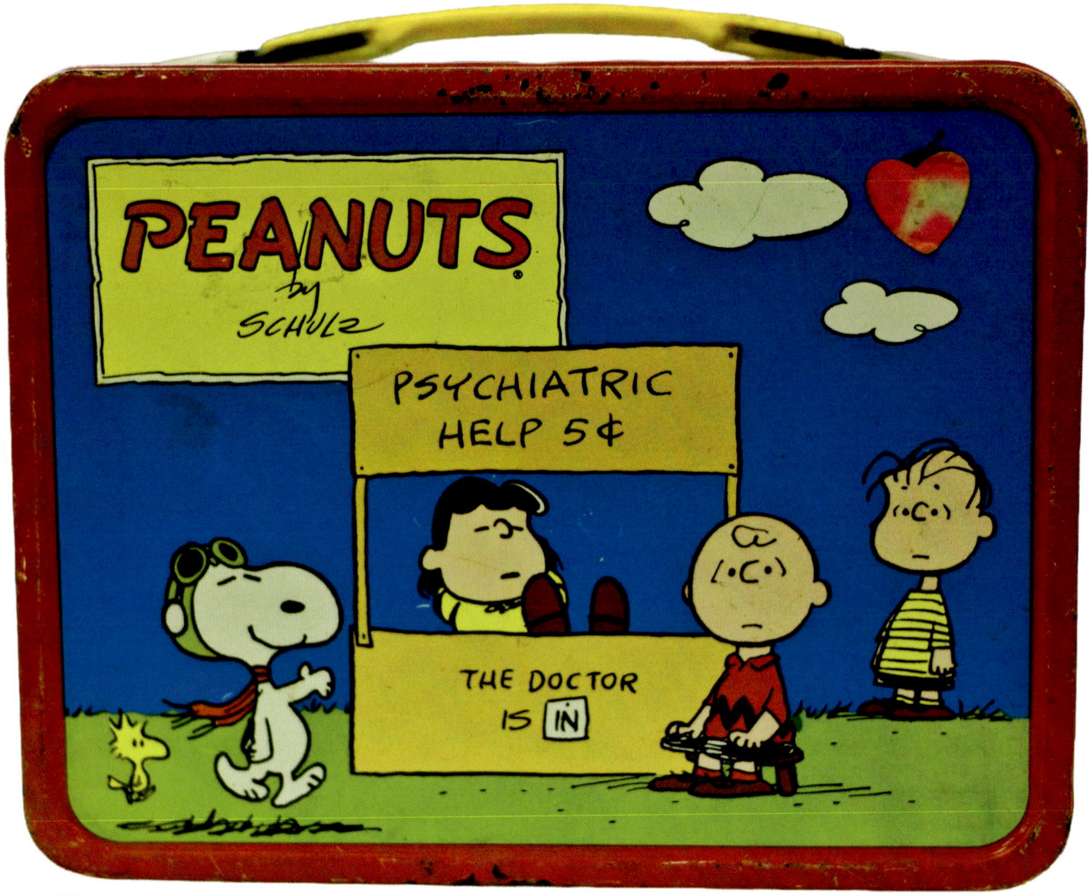

1950 CHILDREN'S LUNCH BOX

In 1950, Aladdin Industries revolutionized children's dining with the first character-themed lunch box, featuring Hopalong Cassidy. This innovation transformed a simple container into a cultural phenomenon spanning generations. These durable metal boxes, featuring Roy Rogers, Howdy Doody, and countless other characters, became coveted status symbols in American schoolyards. Children eagerly anticipated new designs, transforming lunch-carrying into personal expression and cultural identity. Inside, mothers packed wholesome meals: sandwiches, fresh fruit, homemade cookies, and milk in matching thermoses. This daily ritual connected home and school, linking family care with peer interaction. During lunch periods, cafeterias became impromptu galleries where children proudly displayed their themed boxes. This showcase fostered social exchanges—trading treats, comparing acquisitions, and navigating the complex hierarchies these objects helped establish. The lunchroom antics reflected postwar prosperity, enabling families to invest in specialized children's items. Lunch boxes embodied parents' ability to provide small luxuries that enhanced belonging and joy. Safety concerns and production costs prompted a shift from metal to plastic in the 1980s, but the core concept endured. Today's boxes continue this legacy, featuring contemporary characters while serving the same fundamental purposes. The character lunch box evolution illustrates how ordinary objects transcend utility to become powerful symbols of childhood, family care, and cultural participation—finding meaning in our daily rituals.

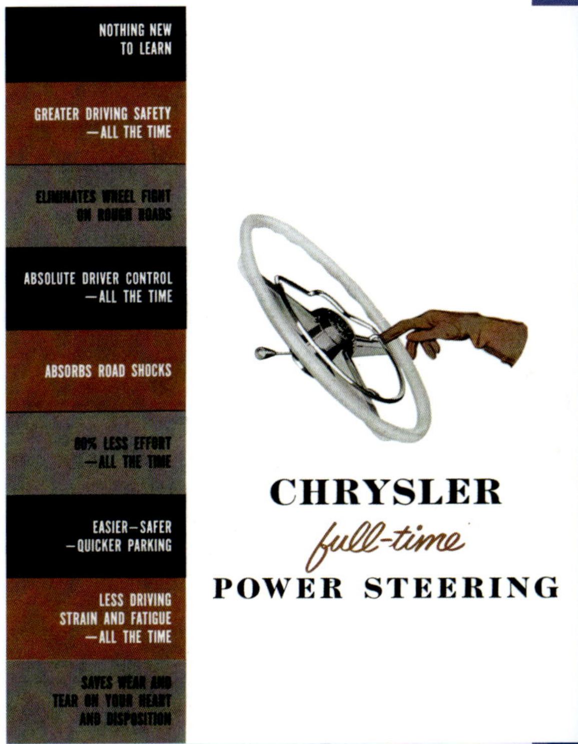

1951 POWER STEERING

Francis W. Davis (1887–1978) invented the first practical power steering system in the 1920s while working for truck manufacturer Pierce-Arrow. His system used hydraulic power from engine-driven pumps to assist drivers in turning the wheel, dramatically reducing the physical effort required to steer heavy vehicles. Initially developed for commercial trucks and buses, power steering remained a luxury feature for decades due to cost and complexity. In 1951, Chrysler became the first automaker to offer it as an option on passenger cars, followed by other manufacturers throughout the 1950s. The technology transformed driving from a physically demanding task into an effortless experience, and helped revolutionize the modern automobile industry.

1952 MR. POTATO HEAD

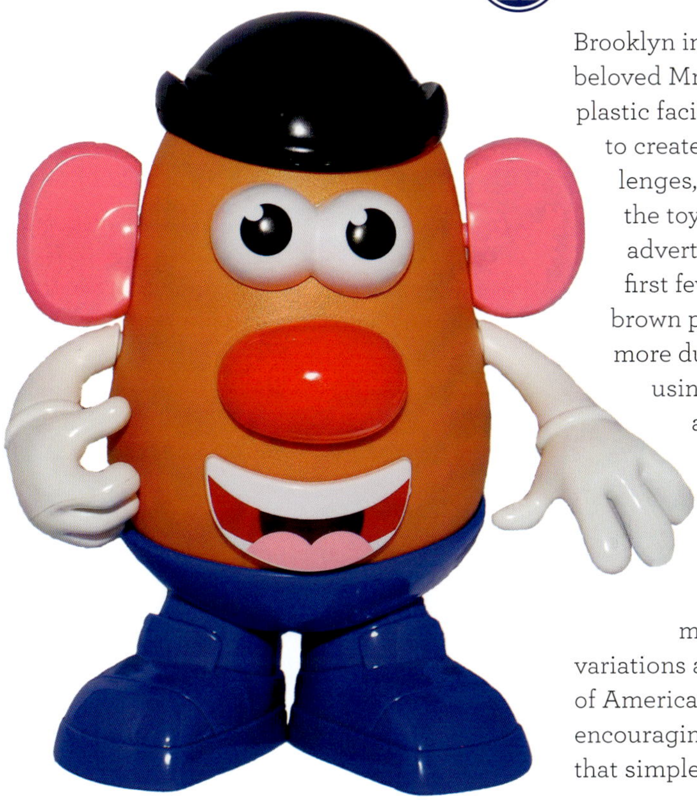

Brooklyn inventor George Lerner (1922–1995) created the beloved Mr. Potato Head in 1949. His original concept used plastic facial features that children stuck into real potatoes to create characters. After initial manufacturing challenges, Hasbro acquired the rights in 1952 and brought the toy to market. Mr. Potato Head became the first toy advertised on television, earning over $4 million in its first few months. In 1964, Hasbro introduced the iconic brown plastic potato body, making the toy safer and more durable while eliminating hygiene concerns from using real potatoes. The toy's popularity endured across generations, reaching new heights as a starring character in Pixar's Toy Story franchise (1995–2026). This exposure introduced Mr. Potato Head to new audiences and solidified its cultural significance. Today, Mr. Potato Head remains a testament to creative play. Its mix-and-match concept has spawned numerous variations and inspired other toys, establishing it as one of America's most enduring playthings. The toy continues encouraging children's creativity and imagination, proving that simple ideas often have the most lasting impact.

1953 *I LOVE LUCY*

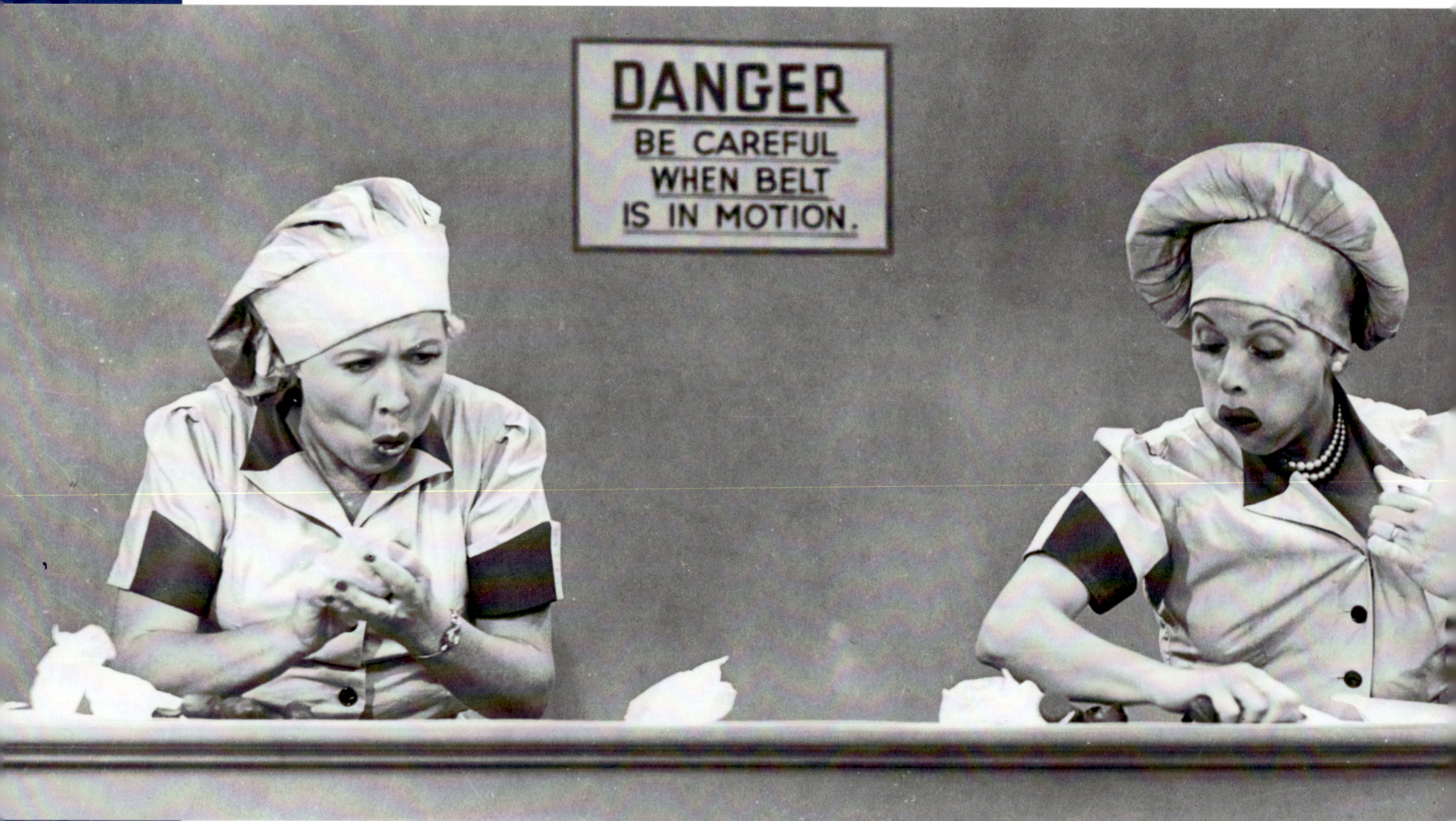

I Love Lucy, starring Lucille Ball and Desi Arnaz, became the most popular show in America for four of its six prime-time seasons. The show was revolutionary in multiple ways. First, it featured a woman as the main character, which was rare during that era. Second, Ball's portrayal of Lucy Ricardo challenged conventional gender roles of the time, presenting a female character who was ambitious, mischievous, and often disobedient to her husband. Finally, *I Love Lucy* was among the first sitcoms to be filmed rather than broadcast live, allowing for higher production values and better distribution. January 19, 1953, was a landmark date for the show. On that day, the episode "Lucy Goes to the Hospital," showing Lucy giving birth to her first child, aired the day before the inauguration of Dwight D. Eisenhower as US president. It was a massive television event, drawing more viewers than any other television episode to date—even more than the inauguration the following day. *I Love Lucy* continues to be held in high esteem by television critics and remains perennially popular. It established many conventions of television comedy that persist today and demonstrated the commercial viability of filmed television production, influencing the medium's future development for decades to come.

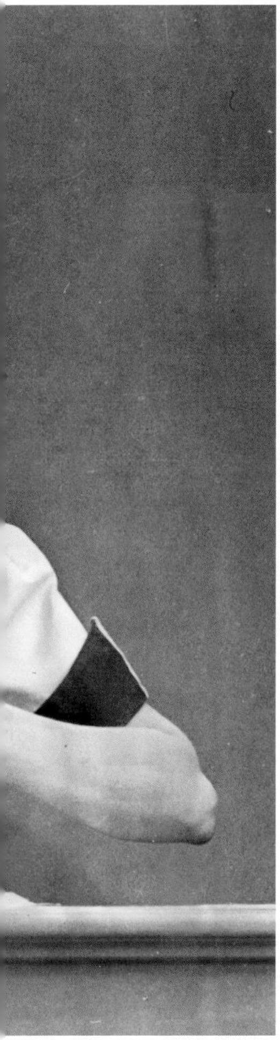

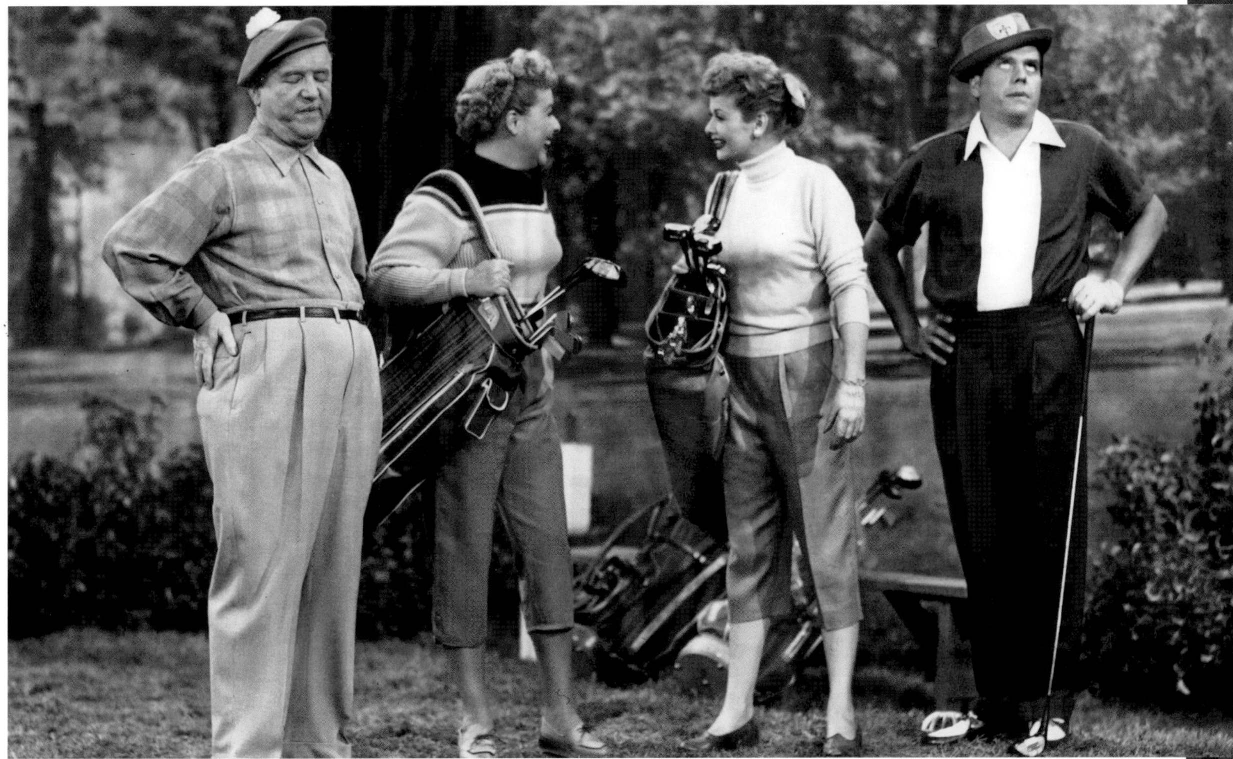

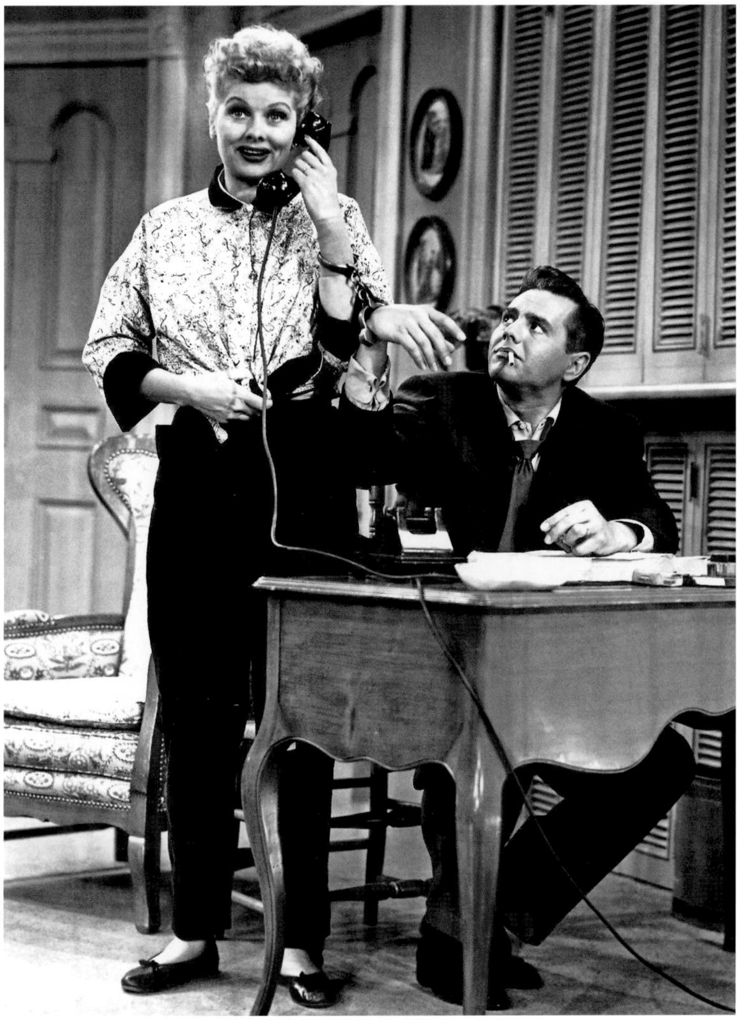

FAR LEFT: Lucy and Ethel in one of their finest bits, as chocolate makers. **TOP:** Lucy and Desi with costars Vivian Vance and William Frawley (Fred and Ethel Mertz). **BOTTOM:** Lucy getting Ricky into another fine mess.

1954 "THAT'S ALL RIGHT"

"That's All Right" stands as one of the most important recordings in music history, marking Elvis Presley's (1935–1977) breakthrough and, as many consider, the birth of mainstream rock and roll. Recorded on July 5, 1954, at Sun Studio in Memphis, Tennessee, the song emerged spontaneously during a break in the recording session when Elvis began ad-libbing with Arthur Crudup's blues song, transforming it with his unique vocal style and rhythm. Producer Sam Phillips recognized something remarkable in Elvis's interpretation—a white performer who could authentically channel the emotion and energy of African American blues and gospel music. They captured Elvis's song on the recording, a sound that was both familiar and entirely new. The song represented a cultural moment, bridging racial divides in music during the segregated 1950s South. When released as Elvis's first single, it received airplay on both country and R&B radio stations, demonstrating its crossover appeal. "That's All Right" essentially launched Elvis's career as the King of Rock and Roll and established a template for countless rock musicians who followed, forever changing popular music.

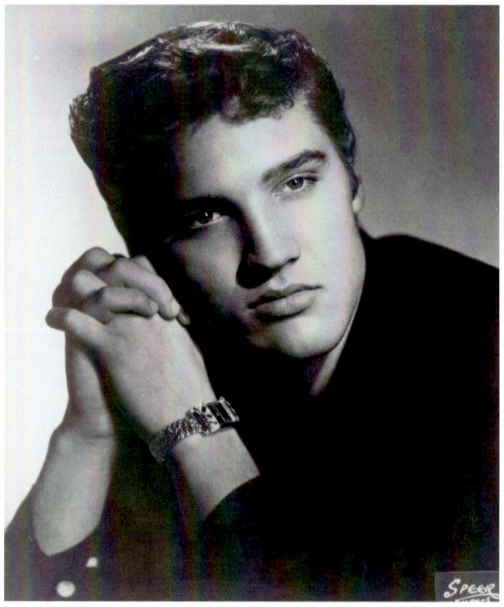

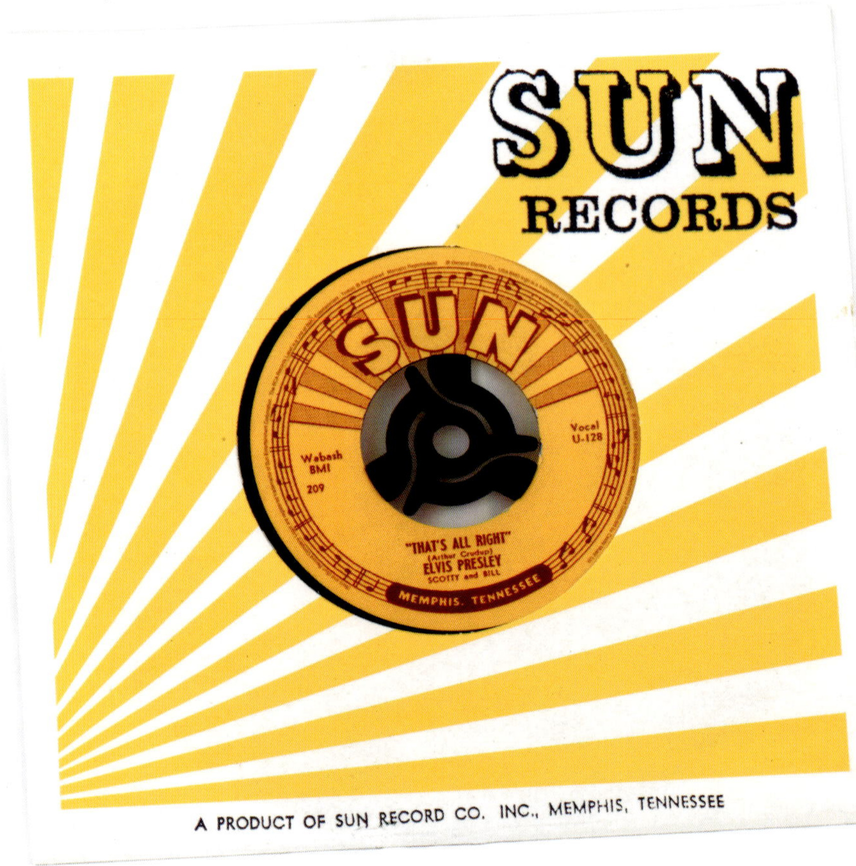

Elvis, c. 1954, and the record that made him a star.

1955 POLIO VACCINE

Polio is a highly contagious disease that typically affects children under the age of five, attacking the human nervous system and potentially leading to paralysis and even death. The polio vaccine stands as one of medicine's greatest triumphs, fundamentally transforming global health in the twentieth century. Before its development, polio threatened communities worldwide. The disease reached epidemic proportions in the 1940s and early 1950s, with the United States experiencing over fifty thousand cases annually at its peak. Jonas Salk (1914–1995) developed the inactivated vaccine, which was introduced in 1955. The vaccine's impact was immediate and dramatic. Polio cases plummeted by over 99 percent within decades, and the disease was eliminated from most of the world. Beyond its direct medical impact, the polio vaccine established a blueprint for modern vaccination programs and demonstrated the power of international cooperation in combating disease. It remains a powerful symbol of scientific achievement and public health success.

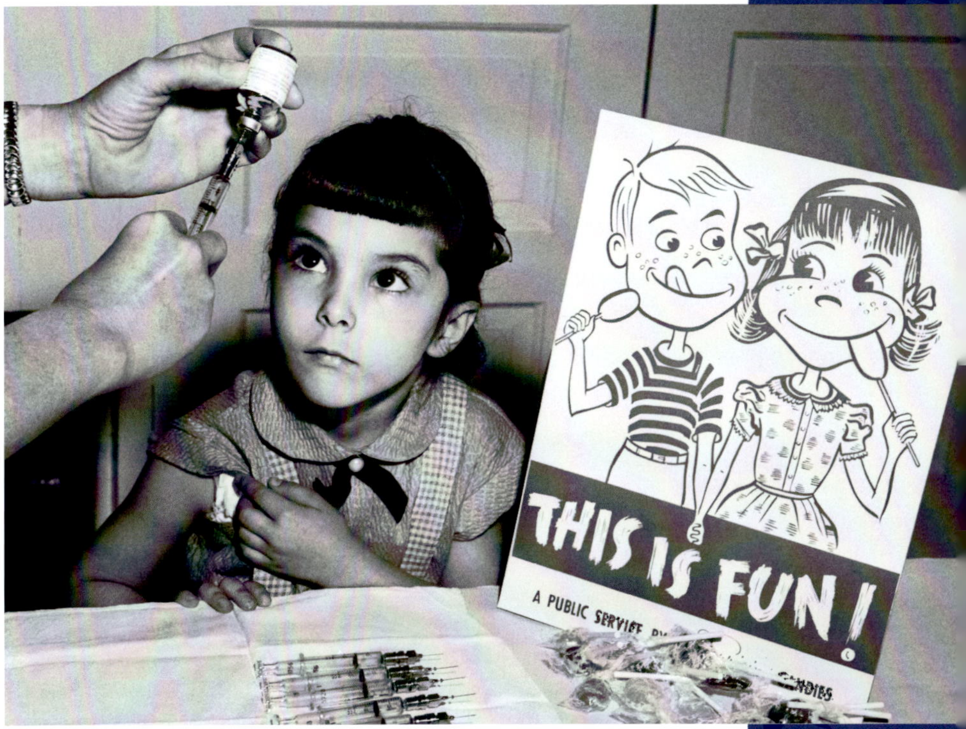

A young girl getting her polio shot, and clearly not believing the sign, c. 1955.

1955 Disneyland

Disneyland opened on July 17, 1955, in Anaheim, California, under Walt Disney's direct supervision. The park revolutionized the amusement industry by creating an immersive storytelling experience rather than just individual attractions. Walt Disney famously declared that "Disneyland will never be completed. It will continue to grow as long as there is imagination left in the world." Though Disney didn't invent theme parks, Disneyland set the standard for what they could and should be. The park's legacy extends far beyond entertainment, inspiring countless theme parks worldwide and establishing Disney as a cultural phenomenon.

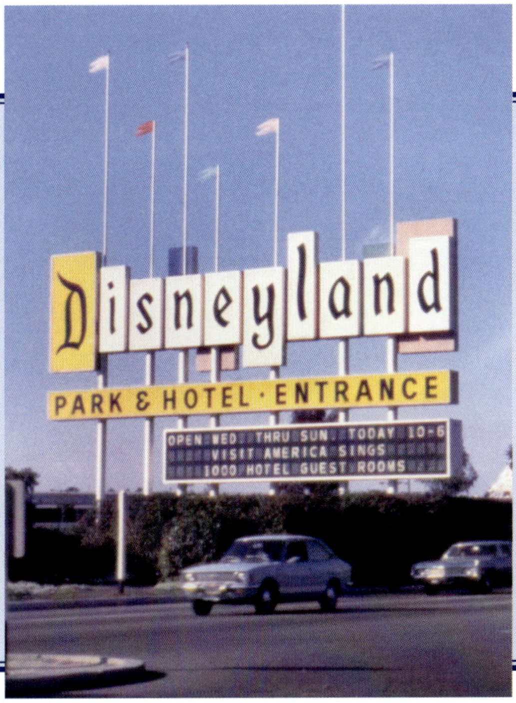

1956 THE ROLODEX

The Rolodex was invented in 1956 by American entrepreneur Arnold Neustadter (1910–1996) and his Danish business partner, engineer Hildaur Neilson. The rotating card file system, with its distinctive cylindrical design and removable cards, became the standard tool for organizing business contacts, phone numbers, and addresses. The device's genius lay in its simplicity and efficiency. Unlike traditional address books, the Rolodex allowed users to easily add, remove, and reorganize contacts without rewriting entire sections. Its quick-flip mechanism made finding information fast and intuitive, whereas the standardized card format encouraged systematic organization. For decades, the Rolodex dominated office desks worldwide, becoming synonymous with professional networking. The phrase "let me check my Rolodex" entered common jargon, and having an extensive Rolodex became a mark of business success and connectivity. Though rendered largely obsolete by smartphones, the Rolodex's legacy endures.

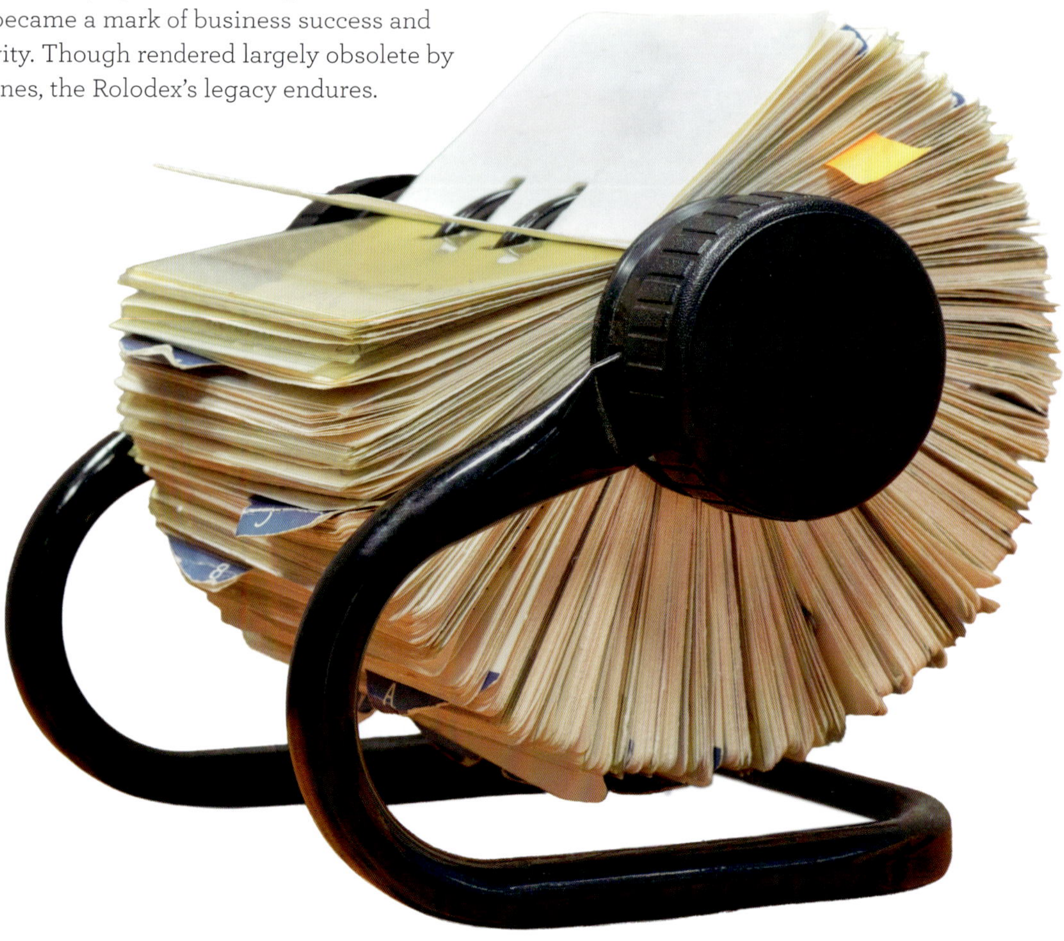

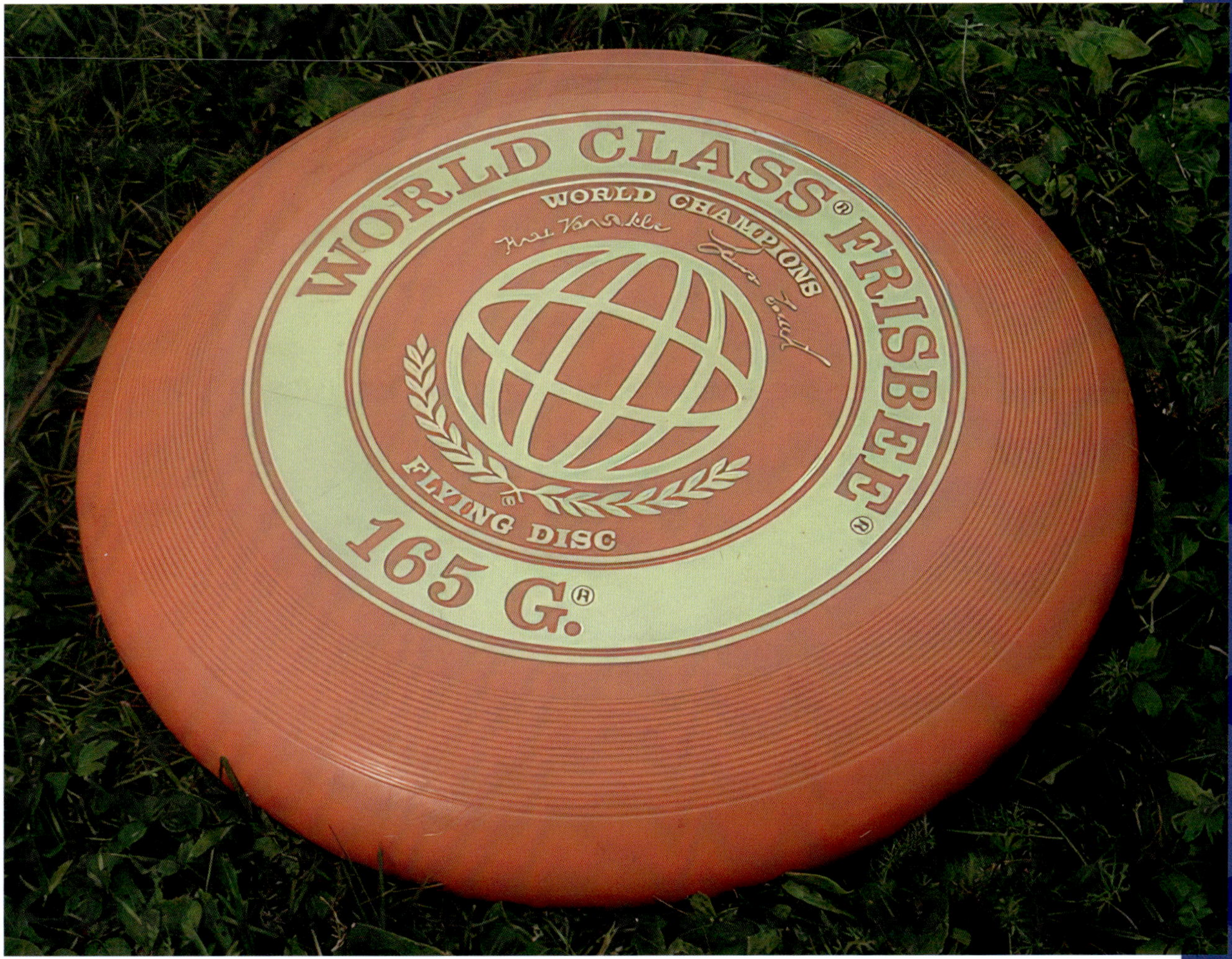

1957 THE FRISBEE®

The Frisbee® emerged from an unlikely origin story. Legend has it that in the 1940s, university students (claims to this story have been made by both Yale University and Middlebury College alumni) discovered that empty pie tins from the Frisbie Pie Company made excellent flying discs when thrown across campus. Tossing the metal tins became a popular activity, with students shouting "Frisbie!" as a warning to those in the projectile's path. The modern plastic Frisbee was born in 1957 when Walter Morrison (1920–2010) partnered with Wham-O toy company to mass-produce his "Pluto Platter." The company later renamed it the "Frisbee" to honor its collegiate roots, though they altered the spelling. Morrison's design featured the aerodynamic properties that made the disc fly smoothly and predictably. The toy's popularity spawned organized sports like Ultimate Frisbee, which combines elements of football, basketball, soccer, and disc golf, played on courses worldwide. Today, millions of Frisbees are sold annually, making it one of the most recognizable and enduring toys in American history.

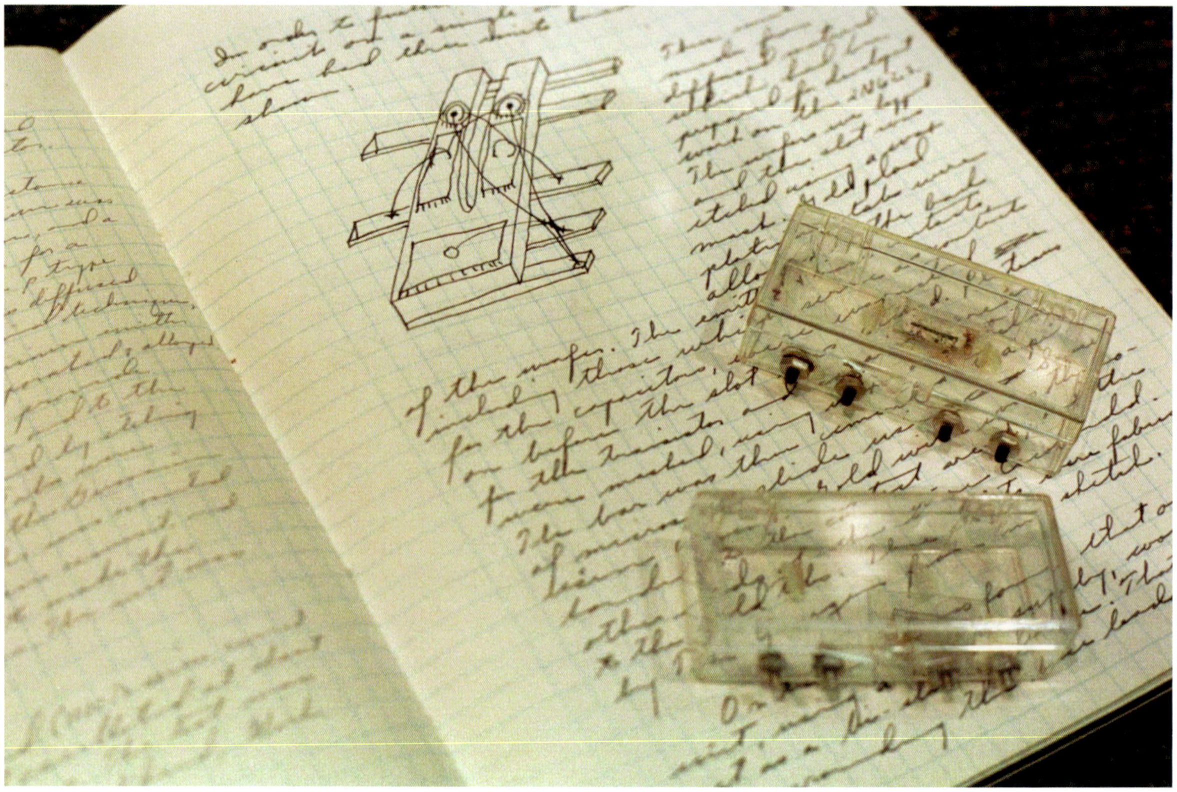

Inventor Jack Kilby's notebook and the two original integrated circuits.

1958 THE MICROCHIP

The microchip, also called an integrated circuit, was invented in 1958 by Jack Kilby (1923–2005) of Texas Instruments and Robert Noyce (1927–1990) of Fairchild Semiconductor. Their breakthrough involved etching multiple electronic components onto a single piece of semiconductor material, typically silicon, rather than connecting separate components with wires. This innovation solved the "tyranny of numbers" problem that plagued early electronics, where complex circuits required thousands of individual connections that were prone to failure and difficult to manufacture reliably. The microchip gave birth to personal computers, smartphones, and countless other digital devices that define modern life. Beyond personal technology, microchips revolutionized industries from healthcare to transportation. They enabled life-saving medical devices like pacemakers and MRI machines, transformed manufacturing through precision automation, and made possible global communications networks, including the Internet. The microchip fundamentally reshaped how humans communicate, work, learn, and interact with the world. By creating the digital age and laying the foundation for modern computing and artificial intelligence, the microchip stands as arguably the most transformative invention of the twentieth century—one that continues to drive innovation and progress in the twenty-first century.

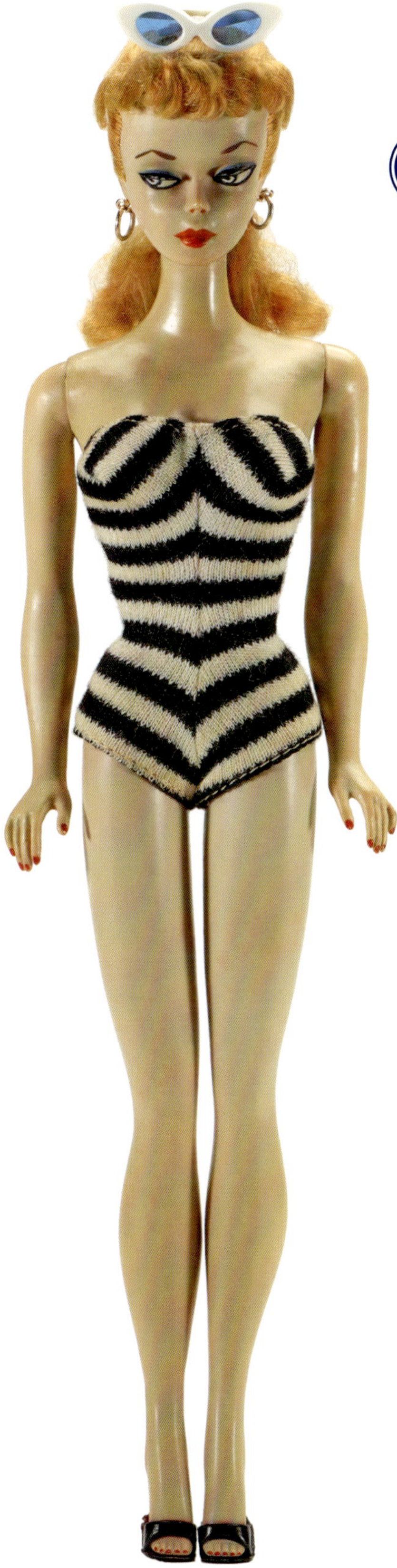

1959 BARBIE DOLL

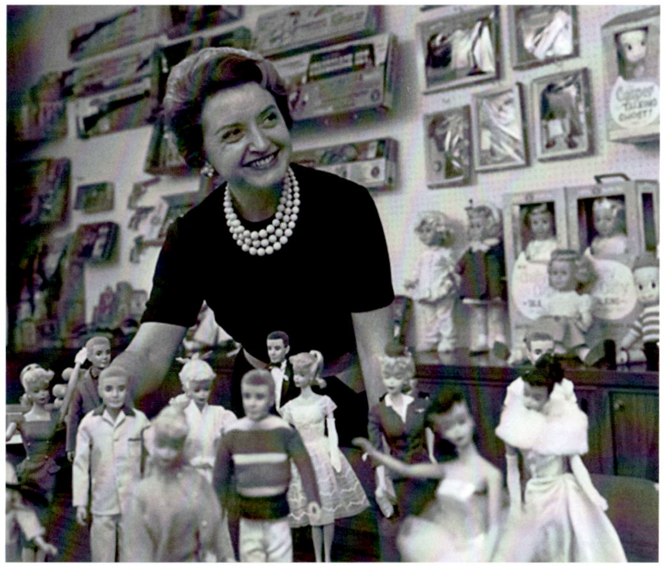

Ruth Handler (1916–2002), cofounder of Mattel, invented the Barbie doll in 1959 after observing her daughter playing with paper dolls and imagining them as adults. Inspired by the German Bild Lilli doll, Handler created a three-dimensional adult woman toy named after her daughter Barbara. Barbie debuted at the American International Toy Fair in New York City on March 9, 1959. Barbie revolutionized the toy industry by introducing an adult-figured doll with extensive wardrobes and accessories—a radical departure from the baby dolls that dominated the market. Over decades, she has pursued more than two hundred careers, from astronaut to president, challenging traditional gender expectations and encouraging girls to imagine possibilities beyond conventional boundaries. However, Barbie has attracted criticism regarding body image and beauty standards. Her anatomical proportions, if translated to human scale, would be physically unattainable, sparking debates about her potential influence on children's self-perception. Critics argue her unrealistic figure may contribute to unhealthy beauty ideals. Despite controversies, Barbie's commercial success remains undeniable. She continues ranking among history's best-selling toys, demonstrating enduring appeal across generations and cultures. Her legacy reflects both the power of imaginative play and the complex relationship between toys and societal values.

ABOVE: The original 1959 Barbie.
RIGHT: Barbie inventor Ruth Handler with her creations in 1961.

1960 *TO KILL A MOCKINGBIRD*

Harper Lee's (1926–2016) *To Kill a Mockingbird*, published in 1960, emerged from the author's childhood and her observations of racial injustice in the Depression-era South. Written during the civil rights movement, the novel channels Lee's experiences growing up in Monroeville, Alabama, with characters like Atticus Finch reportedly inspired by her own father. The novel's publication coincided with America's growing civil rights consciousness, earning Lee the Pulitzer Prize for Fiction in 1961. Its portrayal of moral courage through Atticus Finch's defense of Tom Robinson resonated powerfully with readers confronting contemporary racial tensions. For decades, *To Kill a Mockingbird* was widely taught in schools as an anti-racist text. However, modern scholarship has increasingly critiqued its white-savior narrative and questioned whether it truly serves anti-racist education. Despite these debates, the novel continues to influence American literature and legal culture, remaining one of the most frequently assigned books in high school curricula.

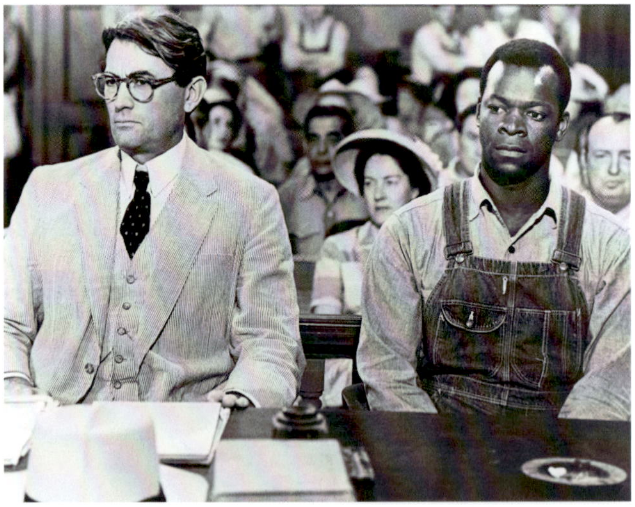

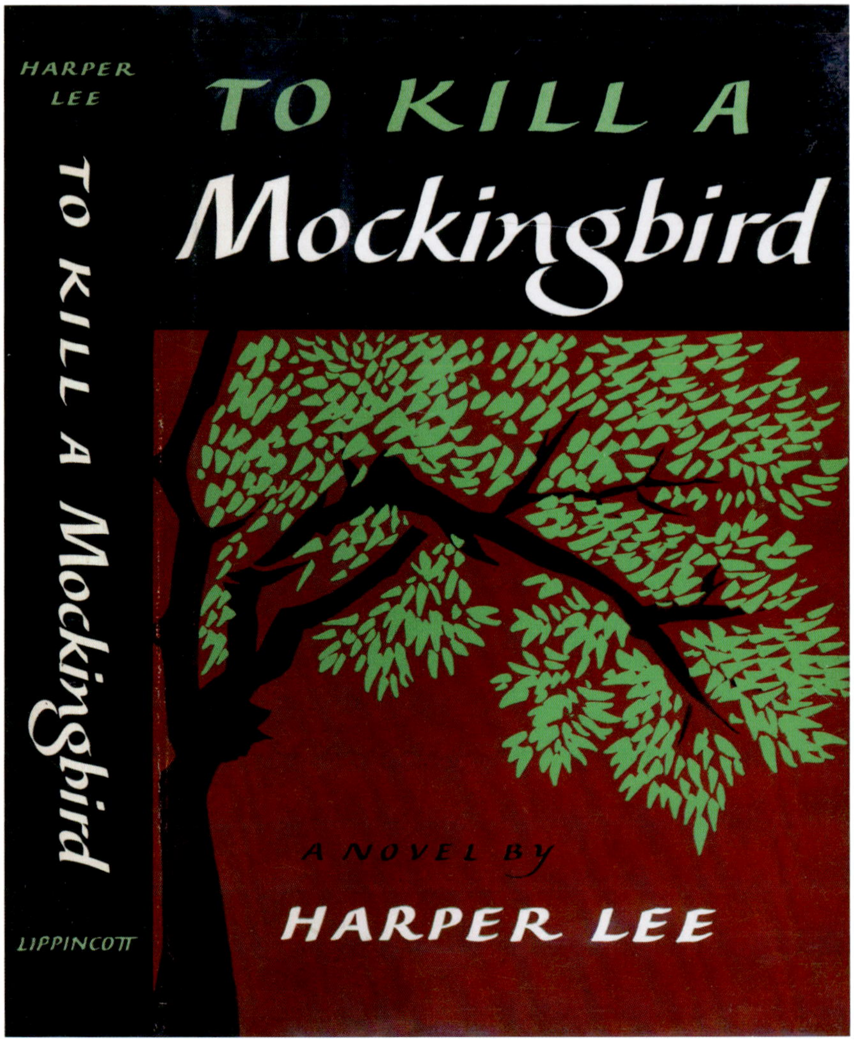

ABOVE: Gregory Peck as Atticus Finch and Brock Peters as Tom Robinson in the 1962 movie.

1961 THE PEACE CORPS

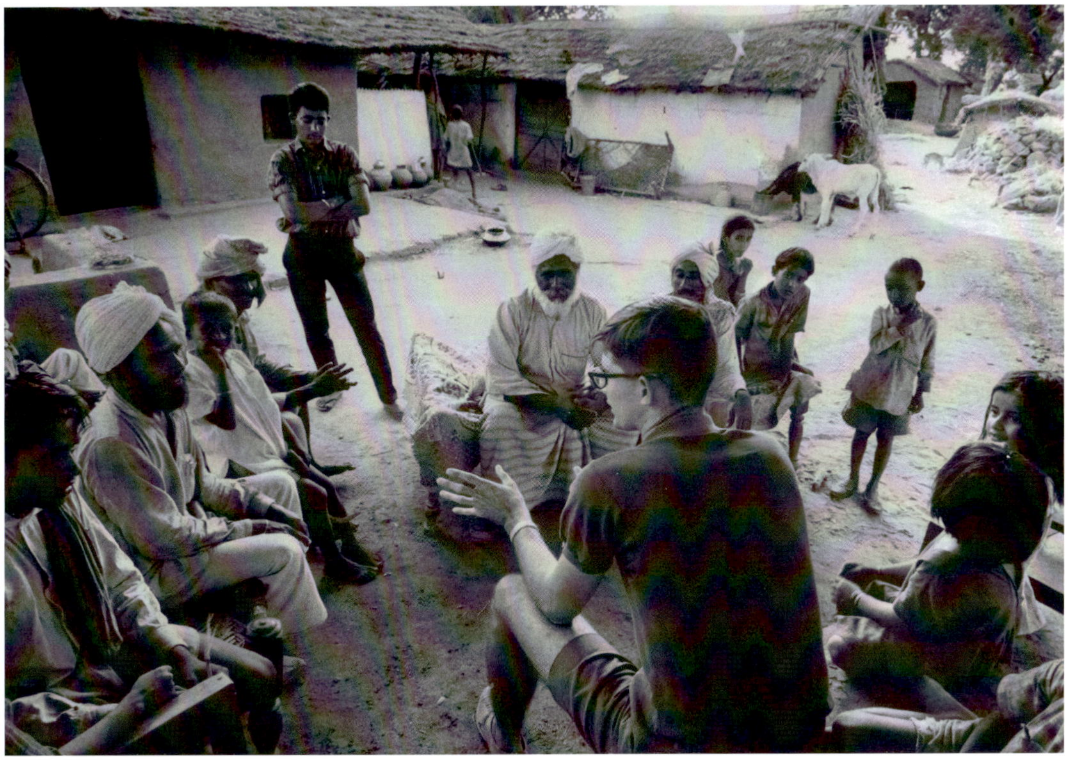

The Peace Corps was founded in 1961 by President John F. Kennedy (1917–1963), who challenged Americans to serve their country by living and working in developing nations. The idea emerged from Kennedy's campaign promise to create a program that would promote world peace and friendship through voluntary service. Congress officially established the Peace Corps in March 1961, with Kennedy's brother-in-law R. Sargent Shriver as its first director. The program sends American volunteers to work on projects involving education, health, agriculture, and community development in host countries for typically two years. Volunteers live at the grassroots level, learning local languages and immersing themselves in local cultures while sharing their skills and knowledge.

Since its inception, the Peace Corps has fostered cross-cultural understanding, provided valuable development assistance, and created lasting relationships between the United States and host nations. Many volunteers have gone on to careers in public service, international development, and education, bringing global perspectives to American society. The organization continues to adapt to changing world needs while maintaining its core mission of promoting peace through service.

Peace Corps volunteer George Beebe talks with villagers in India about the School Partnership Program, 1968.

1962 TELSTAR 1

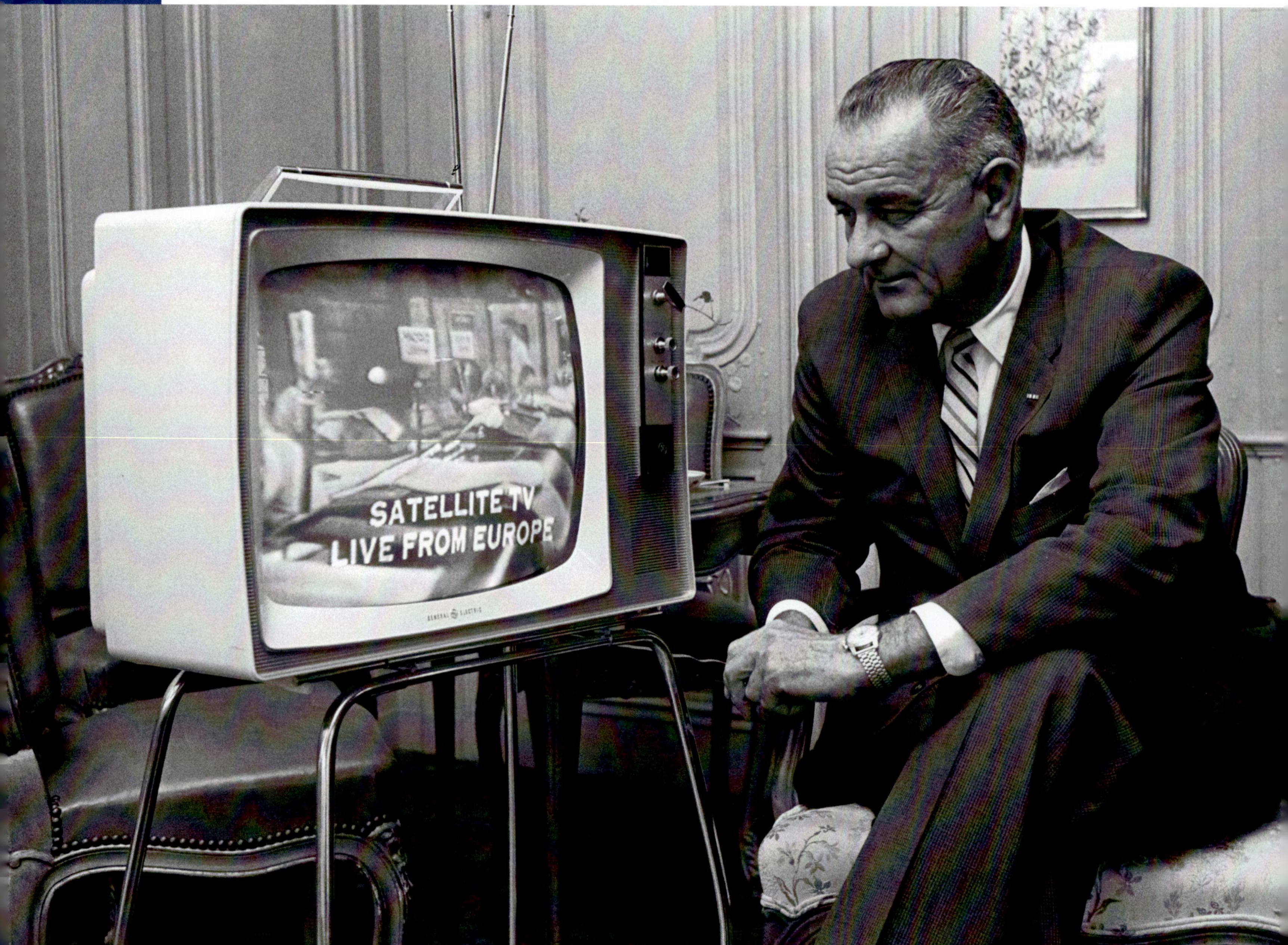

Telstar 1, launched on July 10, 1962, marked a revolutionary leap in global communications as the first privately sponsored satellite capable of relaying live television signals between continents. Developed by Bell Telephone Laboratories and launched by NASA, the beach ball–sized satellite weighed 170 pounds and orbited Earth every 158 minutes, providing brief windows for transatlantic communication. The satellite's first transmission occurred on July 23, 1962, captivating the world as it beamed live images. Millions witnessed the historic moment, including the first live transatlantic television broadcast featuring American and European performers. Though Telstar 1 operated for only seven months before radiation from nuclear tests damaged its systems, its impact was profound. It demonstrated the feasibility of satellite communications and paved the way for the modern global telecommunications network. The satellite inspired countless innovations in broadcasting, international news coverage, and eventually the Internet age. Telstar 1 fundamentally changed how people communicate, making real-time global connection possible and shrinking the world in ways previously unimaginable.

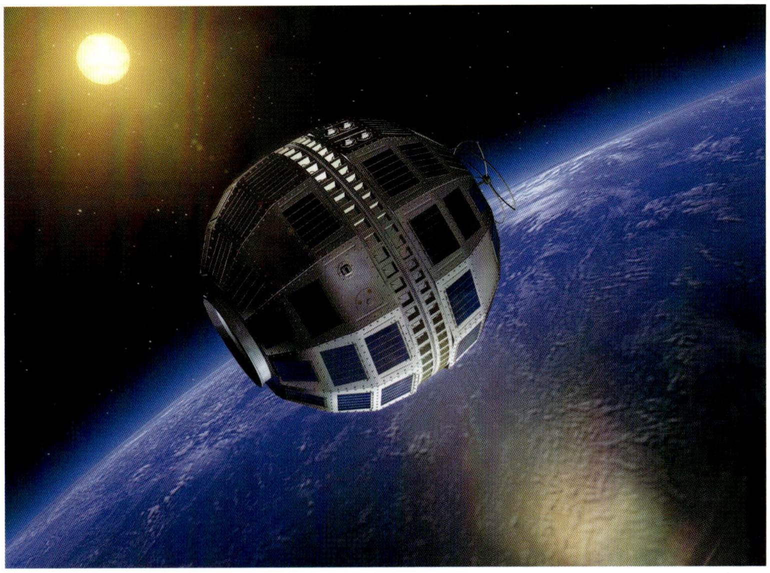

LEFT: Vice President Lyndon Johnson watches the Eurovision Television broadcast relayed by the new communications satellite, Telstar, in his home, July 23, 1962. **ABOVE:** Telstar 1 in orbit, 1962.

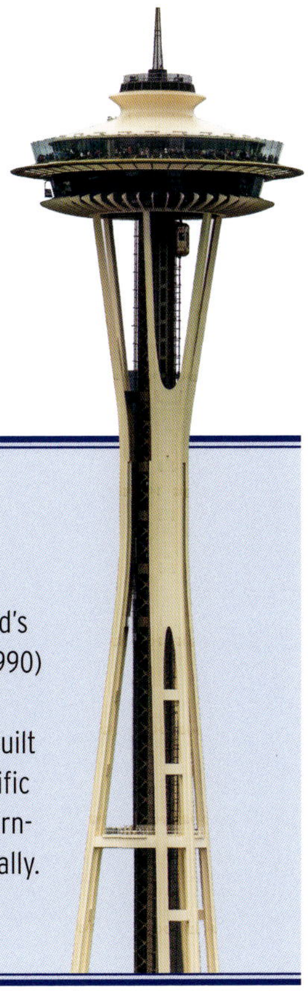

1962 The Space Needle

The Space Needle, Seattle's most recognizable landmark, was constructed for the 1962 World's Fair as a symbol of humanity's space-age aspirations. Designed by Edward E. Carlson (1911-1990) and John Graham Jr. (1908-1991), the 605-foot tower featured a flying saucer-like top and slender profile that captured the era's optimism about technology and space exploration. Built in just four hundred days, the Space Needle is an enduring icon of Seattle's skyline and Pacific Northwest identity. Its legacy extends beyond architecture, representing mid-century modernism and serving as a cultural touchstone that continues attracting millions of visitors annually.

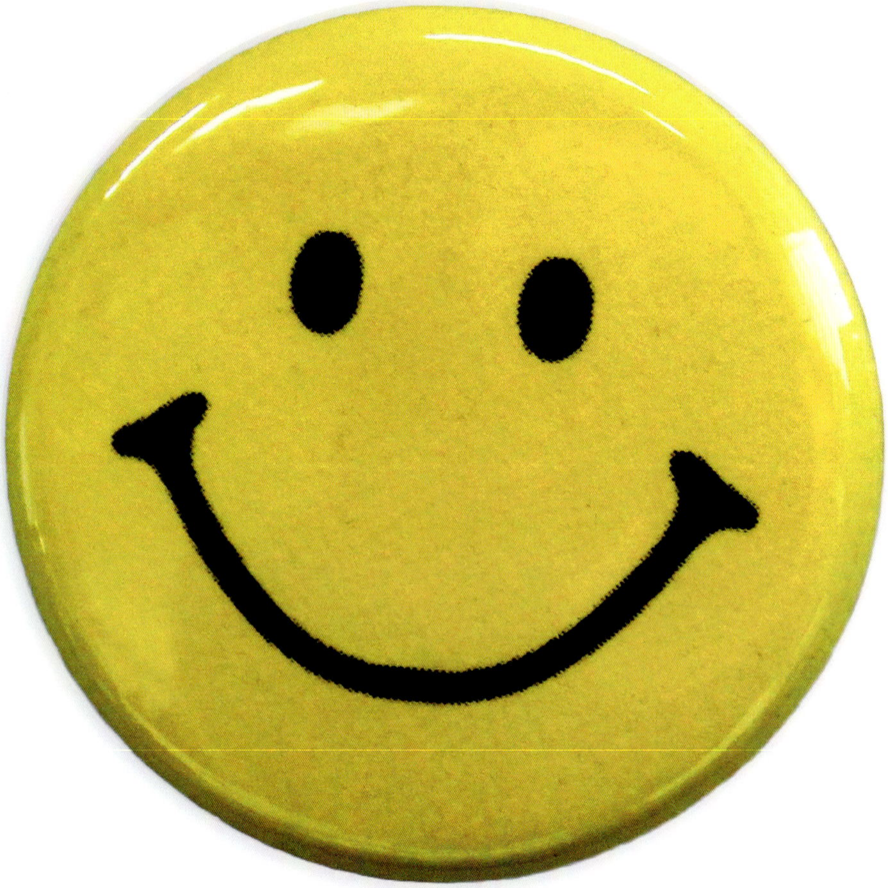

1963 THE SMILEY FACE

The smiley face icon was created in 1963 by Harvey Ball (1921–2001), a commercial artist from Worcester, Massachusetts. Ball was commissioned by State Mutual Life Assurance Company to design a symbol that would boost employee morale during a period of corporate reorganization. Working for just $45, Ball created the simple yellow circle with two black dots for eyes and a curved line for a smile in approximately ten minutes. The design's genius lay in its simplicity. Ball deliberately kept it basic, believing that complexity would diminish its emotional impact. The company initially used it on buttons, posters, and desk cards, distributing thousands to employees. After the icon was appropriated, some say by the owners of a greeting card shop, the smiley face became a cultural phenomenon in the 1970s, adopted by the counterculture movement and later commercialized widely. The design evolved into digital emoticons and emojis, becoming fundamental to online communication. Ball's creation inadvertently established the visual language of digital emotion, influencing how billions express feelings in text messages, social media, and digital platforms. Today, it represents one of the most recognizable symbols globally, transcending language barriers and cultural differences.

1964 POP-TARTS

Pop-Tarts were invented in 1964 by Kellogg's as a convenient breakfast pastry for busy Americans. The concept emerged during America's convenience food revolution, when hectic lifestyles demanded quick meal solutions requiring minimal preparation. Originally marketed as "Fruit Scones," these rectangular pastries featured simple fruit fillings in flaky pastry shells, engineered to be heated in toasters. The initial lineup included strawberry, blueberry, brown sugar-cinnamon, and apple currant, eventually expanding to dozens of varieties from chocolate chip cookie dough to s'mores. Beyond breakfast, Pop-Tarts evolved into a cultural phenomenon, appearing in art installations, inspiring Internet memes, and becoming emblematic of American processed food culture. The toaster pastries have maintained relevance across generations, adapting to changing tastes while preserving their core identity as grab-and-go breakfast food. Today, Pop-Tarts remain a billion-dollar product line, representing both mid-century food technology innovation and America's enduring relationship with convenience foods. Their success demonstrates how a simple concept—portable, shelf-stable pastries that taste good hot or cold—can become deeply embedded in a nation's culinary landscape.

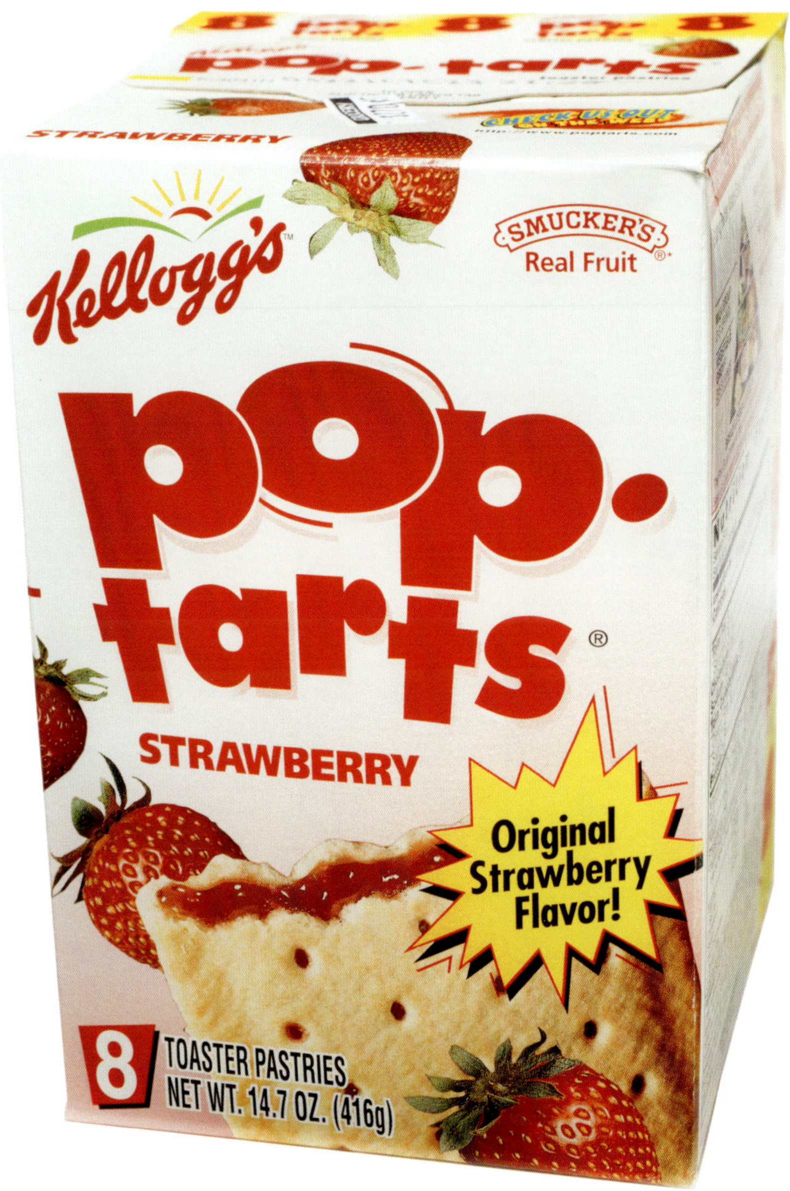

1965 KEVLAR

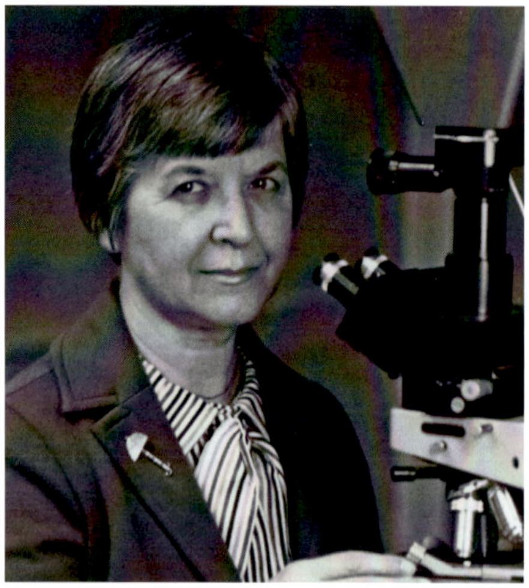

ABOVE: Kevlar inventor Stephanie Kwolek.
BELOW: Maud Fontenoy becomes the first woman to row solo across the North Atlantic Ocean in a boat made of cedar, fiberglass, and kevlar, 2005.

Kevlar was invented in 1965 by chemist Stephanie Kwolek (1923–2014) while she was researching lightweight yet strong fibers for tire reinforcement at DuPont. Working with polymer solutions that appeared cloudy and unpromising, Kwolek chose to spin one particular solution that her colleagues might have discarded. This decision led to the discovery of a revolutionary material five times stronger than steel by weight. The resulting fiber has exceptional tensile strength while maintaining flexibility and lightness. Kevlar's most famous application is in bulletproof vests, which have saved countless lives since their introduction in the 1970s. But its impact extends far beyond personal protection. The material has transformed industries from aerospace to sports equipment, appearing in everything from aircraft components and racing sails to tennis rackets and smartphone cases. Today, the story of Kevlar's invention demonstrates how fundamental research can yield unexpected discoveries with profound real-world applications. Kwolek's perseverance with an unpromising-looking solution ultimately created one of the most important synthetic materials of the modern era.

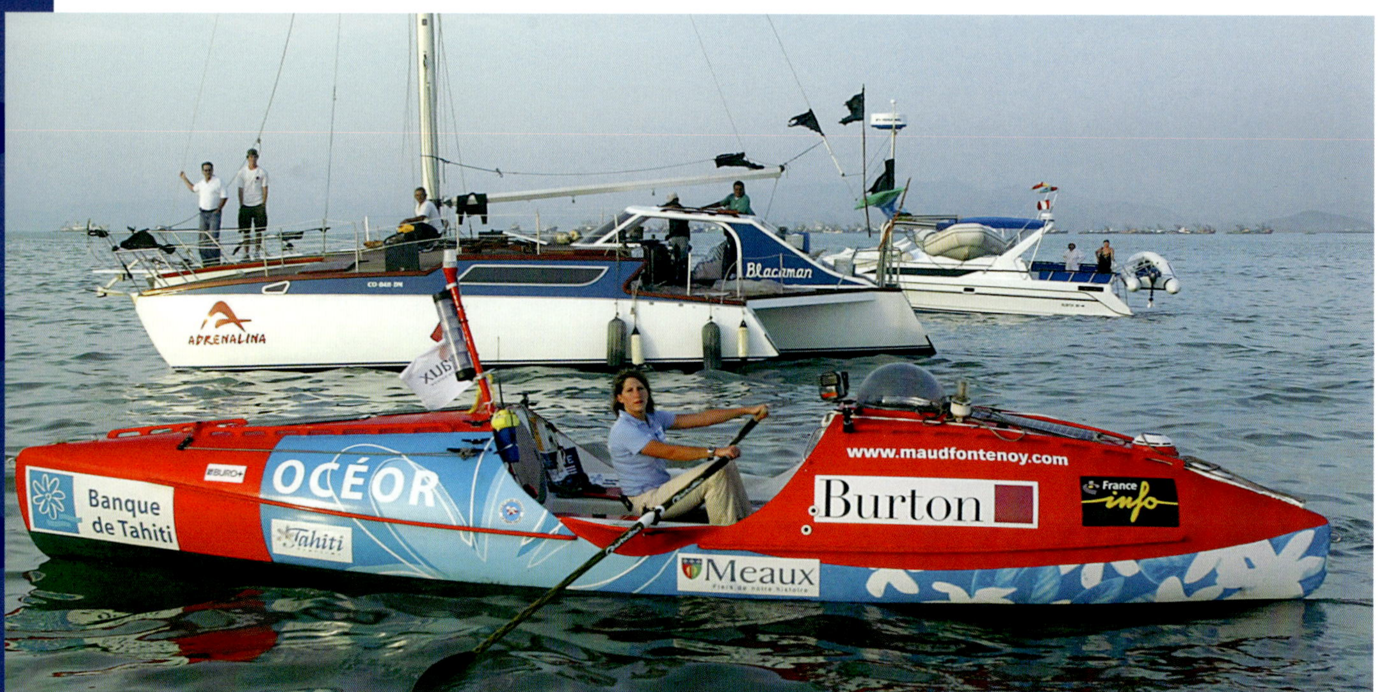

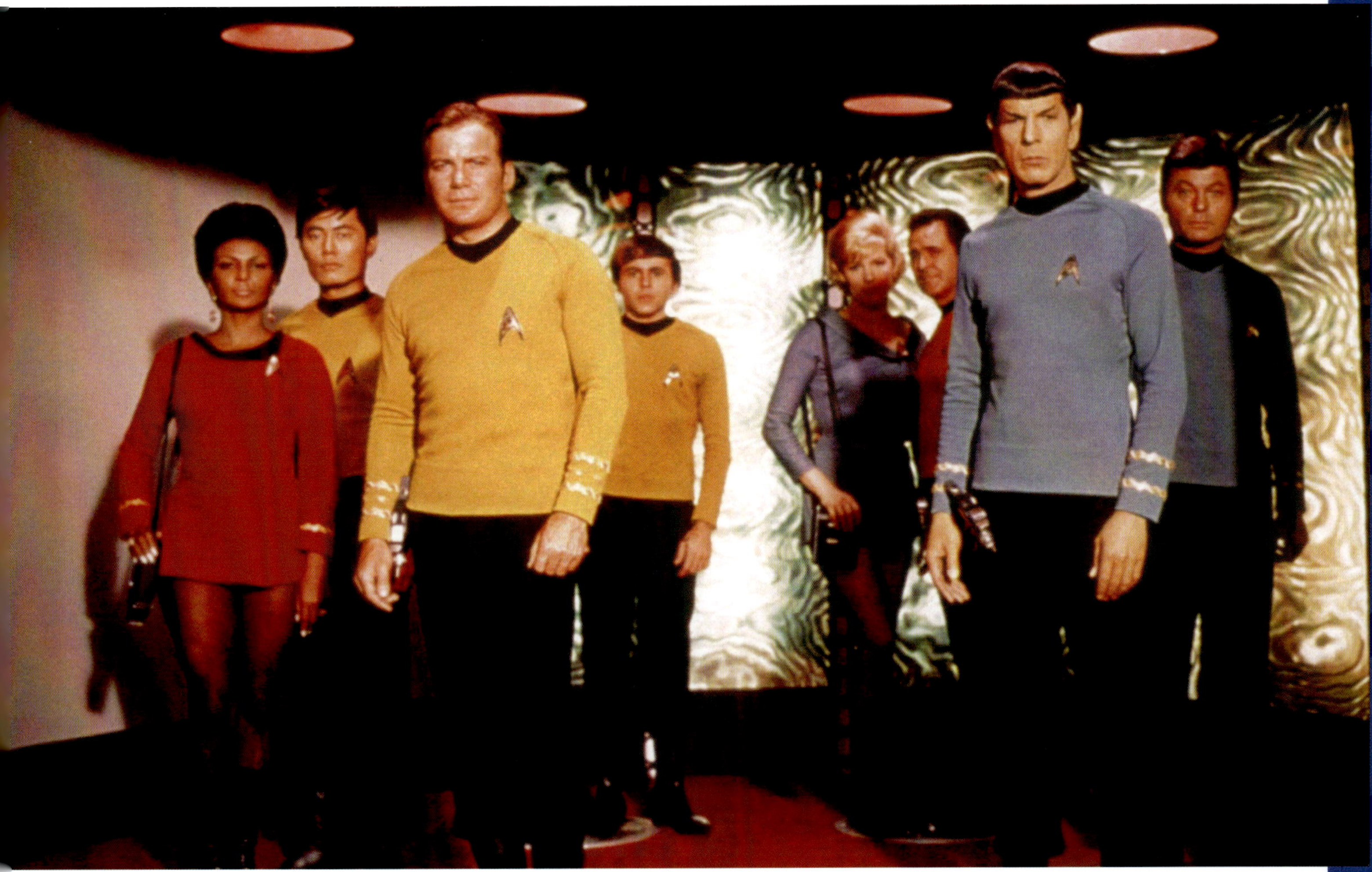

The cast of *Star Trek* ready to transport to their next adventure.

1966 *STAR TREK*

Star Trek emerged from the mind of Gene Roddenberry (1921–1991) in 1966; he pitched it as "Wagon Train to the stars." Kirk, Spock, Dr. McCoy, Uhura, Scotty, and the rest of the crew of the USS *Enterprise* laid the foundation for one of science fiction's most enduring franchises. Roddenberry's creation was groundbreaking for its optimistic vision of humanity's future, depicting a post-scarcity society where diverse species cooperated through the United Federation of Planets. The show tackled contemporary social issues through allegorical storytelling, featuring television's first interracial kiss and promoting themes of equality, exploration, and diplomatic solutions over warfare. Despite cancellation after three seasons, the show spawned multiple series, including *The Next Generation*, *Deep Space Nine*, *Voyager*, and *Discovery*, along with numerous films. It has even influenced real-world technology development, inspiring innovations from cell phones to tablets. The series created a passionate global fan base and established many science fiction conventions still followed today. Its impact on popular culture and its hopeful portrayal of humanity's potential continues to resonate, making *Star Trek* a cultural touchstone that transcends generations.

1967 MICROWAVE OVEN

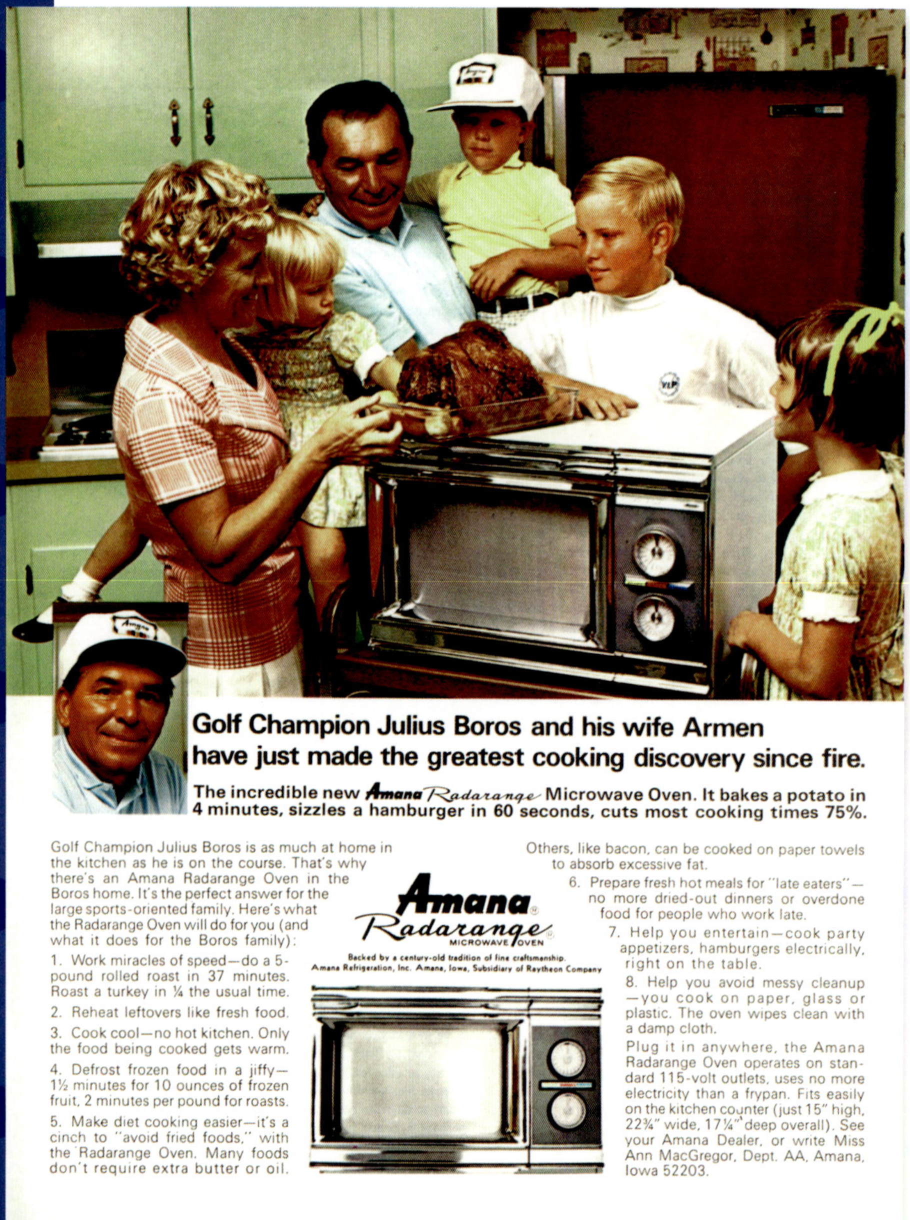

The first countertop microwave oven made its debut in 1967 when Amana Refrigeration, Inc. introduced the Radarange, priced at $495. The name "Radarange" reflected the oven's origins in radar technology. Microwave ovens were originally invented in 1945 by Percy Spencer (1894–1970), an engineer working for Raytheon on radar technology. Spencer discovered the cooking potential of microwaves accidentally when a chocolate bar in his pocket melted while he was working near a magnetron. The first commercial model was released in 1947 and was massive—standing nearly six feet tall, weighing 750 pounds, and costing around $5,000 (equivalent to about $70,000 today). These early units were primarily used in restaurants and industrial settings. The countertop Radarange's launch made microwave cooking accessible to average American households. By 1975, sales of microwave ovens outpaced gas ovens for the first time, demonstrating the technology's rapid adoption. The convenient, time-saving appliance quickly became a symbol of modern living. By the late 1990s, nine out of every ten American households owned a microwave, making it one of the most successful kitchen appliances in history. The Radarange didn't just change how Americans cooked—it fundamentally altered eating habits, meal preparation, and kitchen design, establishing the microwave as an indispensable fixture in modern homes worldwide.

A 1969 ad for the Amana Radarange microwave that appeared in *Sports Illustrated*.

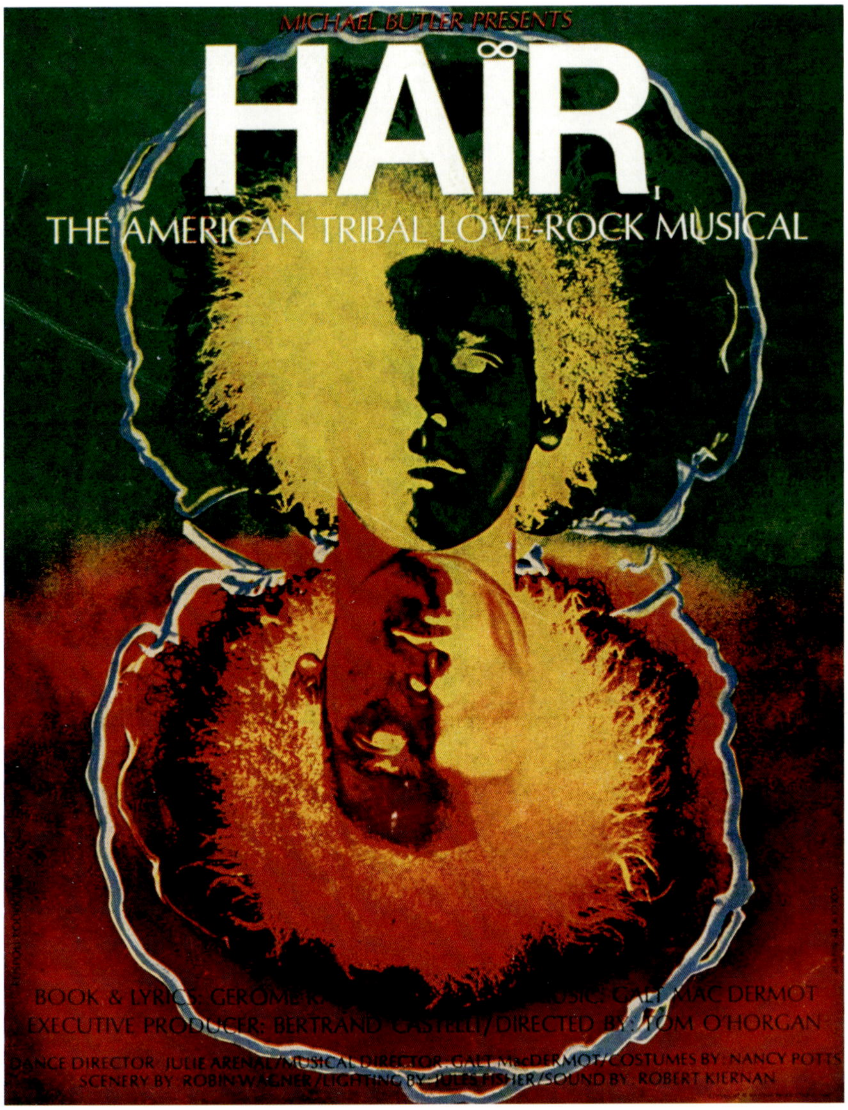

A poster for the original Broadway production, 1968.

1968 *HAIR*

Hair, the groundbreaking musical, premiered off-Broadway in 1967 before moving to Broadway in 1968. Created by Gerome Ragni (1935–1991) and James Rado (1932–2022) with music by Galt MacDermot (1928–2018), the show emerged from the heart of the counterculture movement and became the first rock musical on Broadway. Set against the backdrop of the Vietnam War, *Hair* followed a group of hippies in New York's East Village, exploring themes of peace, love, sexual liberation, and anti-war sentiment. The production was revolutionary for its time, featuring full-frontal nudity, drug references, and rock music on stages traditionally reserved for more conservative fare. *Hair* helped legitimize rock music as a viable Broadway genre while addressing social issues that mainstream entertainment had largely avoided. Its influence can be seen in later concept musicals and rock operas that followed. The 1979 film adaptation introduced the story to new generations, and songs like "Aquarius" and "Let the Sunshine In" became cultural anthems that continue to resonate today. The musical is widely credited with paving the way for later composer-lyricists like Andrew Lloyd Webber and attracting a new, younger audience to theater, fundamentally changing Broadway's demographic and artistic possibilities.

1969 THE APOLLO 11 LUNAR MODULE

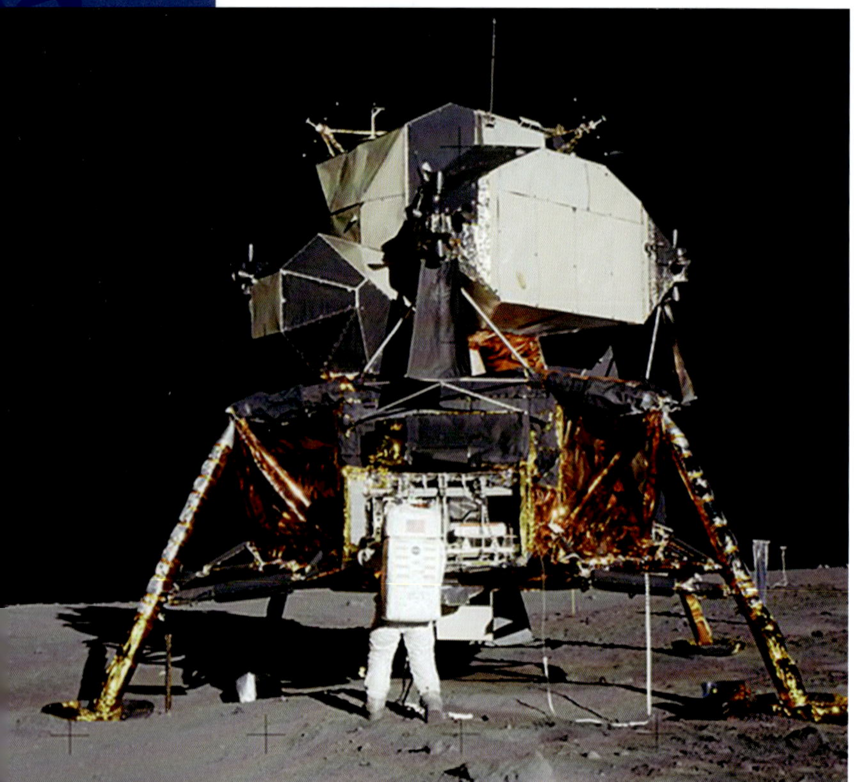

The Apollo 11 mission launched on July 16, 1969, carrying the lunar module *Eagle* on humanity's first crewed moon landing. Eagle consisted of two stages: a descent stage for landing and an ascent stage for returning to orbit. On July 20, 1969, *Eagle* separated from command module *Columbia* with Neil Armstrong and Buzz Aldrin aboard. Despite dangerously low fuel levels, Armstrong successfully piloted *Eagle* to a safe landing in the Sea of Tranquility. Armstrong became the first human to step onto the lunar surface, followed by Aldrin, in a moment broadcast live to millions worldwide. *Eagle* demonstrated the feasibility of lunar exploration and inspired generations of scientists, engineers, and dreamers. The descent stage remains on the moon as a permanent monument bearing a commemorative plaque: "Here men from the planet Earth first set foot upon the Moon July 1969 AD. We came in peace for all mankind," signed by the three Apollo 11 crew members and President Nixon.

Buzz Aldrin outside the lunar module.

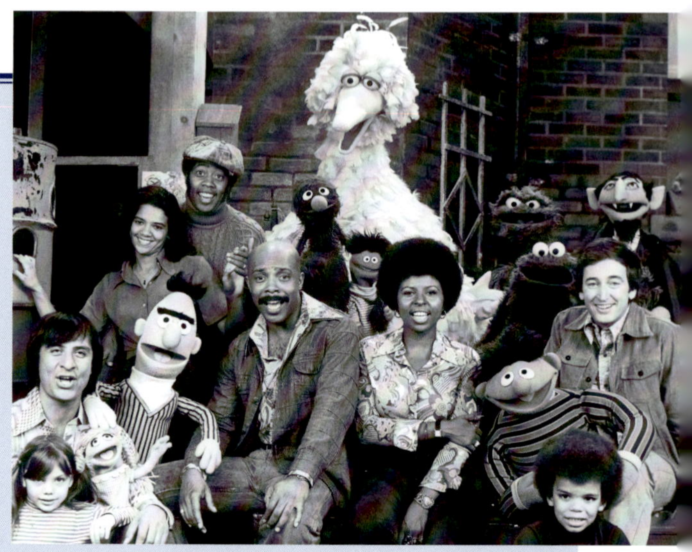

1969 *Sesame Street*

Sesame Street premiered on November 10, 1969, revolutionizing children's television with its innovative blend of education and entertainment. Created by Joan Ganz Cooney (1929-) and Lloyd Morrisett (1929-2023), the show addressed urban childhood experiences while teaching basic literacy and numeracy skills. Its diverse cast, including iconic characters like Big Bird, Elmo, and Cookie Monster, broke new ground in representation and inclusion. The series has profoundly influenced early childhood education, with research demonstrating its positive impact on school readiness and academic achievement. Since its inception, *Sesame Street* has expanded globally, adapting to local cultures while maintaining its core educational mission. It has won numerous awards and continues to evolve with digital platforms, cementing its status as a cornerstone of quality children's programming worldwide.

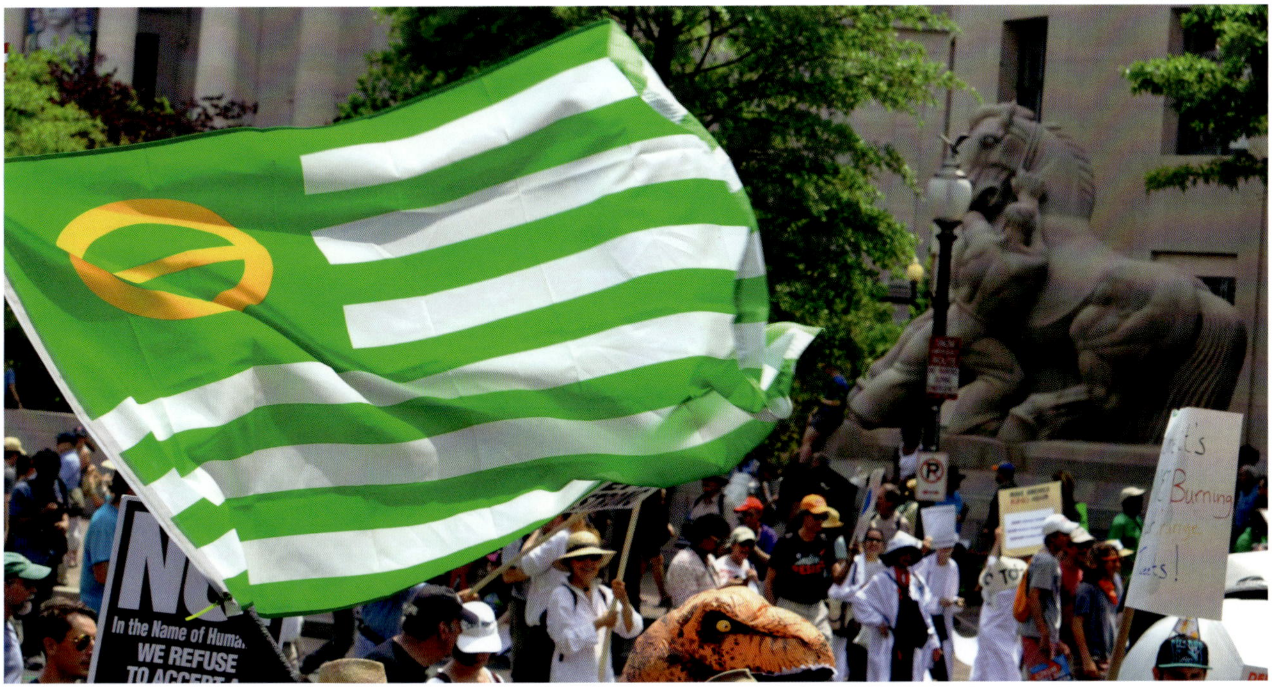

1970 EARTH DAY FLAG

The Earth Day flag was created and copyrighted by peace activist John McConnell (1915–2012) to accompany a celebration in March 1970 in San Francisco, California. However, there's also a significant story about Lanphier High School in Springfield, Illinois, and their historic role in the first Earth Day celebrations. On the first Earth Day, April 22, 1970, dozens of Lanphier students marched behind their flag to the statehouse, picking up litter as they went, to highlight the need for environmental reform. This homemade flag was created in the classroom of science teacher Ray Bruzan (1946–), who had transformed Room 308 into the school's "Environmental Action Center" during the 1969–70 school year. The sixty-one-by-thirty-seven-inch flag features green stripes symbolizing a clean earth, white stripes symbolizing clean air, and the Greek letter theta, representing death. The design was based on one created by Ron Cobb, cartoonist for the *Los Angeles Free Press*. The flag became part of the Smithsonian Institution's permanent collection in 1995 as part of their twenty-fifth anniversary Earth Day exhibit.

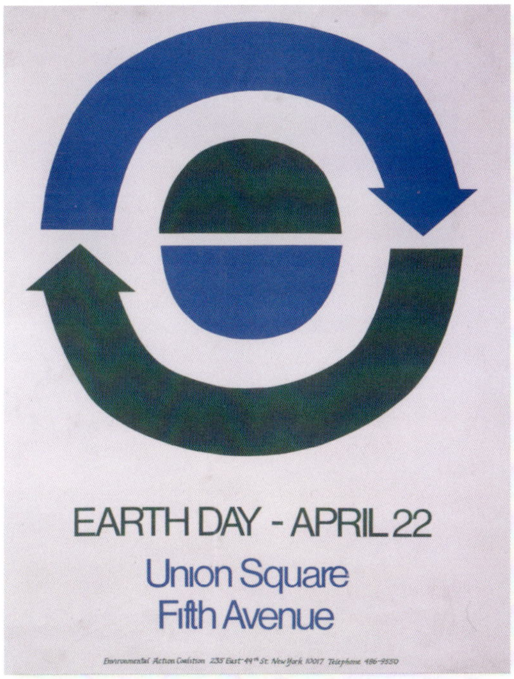

A poster for the first Earth Day, 1970, and the flag that is still used today.

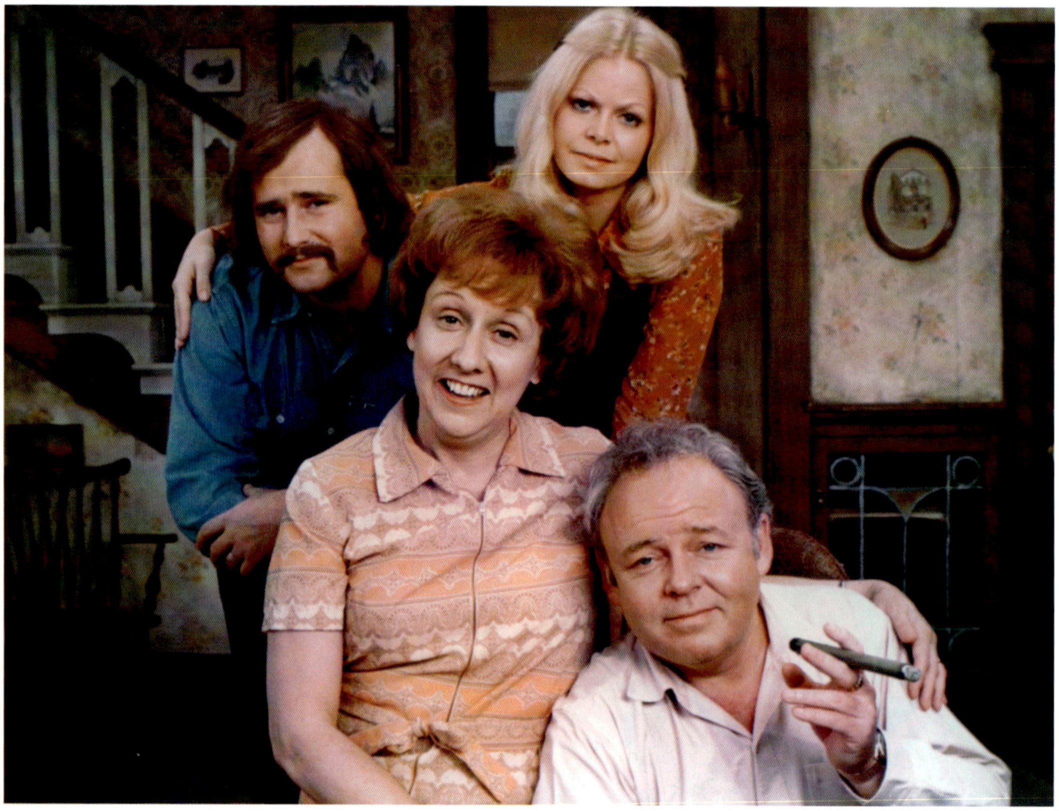

1971 *ALL IN THE FAMILY*

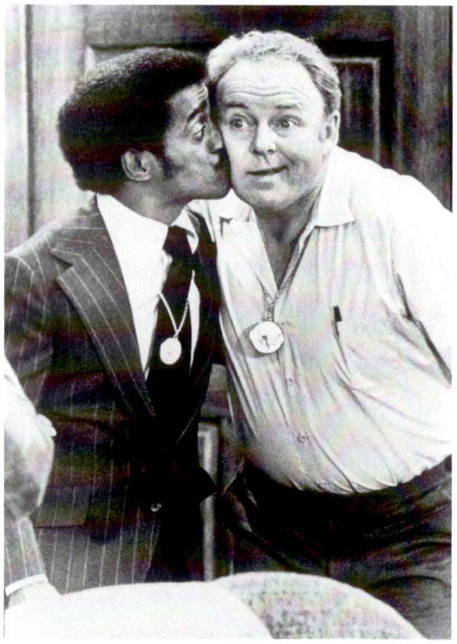

TOP: The family: Archie (Carroll O'Connor) and Edith Bunker (Jean Stapleton), son-in-law Mike Stivic (Rob Reiner), and daughter Gloria (Sally Struthers). BOTTOM: Sammy Davis Jr. shocks the bigoted Archie with a kiss in the iconic episode "Sammy's Visit."

Created by Norman Lear (1922–2023) and adapted from the British sitcom *Till Death Us Do Part*, *All in the Family* premiered on CBS in January 1971. The show centered on Archie Bunker, a working-class bigot from Queens, his naïve but kindhearted wife Edith, and his liberal daughter, Gloria, and son-in-law, Mike, creating explosive dinner-table debates that reflected America's cultural tensions. Lear's groundbreaking series shattered television taboos by directly addressing racism, sexism, homophobia, and political divisions with unprecedented frankness. The show's courage to tackle controversial subjects like the Vietnam War, abortion, and civil rights marked a seismic shift from the sanitized family comedies of the 1950s and 1960s. The series dominated ratings for five consecutive years and spawned multiple spin-offs, including *Maude* and *The Jeffersons*. Its legacy transformed television comedy, proving audiences craved authentic social commentary alongside humor. *All in the Family* opened doors for future shows to examine difficult topics, fundamentally changing how television could function as a mirror for American society's struggles and contradictions.

1972 *PONG*

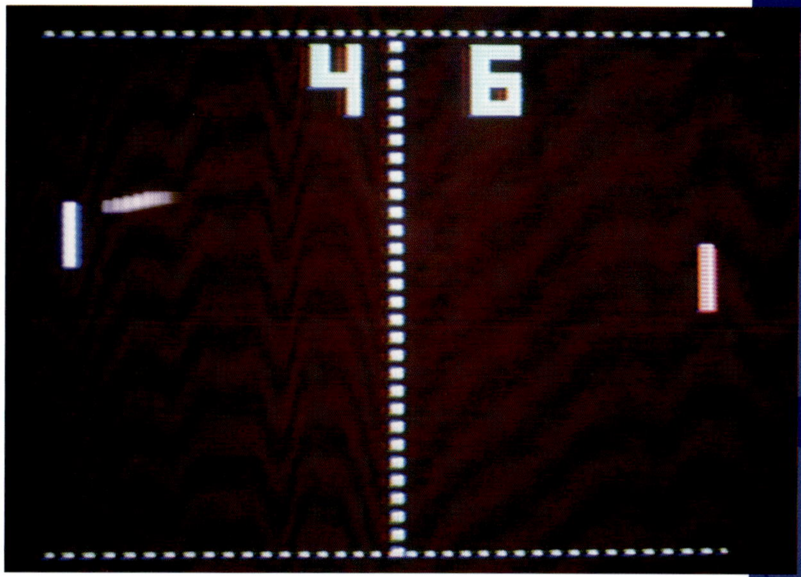

In 1972, Atari created *Pong*, a simple tennis simulation that forever changed entertainment. The game featured two white paddles and a square ball bouncing across a black screen, with players controlling paddles through rotating knobs. What began as an internal training exercise became the first commercially viable video game, birthing an entire industry. Released as an arcade cabinet, *Pong*'s genius lay in its immediacy—players needed no manual or tutorial. This intuitive design helped legitimize video games beyond technical hobbyists, making them accessible entertainment for everyone. The game's commercial success demonstrated that interactive entertainment could generate substantial profits, attracting venture capitalists and entrepreneurs to what was previously a novelty market. *Pong* catalyzed the arcade explosion of the mid-1970s, transforming establishments into cultural gathering places where quarters clinked and high scores became badges of honor. Atari's 1975 home console version marked the first time families could experience video games in their living rooms, representing a seismic shift in entertainment consumption. Beyond commercial impact, *Pong* established fundamental design principles still defining gaming today: competitive multiplayer interaction, score-driven objectives, and real-time responsiveness. These elements influenced everything from *Space Invaders* to modern esports. In its elegant simplicity, *Pong* didn't just launch an industry—it created the DNA of interactive entertainment itself.

The console that controlled the paddles, and the simple—but groundbreaking—screen.

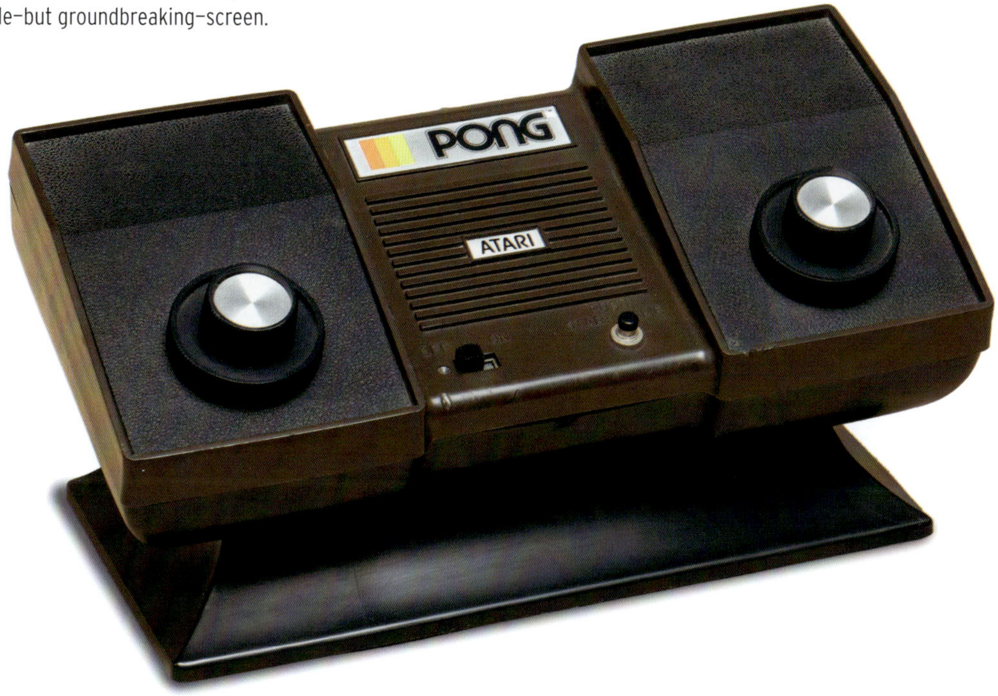

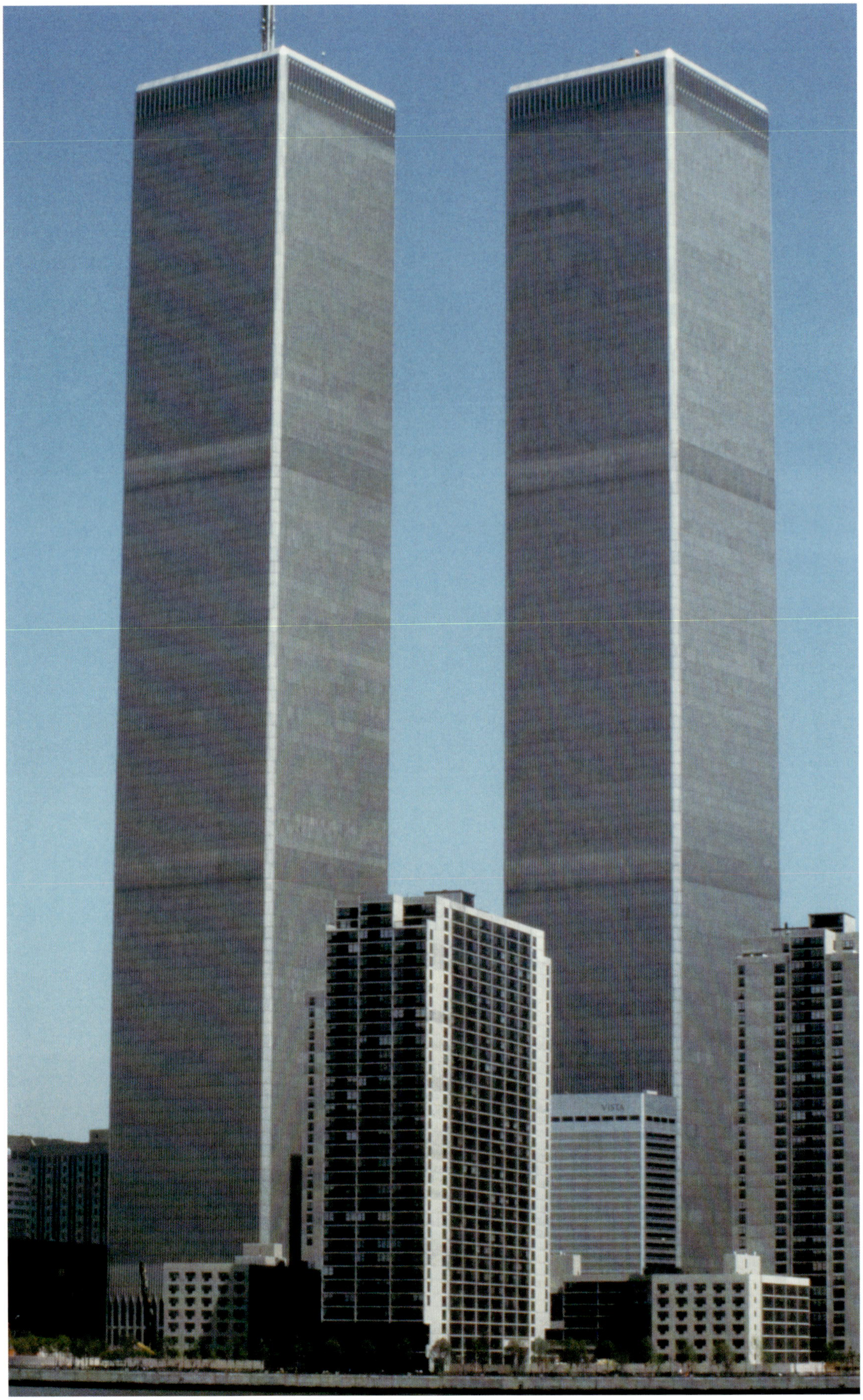
VISTA

1973 THE WORLD TRADE CENTER

After nearly a decade of construction, the original World Trade Center officially opened on April 4, 1973. Designed by architect Minoru Yamasaki (1912–1986), the Twin Towers briefly held the distinction of being the world's tallest buildings, with the North Tower reaching 1,368 feet and the South Tower standing at 1,362 feet. The building complex quickly became an iconic symbol of New York City's skyline. Beyond their architectural significance, the towers represented the economic power and ambition of the United States during the latter half of the twentieth century. The World Trade Center housed thousands of workers and attracted hundreds of visitors daily, serving as a hub of international business, finance, and tourism. However, the towers' prominence also made them targets. Following a 1993 bombing and the devastating September 11, 2001, attacks that claimed nearly three thousand lives, the World Trade Center's legacy became forever intertwined with tragedy, and also resilience. Today, the rebuilt site serves as both a memorial to those lost and a testament to renewal. The new World Trade Center complex, including One World Trade Center and the National September 11 Memorial and Museum, honors the past while symbolizing hope and the enduring spirit of New York City and America.

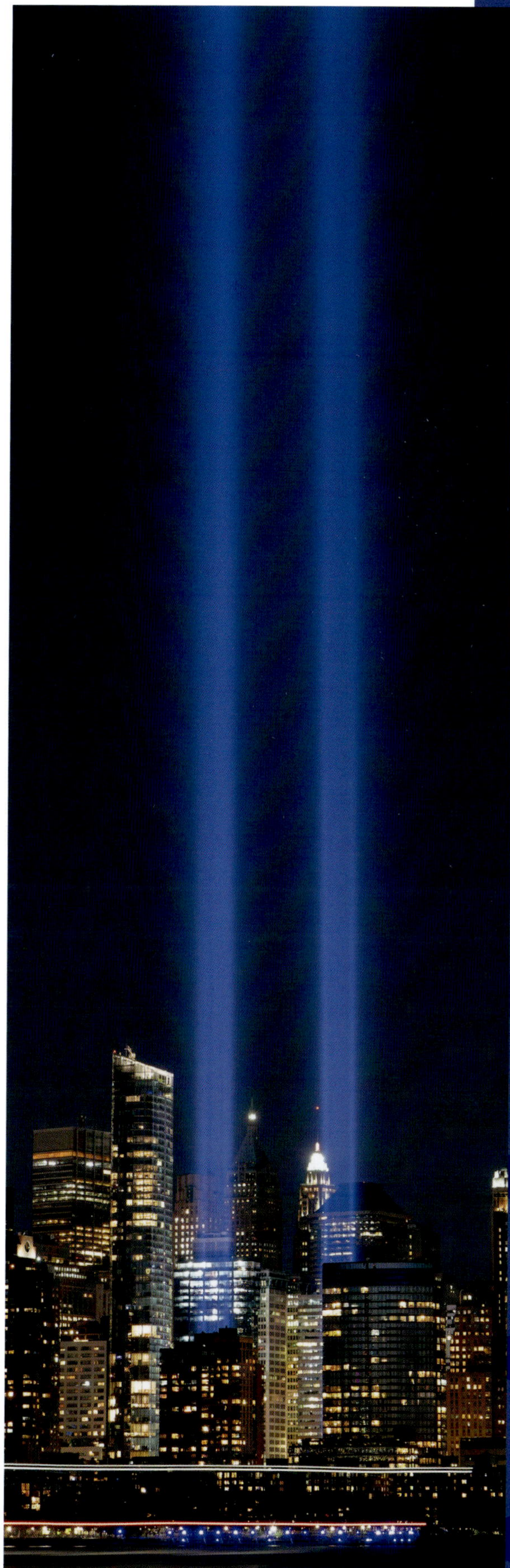

LEFT AND RIGHT: The towers as they stood in 1973, and the memorial lights that are now shown from the footprint of the towers every year on the night of September 11.

1974 THE BARCODE

The barcode's journey from concept to commercial reality spans decades. The technology for barcodes was first patented in 1949, but engineers developed the UPC version of the code in the 1970s to address the need for greater efficiency in grocery stores. The pivotal moment came on June 26, 1974, when a checkout clerk at a Marsh supermarket in Troy, Ohio, became the first person to scan a grocery item—a pack of Wrigley's Juicy Fruit chewing gum—using a Universal Product Code (UPC). The barcode transformed retail operations and supply-chain management worldwide. What began as a solution for grocery store efficiency evolved into a universal system for tracking products, managing inventory, and streamlining logistics across multiple industries. Barcode technology enabled just-in-time manufacturing, improved inventory accuracy, and facilitated the growth of modern retail giants. Today's QR codes and advanced scanning technologies trace their lineage directly to that first scanned pack of gum in Ohio, demonstrating how a simple pattern of black and white lines revolutionized how we track and manage physical goods globally.

1974 Post-it Notes

Post-it Notes were invented in 1968 by Spencer Silver (1941–2021), a 3M scientist attempting to create a superstrong adhesive. Instead, he developed a weak, removable adhesive that seemed useless. In 1974, colleague Art Fry realized it could solve his problem of bookmarks falling from his church hymnal. Using Silver's adhesive, Fry created removable bookmarks that stayed put without damaging pages. 3M recognized the potential and launched Post-it notes nationally in 1980. The iconic yellow color came from available scrap paper. This "failed" experiment became a revolutionary office product, demonstrating how accidents can create transformative innovations.

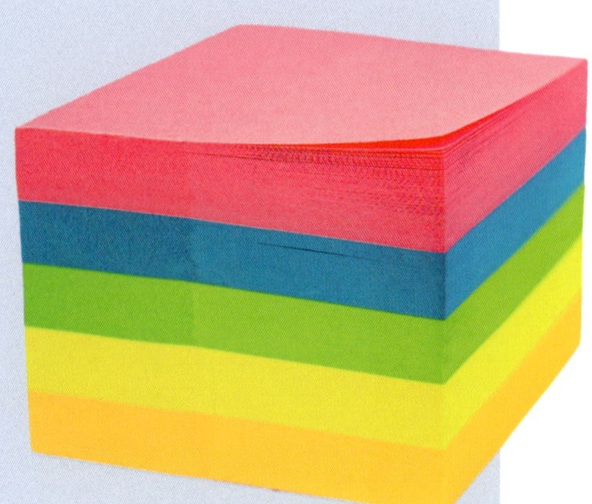

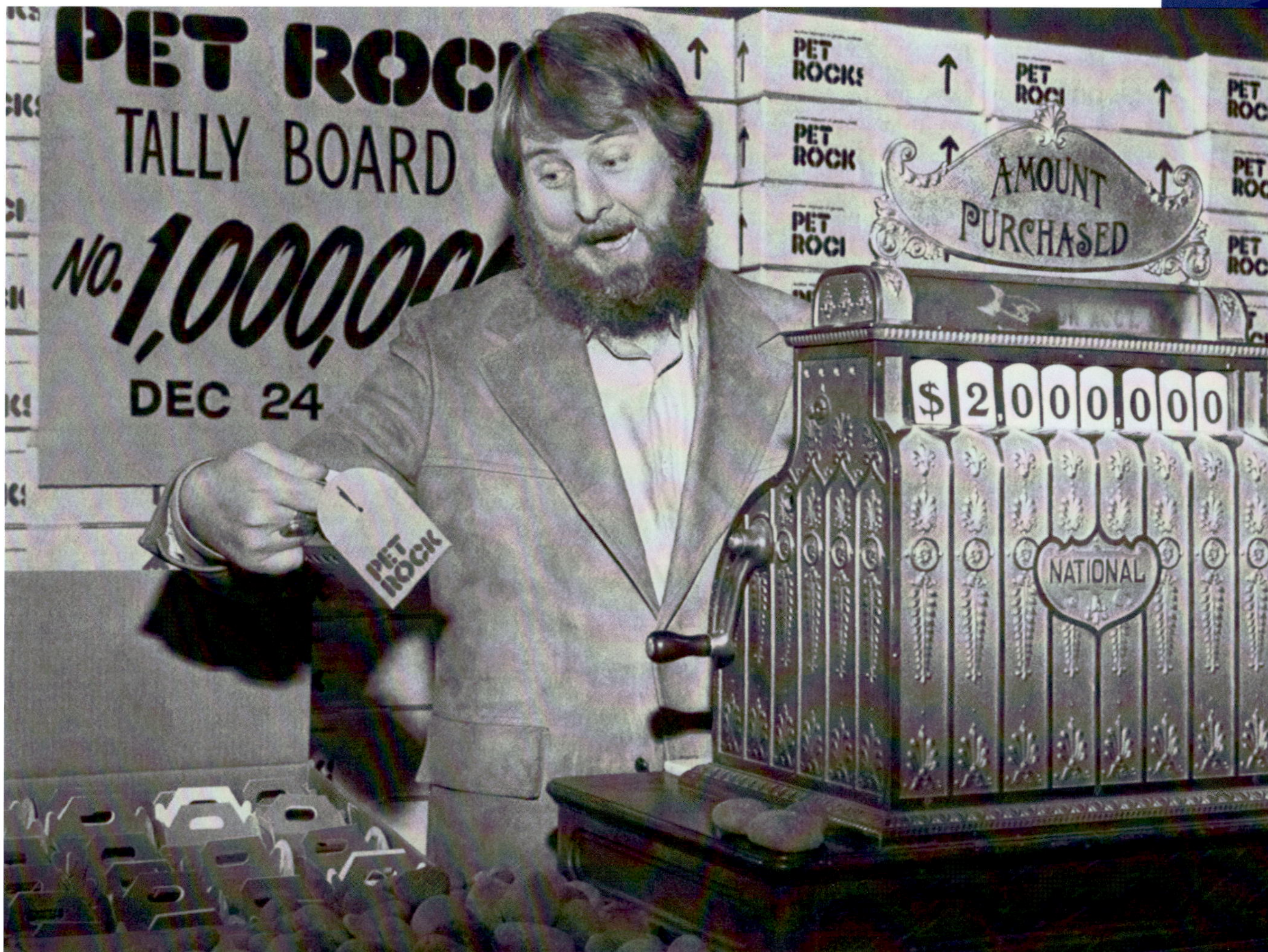

Creator Gary Dahl gets ready to pack up the one millionth Pet Rock sold, Christmas 1975.

1975 THE PET ROCK

Invented by advertising executive Gary Dahl (1936–2015) in 1975, the Pet Rock became one of the most bizarre and successful novelty items of the 1970s. The idea supposedly came to him during a conversation with friends who were complaining about their domestic animals. Dahl jokingly suggested that a rock would be the perfect pet—it wouldn't need feeding, walking, or veterinary care. Dahl packaged ordinary smooth stones from Rosarito Beach in Baja California, Mexico, in cardboard carriers with air holes, complete with a humorous thirty-two-page training manual titled "The Care and Training of Your Pet Rock." Pet Rocks became a cultural phenomenon during the 1975 Christmas season. Dahl sold over one and a half million units, earning him millions of dollars in just a few months. The fad burned out quickly by early 1976, but the Pet Rock remains an enduring symbol of American consumer culture and marketing ingenuity. It demonstrated how clever packaging and humor could transform the most mundane object into a must-have item, inspiring countless novelty products and becoming a cultural touchstone for absurd but successful business ideas.

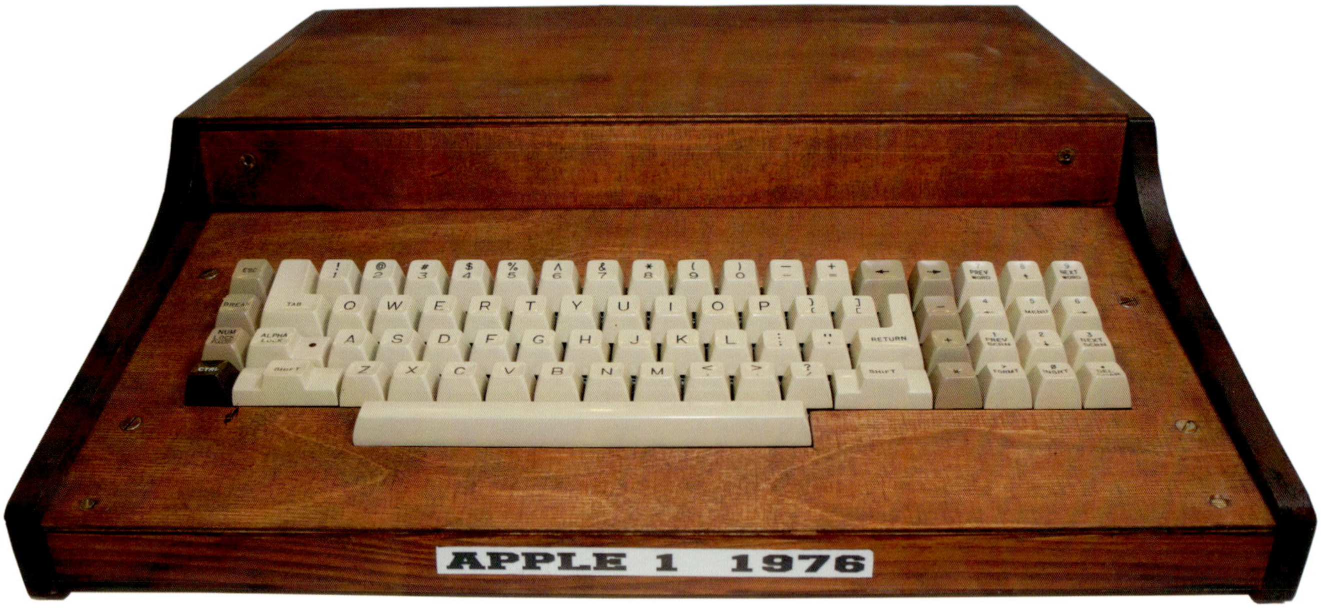

The Apple I was just a single-board computer kit; users had to build their own cases, like the one seen here.

1976 APPLE COMPUTER

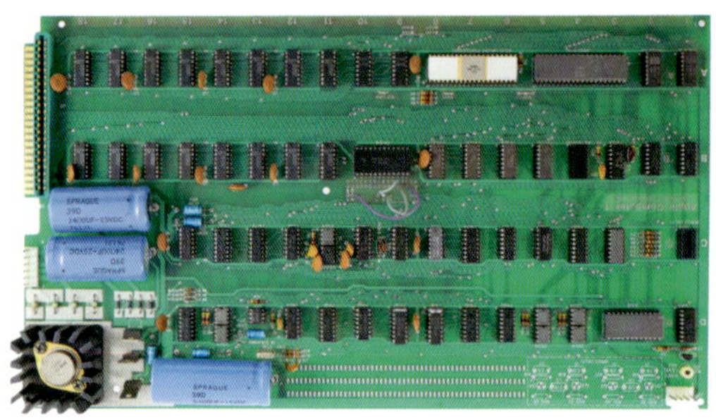

Apple Computer, Inc. was founded in 1976 when Steve Jobs (1955–2011) and Steve Wozniak (1950–) established the company in Jobs's garage in Los Altos, California. The Apple I, their first product, was a revolutionary single-board computer kit that marked an important milestone in the personal computing era. The company gained significant momentum with the release of the Apple II in 1977, which transformed how people interacted with technology by bringing computing power to homes, schools, and small businesses worldwide. Unlike the Apple I, which required users to supply essential components like a case and power supply, the Apple II was delivered as a complete, ready-to-use consumer product. Apple became one of the first successful personal computer companies and played a crucial role in popularizing user-friendly design principles in computing. Apple contributed to the development and adoption of graphical user interfaces, making these interfaces accessible and intuitive for everyday consumers. The success of these early computers established the foundation for Apple Inc., setting the stage for future groundbreaking products including the Macintosh, iPod, iPhone, and iPad. The principles of user-centered design and technological innovation established during this formative period continue to influence the company's approach today.

1977 ALEX HALEY'S *ROOTS*

Based on a 1976 novel by Alex Haley (1921–1992), the miniseries *Roots* premiered on ABC on January 23, 1977, drawing approximately 130 million television viewers. This groundbreaking eight-night broadcast followed Kunta Kinte from his village life in West Africa through his capture by slave traders to his enslavement in the United States. Night after night, viewers witnessed the experiences of Kunta Kinte's descendants and the often brutal conditions they endured as enslaved people on plantations across generations, leading up to the Civil War and eventual emancipation. Haley's novel, *Roots: The Saga of an American Family*, drew from his research into his own family's origins, giving it autobiographical elements. The miniseries had a profound and lasting legacy. First, it marked a significant departure from the typical roles offered to Black actors at the time—instead of minor, stereotypical characters, *Roots* provided substantial parts that helped launch the careers of LeVar Burton, who portrayed the young Kunta Kinte, and numerous other Black actors. Second, it fundamentally changed how Americans, particularly white Americans, understood slavery. Finally, it produced a massive surge in genealogical research, particularly among African American families seeking their roots, and fostered an unprecedented appreciation for Black history.

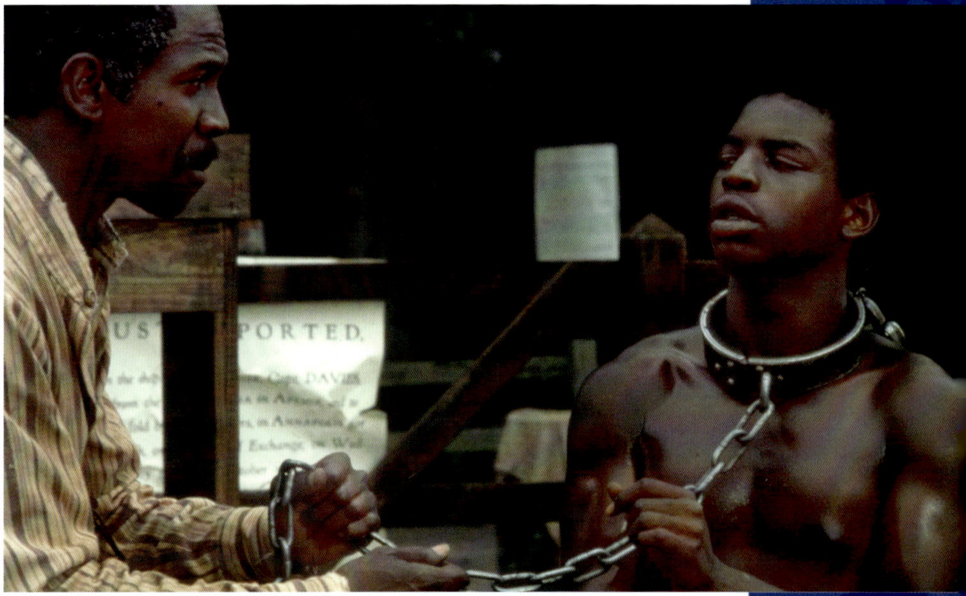

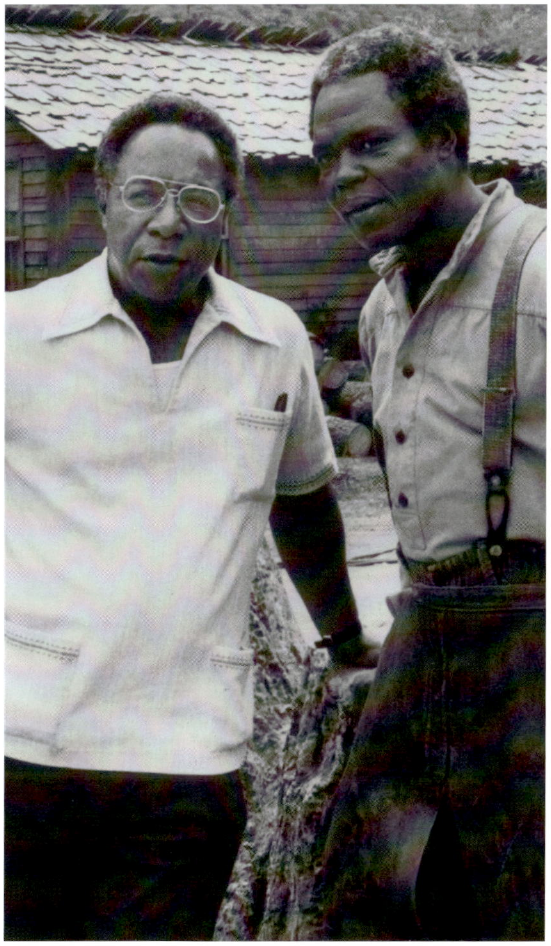

TOP: Louis Gosset, Jr. (l) as Fiddler and LeVar Burton (r) as Kunta Kinte. **RIGHT:** Georg Stanford Brown (r, as Tom Harvey) talks with Alex Haley between takes.

1978 GENETIC ENGINEERING

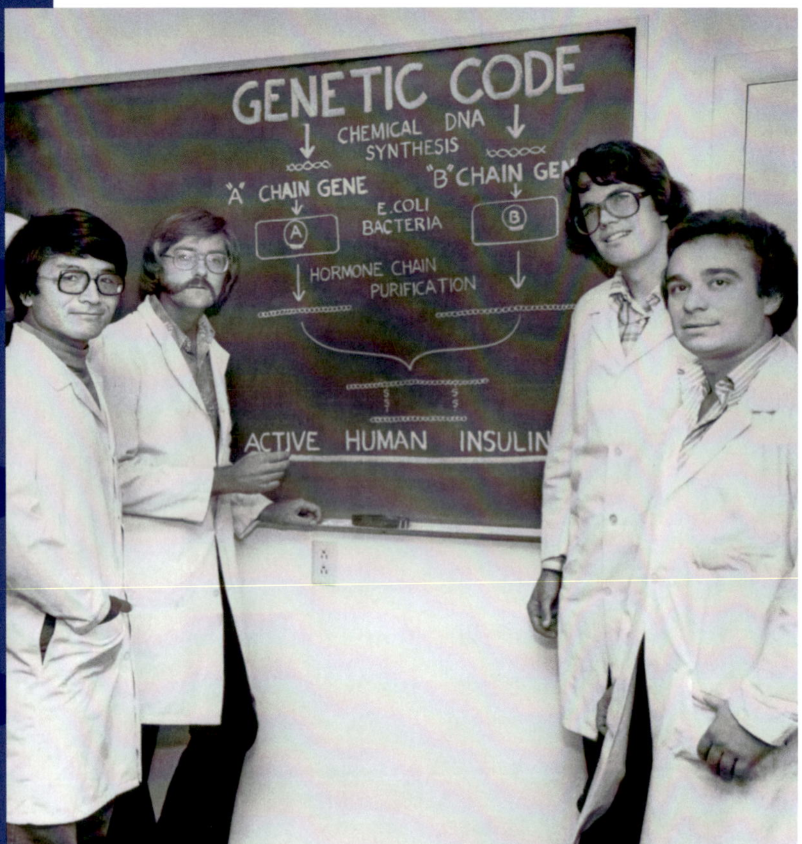

Scientists (l-r) Keiichi Itakura, Arthur D. Riggs, David V. Goeddel, and Roberto Crea explain how they "tricked" *E. coli* bacteria into producing insulin.

The first genetically engineered insulin emerged in 1978 through a collaboration between Genentech and City of Hope National Medical Center. This transformative achievement received FDA approval in 1982 and entered the market as Humulin, manufactured by Eli Lilly and Company. Prior to this breakthrough, diabetics depended on insulin extracted from pig and cow pancreases—a process that was both costly and problematic. This animal-derived insulin frequently triggered allergic reactions in patients and presented supply-chain limitations that affected availability and pricing. The newly developed synthetic insulin was chemically identical to human insulin, effectively eliminating adverse immune responses while ensuring a consistent, reliable supply of the essential medication. This advancement represented far more than a medical improvement—it marked the birth of the modern biotechnology industry and demonstrated that complex human proteins could be successfully mass-produced using recombinant DNA technology. The success of genetically engineered insulin created a blueprint for developing other critical substances, including growth hormones, clotting factors, and innovative cancer treatments. Today, nearly all insulin used globally is genetically engineered, benefiting millions of diabetics worldwide. More significantly, this achievement laid the groundwork for contemporary biotechnology and personalized medicine, creating an industry that continues to revolutionize healthcare through precision treatments and targeted therapies.

1979 ULTRALIGHT SPORTS WHEELCHAIR

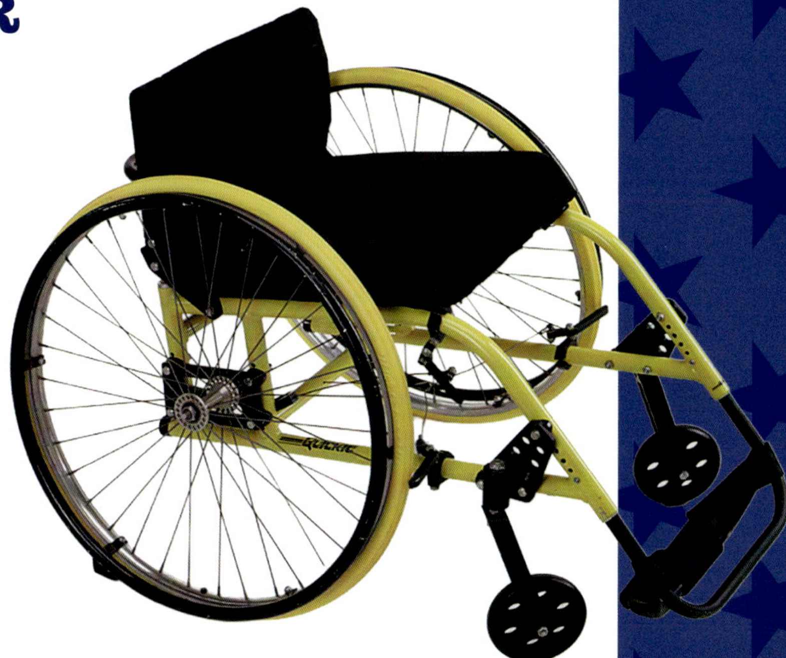

The first ultralight sports wheelchair was invented in 1978 by Marilyn Hamilton (1949–) and two associates, revolutionizing mobility for athletes with disabilities. Hamilton, a former tennis player who became paralyzed in a hang-gliding accident, was frustrated with the heavy, cumbersome wheelchairs of the era. Drawing inspiration from hang-gliding technology, the trio developed a lightweight aluminum frame that weighed significantly less than traditional steel chairs while maintaining structural integrity. Their creation, produced by the company Motion Designs (later rebranded as Quickie) in 1979, featured aircraft-grade aluminum tubing, quick-release wheels, and a rigid frame that dramatically improved performance and maneuverability. The wheelchair weighed around twenty-six pounds compared to the fifty-pound steel alternatives, enabling users to navigate more efficiently and participate in sports previously thought impossible for wheelchair users. The ultralight sports wheelchair sparked a transformation in Paralympic sports and everyday mobility. The design became the foundation for modern racing wheelchairs, basketball chairs, and tennis-specific models. Hamilton's invention didn't just change equipment—it changed perceptions about disability, demonstrating that with proper tools, physical limitations need not define one's potential. Her legacy continues today with advanced carbon-fiber designs that trace their lineage back to that groundbreaking 1979 model.

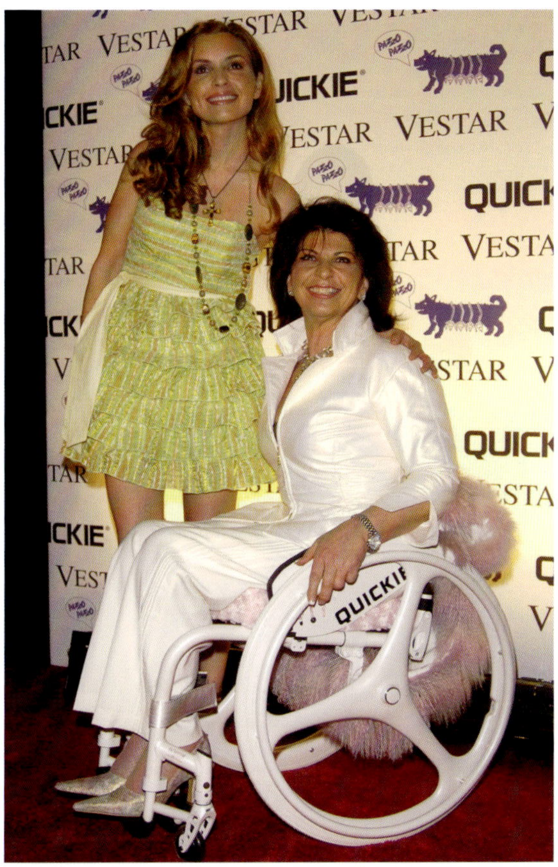

Inventor Marilyn Hamilton with her custom Quickie at a charity event, and the original Quickie.

1980 ROLLERBLADES

Inline skates were invented in 1980 by Scott Olson (1960–), a hockey player from Minneapolis, Minnesota, who wanted to train during the off-season. Olson discovered a pair of vintage inline skates in a sporting goods store and recognized their potential for hockey training. He refined the design, founding Rollerblade Inc. in 1983 to manufacture and market these revolutionary skates. The design of Rollerblades featured wheels arranged in a single line, mimicking the blade of an ice skate more closely than traditional roller skates. This configuration provided better speed, maneuverability, and a more natural skating motion for ice hockey players and recreational users alike. The 1980s and 1990s saw inline skating—or "Rollerblading"—explode in popularity, becoming a mainstream recreational activity and competitive sport. The invention spawned entire industries around inline hockey, aggressive skating, and fitness skating. Rollerblades influenced urban transportation, with many cities developing skate-friendly infrastructure. Today, inline skating remains popular for fitness, recreation, and sport. The technology has evolved to include specialized designs for different activities, from casual cruising to extreme sports. Olson's innovation fundamentally changed how people think about wheeled recreation, creating a lasting impact on sports and urban mobility.

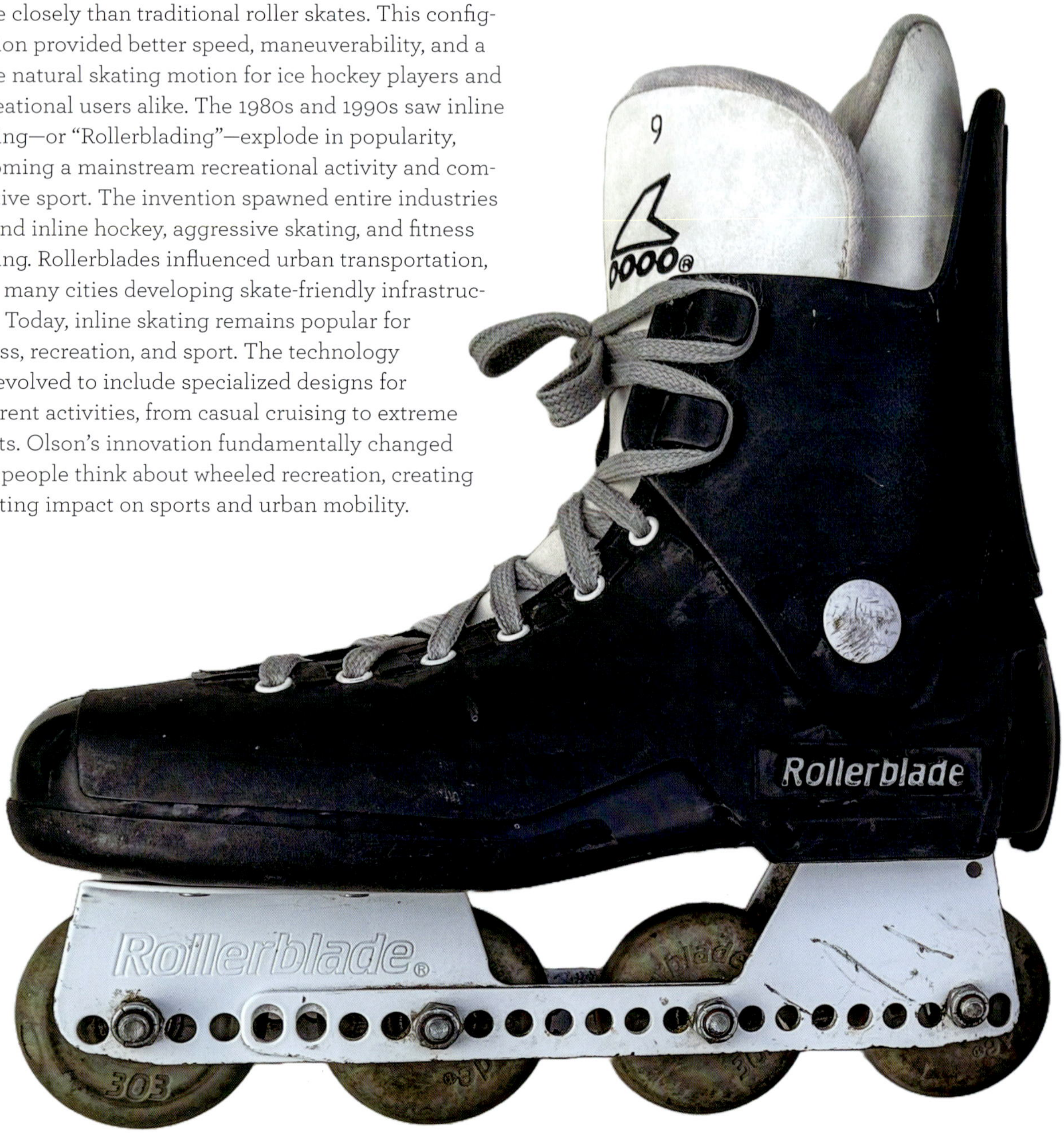

1981 MTV

Music Television—colloquially known as MTV—launched on August 1, 1981, with the prophetic song "Video Killed the Radio Star" by a band called the Buggles. Created by Warner Communications and American Express, the new cable channel revolutionized how music was consumed and marketed. MTV's concept was simple yet groundbreaking: play music videos twenty-four seven, hosted by charismatic video jockeys (VJs) who became cultural icons themselves. The channel's impact extended far beyond music. MTV shaped youth culture, fashion, and language throughout the 1980s and 1990s. It launched many artists' careers, from Madonna to Nirvana, and created the template for modern celebrity culture. Shows like *Total Request Live* made audience participation central to programming, whereas *The Real World* pioneered reality television. MTV's legacy is complex. Though it democratized music discovery and created a global youth culture, critics argue it prioritized image over substance and contributed to the decline of album-oriented rock. The channel's evolution away from music videos toward reality programming marked the end of an era, but its influence on media, marketing, and popular culture remains undeniable, fundamentally changing how entertainment reaches mass audiences.

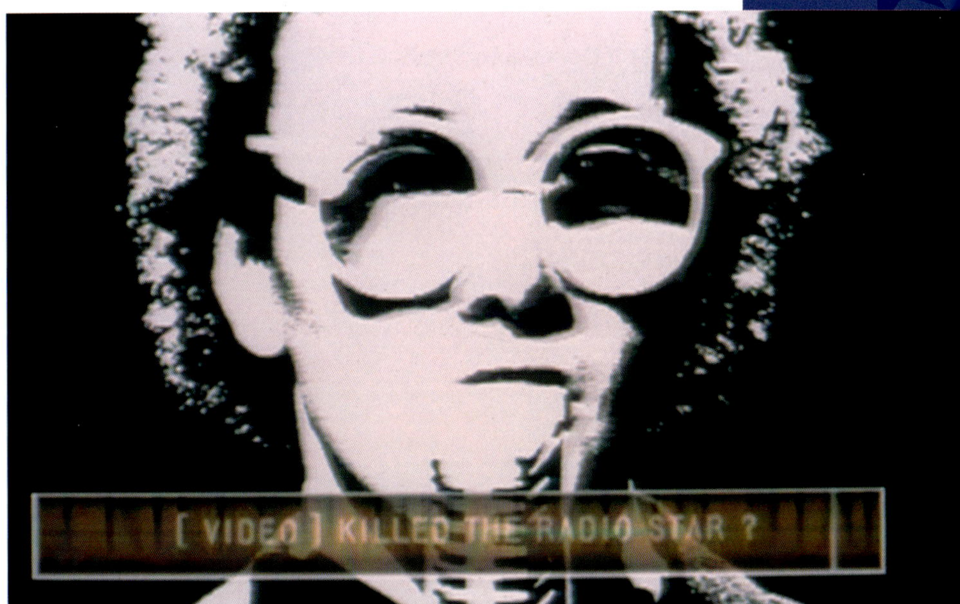

The first video MTV ever played, "Video Killed the Radio Star" by the Buggles. It was also the very last video MTV played when it went off the air on December 31, 2025.

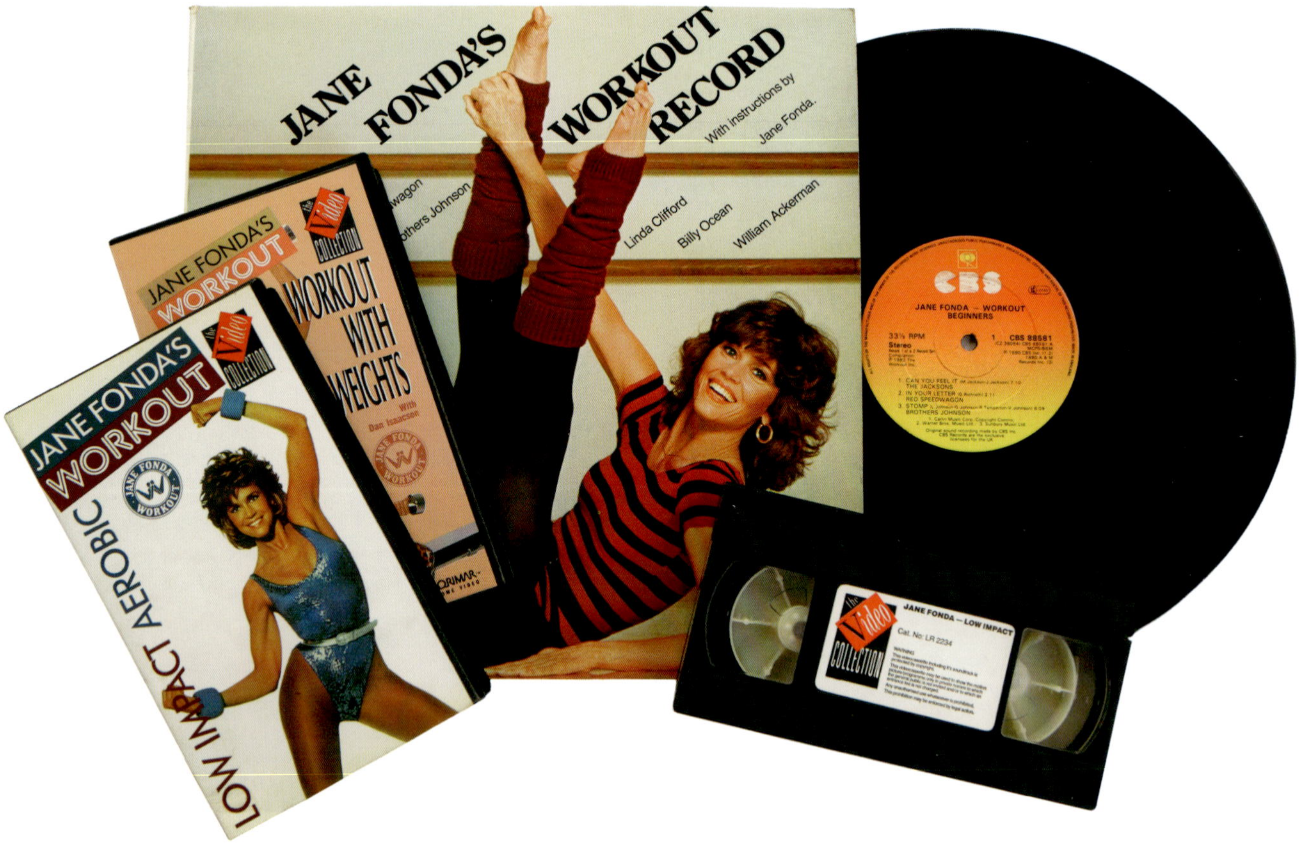

Jane Fonda's 1982 workout tape spawned several sequels and started an entire industry.

1982 HOME WORKOUT VIDEOS

The home workout video revolution began when actress Jane Fonda released *Jane Fonda's Workout* in 1982, which became the best-selling VHS tape of all time. This groundbreaking release capitalized on the emerging home video market and the growing fitness craze, making exercise accessible to millions who couldn't afford gym memberships or preferred working out privately. The format exploded throughout the 1980s and 1990s, with fitness personalities like Richard Simmons, Denise Austin, and Billy Blanks creating diverse offerings from aerobics to Tae Bo. These videos democratized fitness instruction, bringing professional-quality workouts into living rooms worldwide while building massive personal brands for their creators. In the 1980s (and beyond), how you looked while you were exercising was a big part of the fitness fad. Spandex revolutionized 1980s workout gear, transforming exercise clothing from loose-fitting garments into formfitting, stretchy attire that became both functional and fashionable. Invented in 1958 by chemist Joseph Shivers at DuPont, spandex didn't reach mainstream fitness fashion until the 1980s fitness boom. Jane Fonda's 1982 workout video became the catalyst for spandex's dominance in fitness wear.

1983 THE INTERNET

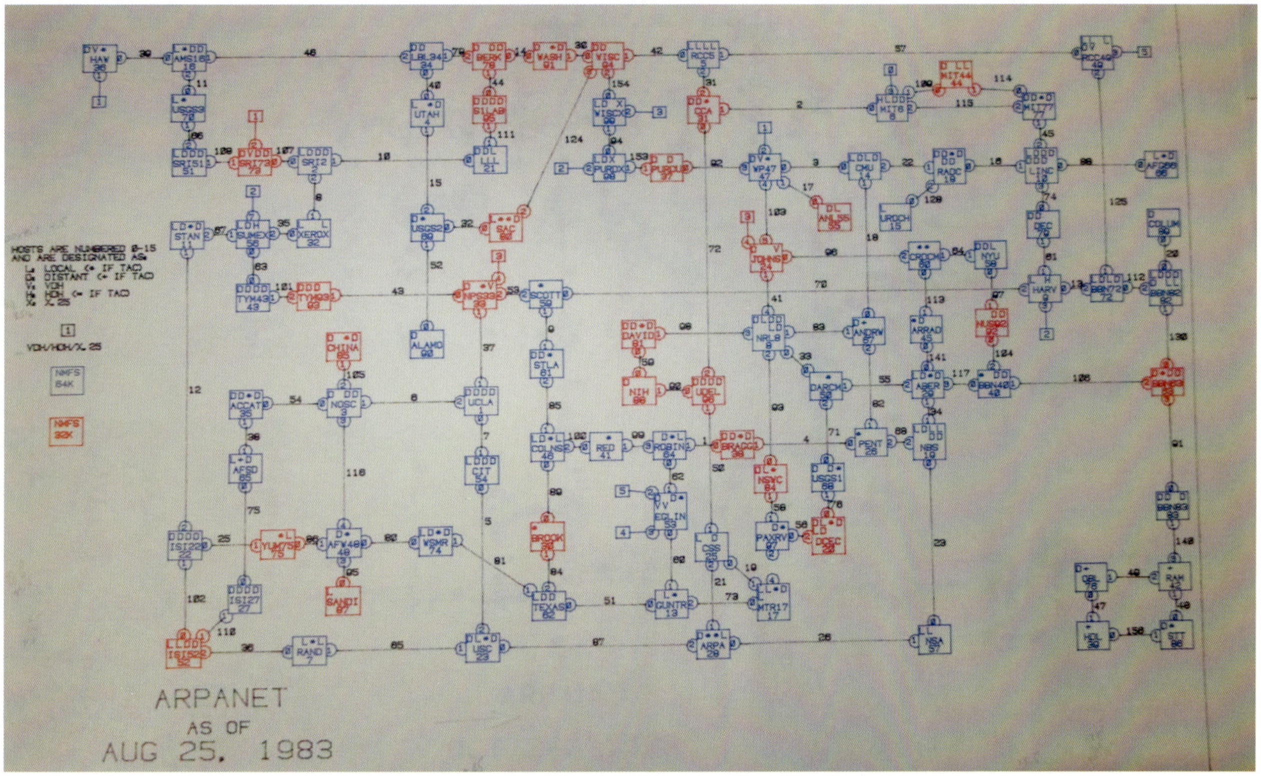

January 1, 1983, marks what many consider the official birthday of the Internet. The Internet evolved from the US Department of Defense's ARPANET project, launched in the late 1960s to create a decentralized communication network during the Cold War. Throughout the 1970s, computer scientists developed the TCP/IP protocol suite, establishing foundational standards that continue to govern Internet communication today. The World Wide Web, invented by Tim Berners-Lee (1955–) at CERN in 1989, transformed the Internet from a specialized research tool into a global information system accessible to the general public through a user-friendly interface. The Internet's impact has been transformative across multiple domains. It revolutionized commerce through e-commerce platforms and digital marketplaces, fundamentally reshaping global trade. Communication became instantaneous and borderless via email, social media, and messaging platforms connecting people worldwide in real time. Information access was democratized, transforming education, journalism, and knowledge sharing while also introducing challenges around misinformation and digital inequality. Today, the Internet functions as essential infrastructure supporting remote work, telemedicine, streaming entertainment, and smart-city technologies. It has fundamentally altered human interaction, learning, work, and social organization, creating unprecedented opportunities for global connection while generating new forms of social and economic disparity.

The original 1983 map showing the network infrastructure of ARPANET.

1984 SPACE SHUTTLE *DISCOVERY*

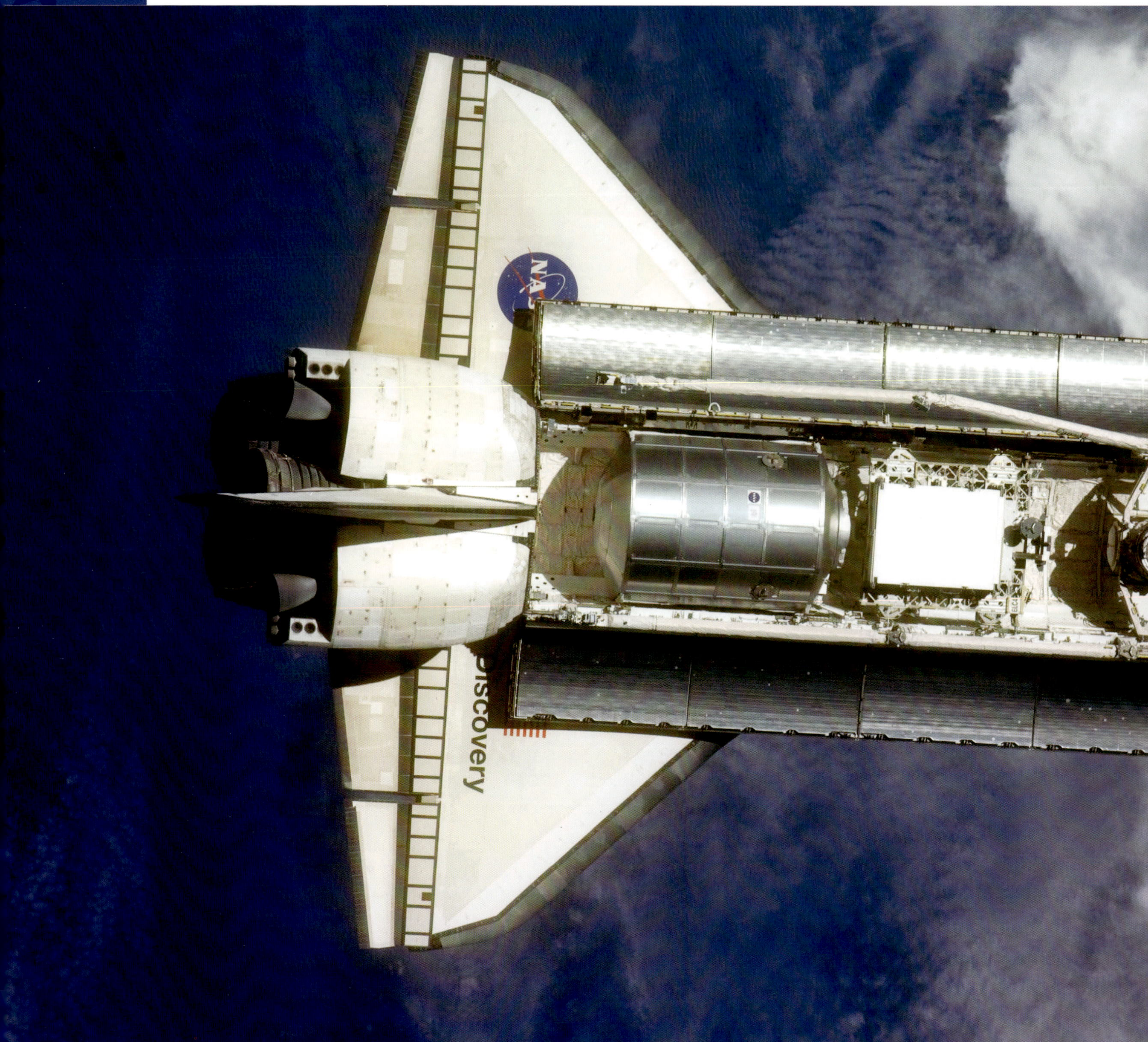

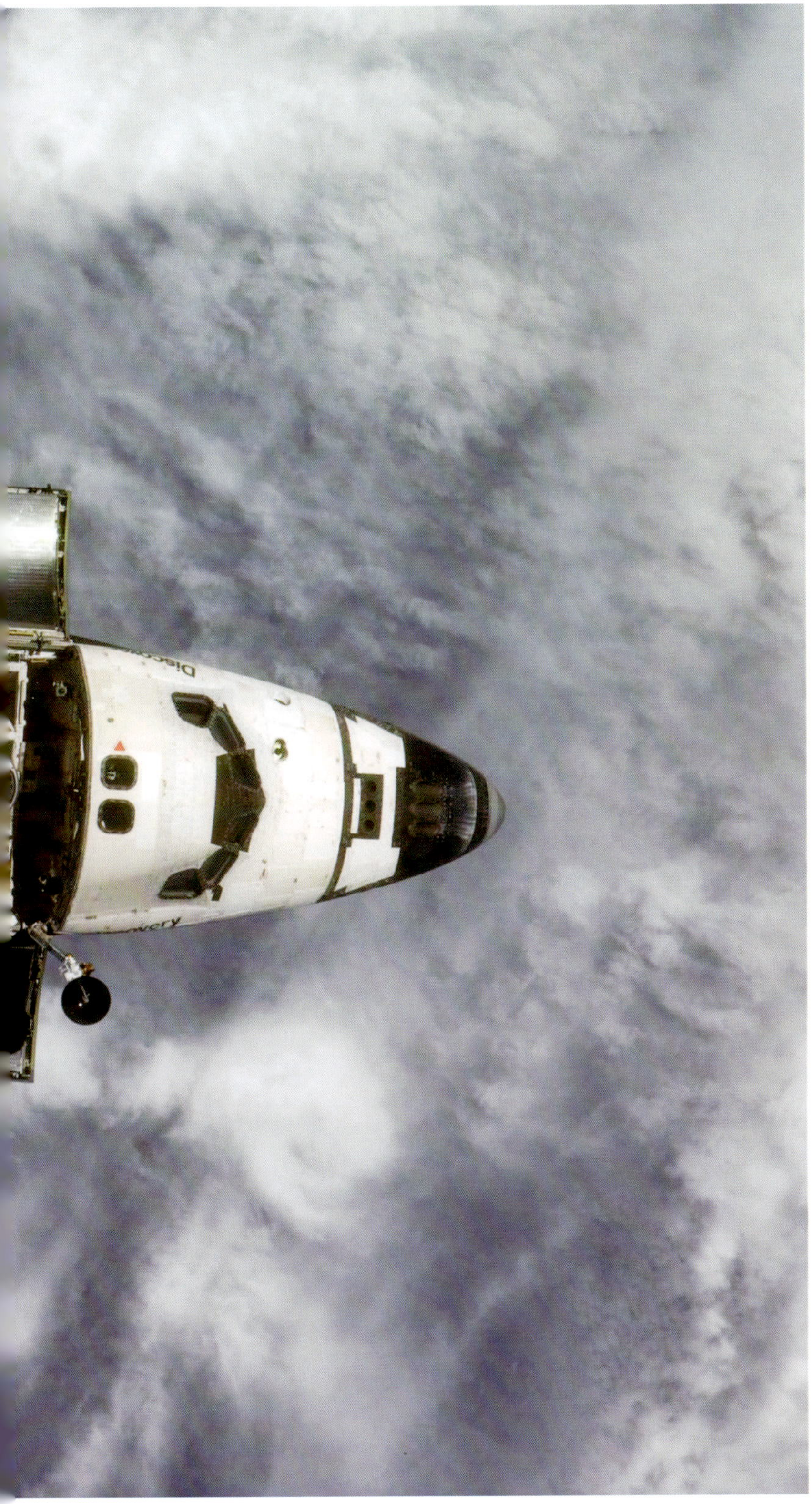

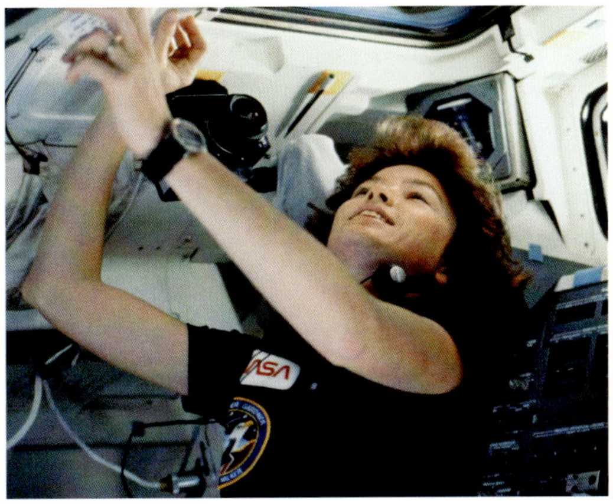

The Space Shuttle *Discovery* is one of NASA's most accomplished satellites, completing thirty-nine missions during its twenty-seven years in operation starting in 1984. As the third working shuttle in the fleet, *Discovery* carried more crew members into space than any other spacecraft, totaling 252 astronauts across its missions. *Discovery*'s achievements include deploying the Hubble Space Telescope in 1990, which revolutionized our understanding of the universe. The shuttle also played crucial roles in constructing the International Space Station, delivering vital components and crew rotations. Notable missions included the return-to-flight operations after both the *Challenger* and *Columbia* disasters, demonstrating NASA's resilience and commitment to space exploration. *Discovery* became a symbol of international cooperation, carrying astronauts from multiple nations and fostering collaborative relationships in space. The shuttle's final mission in 2011 marked the end of the Space Shuttle Program, but its legacy continues to inspire future generations of explorers and engineers.

LEFT: The Space Shuttle *Discovery* in orbit, c. 1984.
ABOVE: Astronaut Anna L. Fisher near the aft flight deck of *Discovery*, November 12, 1984.

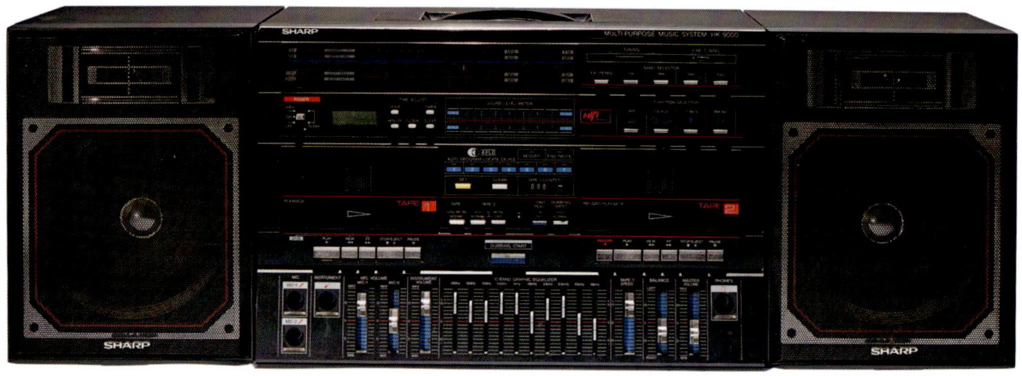

1985 BOOMBOX

The Sharp HK-9000 boombox stands as one of the most iconic portable music systems of the 1980s. First released in 1985, this massive three-piece unit was loaded with features, including dual tape decks, detachable speakers, a ten-band graphic equalizer, AM/FM radio, and two-fader microphone mixing capabilities. Its impressive size and power made it one of the largest boom-boxes of the era, and additional features like adjustable tape speed, song search, Dolby B noise reduction, and multiple input options for microphones and guitar set it apart from competitors. The HK-9000 was famously used by hip-hop pioneer Fab 5 Freddy, cementing its place in music history and earning it a spot in the Smithsonian's National Museum of American History. Among collectors and enthusiasts, the HK-9000 is remembered as exceptionally loud and powerful. Its distinctive design and robust construction have made it a highly sought-after piece among vintage audio collectors, symbolizing the golden age of street culture and portable music.

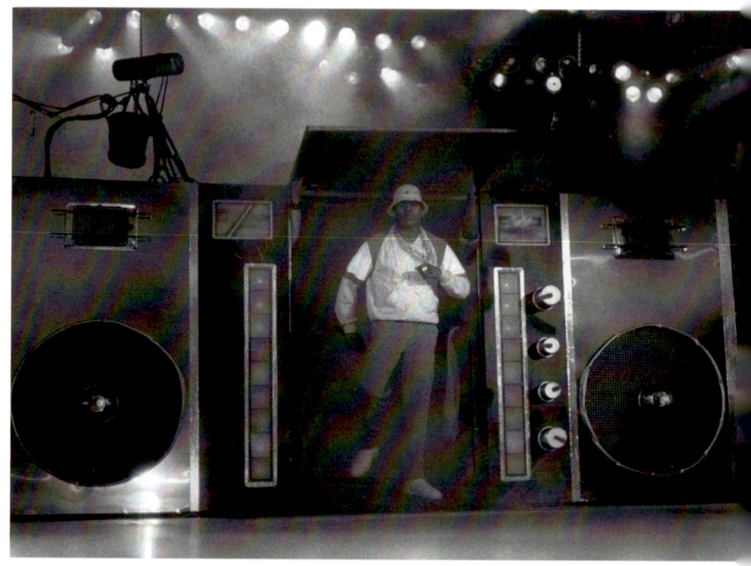

RIGHT: LL Cool J enters the stage through a giant replica of the boombox, c. 1987.

1986 AMERICAN GIRL DOLLS

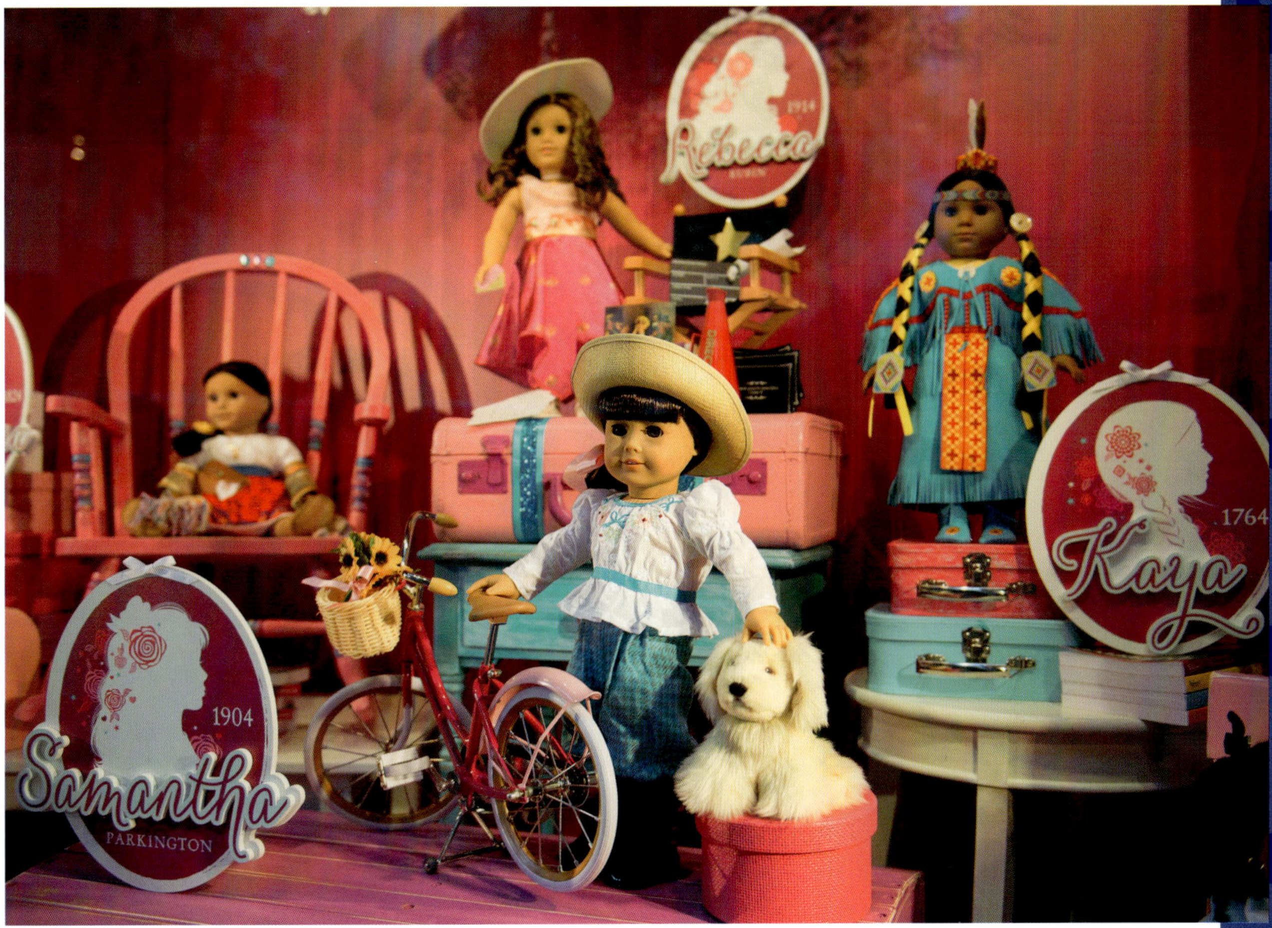

A window display of American Girl dolls in Chicago, Illinois.

American Girl dolls, launched by Pleasant Company in 1986, revolutionized the toy industry by combining high-quality dolls with rich historical narratives. The brand began with three characters: Kirsten, Samantha, and Molly, each representing different eras in American history and accompanied by detailed books that brought their stories to life. What set the American Girl line apart was its educational approach to play. Each doll came with historically accurate accessories, clothing, and furniture that reflected their time period, teaching children about American history through immersive storytelling. The eighteen-inch dolls weren't just toys—they were windows into the past, helping young girls understand historical events through relatable characters their own age. The American Girl dolls fostered literacy through their book series, promoted historical awareness, and created a sense of connection between generations. Over time, the dolls became cultural touchstones, with their own magazine, stores, and even a dedicated café experience. Today, American Girl continues to evolve, introducing contemporary characters that address modern issues while maintaining their commitment to empowering young girls through storytelling and imaginative play.

1987 THE AIDS MEMORIAL QUILT

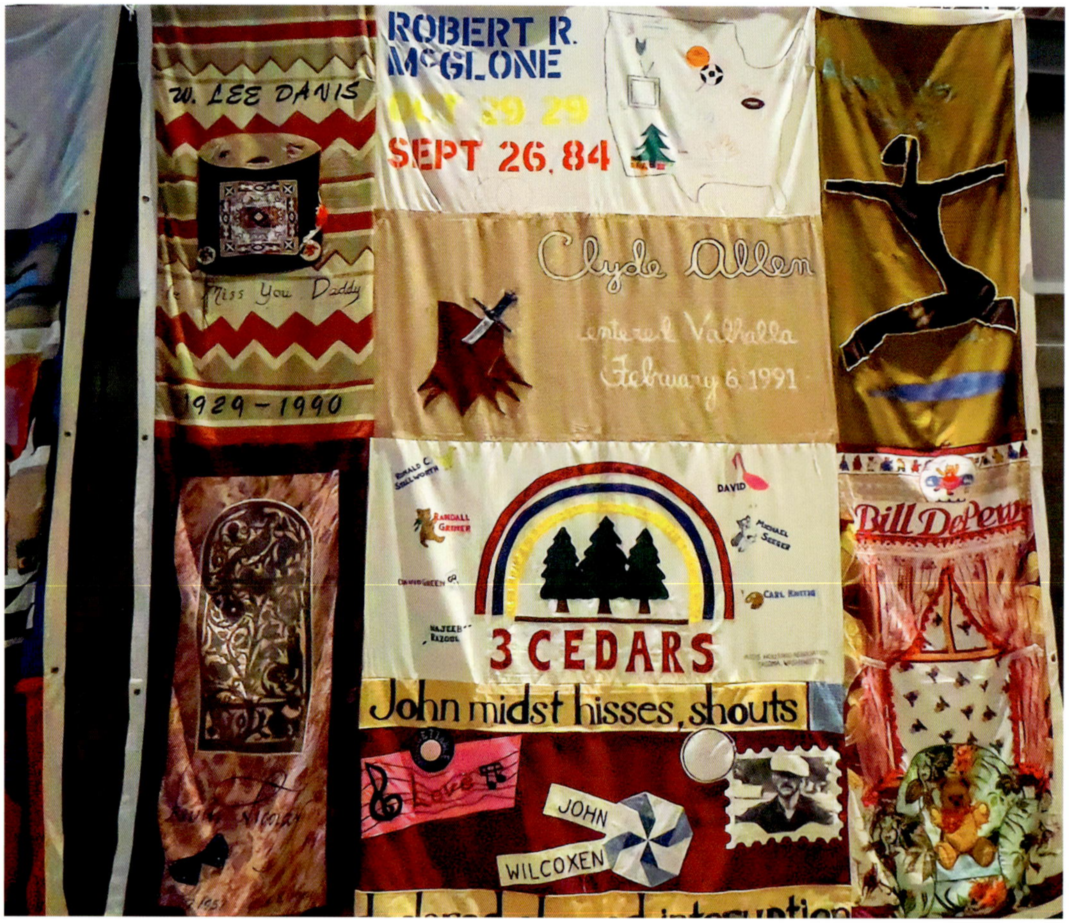

The AIDS Memorial Quilt stands as one of the most powerful grassroots memorials in American history. Conceived by activist Cleve Jones (1954–) in 1985 and first displayed in 1987, the quilt transformed individual grief into collective action during the height of the AIDS crisis. Each three-by-six-foot panel commemorates someone who died from AIDS-related complications, created by loved ones who poured their memories, pain, and love into fabric. The quilt's impact extended far beyond remembrance. It humanized a disease that had been stigmatized and politicized, forcing society to confront the faces behind the statistics. When displayed on the National Mall in Washington, DC, the quilt covered more area than a football field, making the scale of loss impossible to ignore. It became a catalyst for AIDS awareness, funding, and policy changes. Today, with more than forty-eight thousand panels representing more than 105,000 individuals, the quilt continues to educate new generations about the AIDS epidemic while honoring those lost. The quilt serves as proof that art and activism can create profound social change, transforming personal tragedy into a movement for justice and remembrance.

ABOVE: A quilt section with eight panels; at top right is the panel for choreographer Alvin Ailey. **RIGHT:** On October 11, 1996, the full quilt was displayed on the National Mall. It stretched nearly a mile from the Washington Monument to the Capitol. It is the last time the quilt would be displayed in full—it has since become too large for a single display.

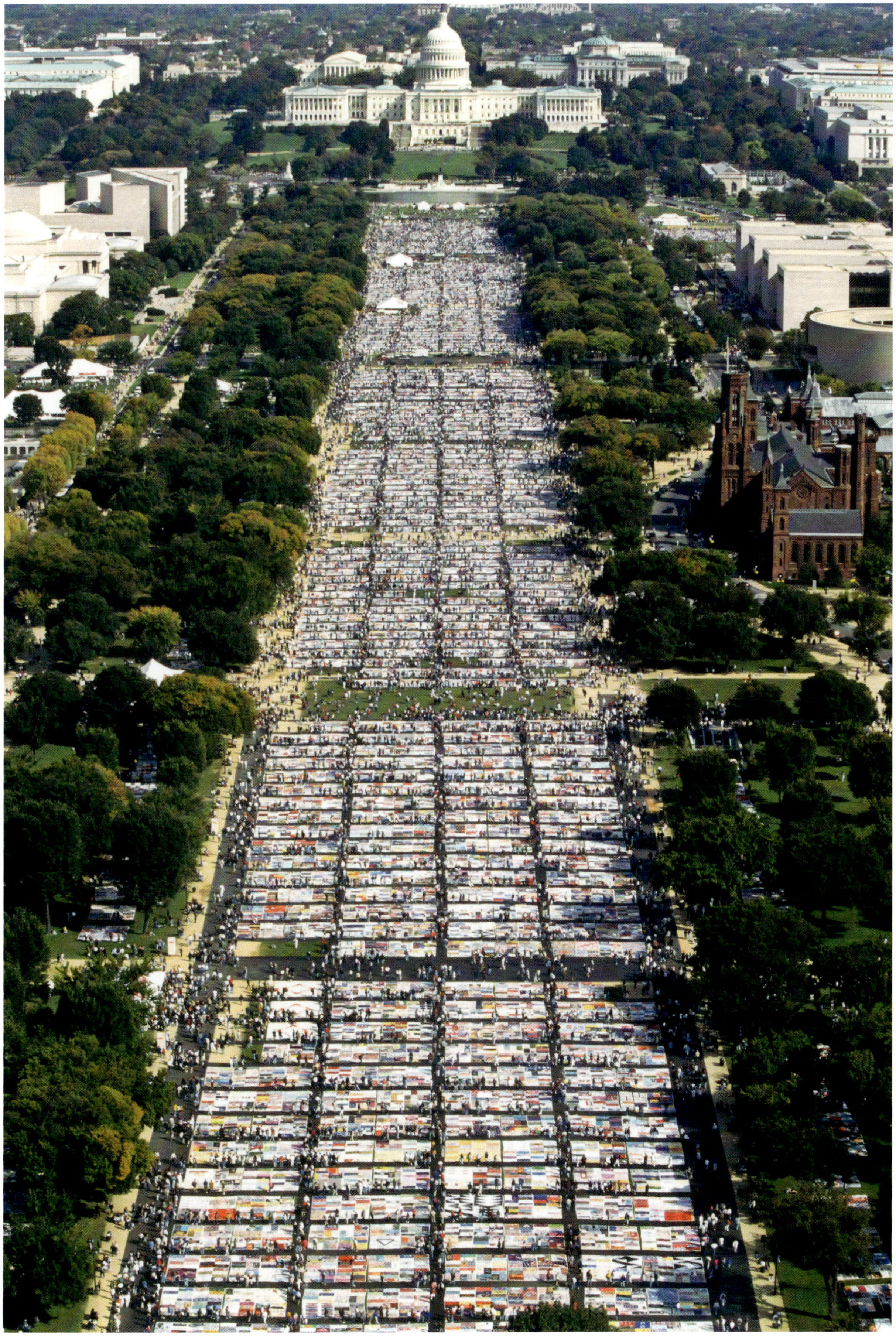

1988 LUNCHABLES

Lunchables—compartmentalized meal kits featuring crackers, processed meats, and cheese—were launched by the Oscar Mayer company in 1988. These tapped into children's desire for autonomy and play, allowing them to assemble their own mini sandwiches. And though it may seem a bit trivial, Lunchables revolutionized the lunch industry by transforming mealtime into an interactive experience. The brand's genius lay in its marketing strategy, positioning lunch as entertainment rather than mere sustenance. With colorful packaging and the tagline "Fun to eat," Lunchables captured the zeitgeist of busy American families seeking convenient solutions. The product ultimately expanded beyond its original format to include pizza kits, nachos, and even dessert combinations. Lunchables normalized processed convenience foods in children's diets and influenced an entire generation's relationship with food. Though the brand sparked debates about nutrition in schools, it simultaneously became a nostalgic touchstone for millennials. Despite criticism over nutritional content, Lunchables endures as a billion-dollar franchise. Their legacy lies in proving that food packaging and presentation could be as important as the food itself, fundamentally changing how companies market to children and families.

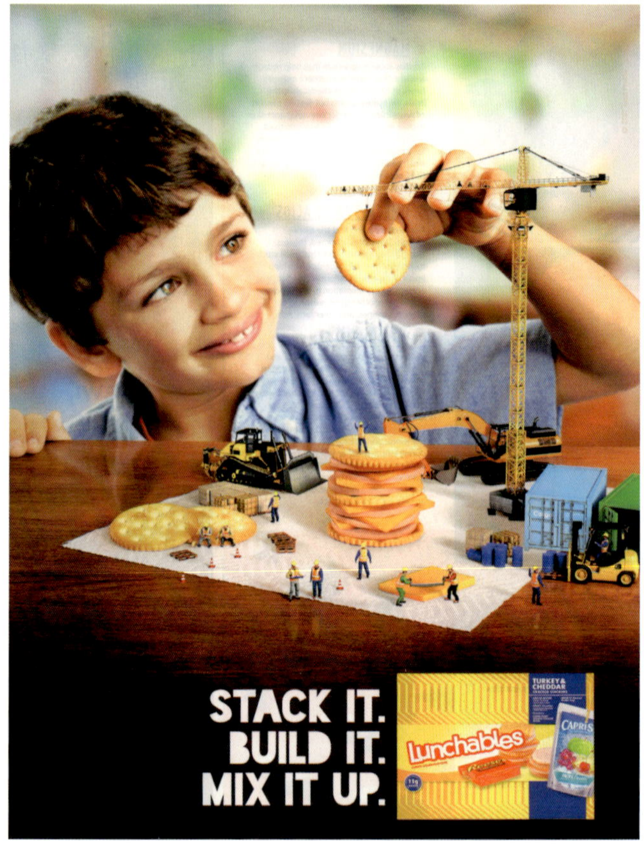

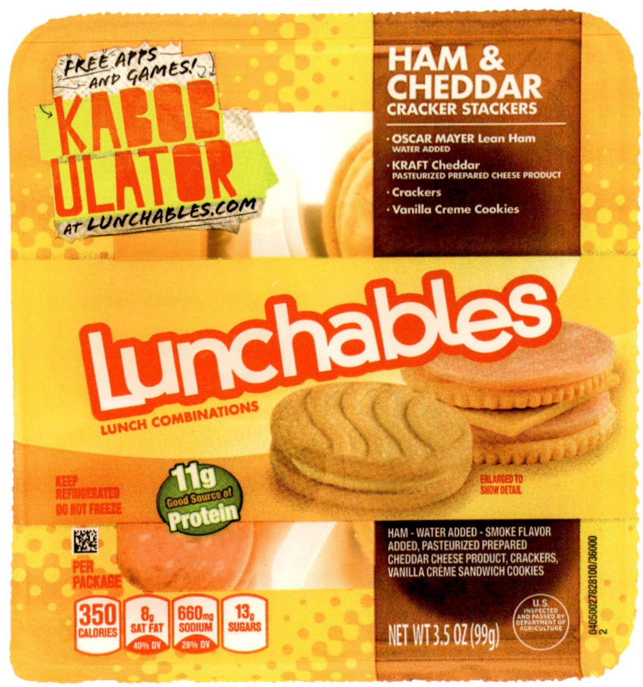

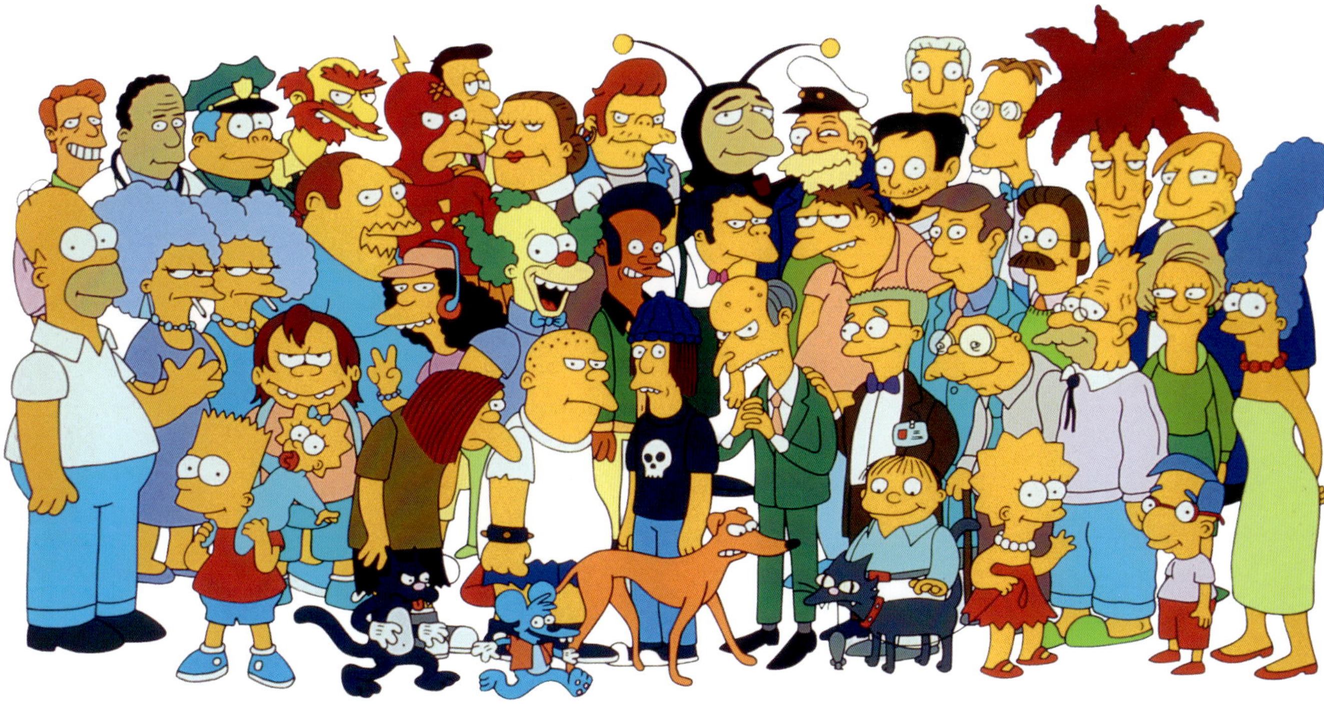

1989 THE SIMPSONS

The Simpsons, which premiered in 1989, will go down in history as one of television's most influential and beloved animated series. Created by Matt Groening (1954–), the show follows the dysfunctional yet lovable Simpson family in the fictional town of Springfield, offering sharp social commentary wrapped in humor that appeals to both children and adults. The series revolutionized TV animation by proving that cartoons could tackle complex social issues while maintaining mass appeal. Its satirical portrayal of American life, from politics to consumerism, has made it a cultural mirror reflecting societal changes for more than three decades—it even contributed new words to the common vernacular, like "D'oh!" and "cowabunga." Its characters, including Homer and Bart Simpson, have become global icons. As the longest-running American sitcom and animated series, *The Simpsons* paved the way for other adult-oriented animated shows like *South Park* and *Family Guy*. Its writers have included comedy legends who went on to create their own successful series, spreading the show's comedic DNA throughout television. The series has maintained cultural relevance through its ability to adapt and comment on contemporary issues, securing its place as a defining piece of American pop culture that continues to influence comedy and animation worldwide.

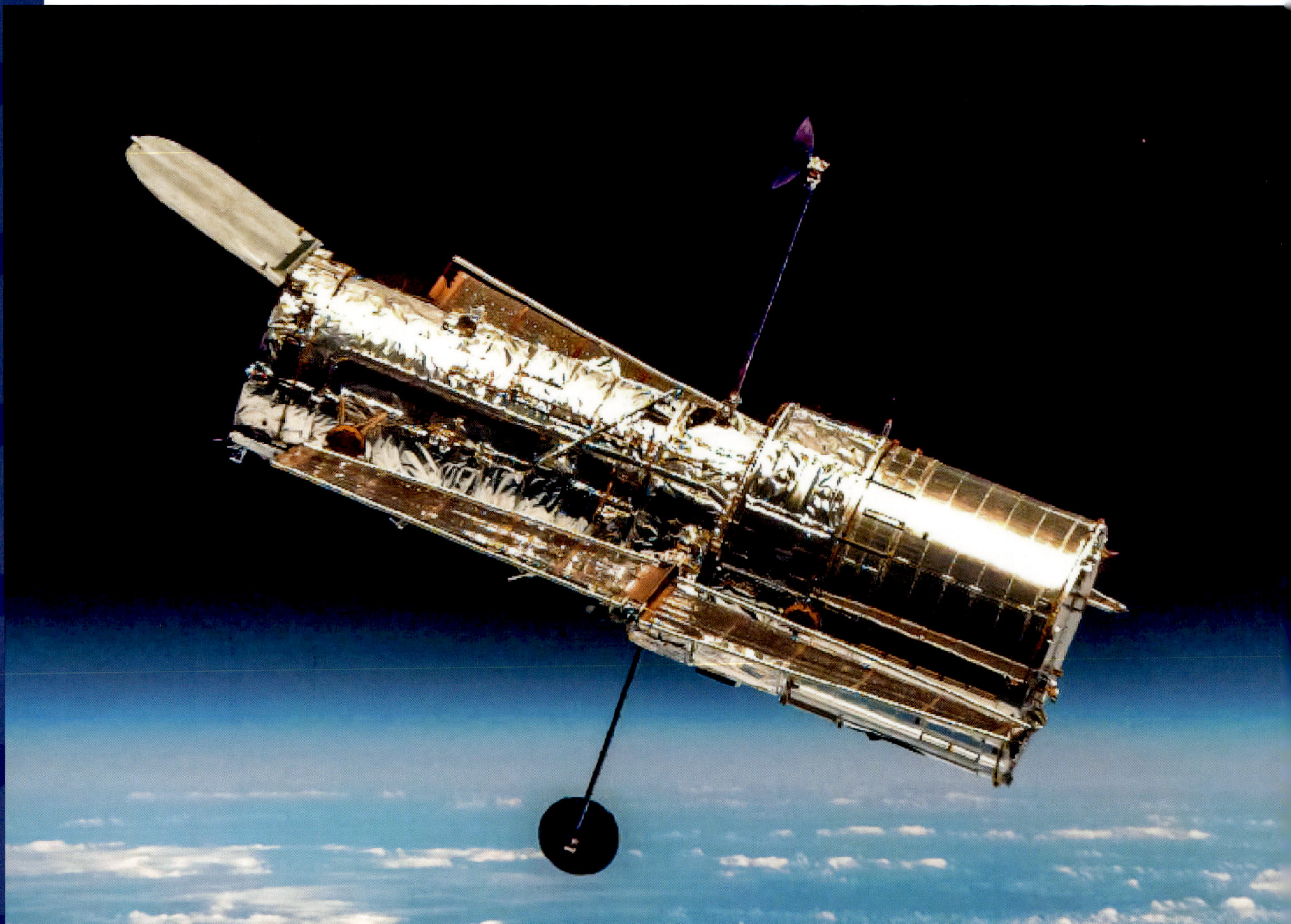

1990 HUBBLE SPACE TELESCOPE

Launched in 1990, the Hubble Space Telescope revolutionized our understanding of the universe by providing unprecedented views of space from beyond Earth's atmosphere. Operating in low orbit over Earth, Hubble has captured stunning images of distant galaxies and other phenomena with remarkable clarity, free from the atmospheric distortion that limits ground-based telescopes. Hubble's scientific achievements include helping determine the age of the universe (13.8 billion years!), providing evidence for dark energy's existence, and discovering that massive black holes exist at the centers of most galaxies. The telescope has also observed the formation of stars and planets, tracked the evolution of galaxies, and revealed the dynamic nature of our solar system. Beyond scientific discovery, Hubble's breathtaking images have inspired countless people to appreciate space in a new way, making astronomy accessible to the general public. The telescope's data has contributed to more than nineteen thousand scientific papers, establishing it as one of the most productive scientific instruments ever created. As Hubble proceeds operating alongside newer telescopes like the James Webb Space Telescope, it continues to fundamentally change how we perceive our place in the universe.

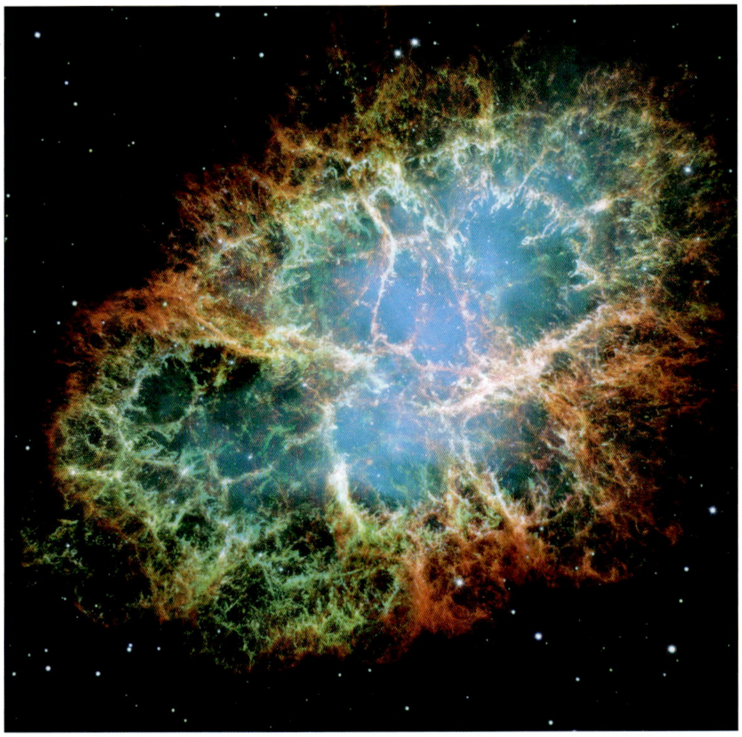

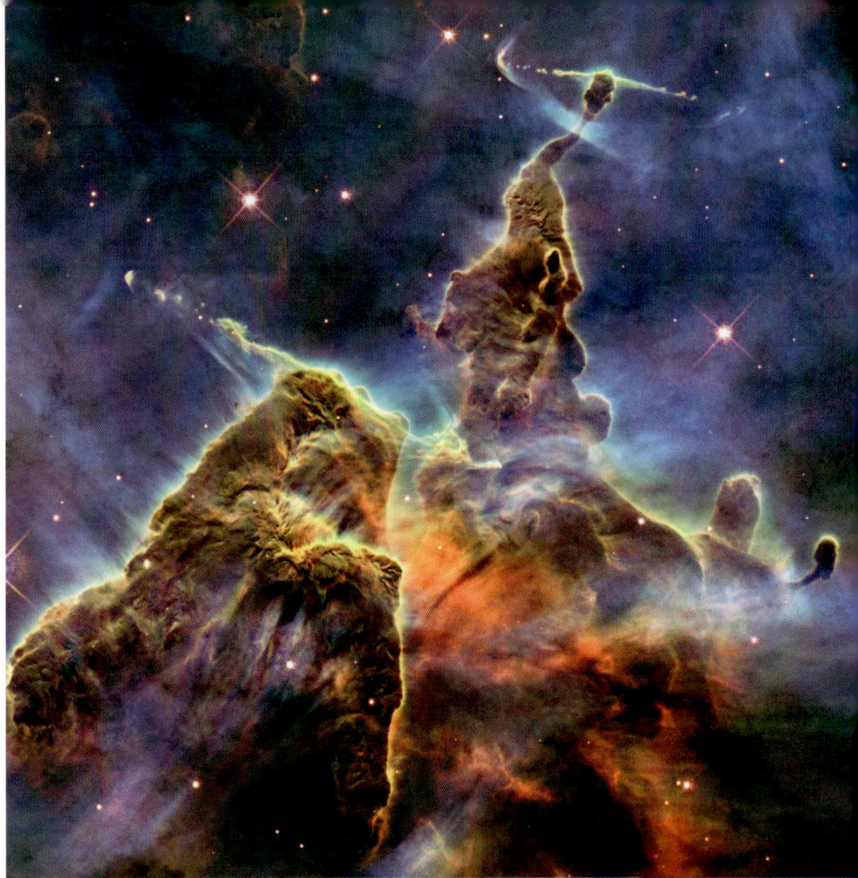

TOP LEFT: The Crab Nebula. **TOP RIGHT:** The Mystic Mountain, three light-years tall, which is being eaten away by the brilliant light from nearby bright stars. **BELOW:** This image of Westerlund 2 became Hubble's twenty-fifth anniversary portrait.

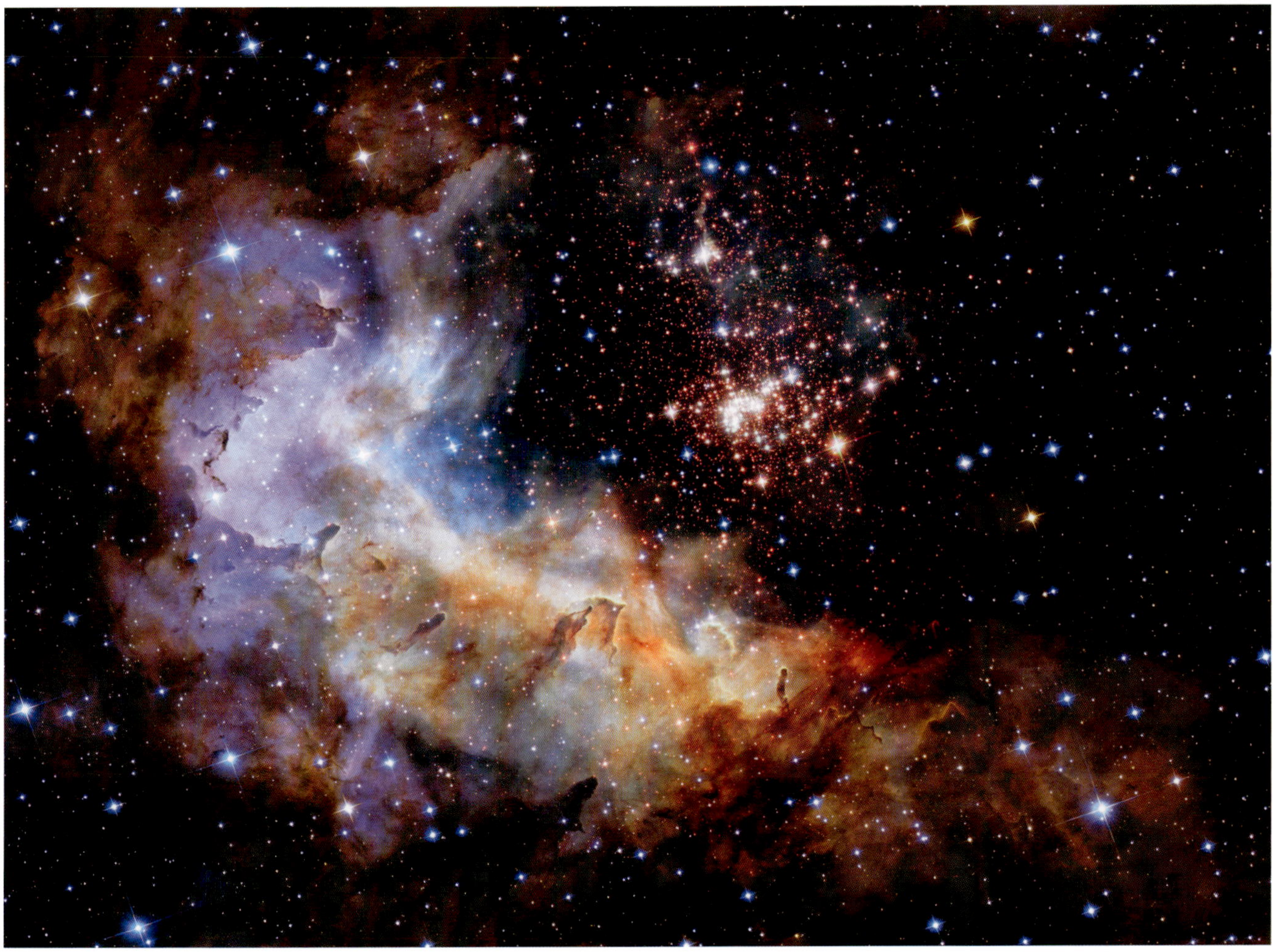

1991 *NEVERMIND*

Released in September 1991, Nirvana's *Nevermind* materially transformed the musical landscape and became one of the most influential albums in rock history. The record marked a seismic shift from the hair metal that dominated mainstream rock to the raw, authentic sound of grunge emerging from Seattle's underground scene. The album's breakout single "Smells Like Teen Spirit" became an anthem for Generation X, capturing the angst and disillusionment of youth in the early 1990s. Kurt Cobain's (1967–1994) emotionally charged vocals, combined with Krist Novoselic's (1965–) driving bass lines and Dave Grohl's (1969–) powerful drumming, created a sound that was both accessible and uncompromisingly honest. *Nevermind* achieved remarkable commercial success, peaking at number one on the *Billboard* 200 and selling more than thirty million copies worldwide. More significantly, it opened the floodgates for alternative rock with bands like Pearl Jam, Soundgarden, and Alice in Chains while establishing Seattle as the epicenter of a cultural revolution. The album gave voice to alienated youth and proved that authentic, noncommercial music could achieve massive mainstream success. The influence of *Nevermind* continues to resonate in contemporary rock music, cementing its legacy as not just a great album but a cultural touchstone that redefined what popular music could be.

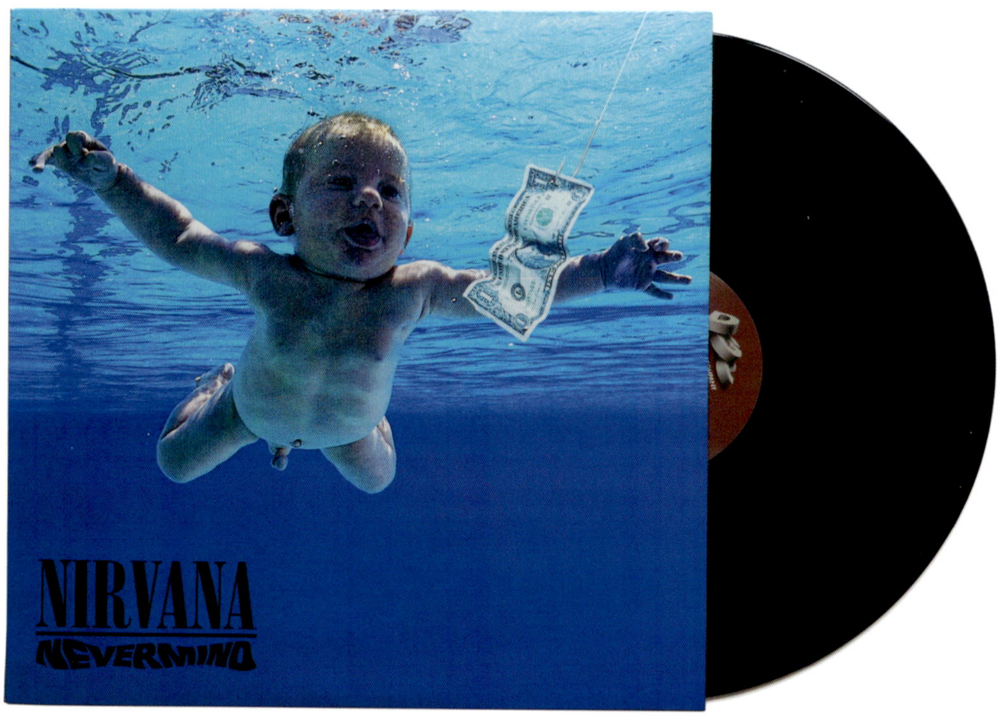

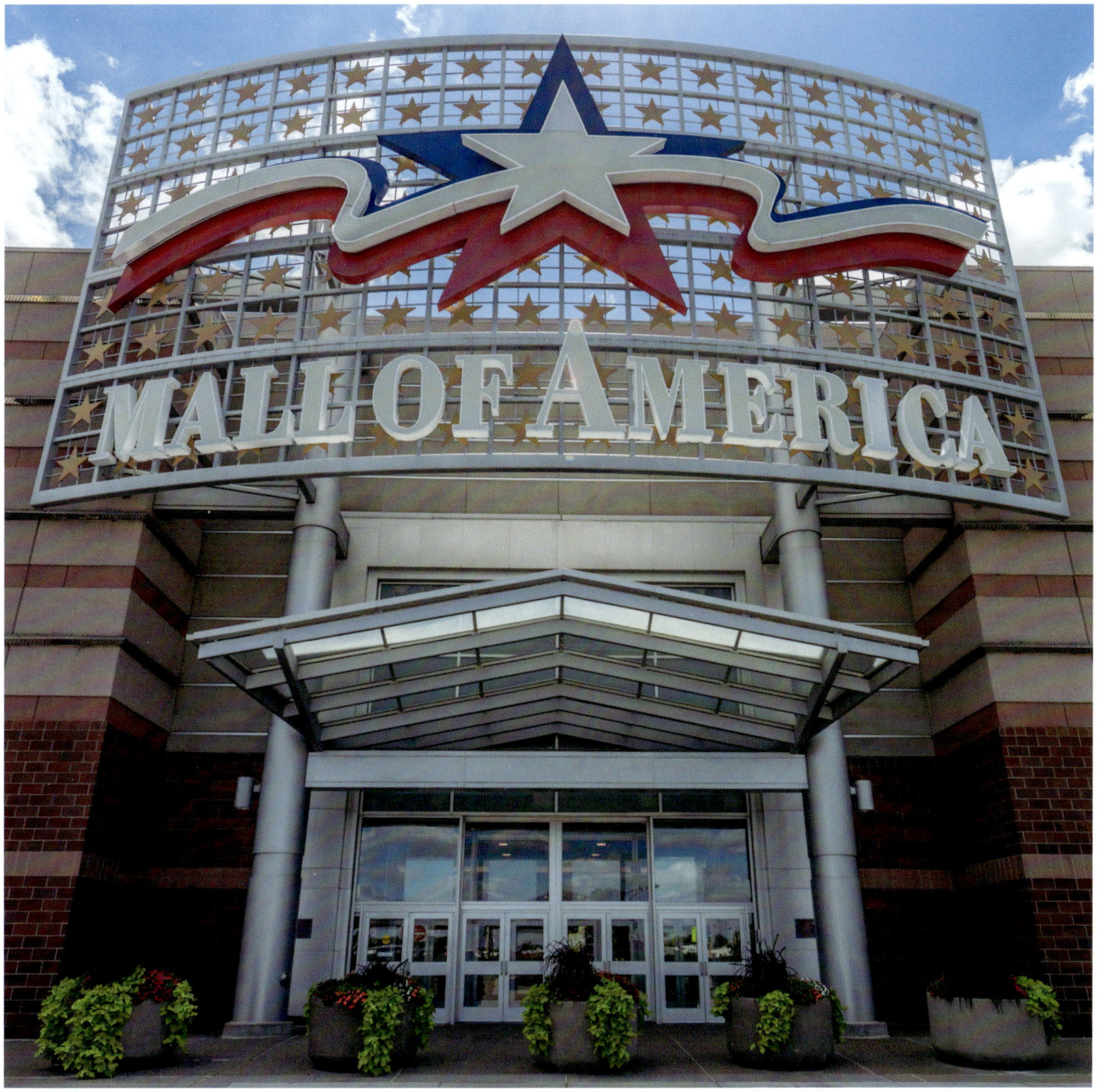

1992 MALL OF AMERICA

The Mall of America—the largest mall in the United States—opened its doors on August 11, 1992, in Bloomington, Minnesota, marking a groundbreaking moment in American retail history. Built on the site of the former Metropolitan Stadium, home of both the Twins and the Vikings, the massive 4.2-million-square-foot complex welcomed its first visitors to a site that included 330 stores and more than 10,000 employees to serve them. The grand opening drew extraordinary crowds, with an estimated 150,000 shoppers. The overwhelming response foreshadowed the mall's lasting appeal as a destination that transcends traditional shopping. Decades later, the Mall of America has become one of the country's top tourist destinations. The mall shook up the standard shopping experience by combining retail with entertainment, featuring attractions like an indoor amusement park and aquarium. Its success helped establish the template for experiential retail destinations, proving that shopping malls can serve as inclusive entertainment complexes rather than mere shopping centers. Today, the Mall of America stands as both an architectural landmark and a cultural icon of American consumer culture.

1993 BEANIE BABIES

Beanie Babies, created by Ty Warner (1944–) in 1993, became one of the most extraordinary toy phenomena of the 1990s. These small, plastic-bean-filled plush toys with their distinctive heart-shaped tags sparked a collecting frenzy that surpassed typical toy markets. What began as simple stuffed animals priced under five dollars evolved into a speculative bubble where rare bears like Princess the Bear could sell for thousands of dollars. The craze was fueled by Warner's brilliant marketing strategy of artificial scarcity—retiring certain animals created urgency and drove collectors to hunt for discontinued pieces. The rise of the online marketplace eBay coincided perfectly with peak Beanie Baby mania, creating a secondary market where prices soared to extraordinary heights. Parents camped outside stores, and the toys became serious investment vehicles for many families. Though the bubble burst by 2000, Beanie Babies left lasting impacts on toy marketing, collectibles culture, and early e-commerce. Beanie Babies demonstrated how manufactured scarcity could create genuine demand and helped establish online auction platforms. Today, they remain a nostalgic reminder of 1990s consumer culture and the power of effective marketing to transform simple products into cultural phenomena.

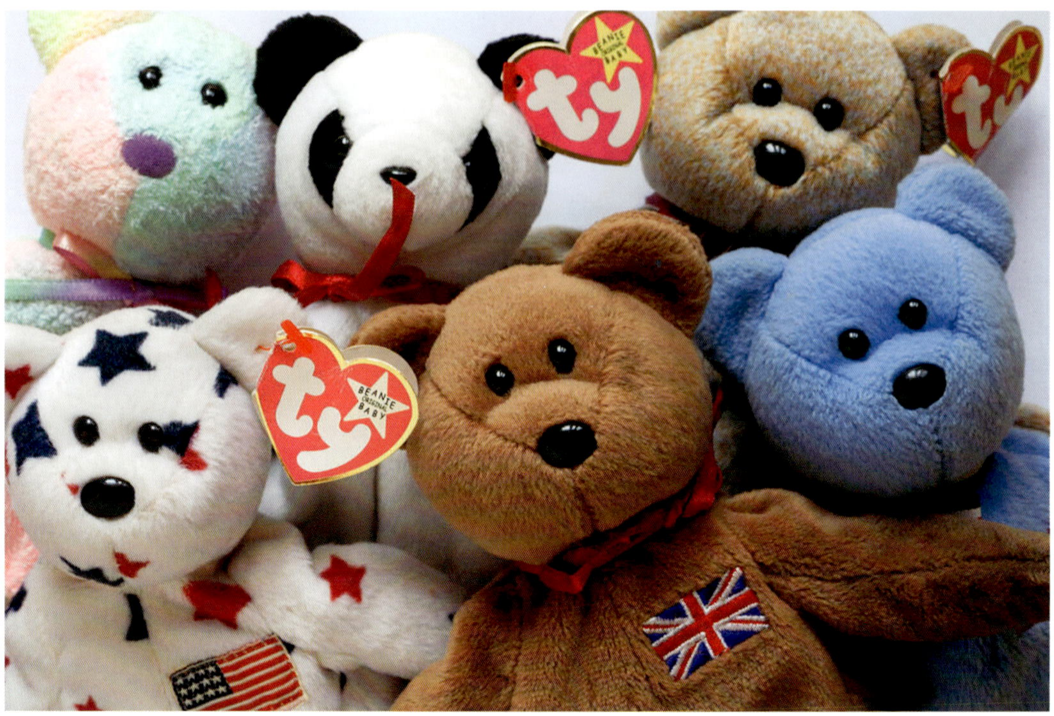

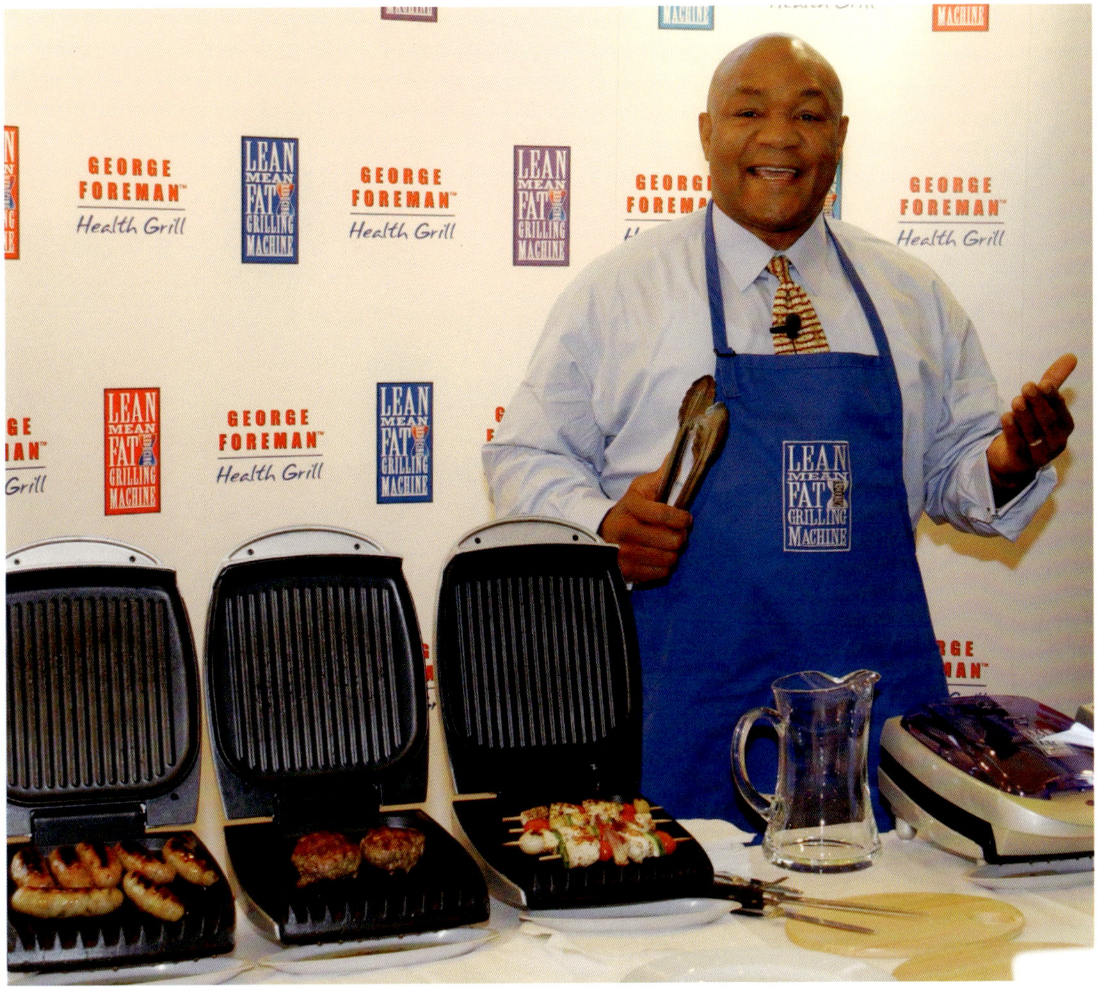

1994 GEORGE FOREMAN GRILL

The George Foreman grill launched in 1994 as the "Lean Mean Fat-Reducing Grilling Machine" and quickly became one of the most successful kitchen appliances in history. The grill changed home cooking by offering an accessible, healthier alternative to traditional grilling methods, featuring a signature sloped surface that drained excess fat. Actually created by inventor Michael Boehm (1944–) of Batavia, Illinois, the grill was promoted through distinctive infomercials featuring the former heavyweight boxing champion, whose jovial personality made the product an instant hit. The grill's success was unprecedented in the kitchen-appliance industry. Sales ballooned from $5 million in 1996 to approximately $200 million in 1999 and more than $300 million in 2000. The phenomenal growth reflected the product's appeal to health-conscious consumers seeking a convenient way to cook with less fat. Since its introduction, more than one hundred million George Foreman grills have been sold worldwide, making it one of the best-selling kitchen appliances of all time. For Foreman personally, the grill deal allowed him to pocket at least $250 million, surpassing his earnings from his illustrious boxing career. The product's enduring popularity demonstrates how effective celebrity endorsement combined with practical innovation can create lasting consumer appeal and cultural impact.

George Foreman demonstrating his "Lean Mean" machine.

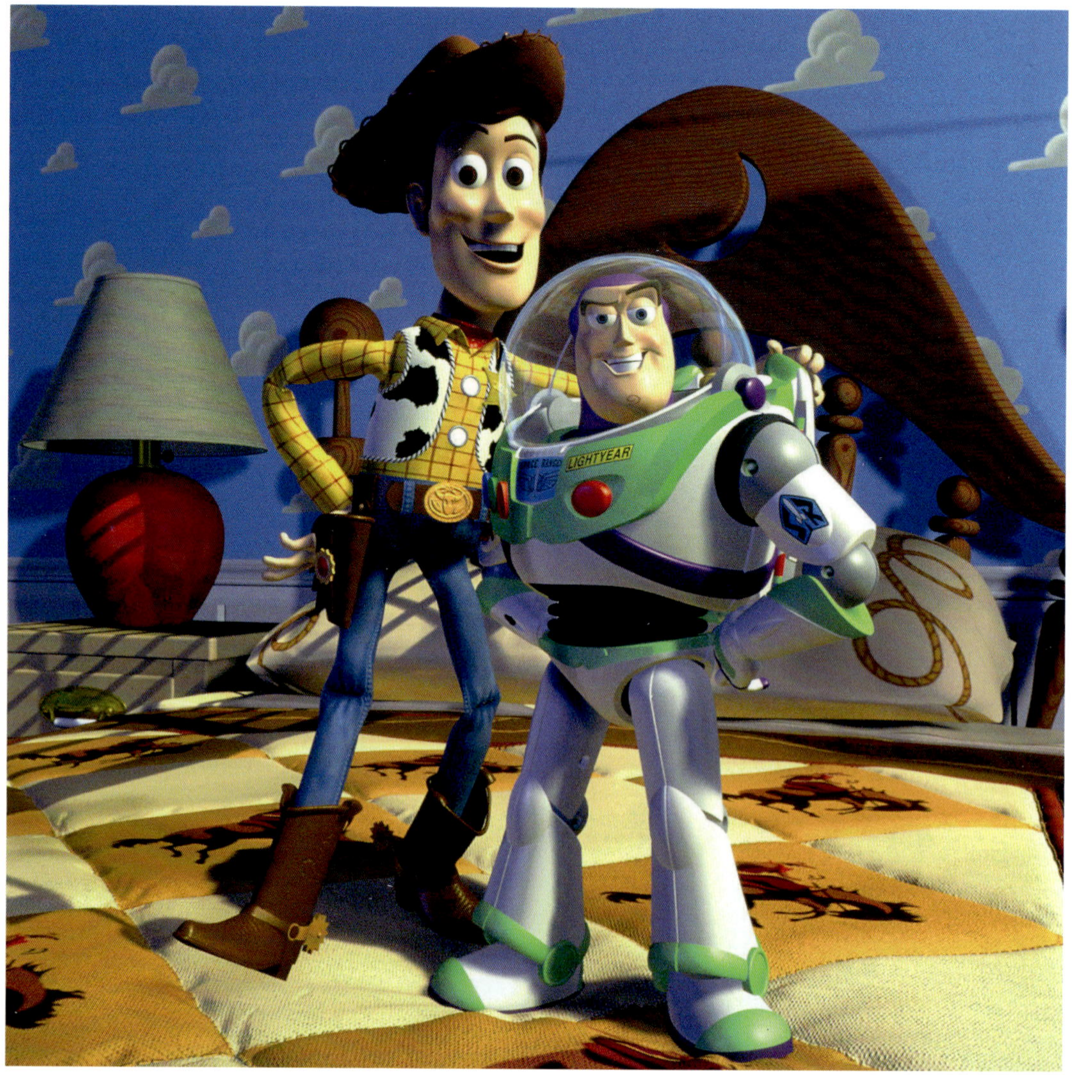

1995 *TOY STORY*

Pixar Animation Studios is a computer animation studio known for creating some of the most beloved animated films in cinema history. But back in 1995, they were a relative unknown in the industry. *Toy Story*, which had its Hollywood premiere on November 19, 1995, is the first fully computer-animated feature film. It not only introduced audiences to beloved characters like Woody and Buzz Lightyear, but it also radically transformed the animation industry forever. The film's technical achievements were unprecedented, proving that computer animation could deliver both visual spectacle and emotional depth. Its success launched Pixar to becoming one of Hollywood's most successful studios, establishing a new standard for animated storytelling that prioritized character development and sophisticated themes alongside cutting-edge technology. The Toy Story franchise has spawned three sequels to date, each pushing creative boundaries while exploring themes of determination, growing up, and friendship. The series demonstrated that animated films could resonate with both children and adults, paving the way for the modern era of family entertainment. More broadly, *Toy Story* catalyzed the industry's shift from traditional hand-drawn animation to digital techniques, influencing many films and inspiring a generation of animators and filmmakers to embrace new technologies while maintaining storytelling excellence.

1996 ELECTRIC CAR

On November 14, 1996, production began on the EV1 by General Motors. That year, GM built over one thousand of the cars and leased most of them to consumers in California, Arizona, and Georgia. It was the first electric car built by a major manufacturer in about ninety years and the first modern electric car designed for a mass market. The EV1 emerged as a response to California's groundbreaking zero-emission vehicle (ZEV) mandate passed in 1990, which required automakers to produce zero-emission vehicles to combat air pollution. The EV1 featured innovative technologies, including the first heat-pump automotive climate system, electro-hydraulic power steering, and power-blended, electro-hydraulic regenerative braking. Despite its technical achievements, GM controversially ended the EV1 program in 1999 and famously recalled and destroyed most of the vehicles, a decision that became the subject of the documentary *Who Killed the Electric Car?* However, the EV1's legacy proved enduring. It transformed how electric cars are perceived, accelerated technological advancements, and instilled valuable lessons for sustainable practices. The EV1 established the foundation for today's electric vehicle revolution, proving that consumer demand for clean transportation existed and inspiring the development of modern EVs like the Tesla Model S and GM's own Chevrolet Bolt EV.

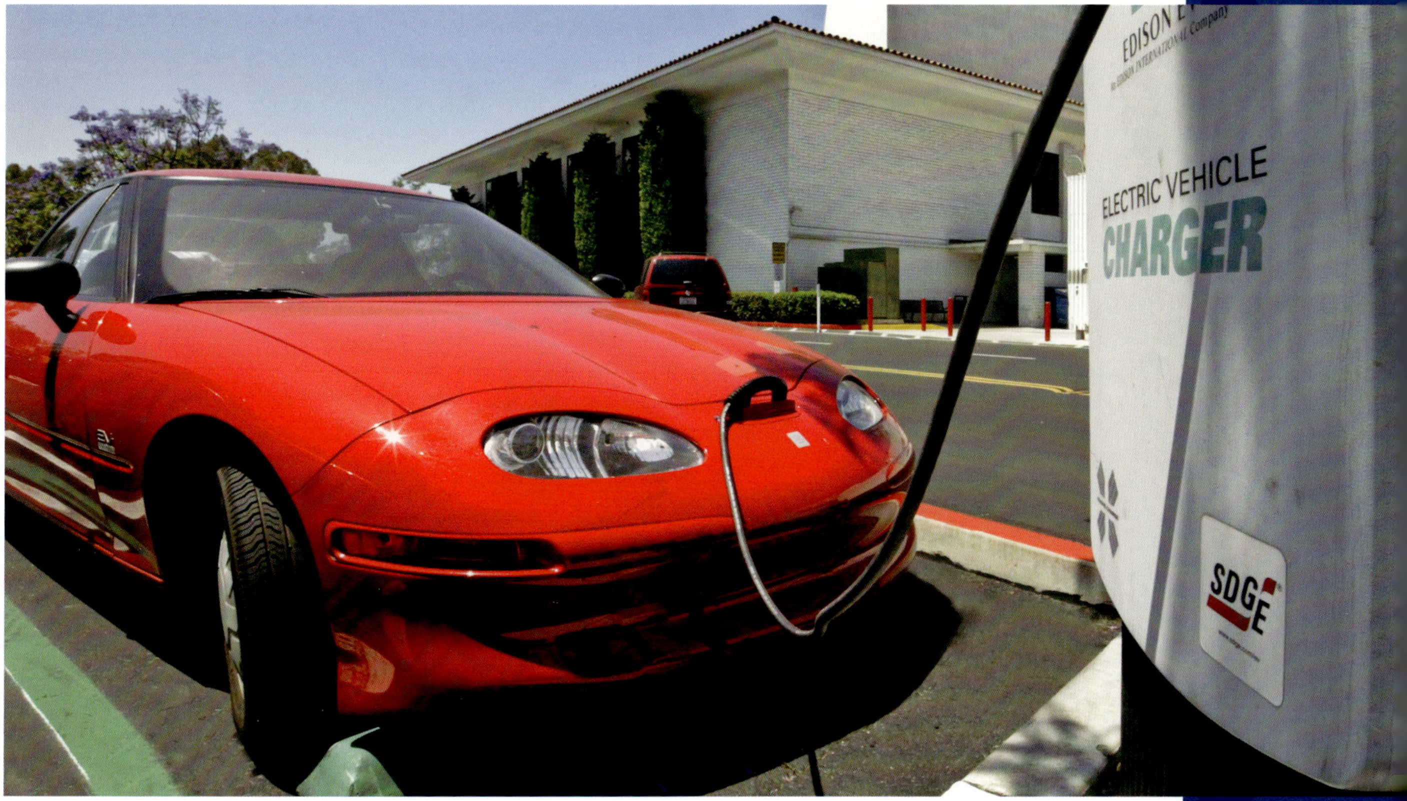

1997 GOOGLE

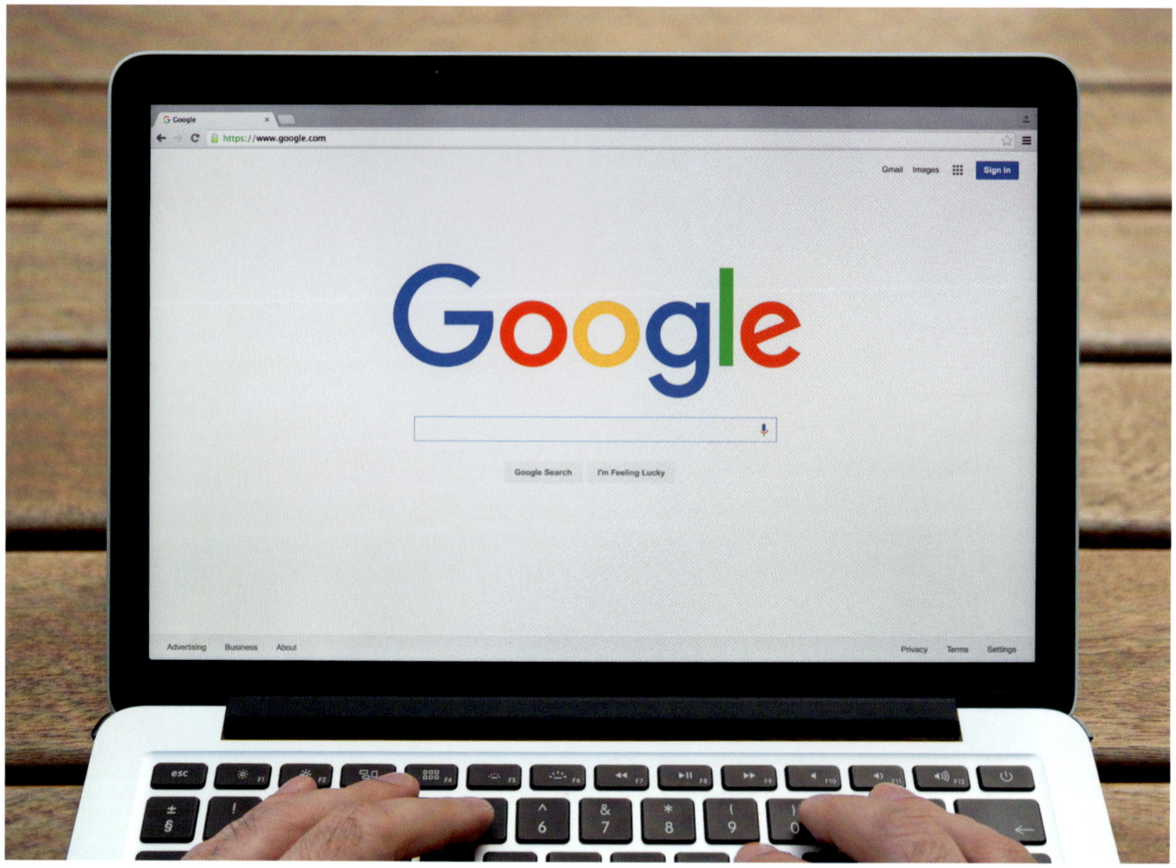

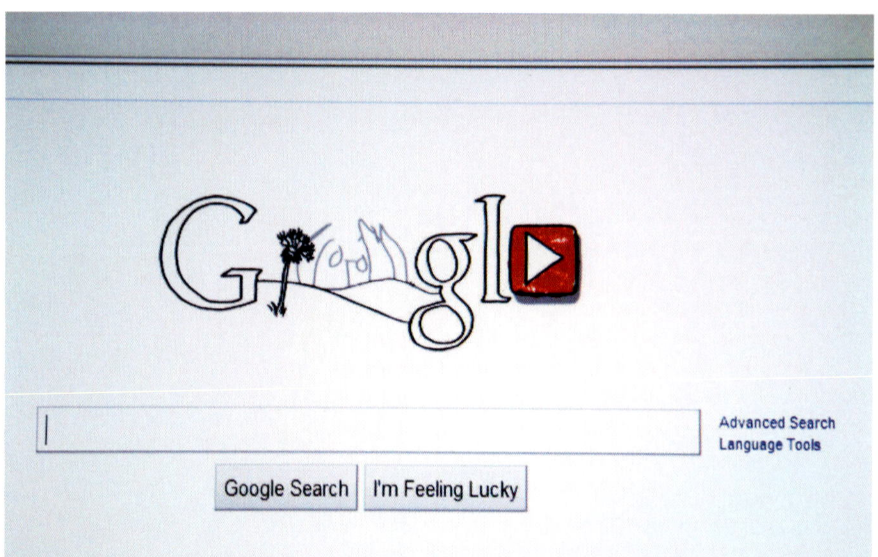

LEFT: A now-iconic feature of the Google homepage is the Google Doodle, a temporary logo that is changed daily, celebrating events, heroes, holidays, and more. The Doodle shown here was created to celebrate what would have been John Lennon's seventieth birthday.

On September 15, 1997, a domain name was quietly registered that would soon fundamentally reshape the Internet landscape. Larry Page (1973–) and Sergey Brin (1973–) secured Google.com nearly a year before officially incorporating their company on September 4, 1998. This simple act of registering a domain marked the beginning of what would become the world's most dominant Internet search engine and technology company. The timing was significant—it occurred just over seven years after Tim Berners-Lee had released the World Wide Web to the public, placing Google at the forefront of the Internet's explosive growth period. Originally conceived as part of a research project called "BackRub" at Stanford University, the Google domain represented the duo's vision for organizing the world's information. Google.com became the entry point to an ecosystem encompassing email (Gmail), mobile operating system (Android), cloud computing, advertising platforms, and many other innovations. The domain transformed from a simple research project URL into one of the most valuable digital assets in the world, answering billions of queries daily and fundamentally changing how we access and processes information.

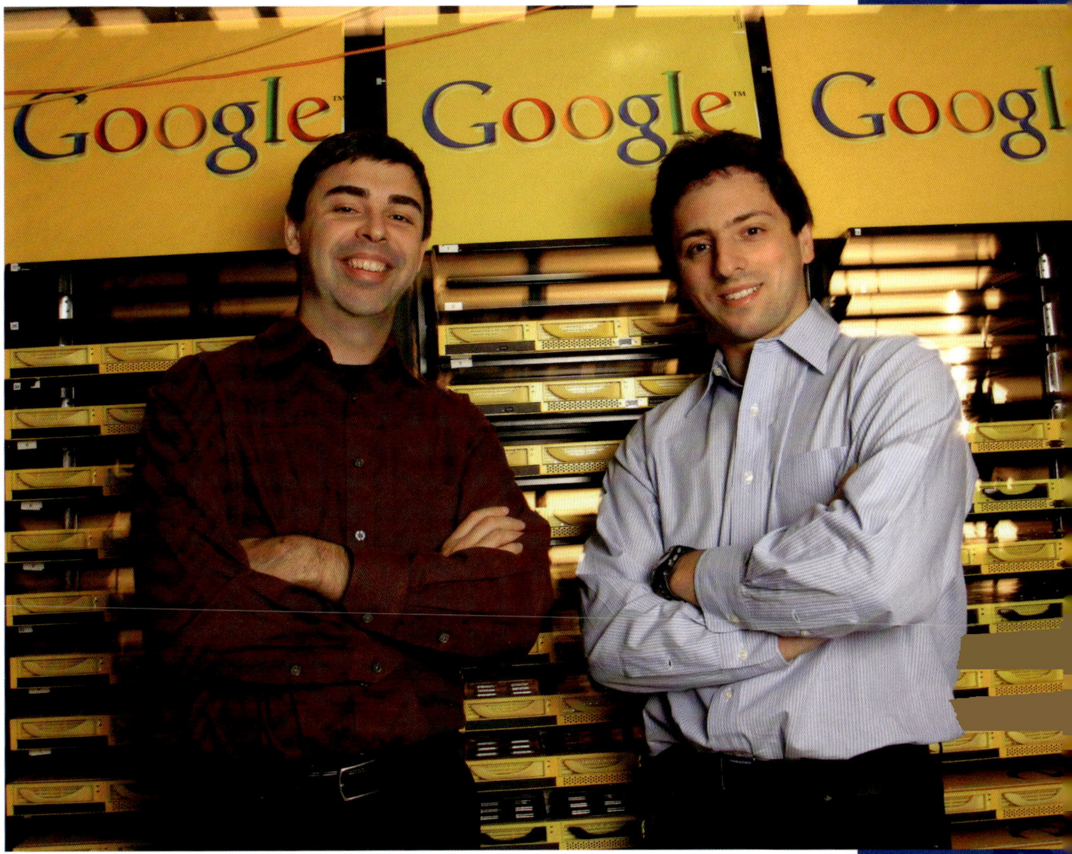

Google's co-founders Larry Page (l) and Sergey Brin (r).

1997 Netflix

Founded by Reed Hastings (1960–) and Marc Randolph (1958–), Netflix launched in 1997 as a DVD-by-mail rental service. The company revolutionized home entertainment by offering unlimited rentals through subscription plans and eliminating the late fees common at brick-and-mortar video rental shops. In 2007, Netflix pivoted to streaming, profoundly transforming how audiences consumed content. Netflix's legacy includes pioneering binge-watching culture, investing heavily in original programming like *House of Cards* and *Stranger Things*, and forcing traditional media companies to adapt. Netflix democratized global content distribution, establishing the template for modern streaming platforms that now dominate entertainment.

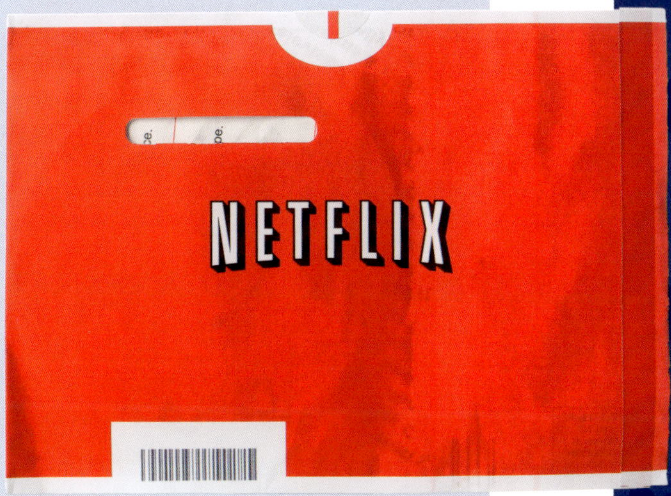

1998 THE FURBY

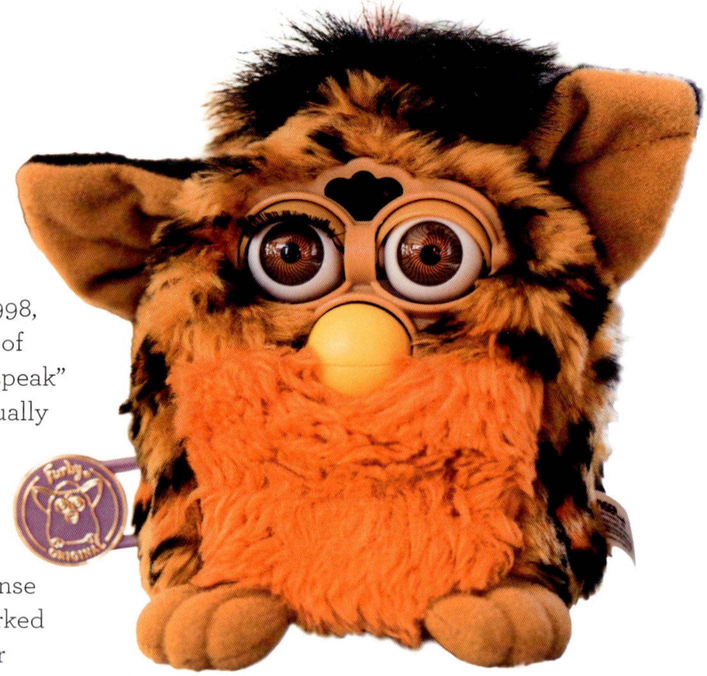

The Furby, launched by Tiger Electronics in 1998, became one of the most iconic electronic toys of the late 1990s. The owllike robotic pet could "speak" in its own language called "Furbish" and gradually "learn" English through interaction with its owner. The toy's sophisticated animatronics, including moving eyes, beak, and ears, combined with its ability to respond to touch, sound, and light, created an unprecedented sense of companionship for children. The Furby sparked the first major electronic pet craze, selling over forty million units in its first three years. Its success paved the way for subsequent interactive toy phenomena like Tamagotchis and modern smart toys. The Furby's seemingly lifelike behavior raised questions about emotional attachment to technology that remain relevant today.

1999 *SPONGEBOB SQUAREPANTS*

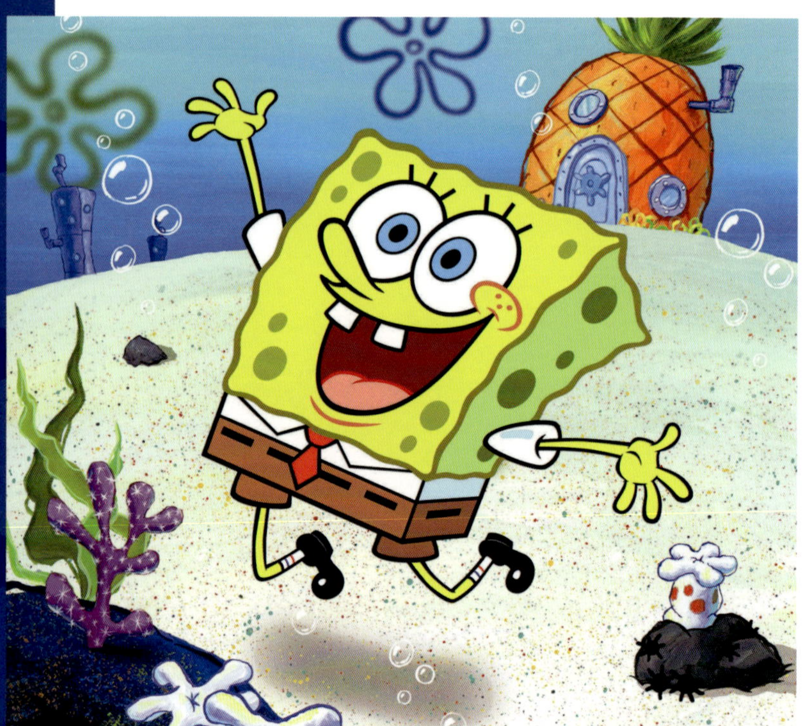

SpongeBob SquarePants has become one of the most long-lasting and influential animated series in television history since its debut on the Nickelodeon network in 1999. Created by marine biologist Stephen Hillenburg (1961–2018), the series has transcended its original target audience of children to become a cultural phenomenon embraced by multiple generations. Its absurdist humor, memorable characters, and quotable dialogue have spawned countless Internet memes and references that permeate popular culture. SpongeBob's relentless optimism and work ethic at the Krusty Krab—the fictional fast-food restaurant where he works—have made him an unlikely icon of positivity and determination. With multiple movies, Broadway adaptations, and merchandise worth billions, *SpongeBob SquarePants* has proven that well-crafted animation can create lasting cultural impact across decades.

2000 HUMAN GENOME DRAFT SEQUENCE COMPLETED

The completion of the draft human genome sequence in 2000 marked a pivotal moment in scientific history and established the United States as the global leader in genomics research. This enormous achievement, announced jointly by the publicly funded Human Genome Project and private US company Celera Genomics, marked the culmination of a decade-long international effort with significant American leadership and investment. The project has been transformative for US biotechnology, spurring the creation of countless companies focused on genetic testing and personalized medicine. It laid the foundation for precision medicine initiatives, enabling treatments tailored to individual genetic profiles. The genomic revolution has positioned American institutions at the forefront of breakthrough discoveries in cancer research, rare disease treatment, and gene therapy. The project's success demonstrated America's capacity for large-scale scientific coordination and has maintained US competitiveness in the rapidly expanding field of biotechnology, contributing billions to the economy while advancing human health worldwide.

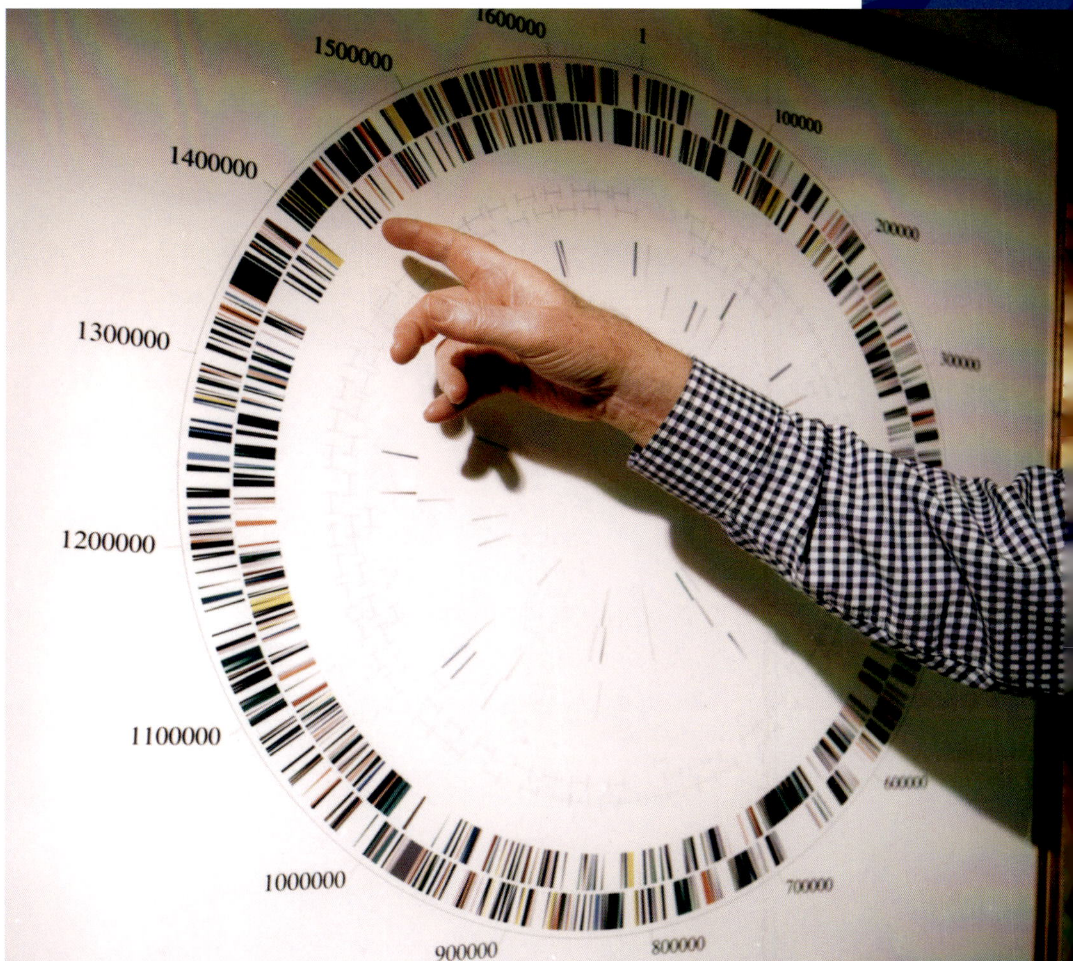

A DNA specialist points to the completed genome sequence.

2001 iPOD

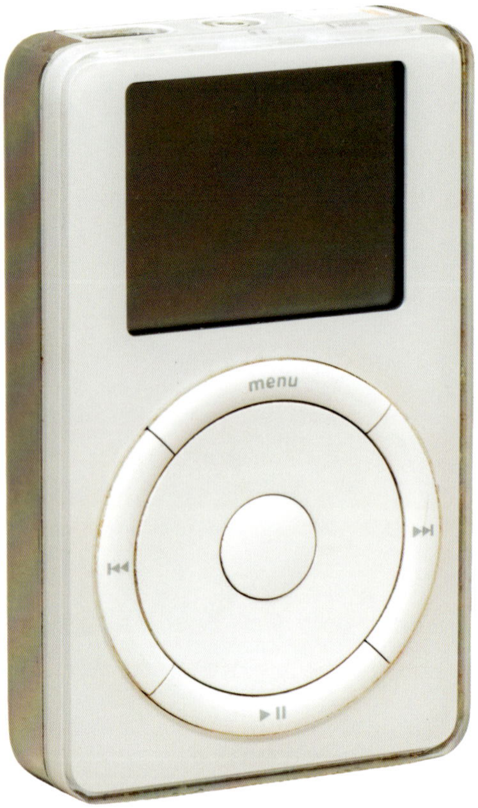

The iPod, launched by Apple in 2001, revolutionized the music industry and consumer electronics landscape. With its sleek design, intuitive scroll wheel, and the tagline "1,000 songs in your pocket," the device made digital music portable and accessible like never before. The iPod's success was enhanced by the iTunes store, which provided a legal and convenient way to purchase and download music, helping combat piracy while giving artists a new revenue stream. The iPod established Apple as a major player in consumer electronics, setting the stage for future innovations like the iPhone and iPad. The device also popularized the concept of premium, design-focused technology products and demonstrated how hardware and software integration could create superior user experiences. The iPod reshaped how people discovered and consumed music, paving the way for playlist culture and, ultimately, streaming services. The iPod's influence on urban culture was profound, with its distinctive white corded earphones—and later earbuds—becoming a fashion statement and status symbol. Though it was discontinued in 2022, the iPod's legacy lives on in every smartphone and streaming platform.

2001 The Segway

The Segway, invented in 2001 by inventor and entrepreneur Dean Kamen (1951-), was a revolutionary two-wheeled, self-balancing personal transporter that promised to transform urban mobility. Despite initial hype and a five-thousand-dollar price tag, it never achieved mainstream adoption due to safety concerns, regulatory restrictions, and impracticality for daily use. The device became more of a novelty for tourists and security personnel than a transportation revolution. However, the Segway's gyroscopic technology and electric propulsion concepts influenced modern electric scooters, hoverboards, and other personal mobility devices. Production ceased in 2020, but its engineering innovations continue to shape today's micromobility landscape.

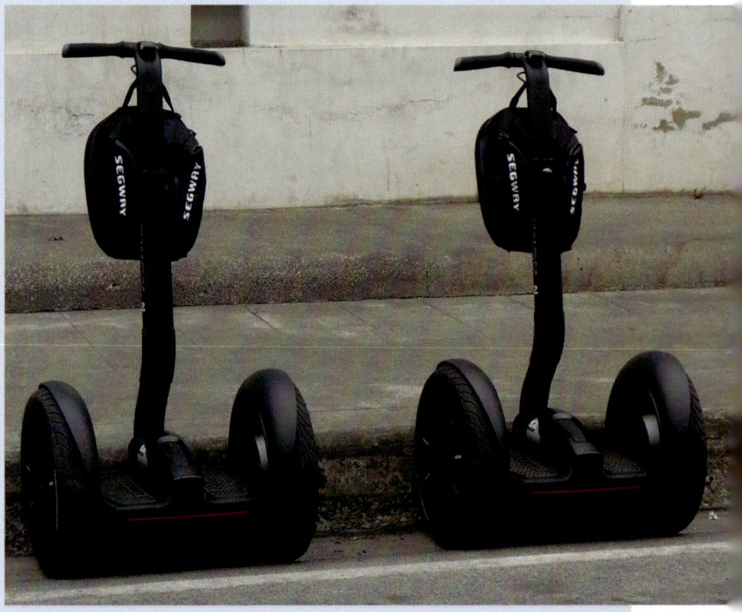

2002 ROOMBA

The Roomba, launched by the company iRobot in 2002, revolutionized home cleaning by introducing the first commercially successful autonomous vacuum cleaner. The disc-shaped robot navigates homes using a combination of sensors, algorithms, and initially random movement patterns, gradually evolving to include mapping capabilities and smart home integration. It demonstrated that robots could be helpful household companions rather than threatening entities, paving the way for broader acceptance of domestic automation. The device also sparked countless viral videos of pets riding Roombas and inspired numerous memes, embedding itself in popular culture. Technologically, the Roomba established foundational principles for autonomous navigation, obstacle avoidance, and user-friendly robotics interfaces. Its success inspired an entire industry of robotic home appliances, from lawn mowers to window cleaners. The device also contributed valuable data about indoor environments and human-robot interaction patterns. But beyond its practical impact, the Roomba became a cultural phenomenon and gateway device for robotics in everyday life.

2003 ICE BIKE

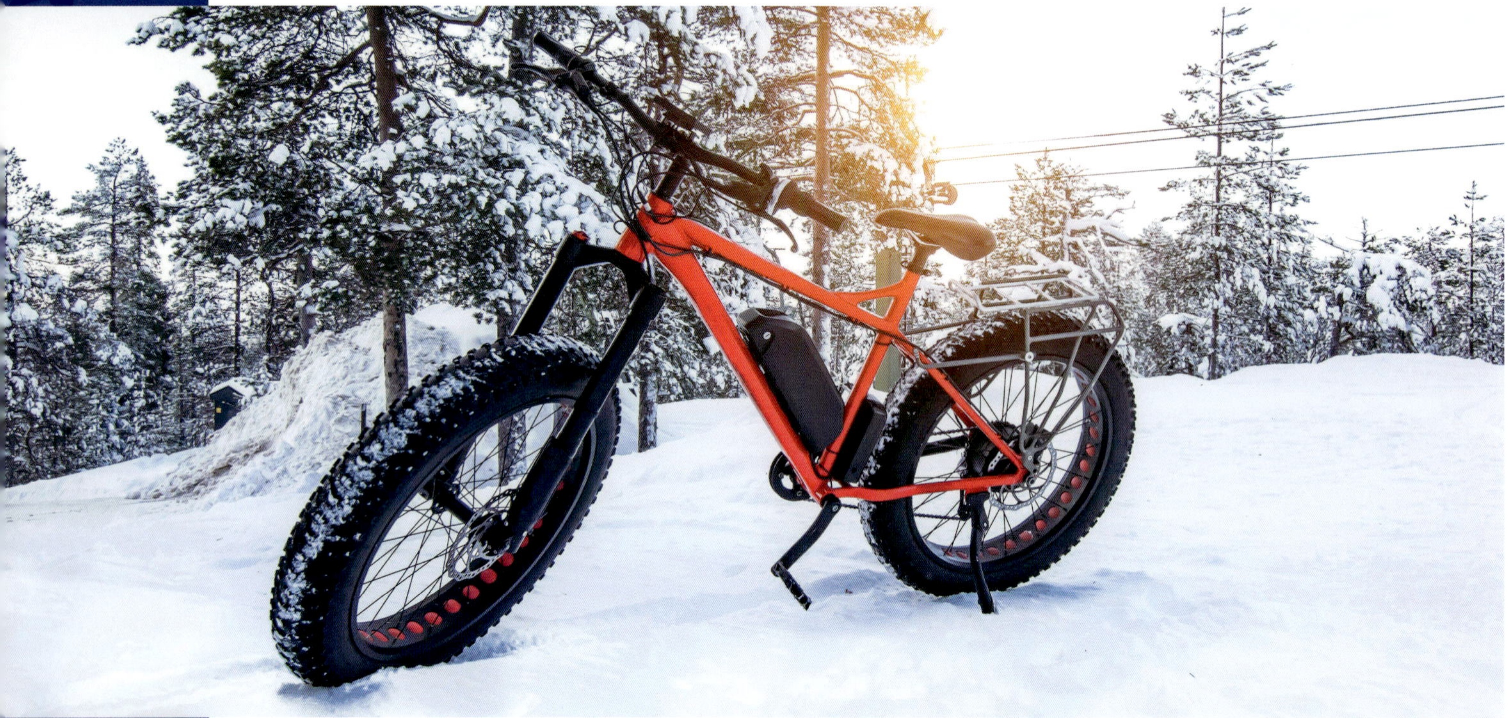

The ice bike was invented in 2003 by Dan Hanebrink (1938–2018), a visionary mountain-bike designer and aeronautical engineer known for pioneering suspension systems and fat tires in mountain bikes during the 1980s and 1990s. The ice bike was born from adventurer Doug Stoup's need for faster transportation during his Antarctic expeditions, as traditional skiing proved too slow for polar exploration. Hanebrink's ice bike omitted plastic components to prevent freezing and shattering in extreme temperatures, and featured super-fat, low-pressure tires designed specifically for Antarctic conditions. The bike was recognized as one of *TIME* magazine's Best Inventions of 2003, cementing its place in innovation history. Hanebrink's ice bike demonstrated how specialized bicycle design could conquer previously impossible terrains, inspiring future developments in extreme-condition cycling. His innovation proved that with proper engineering, bicycles could function in Earth's most challenging environments. The ice bike helped create today's Extreme Terrain Bike, which works well in diverse environments, including snow, sand, mud, pavement, and traditional mountain bike trails. Many believe it to be the ultimate all-terrain cycle.

Hanebrink's ice bike, nicknamed "fat bike" because of the tires.

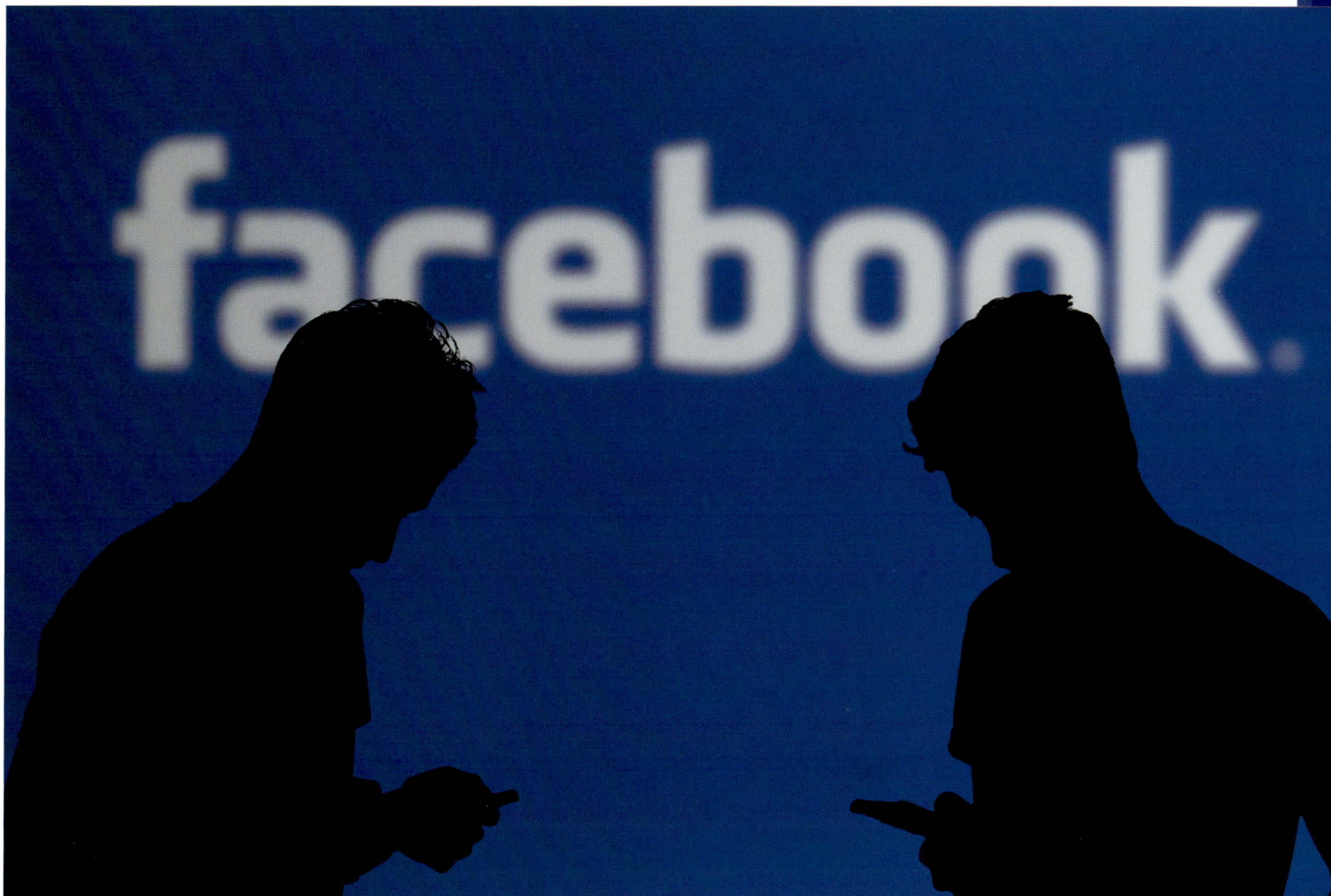

2004 FACEBOOK

Facebook was invented in 2004 by Mark Zuckerberg (1984–), along with co-founders Eduardo Saverin, Andrew McCollum, Dustin Moskovitz, and Chris Hughes, while they were students at Harvard University. Originally called "TheFacebook," it began as a social networking site exclusively for college students before expanding to high schools and eventually the general public in 2006. The platform transformed how people connect and share information online, becoming the blueprint for modern social media. Facebook introduced features like the News Feed, Timeline, and Like button that ultimately became the standard across other social media platforms.

Facebook's legacy is profound and complex. Though the platform connects billions of people worldwide, enabling long-distance relationships, community building, and global communication, it also raises some concerns about privacy, data collection, and the spread of misinformation. In addition, Facebook has been scrutinized for its role in political polarization and its impact on mental health, particularly among young users. Today, Facebook (now Meta) continues to shape digital communication while grappling with regulatory challenges and evolving toward virtual reality, cementing its position as one of the most influential technological innovations of the twenty-first century.

2005 YOUTUBE

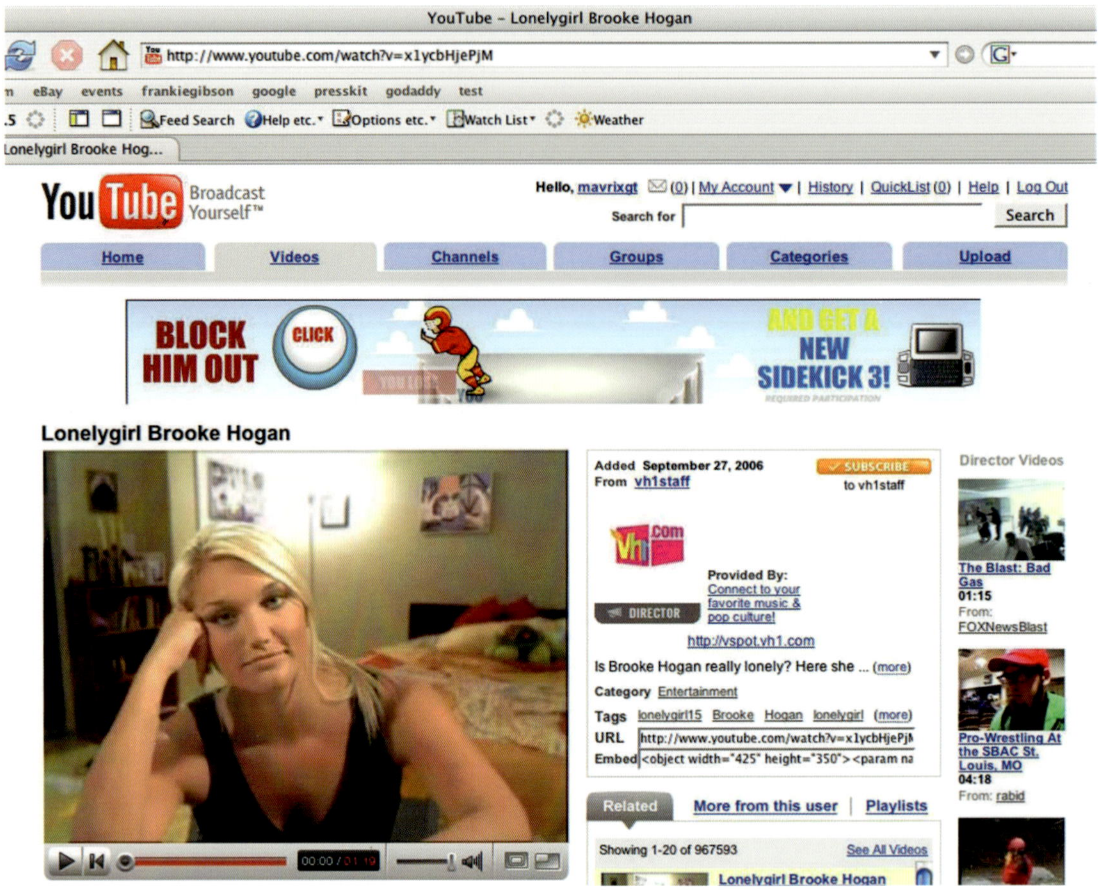

Videos from Lonelygirl Brooke Hogan was one of the early successes of YouTube.

YouTube launched on February 14, 2005, when three former PayPal employees—Chad Hurley (1977–), Steve Chen (1978–), and Jawed Karim (1979–)—created a platform to share videos online. The first video, "Me at the Zoo," was uploaded by Karim in April of that year, marking the beginning of a digital revolution. Within months, YouTube exploded in popularity, attracting millions of users who could finally upload, share, and discover video content more easily. Google recognized its potential and acquired the platform for $1.65 billion in October 2006, just twenty months after its launch. YouTube democratized content creation, enabling anyone with a camera to reach global audiences. The platform also birthed new career paths—YouTubers, influencers, and digital marketers—while transforming how we consume news, education, and culture. It became a powerful tool for social movements, political discourse, and cultural exchange. Today, YouTube processes more than five hundred hours of video uploaded every minute, serving billions of users worldwide. Its impact on society, from changing television consumption habits to creating new forms of celebrity, makes it one of the most influential platforms in Internet history.

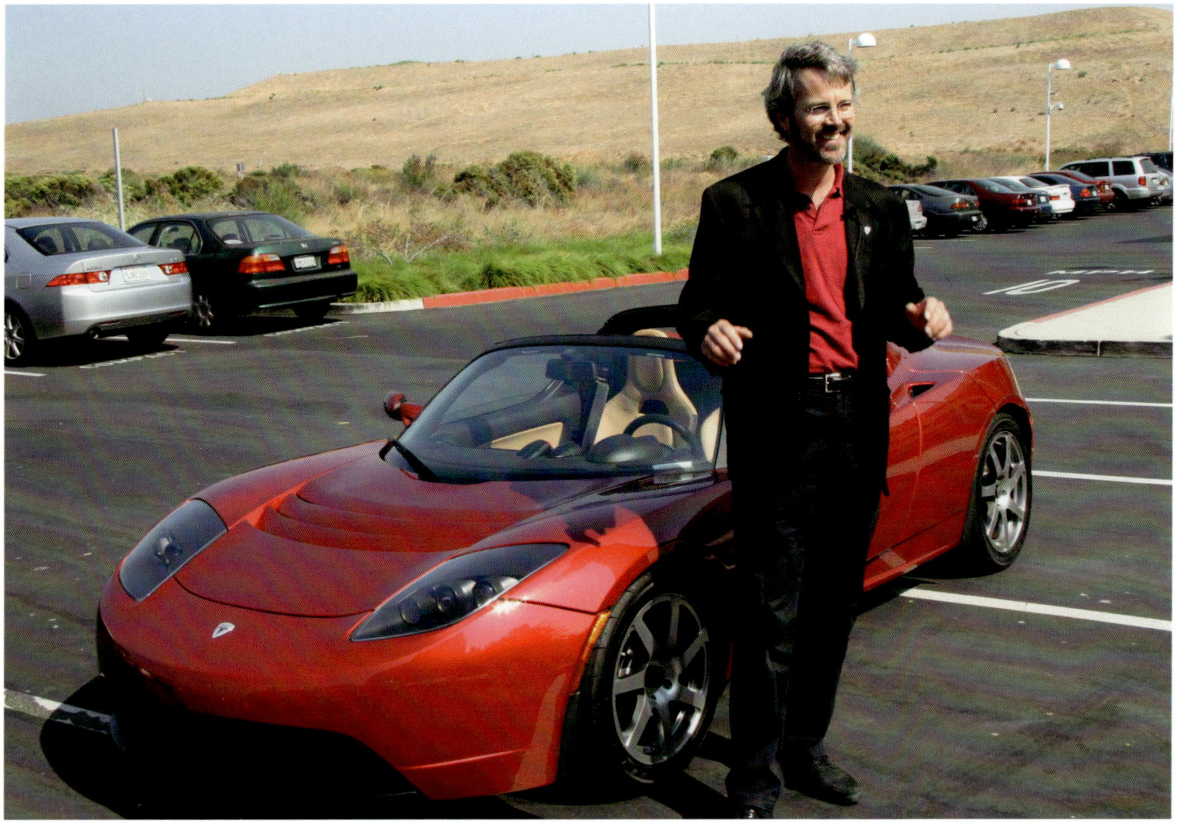

Former Tesla CEO Martin Eberhard with his Tesla Roadster, 2006.

2006 TESLA ROADSTER

The first prototype for the Tesla Roadster was completed in 2006. When the car was unveiled to the public in 2008, it marked a revolutionary moment in automotive history—it was the first highway-capable electric vehicle made in serial production by Tesla Motors. Built on a modified Lotus Elise chassis, the Roadster shattered preconceptions about electric cars being slow and impractical. With its lithium-ion battery pack delivering 250 miles of range and acceleration from 0–60 mph in just 3.9 seconds, it proved that electric vehicles could be both high-performance and desirable. The Roadster served as Tesla's proof of concept that electric vehicles could compete with traditional sports cars and generate the capital and credibility necessary for the company's expansion. The vehicle's success paved the way for the Model S, Model 3, and Tesla's eventual dominance in the electric vehicle (EV) market. More broadly, by making electric vehicles aspirational rather than merely practical, it influenced established automakers to accelerate their own EV development programs, substantially reshaping the future of transportation.

2007 IPHONE

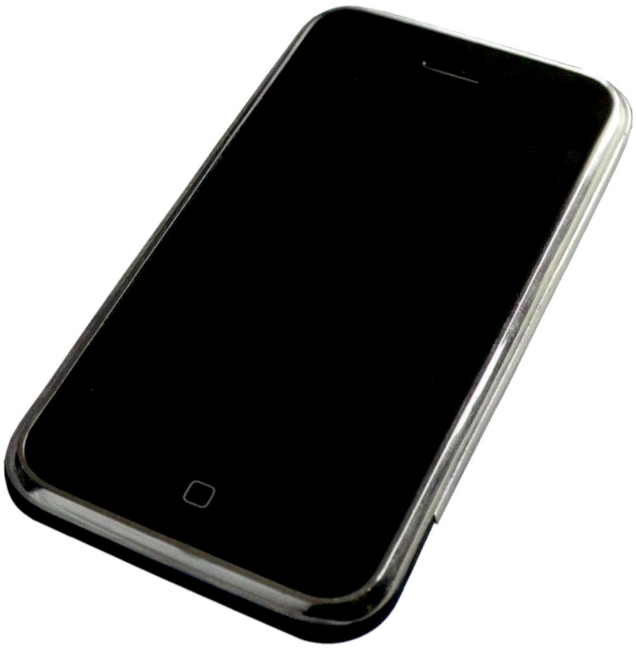

The original iPhone was unveiled by Steve Jobs on January 9, 2007 at the Macworld conference and expo. Jobs famously introduced it as "three revolutionary products in one: a widescreen iPod with touch controls, a revolutionary mobile phone, and a breakthrough Internet communications device." The new device fundamentally transformed the mobile phone industry and modern communication. The first-generation iPhone featured a 3.5-inch multi-touch display, eliminating the physical keyboard that dominated smartphones of the era. It ran on a custom version of Mac OS X, offered visual voicemail (revolutionary at the time), and provided a full web-browsing experience through Safari. The device's intuitive interface and sleek design captivated users. The iPhone established the foundation for the modern smartphone ecosystem, popularized touch-based interfaces, and created the App Store model that revolutionized software distribution. The device sparked fierce competition, accelerated mobile Internet adoption, and radically changed how people interact with technology, communication, and information in their daily lives.

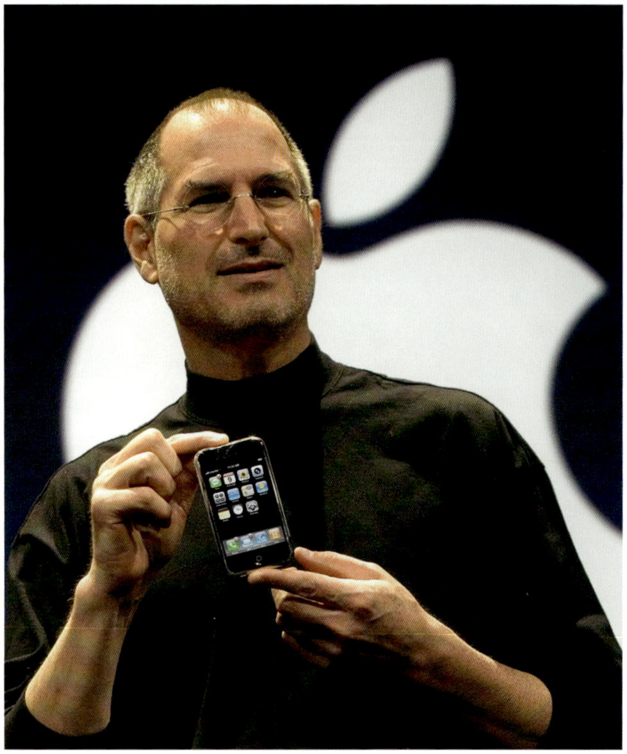

Apple CEO Steve Jobs debuts the iPhone at Macworld, January 9, 2007.

2008 THE SNUGGIE

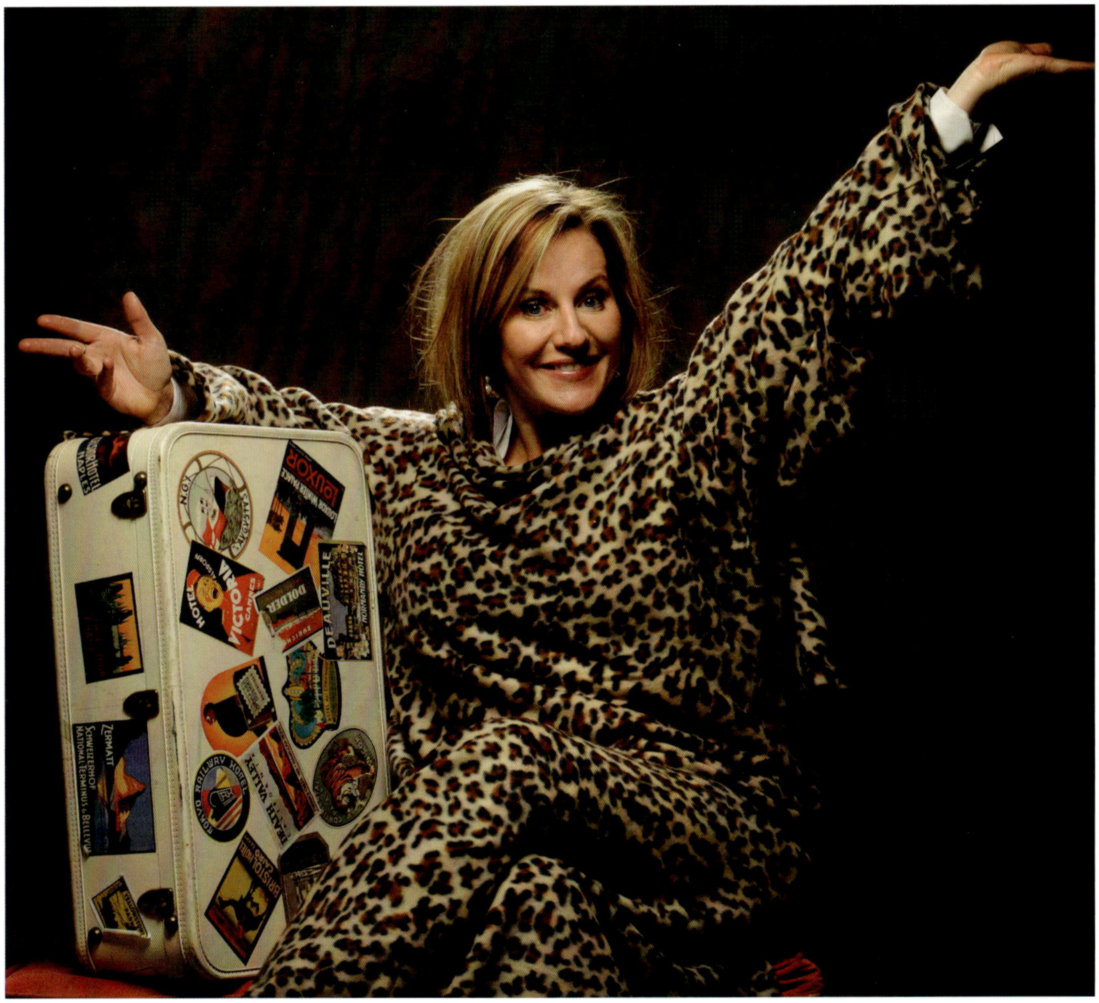

Radio star Kelly Ford shows off her cheetah Snuggie that she takes with her whenever she travels.

The Snuggie, the iconic blanket-with-sleeves, was invented by Scott Boilen (1966–) in 1998, though it didn't achieve widespread fame until Allstar Products Group's aggressive marketing campaign launched in 2008. The simple concept—a fleece blanket with armholes—solved the age-old problem of keeping warm while using your hands for reading, remote controls, or snacking. The Snuggie's infomercial became a phenomenon, spawning countless parodies and memes. Its deliberately cheesy advertising, featuring families awkwardly wearing their Snuggies at sporting events, became part of Internet folklore. Nevertheless, the product sold more than thirty million units worldwide. The Snuggie paved the way for countless similar novelty items and became a cultural touchstone representing both American consumerism and the power of unabashed, campy advertising to create genuine demand for products nobody knew they wanted such as the ShamWow (a cleaning cloth) and Hot Buns (a device for making sock buns in hair).

2009 INTERNATIONAL TRANSGENDER DAY OF VISIBILITY

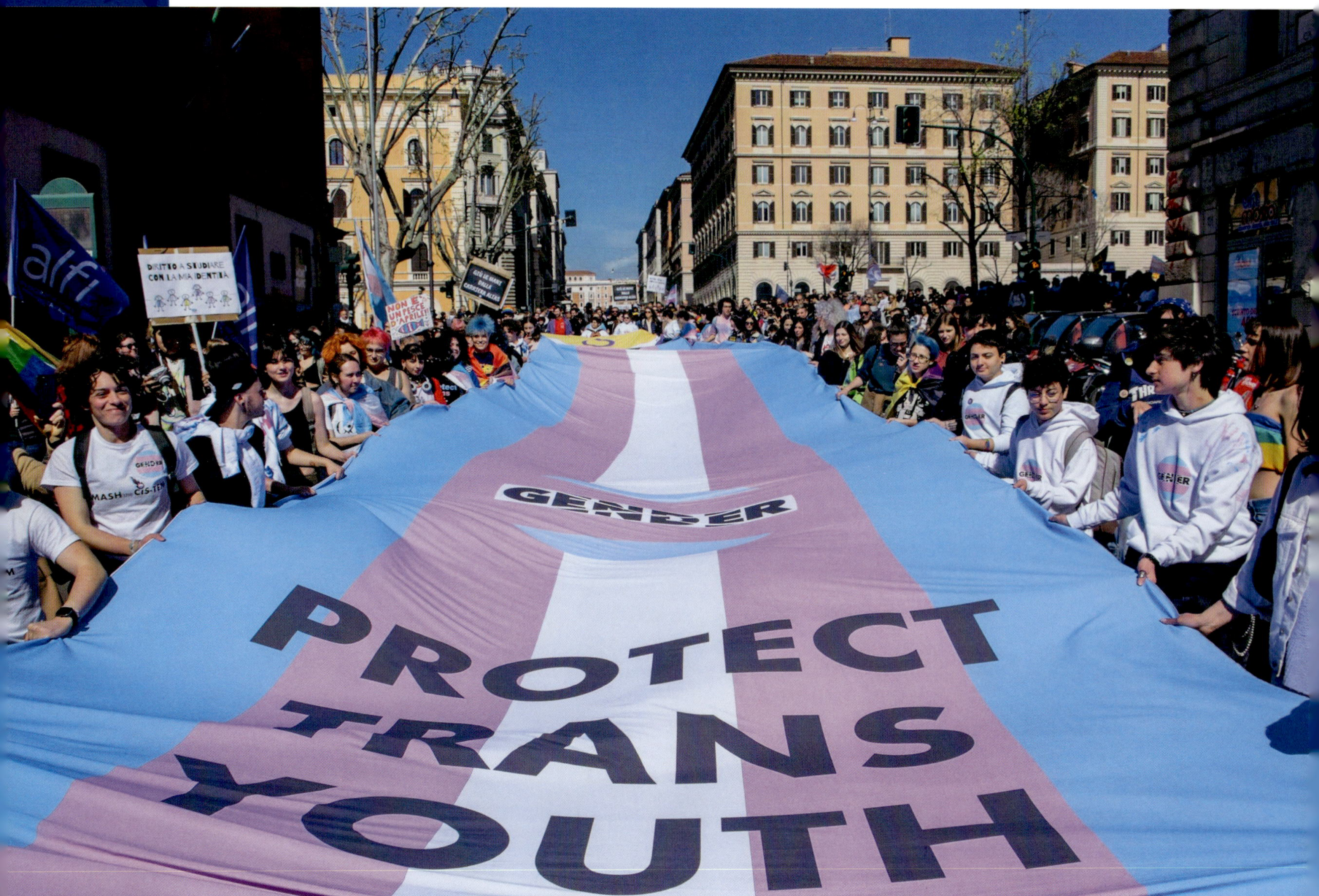

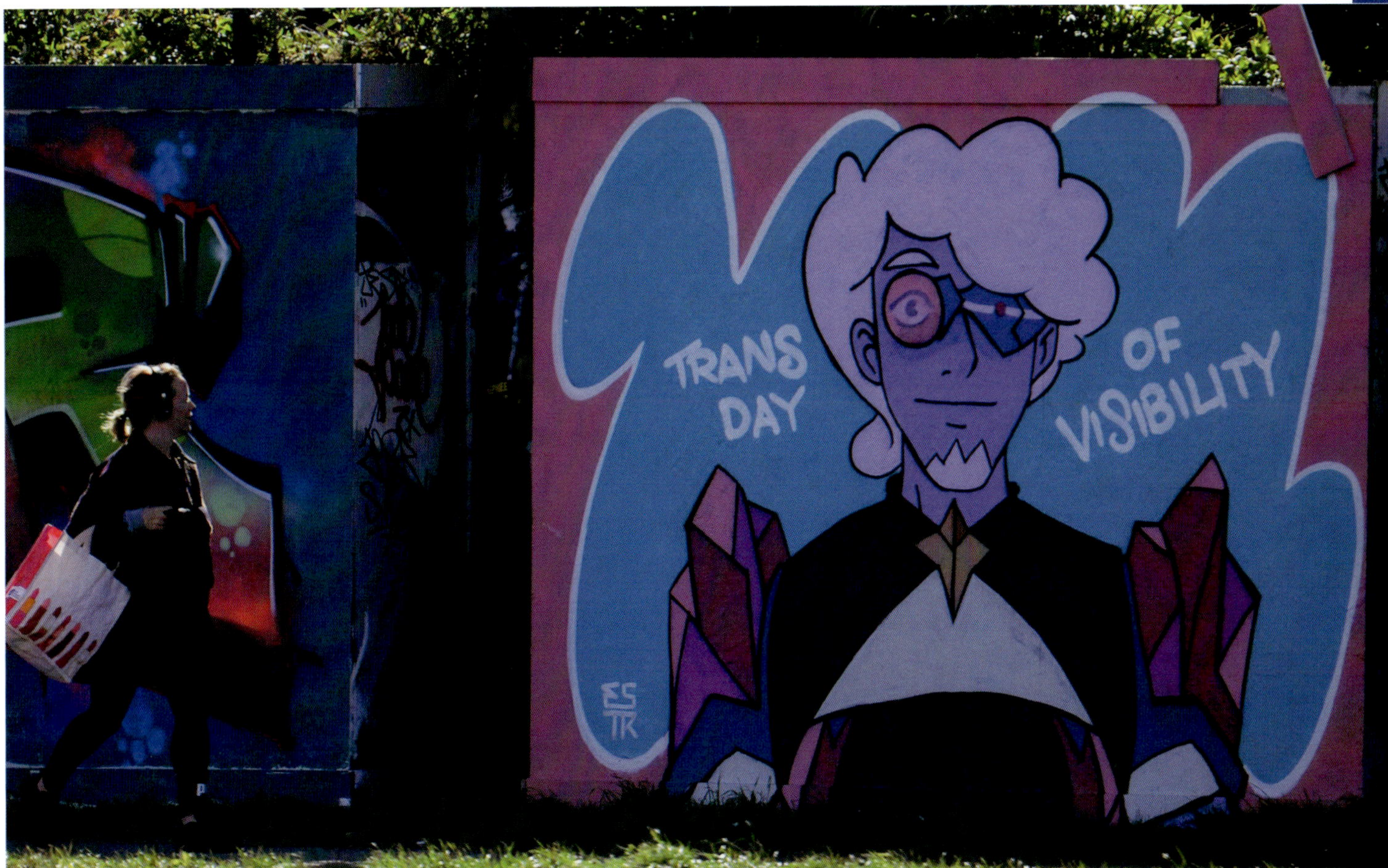

International Transgender Day of Visibility—March 31—was created in 2009 by transgender activist Rachel Crandall-Crocker (1958–), a psychotherapist from Michigan. She founded the day in response to the lack of LGBTQ+ recognition of transgender people and the overwhelming focus on violence in media coverage of the transgender community. Crandall-Crocker was frustrated that the only prominent transgender-centered day was the Transgender Day of Remembrance, which mourned the dead. Having lost her marriage when she came out and her job when she transitioned, Crandall-Crocker hoped to create a day where people could celebrate transgender lives. The observance has grown globally, shifting narratives from tragedy to celebration, empowering transgender individuals to live authentically while building community connections, and raising awareness about transgender experiences, achievements, and ongoing challenges. Although violence remains a critical concern, there are many beautiful stories of transgender joy that are increasingly exposed thanks to efforts like Crandall-Crocker's.

LEFT: A march in Rome on the International Transgender Day of Visibility, 2023. **ABOVE:** A mural of Jewelstar from Netflix's *She-Ra and the Princesses of Power* in Dublin's Grand Canal area painted by Irish artist Emmalene Blake for International Transgender Day of Visibility, 2021.

2010 MALARIA-PROOF MOSQUITO

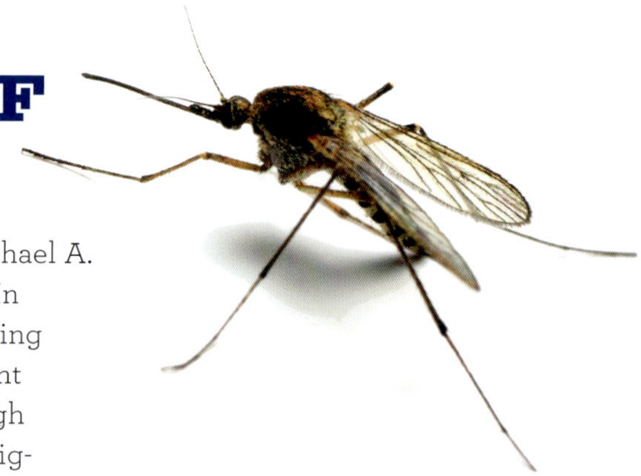

University of Arizona entomologists, led by Professor Michael A. Riehle, reached a significant scientific milestone in 2010. In that year, the research team succeeded in genetically altering *Anopheles* mosquitoes to render them 100 percent resistant to the single-celled parasite *Plasmodium*. The breakthrough involved manipulating the insulin/insulin-growth factor signaling pathway in these disease-carrying mosquitoes. The researchers' long-term vision involves developing genetically engineered malaria-resistant mosquitoes that could reproduce rapidly enough to replace wild populations of malaria-carrying mosquitoes. This research represents a significant advancement in vector control strategies that could potentially prevent millions of malaria infections worldwide.

2011 THE PELOTON

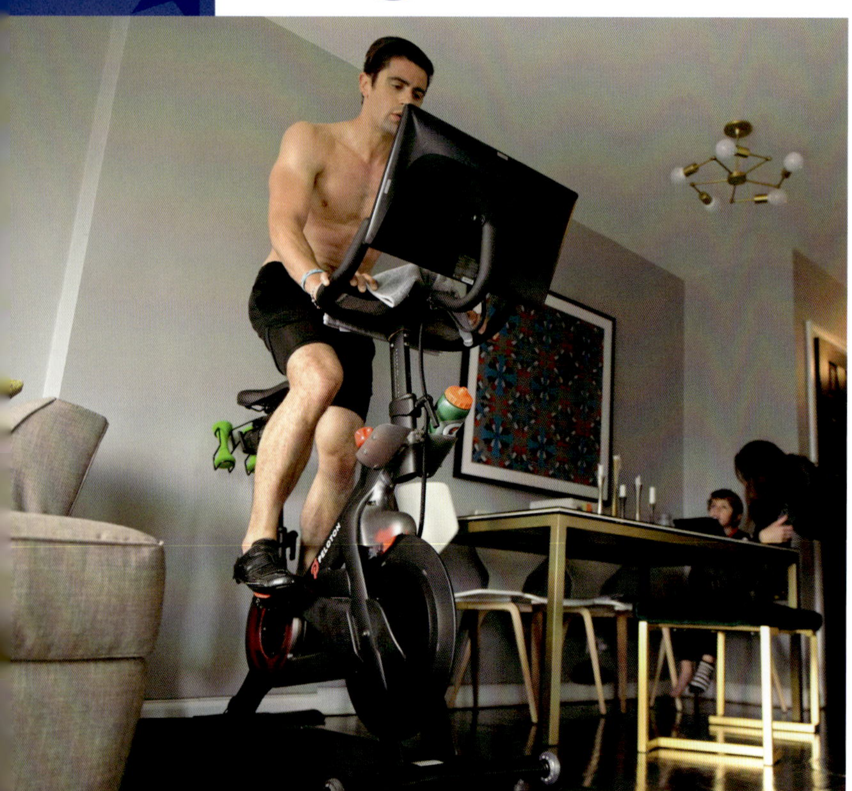

In 2011, after missing a spin class due to a snowstorm, John Foley (1971–), an executive at Barnes & Noble in New York City, approached his coworker, Tom Cortese (1980–), proposing that advanced technology could enable busy individuals to enjoy the benefits of premium indoor cycling classes from the comfort of their own homes. The partners spent the rest of the year putting together key team members and raising funds for their business. Founded in 2012, Peloton rose to become a fitness startup valued at $1 billion by selling exercise bikes with large Internet-enabled touchscreens that played live broadcasts of fitness classes. The company revolutionized home fitness by combining high-quality equipment with interactive technology, creating a social exercise experience that connected users with live and on-demand classes. Peloton's legacy lies in permanently shifting consumer expectations toward interactive, connected fitness experiences.

2012 GENE EDITING

CRISPR-Cas9, discovered in 2012 by American scientist Jennifer Doudna (1964–) and her French colleague, Emmanuelle Charpentier (1968–), revolutionized genetic engineering by providing a precise, programmable tool for editing DNA. Originally identified as part of bacterial immune systems, CRISPR (clustered regularly interspaced short palindromic repeats) uses guide RNAs to direct the Cas9 enzyme to specific DNA sequences, where it makes targeted cuts. This breakthrough earned Doudna and Charpentier the 2020 Nobel Prize in Chemistry and transformed multiple fields. In medicine, CRISPR has enabled treatments for sickle cell disease and beta-thalassemia and shows promise for treating cancer, blindness, and genetic disorders. Agricultural applications include developing drought-resistant crops and improving nutritional content. Research has accelerated dramatically, with CRISPR enabling the rapid creation of disease models and therapeutic targets. However, CRISPR has also raised ethical questions about human genome editing and the potential for unintended consequences. CRISPR continues evolving with new variants like base editors and prime editors, promising even more breakthroughs to come.

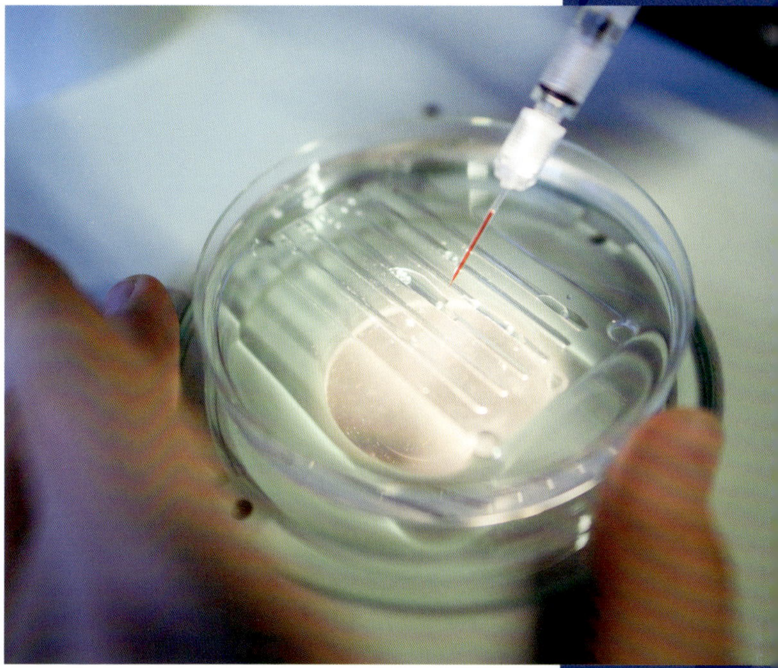

ABOVE: A researcher performs a CRISPR/Cas9 process. **BELOW:** Emmanuelle Charpentier (l) and Jennifer Doudna (r) receive the 2015 award for Technical and Scientific Research at the Princess of Asturias Awards in Spain.

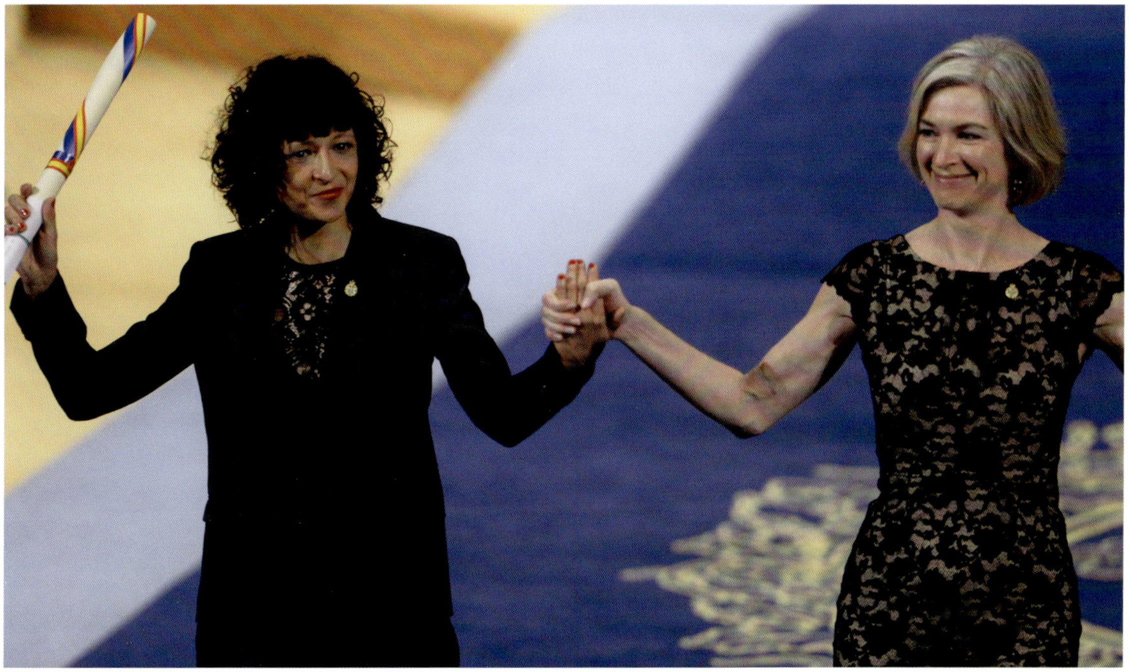

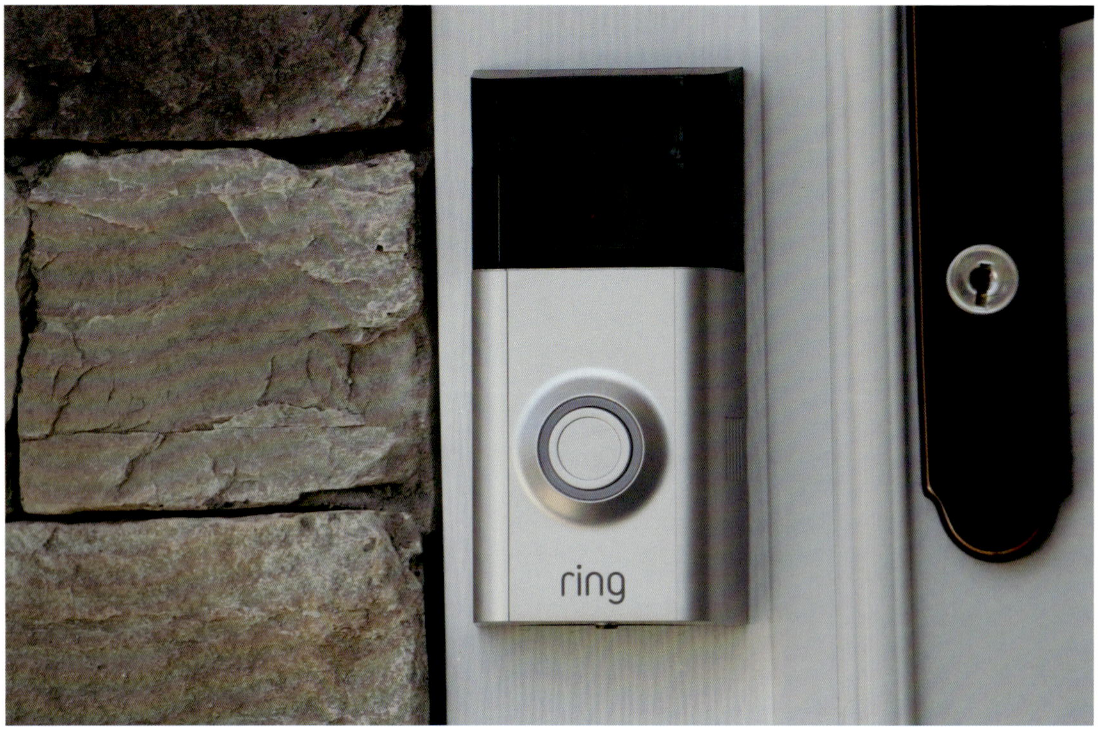

2013 RING VIDEO DOORBELL

The Ring doorbell was created by American entrepreneur Jamie Siminoff (1976–) with a simple yet revolutionary concept: a doorbell that sends live video of visitors directly to the homeowner's smartphone, allowing them to communicate with visitors no matter where they are. After struggling to gain traction and being rejected on the TV show *Shark Tank*, Siminoff's company was officially founded in fall 2013 as a crowdfunded startup called Doorbot. He renamed it Ring the following year, after the company finally began to receive interest from investors. Siminoff's gumption paid off—the company was acquired by Amazon in 2018 for approximately $1 billion and transformed Ring from a garage startup into a dominant force in home security, with the company becoming a fixture in more than ten million American homes. Ring helped create the modern smart-home ecosystem, inspiring countless competitors and innovations. That said, Ring's product lines have faced scrutiny over privacy and information access issues. Nevertheless, the invention is an inspiring example of how perseverance can pay off for entrepreneurs everywhere.

2014 AMAZON ALEXA

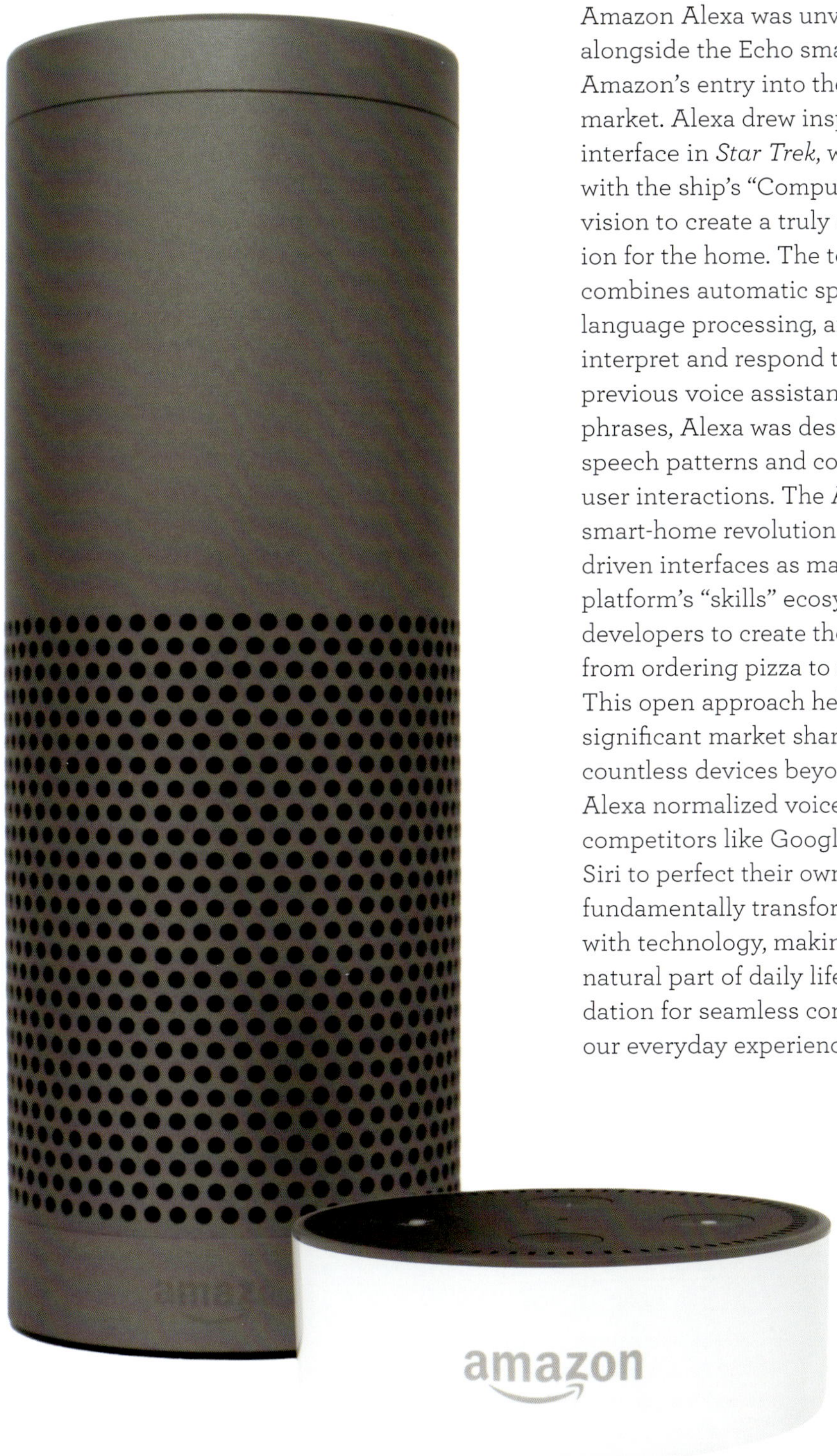

Amazon Alexa was unveiled in November 2014 alongside the Echo smart speaker, marking Amazon's entry into the voice-controlled assistant market. Alexa drew inspiration from the voice interface in *Star Trek*, where characters interacted with the ship's "Computer," reflecting Amazon's vision to create a truly conversational AI companion for the home. The technology behind Alexa combines automatic speech recognition, natural language processing, and cloud computing to interpret and respond to voice commands. Unlike previous voice assistants that required specific phrases, Alexa was designed to understand natural speech patterns and continuously learn from user interactions. The Alexa system catalyzed the smart-home revolution and established voice-driven interfaces as mainstream technology. The platform's "skills" ecosystem enabled third-party developers to create thousands of applications, from ordering pizza to controlling smart lights. This open approach helped Amazon capture significant market share and integrate Alexa into countless devices beyond their Echo speakers. Alexa normalized voice interaction and influenced competitors like Google Assistant and Apple's Siri to perfect their own offerings. The platform fundamentally transformed how people interact with technology, making voice commands a natural part of daily life and establishing the foundation for seamless computing integration in our everyday experiences.

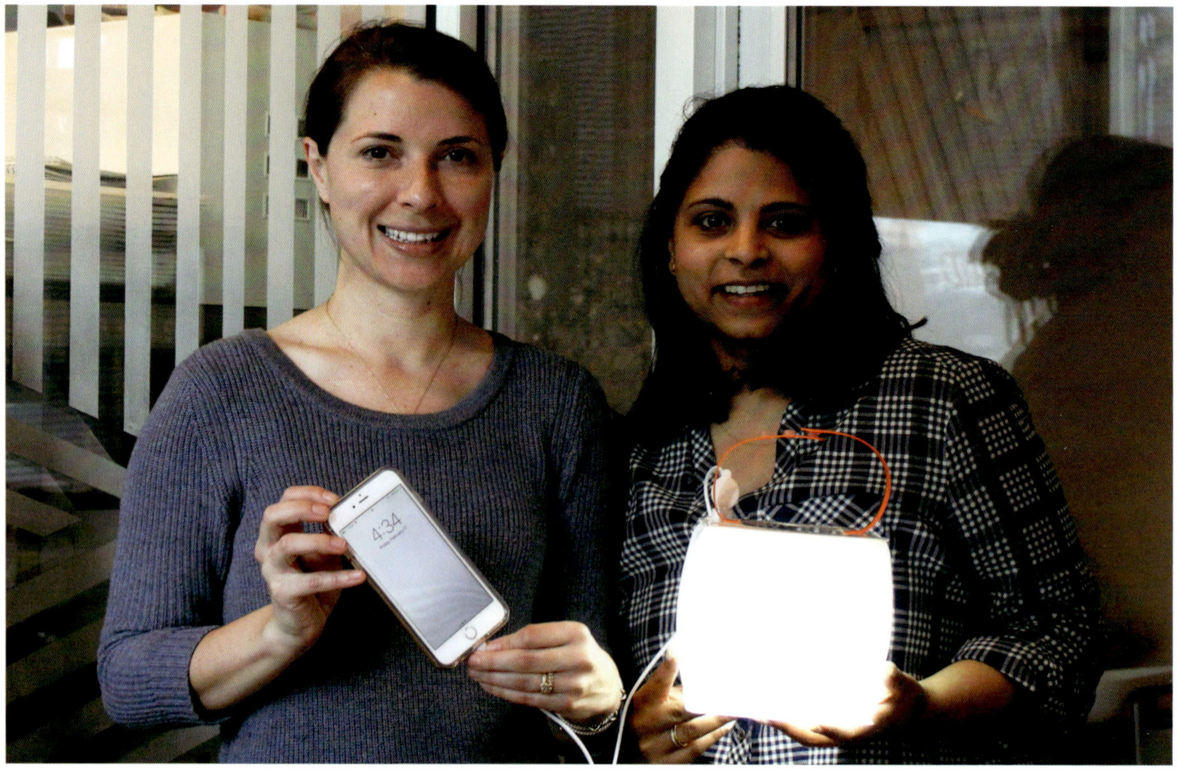

Inventors and co-founders Anna Stork (l) and Andrea Sreshta (r) with their lantern.

2015 LUMINAID SOLAR LIGHT

LuminAID's path to success began with a humanitarian crisis. Founders Anna Stork (1984–) and Andrea Sreshta (1983–) conceived their inflatable solar lantern after witnessing the devastating aftermath of Haiti's earthquake, recognizing the urgent need for portable, sustainable lighting in disaster zones. Before their pivotal *Shark Tank* appearance in February 2015, the duo had already demonstrated market validation through an Indiegogo campaign that raised more than $50,000. This early success set the stage for what would become one of the show's most memorable pitches. On *Shark Tank,* LuminAID captured the attention of all five investors, with each "shark" extending an offer. Stork and Sreshta ultimately accepted Mark Cuban's deal: $200,000 for 15 percent equity in their company. Cuban's investment proved transformative, enabling the founders to expand their product line beyond basic lanterns to include models with phone-charging capabilities. The *Shark Tank* exposure launched LuminAID into extraordinary growth. Through strategic partnerships with humanitarian organizations, the company has distributed their solar lanterns to communities in need across more than one hundred countries worldwide. This dual focus on commercial success and social impact earned LuminAID recognition from ABC as one of *Shark Tank*'s best companies, highlighting how innovative solutions can achieve both financial returns and meaningful social change.

2016 AIRPODS

Apple's AirPods launched on December 13, 2016. The wireless earbuds emerged as Apple's solution to removing the headphone jack from the iPhone 7 (launched a few months earlier on September 7), a decision that initially faced skepticism from consumers who worried about losing the tiny devices. Despite the early criticism, AirPods quickly transformed the audio landscape. Within two years, they became Apple's most popular product. The seamless pairing with Apple devices set a new standard for wireless earbuds. The product line evolved rapidly, with the second generation of AirPods launching in 2019 featuring a more advanced chip and hands-free Siri activation. AirPods Pro followed in October that year with active noise cancellation, and AirPods Max arrived in December 2020 as over-ear headphones. The invention of AirPods spurred widespread adoption of truly wireless earbuds across the industry. They normalized wireless audio consumption across the industry, with other manufacturers soon making their own wireless products.

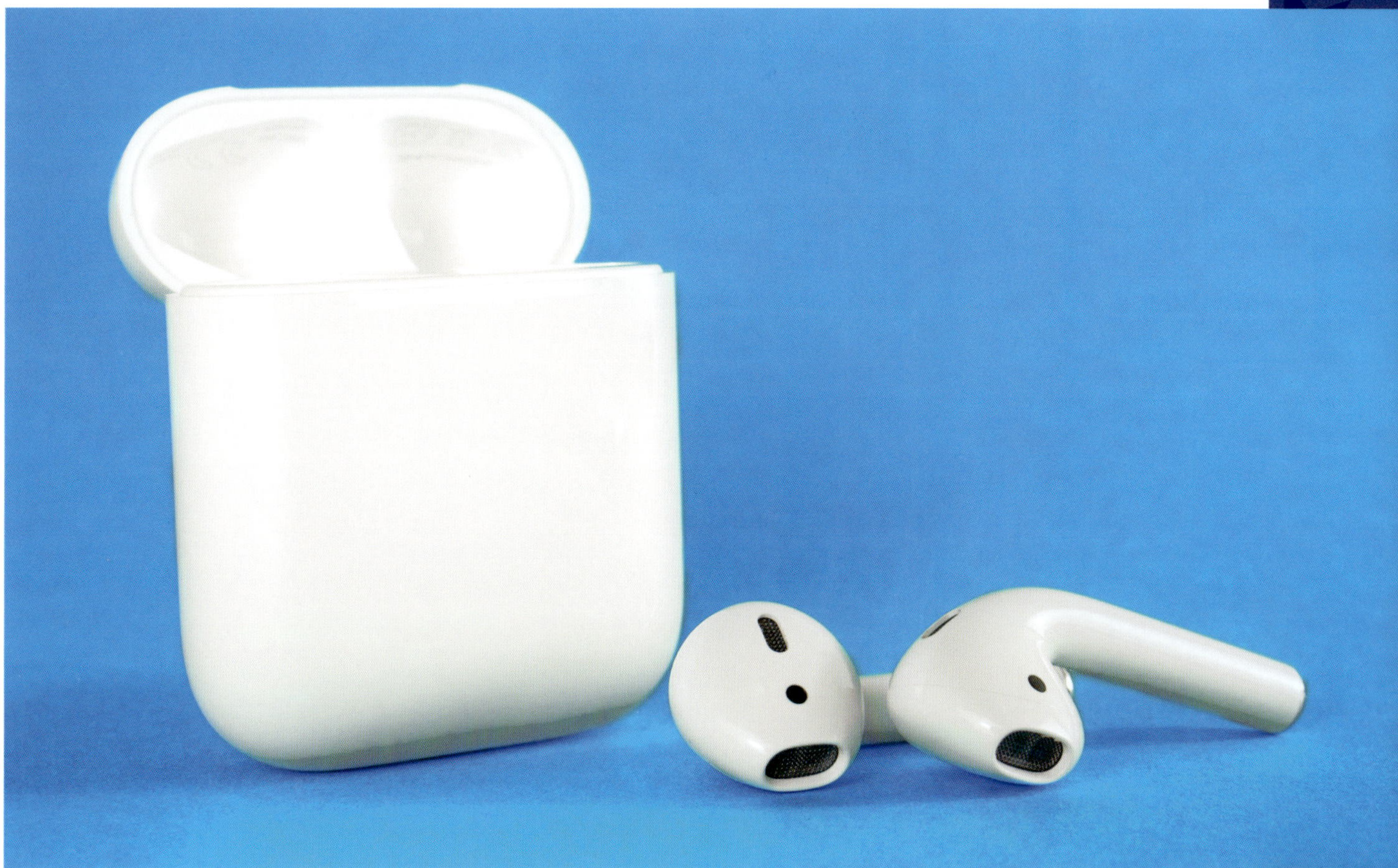

2017 THE REUSABLE ROCKET

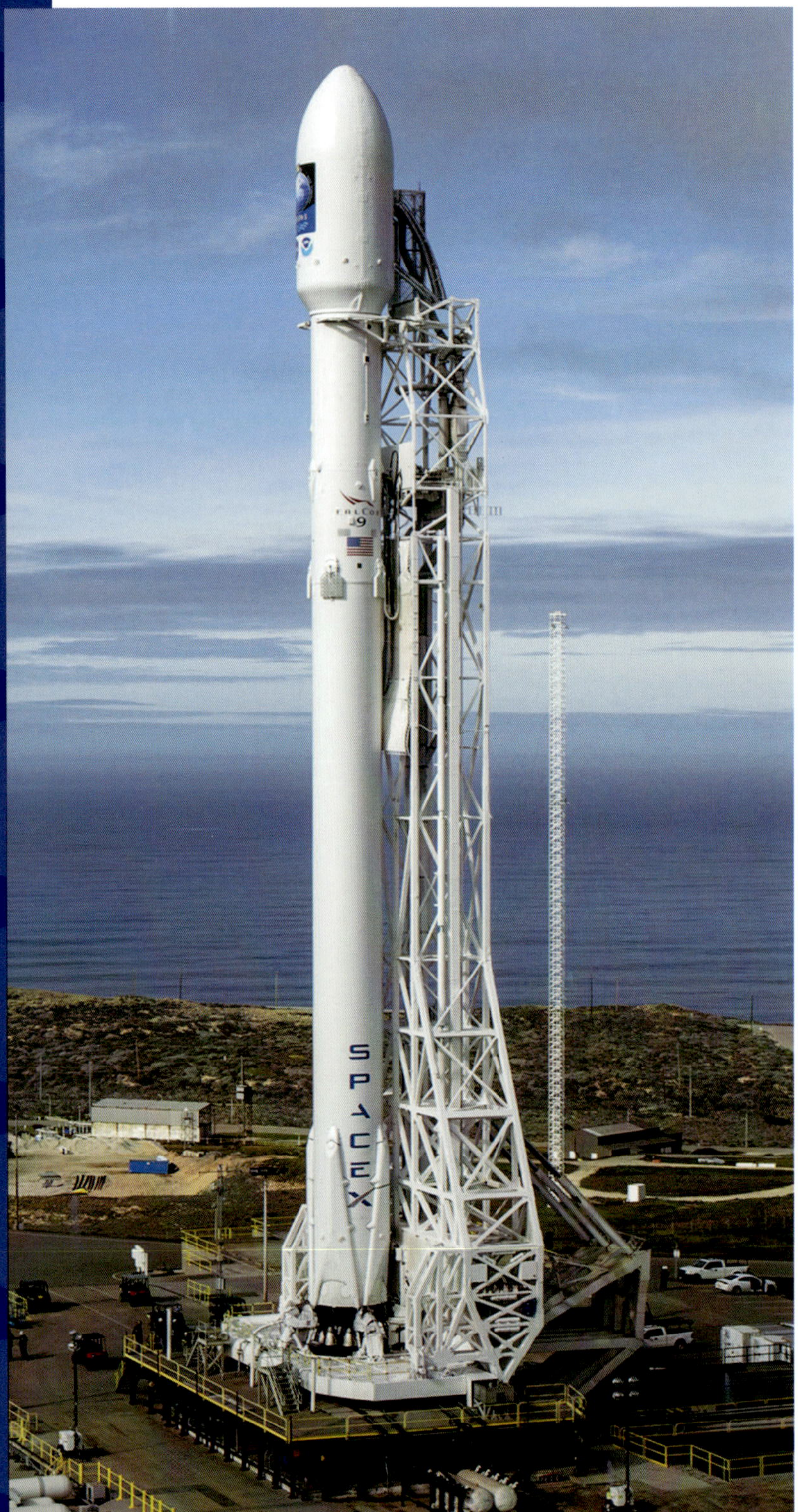

The first relaunch of a previously flown rocket booster occurred on March 30, 2017, with a rocket manufactured by the company SpaceX. On this historic day, SpaceX reused a Falcon 9 rocket booster to launch the SES-10 satellite into orbit from Cape Canaveral Air Force Station in Florida. This event demonstrated that rocket reusability was not just theoretical but a practical reality that could revolutionize the economics of operating in space. Now the most expensive components of a rocket—particularly the first-stage boosters—could be recovered, refurbished, and reflown multiple times rather than being discarded after a single use. SpaceX's second reflight occurred five months later in June, cementing its achievement. These successes validated the reusable rocket model and laid the foundation for the modern space industry's transformation. But the legacy of reusable rockets extends far beyond cost savings. The technology has enabled unprecedented launch frequency, made satellite constellations economically viable, and paved the way for ambitious projects like space tourism and Mars-colonization missions, potentially reshaping humanity's relationship with space exploration.

The Falcon 9 at the Vandenberg Space Force Base.

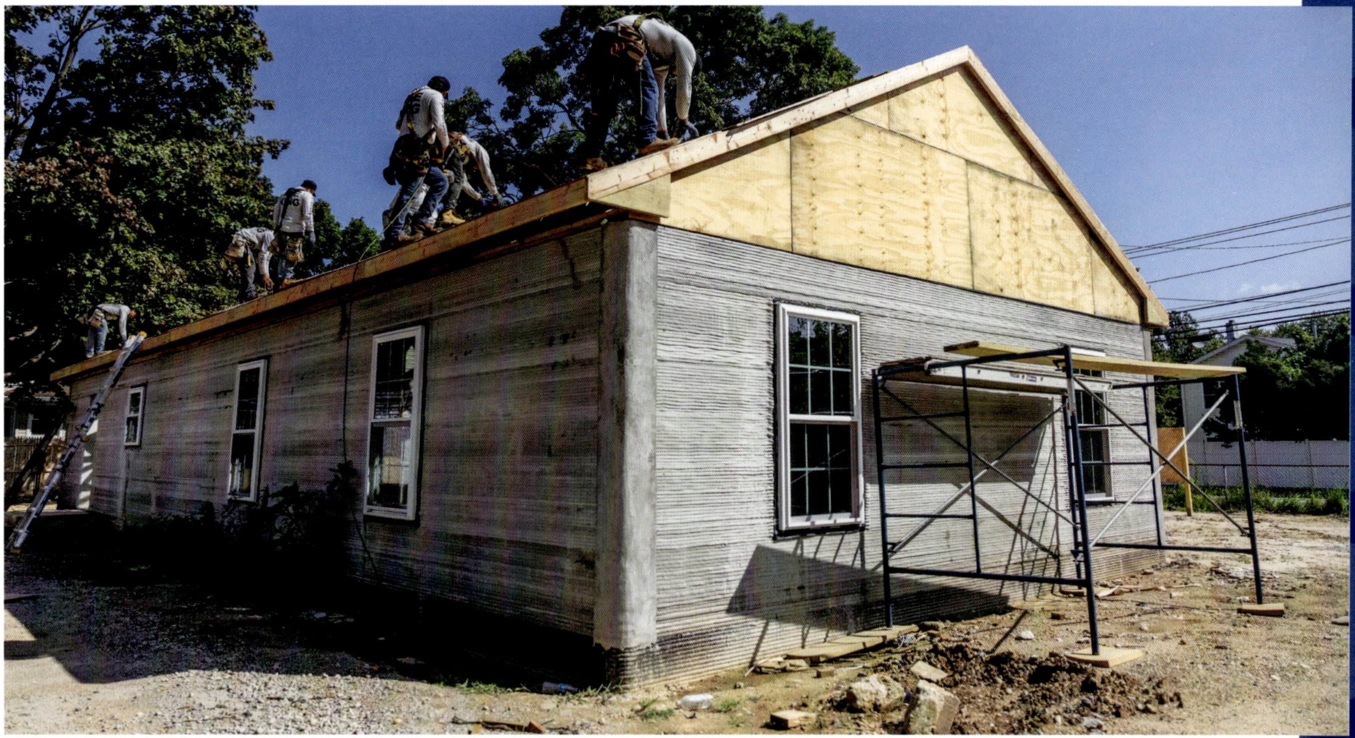

2018 3D PRINTING TO END HOMELESSNESS

The launch of 3D printing as a solution ending homelessness began around 2018, when organizations like New Story, a nonprofit organization, partnered with Austin-based ICON Technology, Inc., to explore rapid housing construction. A man named Tim Shea became the first formerly unsheltered person to move into ICON's 3D-printed home in Austin, Texas, symbolizing a pivotal moment in addressing chronic homelessness through innovative construction technology. The technology's promise lies in its speed and cost-effectiveness. Industrial-sized 3D printers can print an entire house in less than twenty-four hours. The 3D printing machines can build up to 40 percent of homes during the construction process and save money on labor costs. The initiative continues to expand, with a hundred 3D-printed homes by ICON currently under construction for chronically homeless individuals in Texas. This scalable approach represents a paradigm shift in addressing housing shortages, combining technological innovation with social impact to create sustainable solutions for one of society's most pressing challenges.

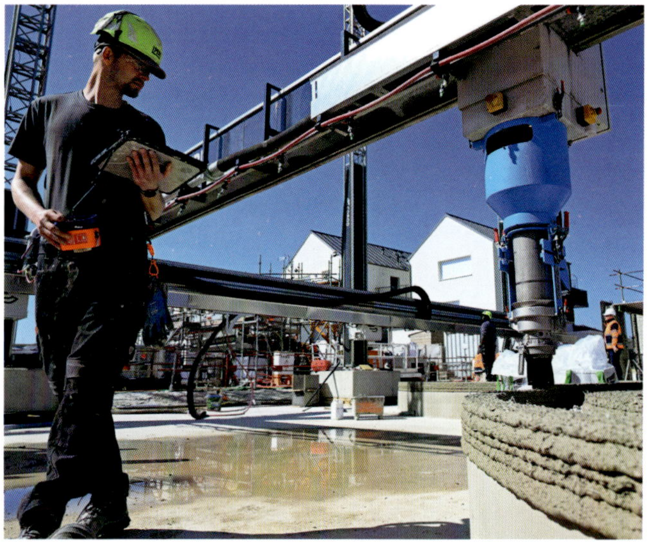

TOP: Habitat for Humanity workers put the finishing touches on a 3D-printed home in Brentwood, New York. **BOTTOM:** A worker supervises the printing of a home in Bezannes, France.

2019 SMART OVEN

American housewares brand KitchenAid launched the Smart Oven+ at the Consumer Electronics Show (CES) in 2019, one of the world's largest and most influential technology trade shows. The Smart Oven+ won a CES Innovation Award in the Smart Home product category and marked the brand's entry into the sector. The oven's defining innovation was its collection of interchangeable powered attachments inspired by KitchenAid's iconic stand mixer, including a grill, steamer, and baking stone that connect to a powered hub at the back of the oven. These attachments expanded traditional oven capabilities to include indoor grilling, steam cooking, and baking stone functionality. The Smart Oven+ was offered in three configurations: single, double, and combination ovens, each featuring smart connectivity through mobile app control. This innovation represented KitchenAid's vision of empowering home cooks with professional-grade versatility while maintaining the brand's signature commitment to quality and creativity. The legacy of the Smart Oven+ lies in its groundbreaking approach to wall-oven design, introducing the first-ever powered attachment system that transformed a single appliance into a multifunctional cooking platform.

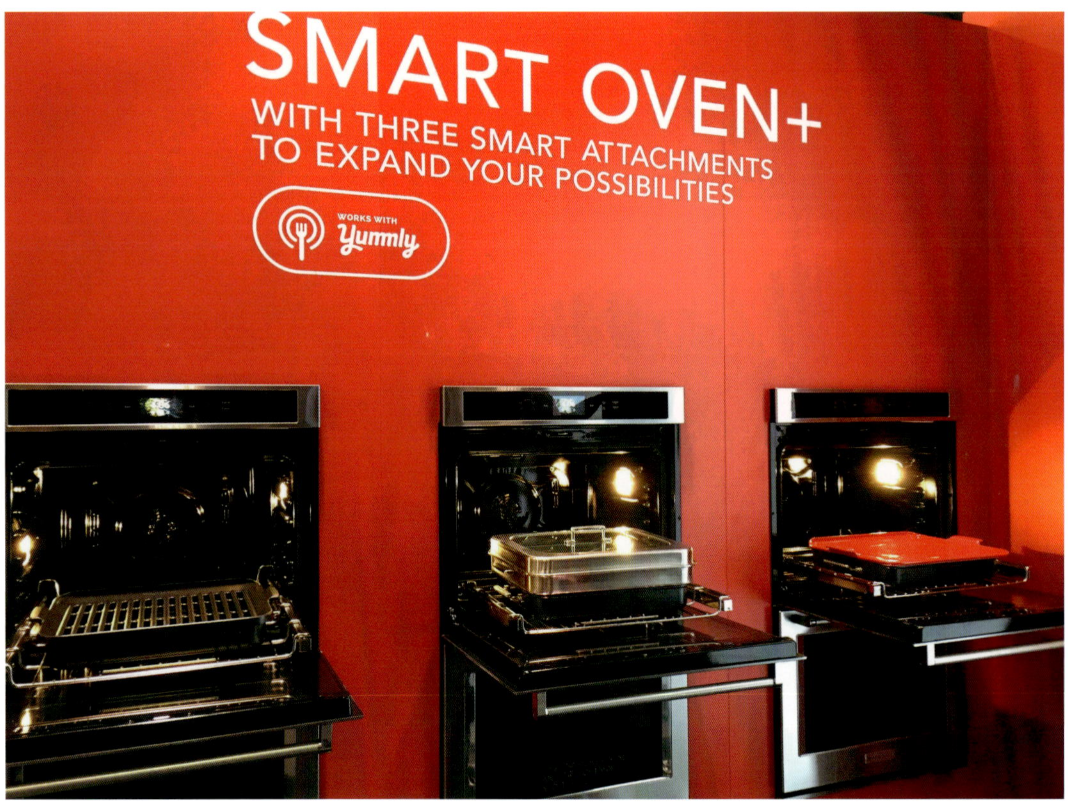

The Smart Oven display at the 2019 Consumer Electronics Show.

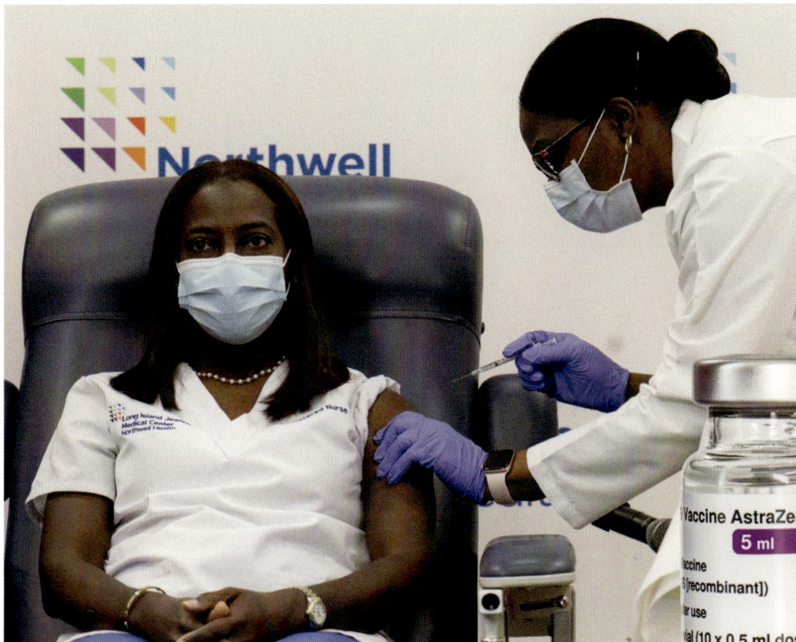

Sandra Lindsay, a nurse at Long Island Jewish Medical Center in Queens, New York, is the first person in the United States to be inoculated with the COVID-19 vaccine, Monday, December 14, 2020. Lindsay became the face of the medical first responders and was awarded the Presidential Medal of Freedom for her work during the epidemic.

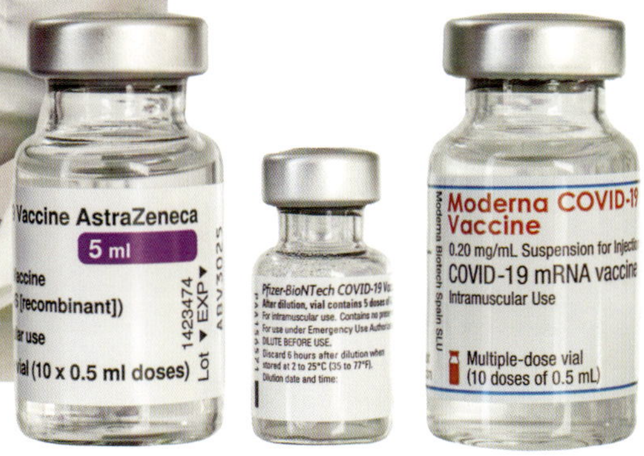

2020 COVID-19 VACCINE

The creation of the COVID-19 vaccine represents one of the most remarkable scientific achievements in modern history. It was developed with unprecedented speed and through global collaboration and innovative technologies. Multiple vaccines emerged within a year of the pandemic's onset, including mRNA vaccines by Pfizer-BioNTech and Moderna. The development process was accelerated by conducting steps concurrently that normally would be conducted sequentially, dramatically compressing the typical vaccine development timeline from years to months. This parallel processing model eliminated traditional delays between development phases without compromising safety standards. Another big factor in the success of the vaccine was a massive financial investment that eliminated typical funding delays. Scientists built upon decades of previous coronavirus research, particularly from SARS and MERS outbreaks. The vaccine's legacy extends far beyond COVID-19 control. It saved millions of lives globally, prevented healthcare system collapse, and enabled economic recovery by reducing severe illness and death rates. The mRNA platform pioneered for COVID-19 has opened new possibilities for cancer treatments, seasonal flu vaccines, and rapid response to future pandemics. Perhaps most significantly, the COVID-19 vaccine development demonstrated humanity's capacity for unprecedented scientific cooperation, regulatory flexibility, and rapid innovation when facing existential threats, establishing new paradigms for vaccine development and global health emergency response.

2021 PERLMUTTER SUPERCOMPUTER

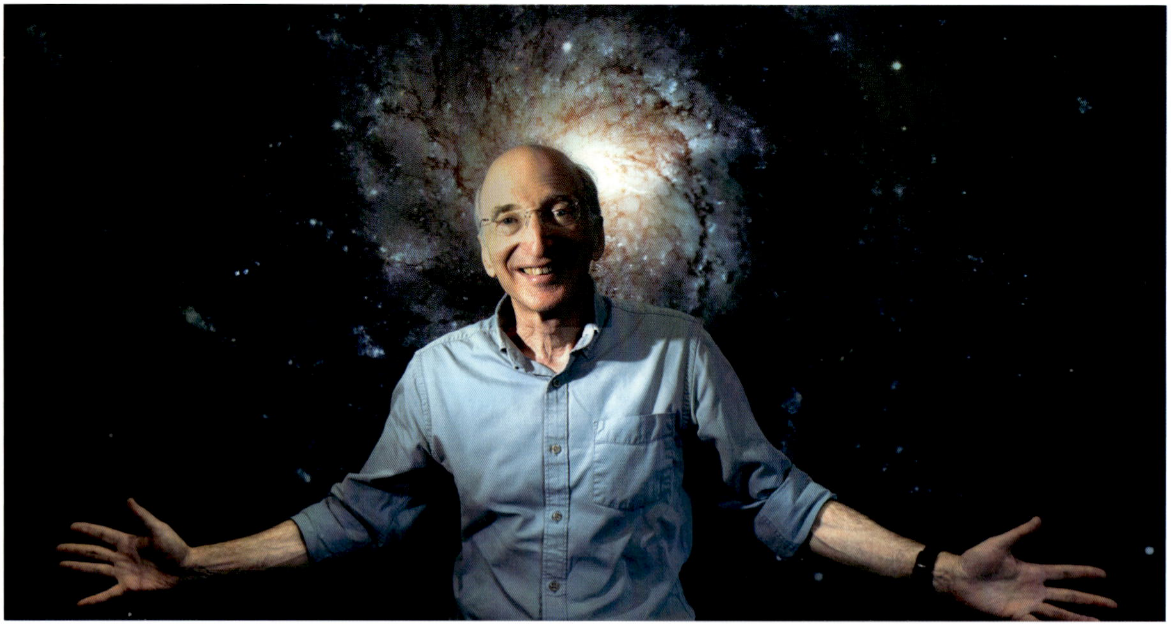

Saul Perlmutter, the Nobel Prize-winning scientist for whom the supercomputer is named.

Perlmutter is a groundbreaking supercomputer developed for the National Energy Research Scientific Computing Center (NERSC) at Lawrence Berkeley National Laboratory. The system is named after Saul Perlmutter (1959–), a Nobel Prize–winning scientist who discovered that the universe is expanding faster and faster. When Perlmutter launched in May 2021, it was called the world's fastest AI supercomputer. It uses more than six thousand NVIDIA A100 graphics cards and can perform nearly four exaflops of calculations per second. (That's an enormous number of calculations.) The computer was built by Cray Inc. using their Shasta design and AMD processors. It was the first supercomputer built specifically to handle both traditional scientific computing tasks and artificial intelligence work on a massive scale. Perlmutter has changed how scientists approach computing by combining AI and traditional high-performance computing. This has led to major advances in understanding climate, studying materials, and exploring space. The system can store thirty-five petabytes of data (equivalent to millions of laptops worth of storage) and processes information incredibly fast. Perlmutter has become the model for future supercomputers that blend traditional scientific simulations with machine learning.

2021 Cabenuva Regimen

The FDA's approval of Cabenuva in January 2021 marked a revolutionary milestone in HIV treatment. It is the first FDA-approved complete regimen for HIV-1 infected adults that is administered once a month via injection. The injection offered virally suppressed adults an alternative to daily oral medications, fundamentally transforming HIV care delivery. This innovation paved the way for long-acting HIV therapies, demonstrating that injectable regimens could maintain efficacy while improving quality of life.

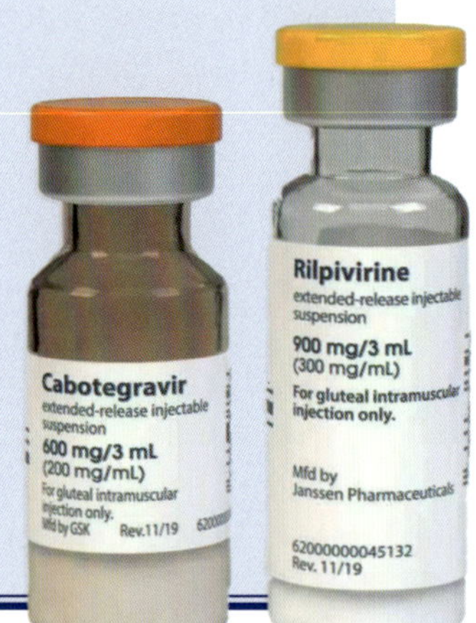

2022 REGENT SEAGLIDER

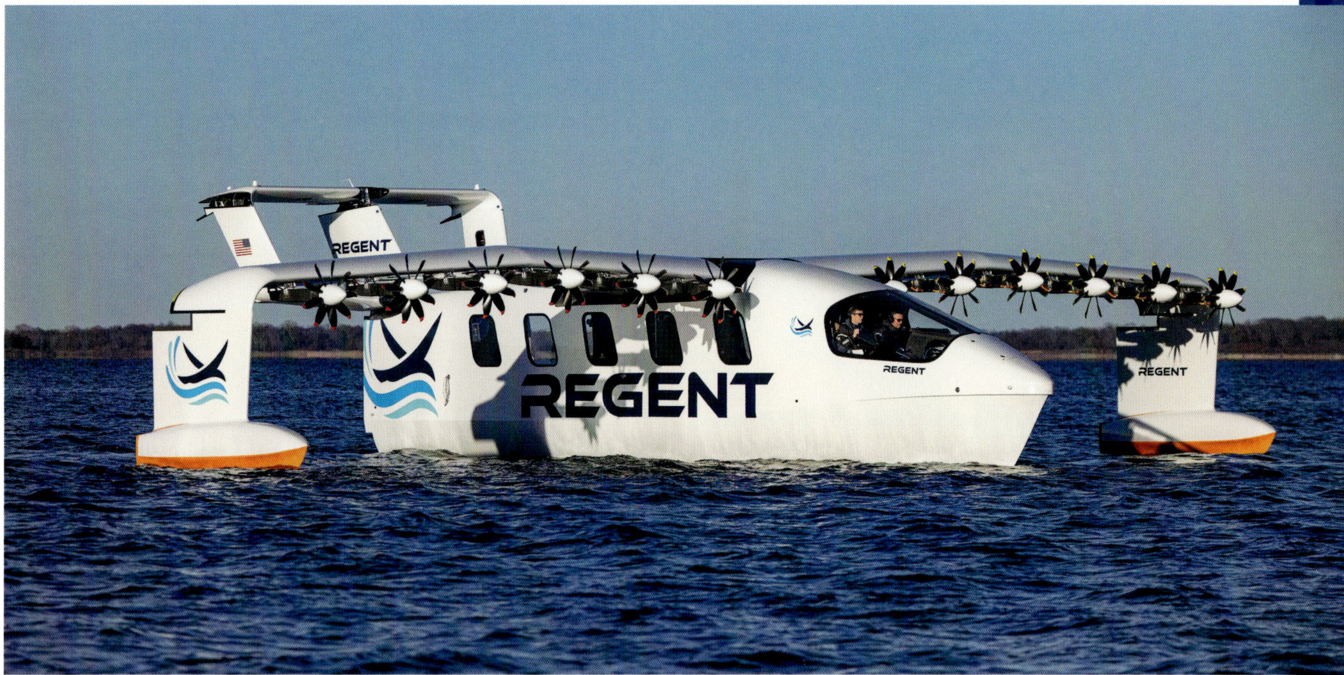

The REGENT Viceroy Seaglider on a test run in Narragansett Bay, August 2025.

The REGENT Seaglider represents a revolutionary approach to coastal transportation. Founded in 2020 by Michael Klinker and Billy Thalheimer, REGENT is a Rhode Island–based company whose name is an acronym for "Regional Electric Ground Effect Nautical Transport." The company's flagship vehicle, the Viceroy, is a proposed electric-powered glider. Its innovation lies in its unique operational method: the twelve-seat Viceroy Seaglider uses retractable hydrofoils during takeoff before transitioning to wing-in-ground-effect flight over coastal waters—a fascinating aerodynamic phenomenon that occurs when an aircraft flies very close to a surface, typically water or land. When a wing operates within approximately one wingspan's distance from the ground, it experiences dramatically improved lift and reduced drag compared to normal flight. In 2022, REGENT successfully demonstrated a quarter-scale prototype of the Seaglider in Narragansett Bay. This proof-of-concept test was crucial for validating REGENT's technology. The company is currently testing full-scale models of the Seaglider for an imminent launch. Reports boast that the vehicle will be able to travel up to 180 miles per hour and will be able to serve small communities that have been unreachable before.

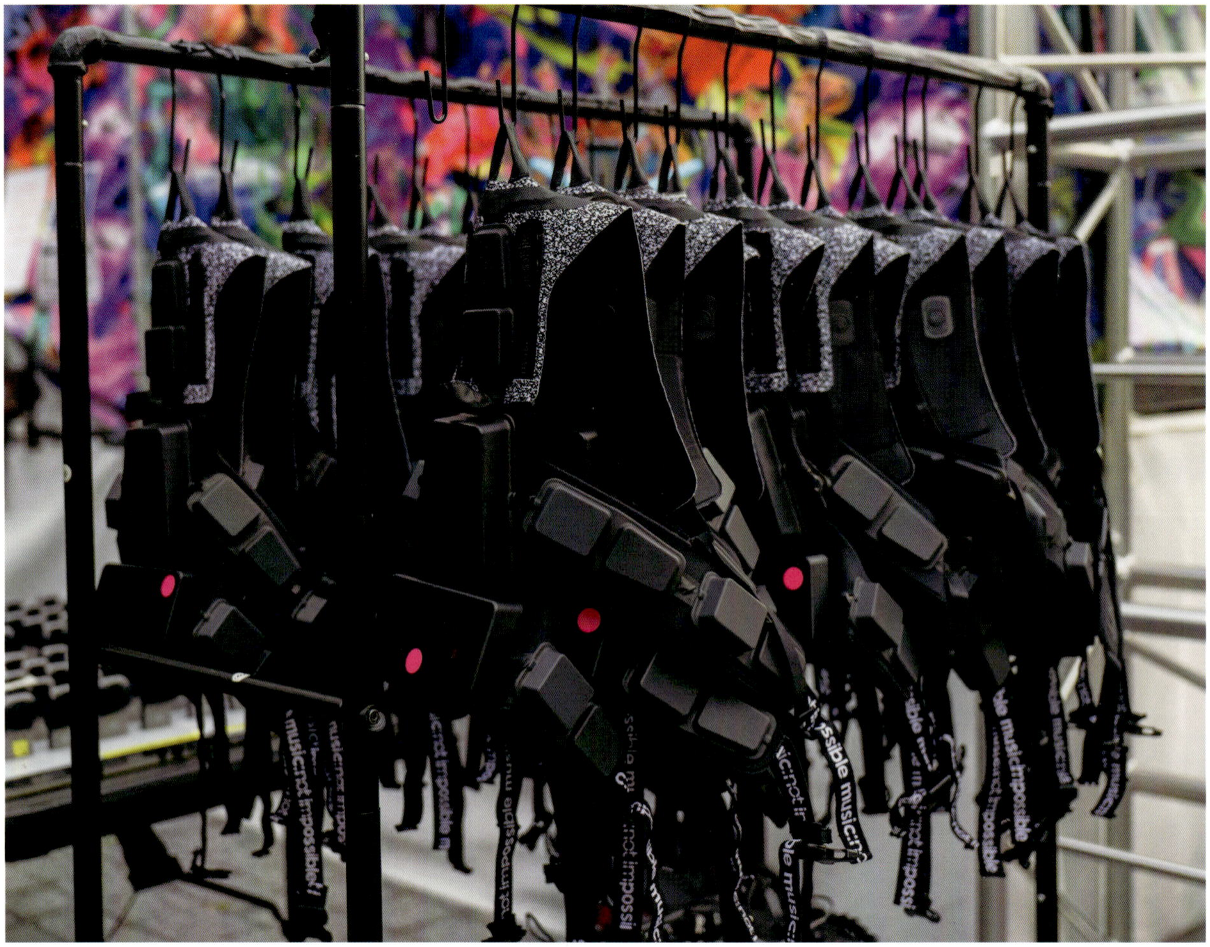

2023 FEELING THE BEATS

Invented in 2023, vibrating haptic suits developed by Music: Not Impossible (of Not Impossible Labs) translate sound from concert venues into vibrations that allow hard of hearing and deaf people to feel music on their skin through a wearable backpack-like vest and wrist and ankle bands. (*Haptics* is the science and technology of touch-based feedback and interaction. The term comes from the Greek word *haptikos*, meaning "pertaining to the sense of touch.") The initial tests on Music: Not Impossible's haptic suit involved strapping cell phone motors to people's bodies, but the technology became increasingly sophisticated. Ultimately, the wearable Vibrotextile technology features multiple haptic "actuators" that map different musical elements to various body parts, creating a comprehensive sensory experience. The Not Impossible haptic suit was recognized as one of *TIME*'s Best Inventions of 2023, highlighting its significant impact on accessibility technology. The suit has been featured at prestigious venues, including Lincoln Center, demonstrating its potential to revolutionize how people experience live music and paving the way for more inclusive entertainment technology.

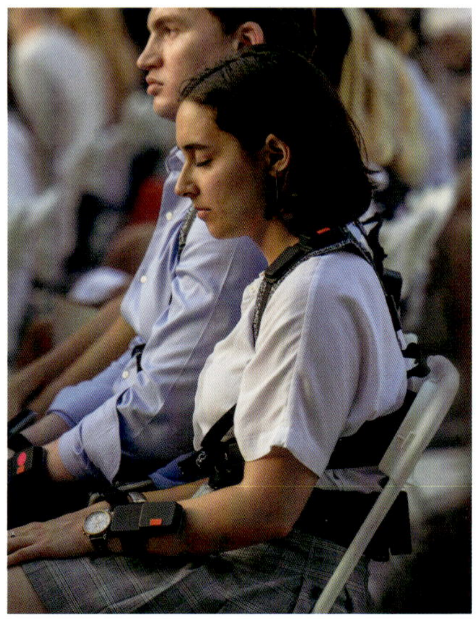

A woman wears a haptic suit at an outdoor concert at Lincoln Center in New York.

2024 EXTRAVEHICULAR SPACE ACTIVITY SUIT

The year 2024 marked a pivotal moment for commercial spaceflight, highlighted by groundbreaking advances in extravehicular activity (EVA) suit technology. At more than four hundred miles above Earth, the crew of SpaceX's Polaris Dawn mission achieved the first-ever commercial astronaut spacewalk using EVA suits developed entirely by SpaceX. These suits represented a significant technological leap forward in spacewalk capabilities. Though based on the familiar black-and-white intravehicular activity (IVA) spacesuits used inside spacecraft, the EVA versions featured critical enhancements designed for the demanding environment of space walks. Key improvements included lightweight construction for reduced bulk, integrated heads-up displays, helmet-mounted cameras, and advanced joint mobility systems that maintained flexibility even under full pressurization. In our current era of renewed space exploration, continued innovation in EVA spacesuits remains crucial for several key reasons: environmental protection against radiation, micrometeoroids, and extreme temperatures; pressure systems that maintain life-sustaining atmospheric conditions; mobility and durability for complex tasks in harsh conditions; life support independence enabling extended operations outside spacecraft; and communication and navigation systems for crew coordination and safety. As space travel continues to expand from government agencies to private enterprise, advanced EVA capabilities will prove essential for the commercial space economy's growth and the safety of civilian astronauts.

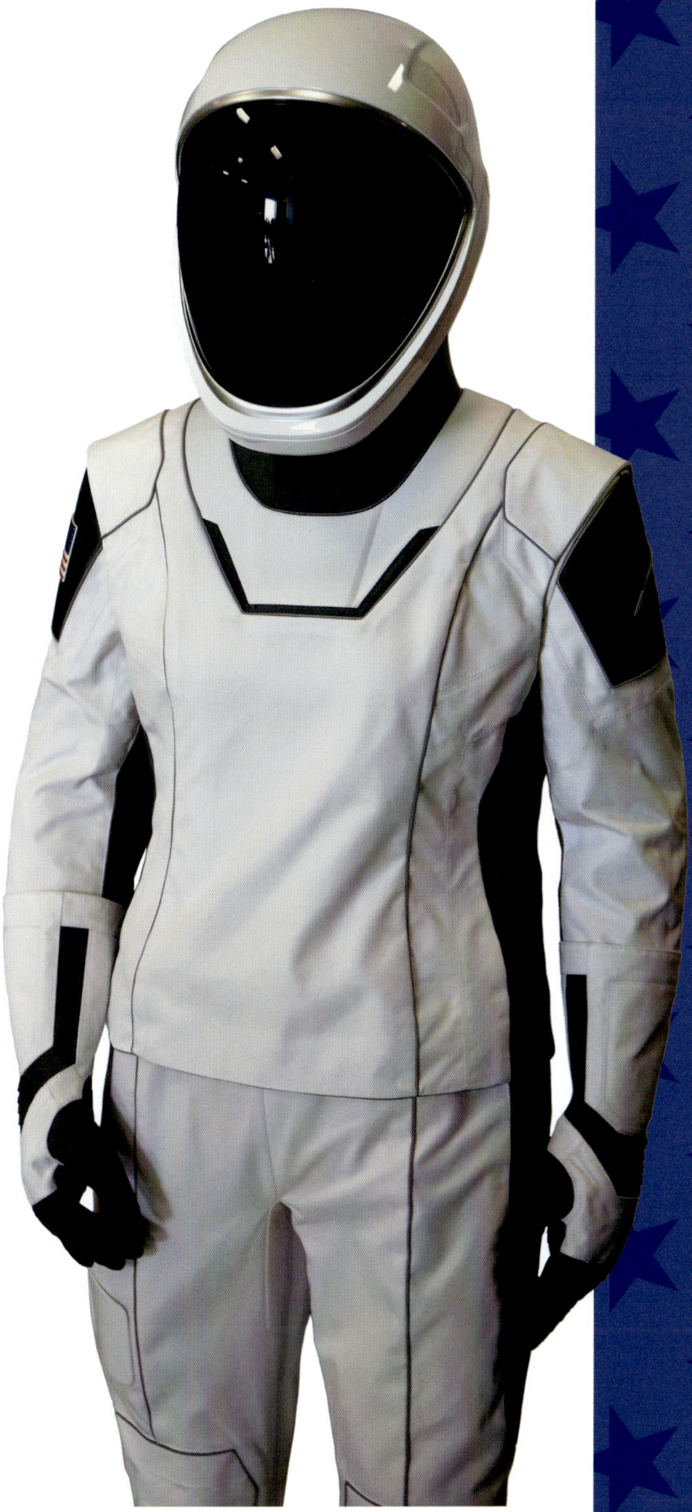

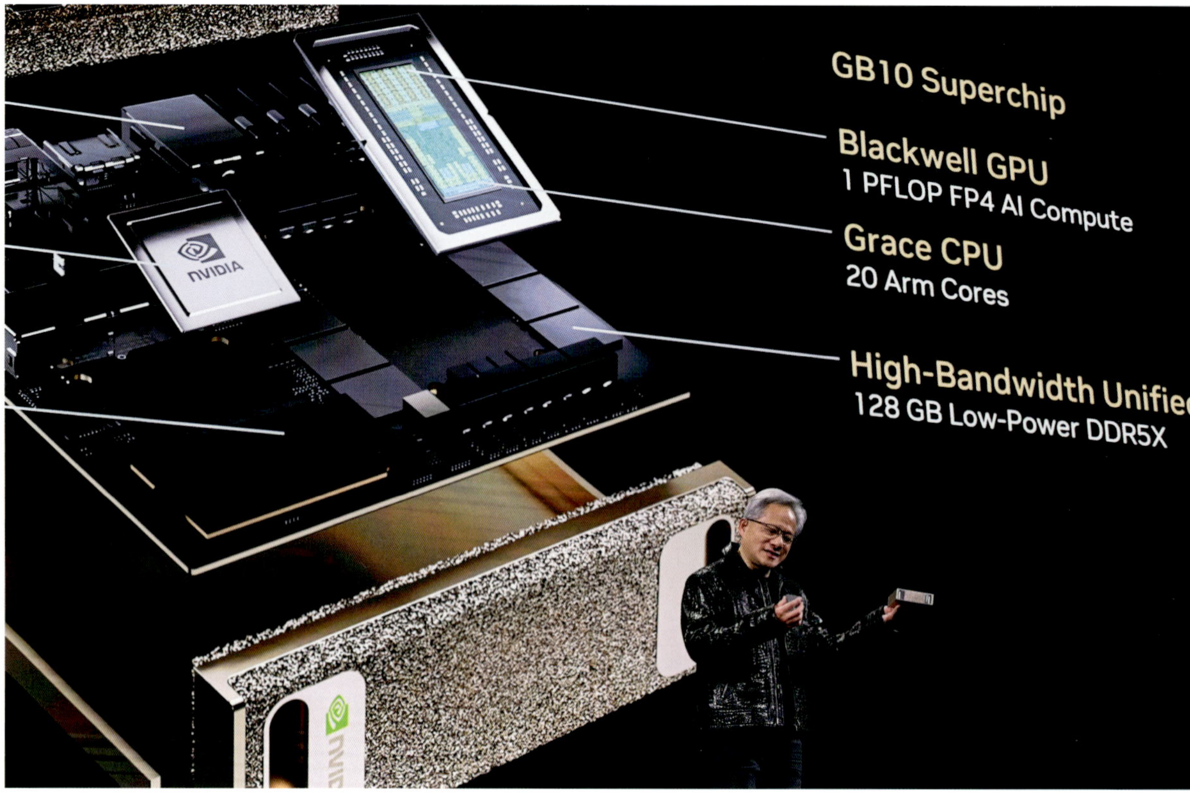

NVIDIA CEO Jensen Huang unveils the next generation of GPU and other software at the 2025 Consumer Electronics Show in Las Vegas.

2025 GRAPHICS PROCESSING UNITS (GPUS)

In 2025, GPUs emerged as perhaps the most consequential hardware pieces in modern computing history. Originally designed to render video game graphics, GPUs became the indispensable foundation of the artificial intelligence revolution. Their ability to perform thousands of parallel calculations simultaneously made them uniquely suited for training the massive networks that power generative Artificial Intelligence (AI) and other advanced technologies. NVIDIA, the Santa Clara, California–based semiconductor company, emerged as the unrivaled architect of this transformation, designing chips that became as essential to twenty-first-century innovation as electricity itself. The American dominance in GPU technology reshaped global power dynamics in 2025. As a response, nations throughout the world recognized that AI supremacy required access to these advanced chips, which elevated semiconductors to the strategic importance of traditional energy resources. In 2025, a massive GPU shortage crystallized a geopolitical reality: controlling this technology meant controlling the future of innovation. Beyond commerce, American GPU innovation accelerated scientific discovery across disciplines. Silicon Valley's chip designers had inadvertently created the most powerful tool for human knowledge expansion since the printing press. GPU's impact extends far beyond the hardware itself. American GPU architecture established the computational foundation upon which the next century of technological advancement would be built, cementing the United States' role as the infrastructure provider of the digital age.

2026 AGENTIC AI

Silicon Valley's constant push for innovation catalyzed the emergence of a new form of artificial intelligence, agentic AI. This new technology is an evolution of generative AI (an artificial intelligence tool that waits for commands, completes one task and stops, and seeks guidance for each next step) into an autonomous form, wherein the technology is able to set its own sub-goals, plan sequences of actions independently, and adapt based on the results. American companies such as Anthropic, Google, Microsoft, OpenAI, and many startups have pioneered systems capable of planning multistep workflows, adapting to obstacles, and executing complex tasks without continuous human oversight. This fundamental shift represents a significant contribution to AI throughout the world.

The concentration of talent, venture capital, and resources in the United States created an insurmountable advantage. The fact that more than 80 percent of leading agentic AI companies are headquartered in America means that the United States set the technical standards and ethical guidelines that govern agentic AI worldwide. This dominance ensured the American values of transparency, accountability, and human oversight became embedded in the technology's DNA. Agentic AI's lasting legacy extends beyond commercial success. Despite skepticism, American agentic AI established the template for human/AI collaboration in the twenty-first century, with the goal to prove that autonomous systems could augment rather than replace human capability, ultimately expanding what individuals and organizations could accomplish.

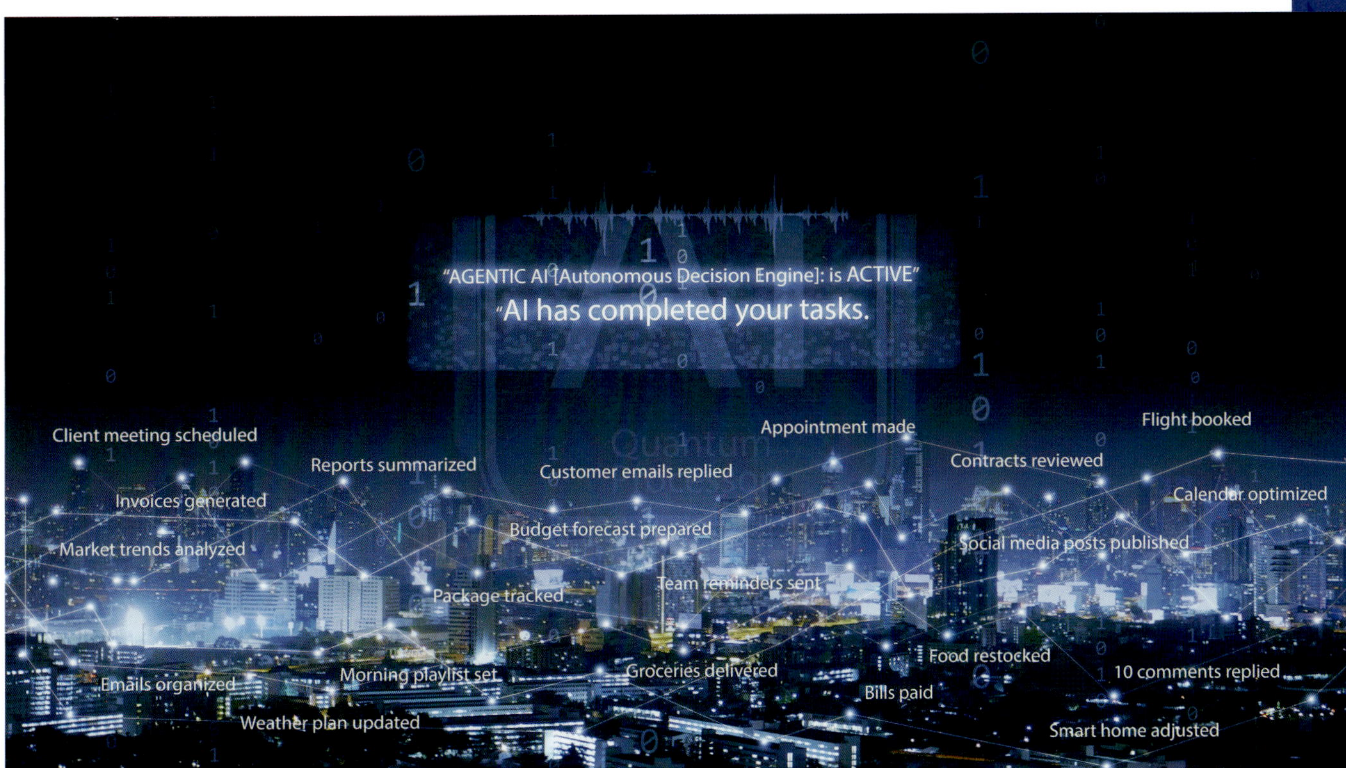

LIST OF OBJECTS BY SUBJECT

AGRICULTURE

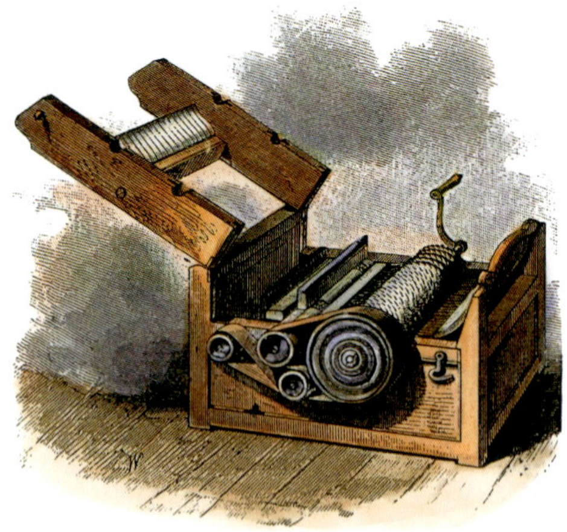

The cotton gin

ARCHITECTURE

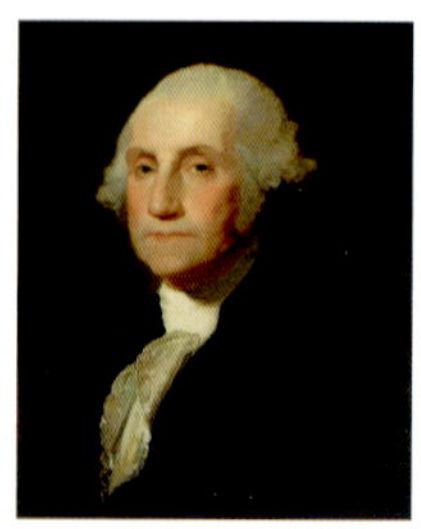

Presidential portraits

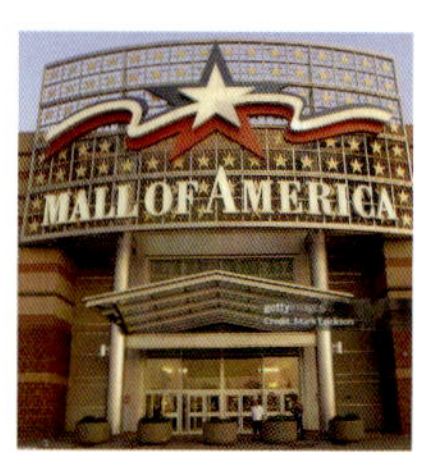

Mall of America

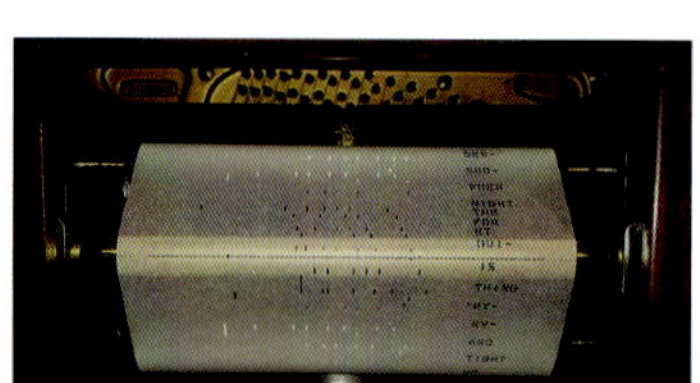

Player piano

THE ARTS

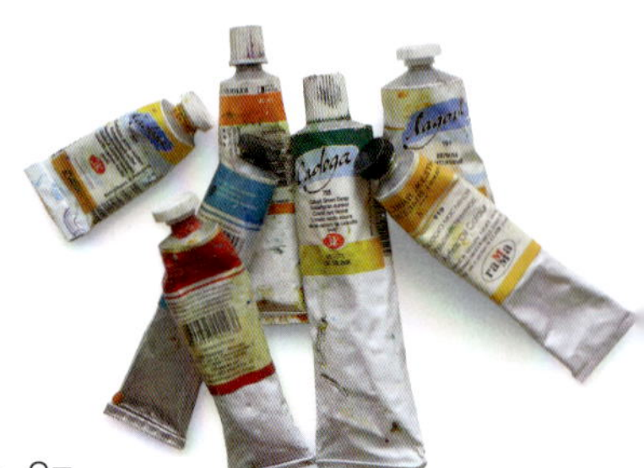

Paint tubes

COMMUNICATION

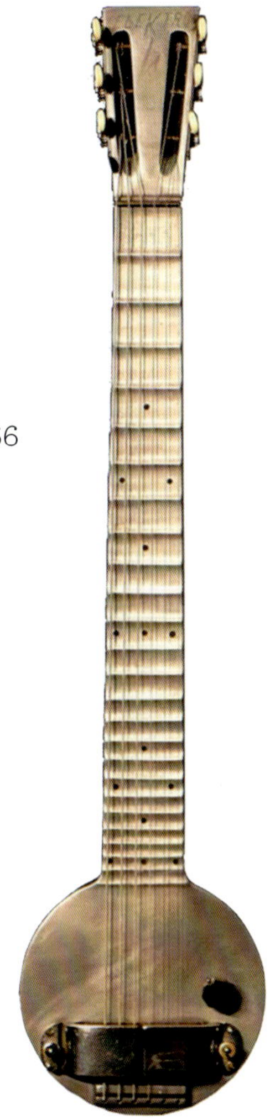
Electric guitar

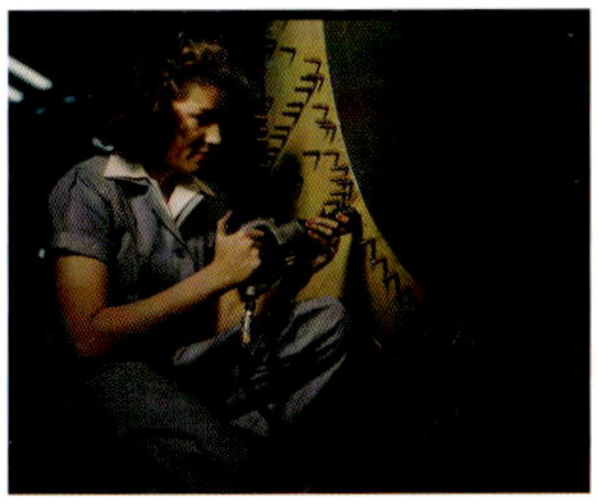
Rosie the Riveter

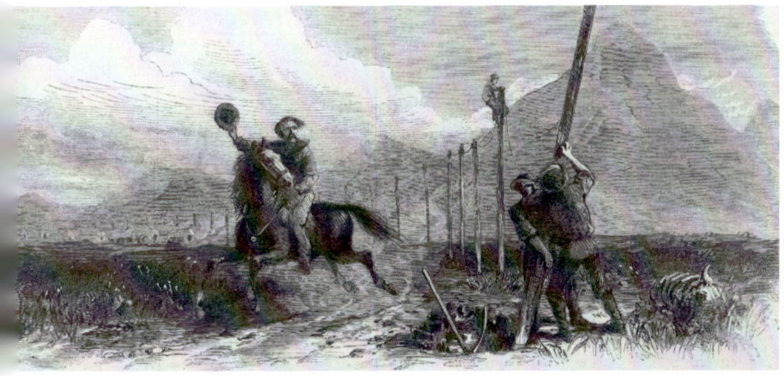
Transcontinental telegraph system

CULTURE

FASHION

FOOD AND DRINK

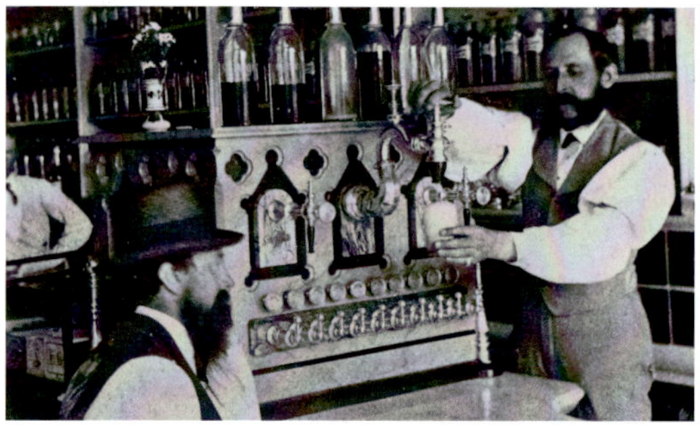

GOVERNMENT AND MILITARY

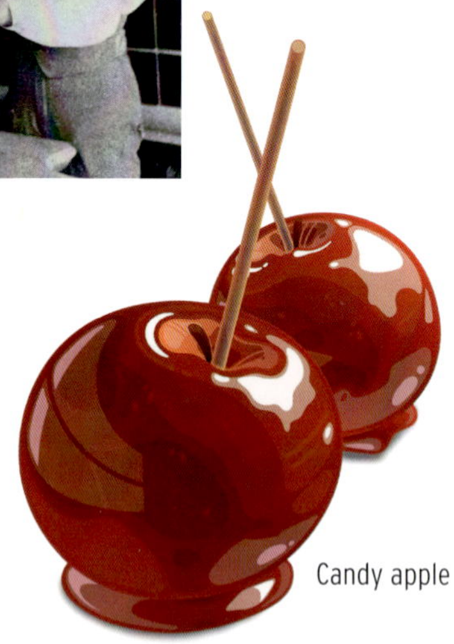

Candy apple

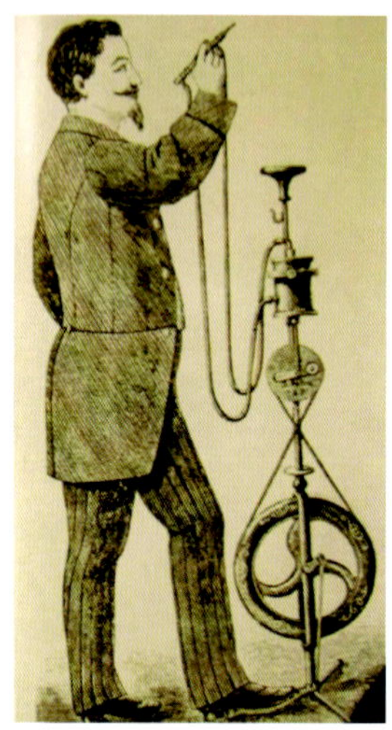

Electric dental drill

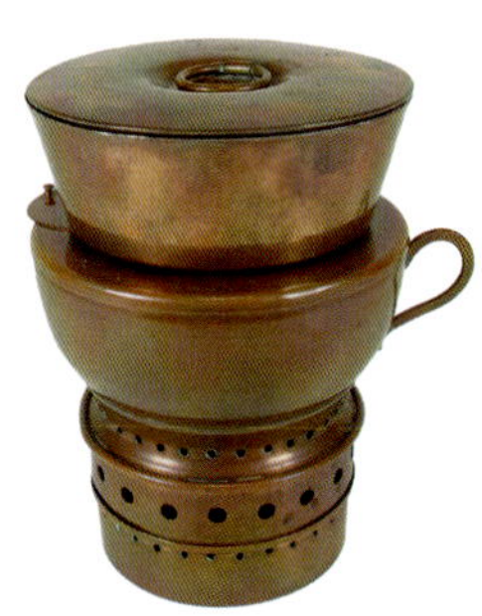

Coffee percolator

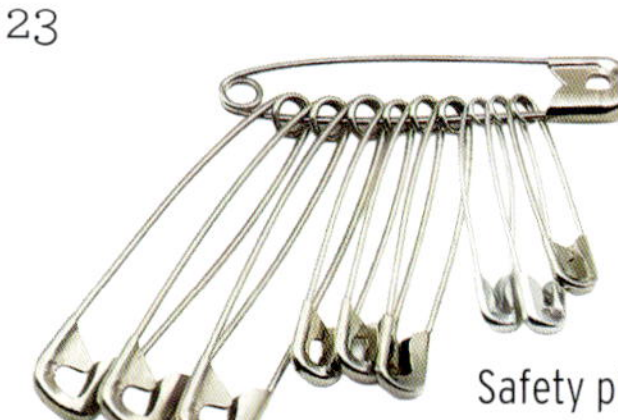

Safety pin

First US coins in circulation

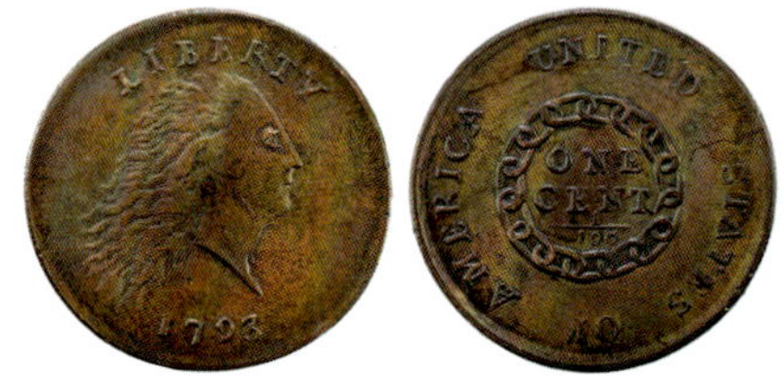

HEALTH AND MEDICINE

IN THE HOME

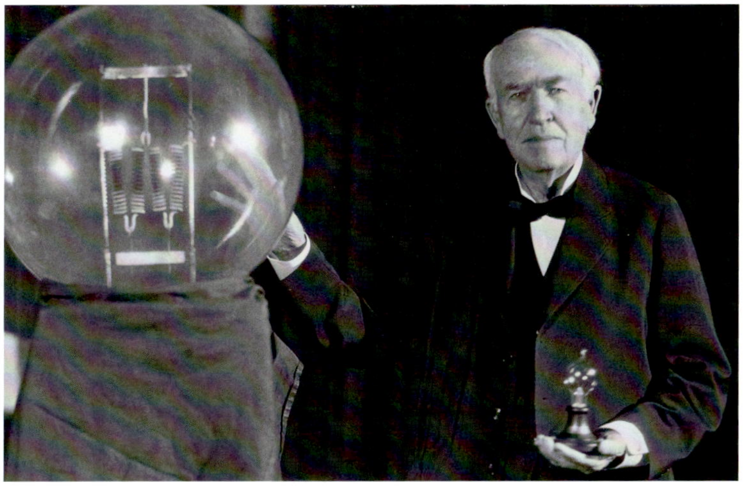

Incandescent light bulb

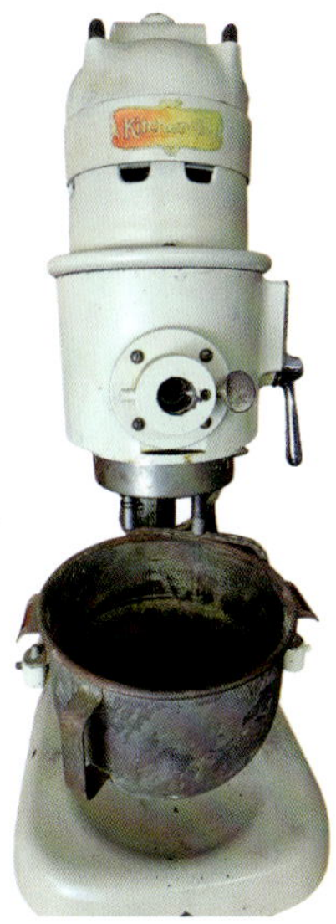

KitchenAid Food Preparer

IN THE WORKPLACE

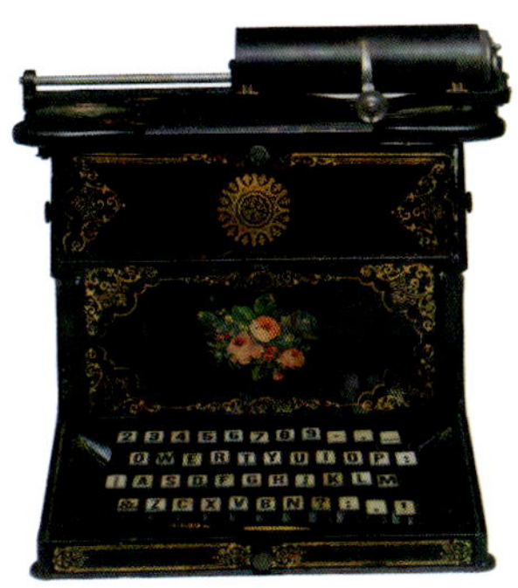

Typewriter

POP CULTURE

SCIENCE AND EXPLORATION

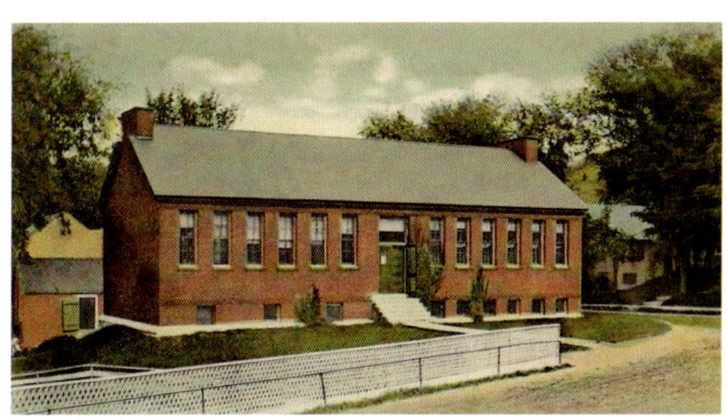

First free public library

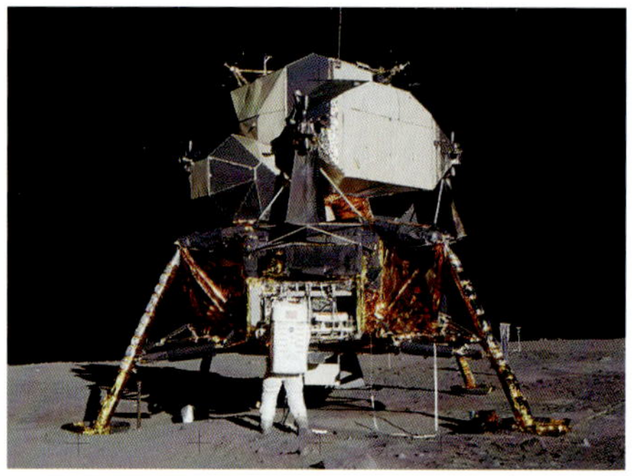
The Apollo 11 lunar module

SPORTS

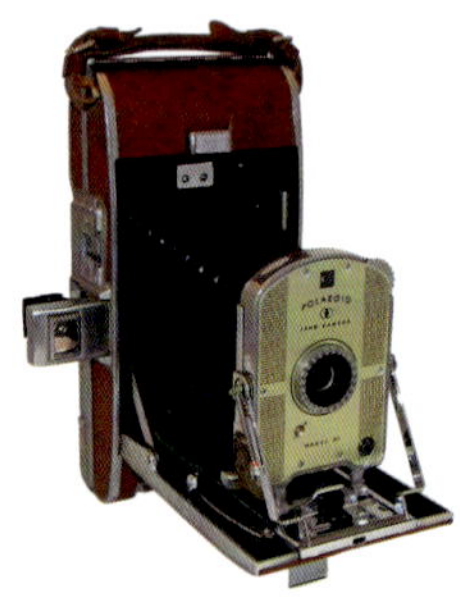
Polaroid instant camera

TECHNOLOGY

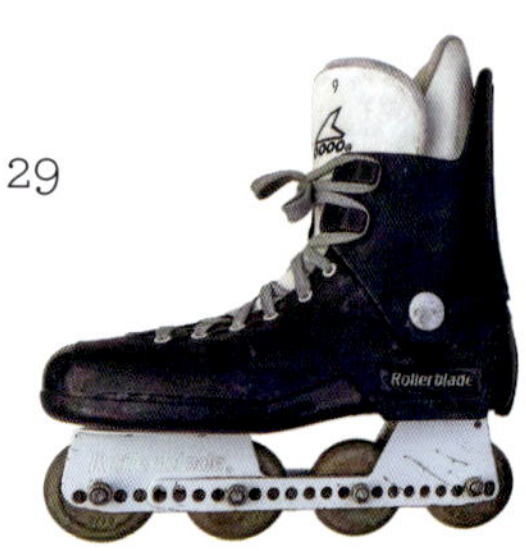
Rollerblades

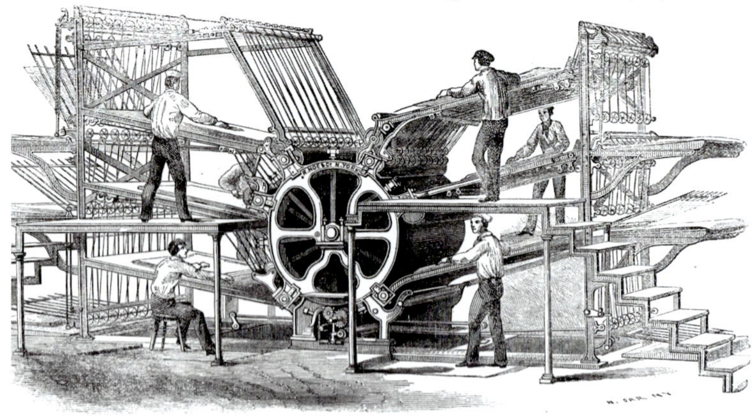
Rotary printing press

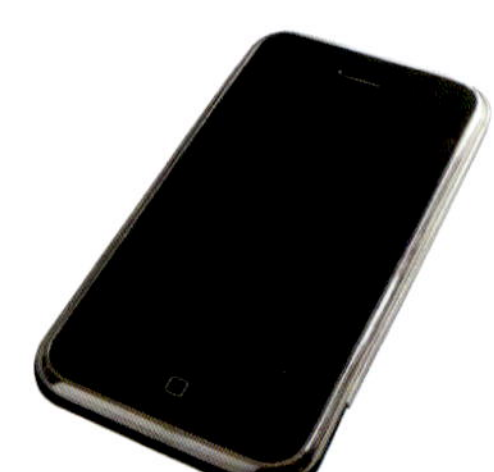
iPhone

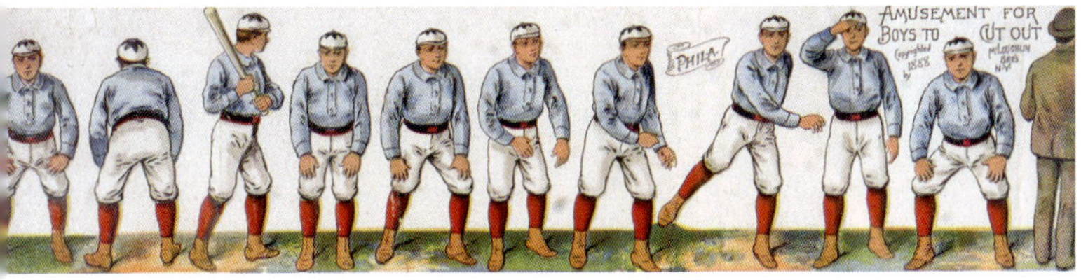

Paper dolls

TOYS AND GAMES

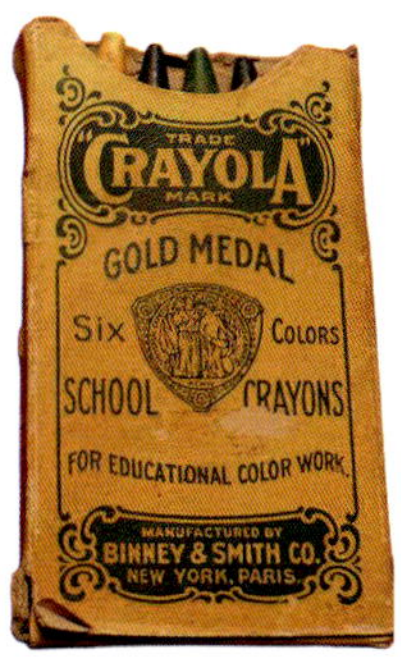

Crayola crayons

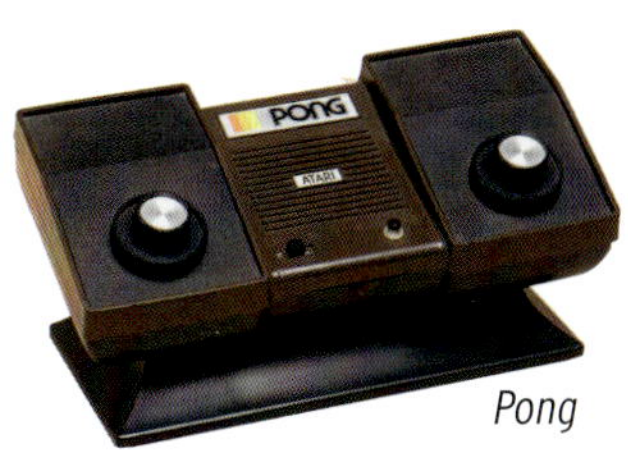

Pong

TRANSPORTATION

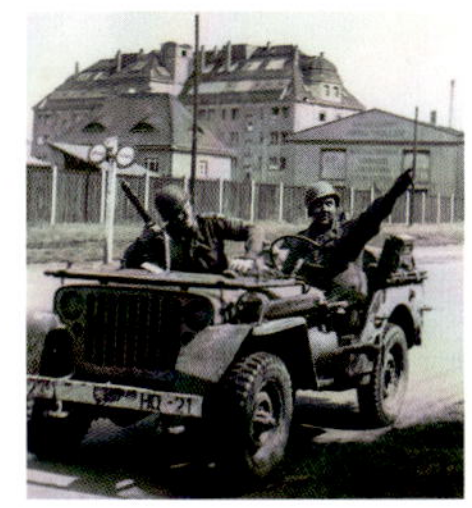

The Jeep

TV, MOVIES, AND THEATER

MTV MUSIC TELEVISION®

Tom Thumb locomotive

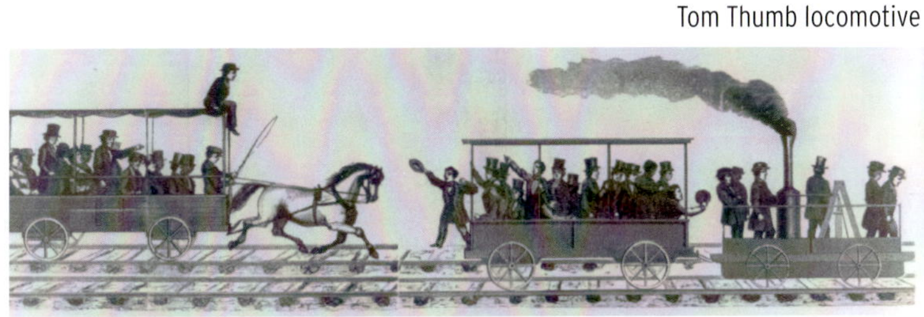

PHOTO CREDITS

T=top; B=bottom; R=right; L=left; M=middle

Archives, American Academy of Arts and Sciences, Cambridge, Massachusetts. Benjamin Hanks, "Mr. Hanks's pneumatic Clock," 1784 [July 1]; [Litchfield, CT]. RG I-C-1: General records. Communications to the Academy – Bound, 1780–1810. Volume 1, p. 56.: 15; Alamy: Alpha Stock 31, Archive PL 217, Book Worm 57, brinkstock 234, Carolyn Jenkins 230, CBW 228, Chronicle 67, ClassicStock 187, Cultural Archive 56, David Tonelson 230, FlixPix 232, Futuras Fotos 142, Glasshouse Images 66–67, Golib Golib Tolibov 256–57, GTCRFOTO 242, Helen Sessions 221, Hera Vintage Ads 50, History and Art Collection 128, Independent Photo Agency Srl 246, INTERFOTO 69, 121, 205, Jim Kidd 162, Jim West 157, Keith Homan 177, Liam Bunce 181, Lynne Sutherland 216, M4OS Photos 234, Malcolm Haines 251, Marmaduke St. John 233, NAPA 240, NASA Archive 218–19, North Wind Picture Archives 29, NurPhoto SRL 247, Patti McConville 200, 224, Penta Springs Limited 112, Photo 12 199, 175, 225, 236, Photology1971 205, Pictorial Press Ltd 58, 138, 173, 198, PictureLux/The Hollywood Archive 204, Pump Park Vintage Photography 123, Science History Images 106B, Stephen Lloyd UK 186, Tanapong Sungkaew 263, The NASA Library 219, The Reading Room 41, 45; UPI 256–57, ZUMA Press, Inc. 182, 213, 222; Boston Public Library Digital Commonwealth 48 ("Harvesting Cranberries on Cape Cod." Card. Tichnor Bros. Inc., Boston, Mass., [ca. 1930–1945]. File: ark.digitalcommonwealth.org/ark:/50959/cn69mp365) 88L; Brooklyn Museum of Art, Gift of Paul Roebling 114; Brooklyn Museum Costume Collection at The Metropolitan Museum of Art, Gift of the Brooklyn Museum, 2009; Funds given by Mrs. Richard Palmer, 1978 41B; Courtesy Everett Collection: 185, 215, 202B, 204B, 211B, Advertising Archive 178, ©Warner Bros Television 211T; CC Image on Flickr: 165 (Jim Griffin), 180 (Joe Haupt), 222 (Ron Cogswell), 238B (David McKelvey); The Commons, Flickr: Internet Archive Book Archive 97B, 68B; 152; Getty Images: ANGELA WEISS 260T, 260B, Anthony Harvey 231, Archive Photos 185T, Bettmann 42B, 74, 94B, 110, 126T, 142T, 143B, 209, 157T, 169T, 171B, 193, 194, 212, CBS Photo Archive 184, Chicago History Museum 146, Corbis Historical 11, Culture Club 51, Cyrus McCrimmon 245, David Paul Morris / Stringer 244, David Pollack 170, Fotosearch 103, FRANCOIS NASCIMBENI, 255B, Graphic House 64T, GraphicaArtis 42T, Harold M. Lambert 206, 145, Heritage Images 113, Hulton Archive 98, 149B, Jean-Christian Bourcart 237, Jim Heimann Collection 144, JP Jazz Archive 148, Kean Collection 13, Kim Kulish 235T, MARCEL MOCHET 198B, Michael Loccisano 248B, Michael Ochs Archives 155, 186T, MIGUEL RIOPA 249B, Movie Poster Image Art 166, Newsday LLC 255T, NurPhoto 262, PAUL BUCK 190, PhotoQuest 151, 153, picture alliance 249T, Raymond Boyd 220, ROBYN BECK 261, Sherman Oaks Antique Mall 171, Silver Screen Collection 192T, Stock Montage 75, Transcendental Graphics 60, Underwood Archives 159, Universal History Archive 141, UniversalImages-Group 63, Visual Studies Workshop 149T; Granger Historical Picture Archive: 116; Internet Archive, Otis Elevator Company, Public Domain Mark 1.0, https://archive.org/details/OtisElevatorCompanyTheOtisElevatorIndustryComprisesLarge_574/page/n15/mode/2up: 189; iStock Photo: Berezka_Klo 208T, Bob Lord 167, Brett_Hondow 188, cafe-de-art 236T, ECummings00 24B, evemilla 197, gojak 128T, Irina Meshcheryakova 253, ivan-96 132B, Jesus Carpio 143T, LICreate 235B, memoriesarecaptured 224B, PictureLake 166, Sjo 158; Sheryl Jaeger, Eclectibles—Fine and Unusual Ephemera: 60; The Jay T. Last Collection of Graphic Arts and Social History, Huntington Digital Library 24T; Photo Courtesy Leland Little Auctions: 44; Library of Congress: 16, 22B, 32, 26, 33, 35T, 40, 46, 58, 61T, 62, 82, 91, 69B, 76B, 76T, 78–9, 81T, 84R, 86T, 87R, 93B, 93T, 94T, 96L, 96R, 99T, 100T, 104T, 104B, 105B, 105T, 106 T, 108R, 114–15, 117, 118, 119T, 120L, 122, 127T, 127B, 129T, 132T, 134, 135T, 137L, 140, 144R, 147, 160, 161, 168TR, 172, 203B; Library of Congress Music Division: 131B; Library of Congress Prints & Photographs Division: 9, 10, 55T, 119B, 128B, Brady-Handy photograph collection 108L, Detroit Publishing Company Collection 49, 107B, Farm Security Administration/Office of War Information Color Photographs 175L, LC-DIG-stereo-1s06666 115B, photograph by Harris & Ewing 35T, photograph by John Margolies, [reproduction number, e.g., LC-MA05-1], 164, 165, Photographs in the Carol M. Highsmith Archive 52T, 52B, West Virginia Collection within the Carol M. Highsmith Archive 43T; Library of Congress, Manuscript Division: 20; Library of Congress, Paradise Valley Folklife Project collection, 1978–1982 (AFC 1991/021), American Folklife Center 30; Library of Congress, Rare Book and Special Collections Division: 10T, 55B, America Singing: Nineteenth-Century Song Sheets 95, Printed Ephemera Collection 23; Licensed under Creative Commons CC0 1.0 Universal Public Domain Dedication, via Wikimedia Commons: 27 (Photograph by Daderot, File:Jefferson cylinder cipher (replica) - National Cryptologic Museum - DSC07717.JPG), 77 (Photograph of Polk Miller by WestCoastDRIP, File:Polk Miller with his banjo.png), 154 (US patent for the traffic signal by Garrett Morgan, File:US1475024-drawings-page-1.png), 213T (Wheelchair designed by Marilyn Hamilton, 1983, The Quickie 1999.0303.1, File:Quickie Tennis Wheelchair NMAH-2000-3254-04.jpg), 254 (Falcon 9 Vertical At Vandenberg Air Force Base, File:2016 Falcon 9 at Vandenberg Air Force Base.jpg); Licensed under the Creative Commons Attribution 3.0 Unported license, via Wikimedia Commons: 102B (Science History Institute, Adams Pepsin Tutti Frutti Gum, File:Adams Pepsin Tutti Frutti Gum.jpg), 135BR (Photograph by Ed Welter, File:Crayola No 6 progression.jpg), 163L (Photograph by the Museum of Making Music at English Wikipedia, File:Elektrofryingpan.jpg), 179R (Photograph by Brigade Piron, File:1947 defibrillator.jpg); Licensed under the Creative Commons Attribution-Share Alike 2.0 Generic license, via Wikimedia Commons: 203T (Photograph by Edward Kimmel, File:Climate March 1090 (34368550605).jpg), 210T (Photograph by Bobo11, File:Museum-Enter-Apple-1-6094652.jpg), 243 (Photograph by Nicki Dugan, File:Martin Eberhard (normalized).jpg), 244 (Photograph by Carl Berkeley, File:iPhone First Generation 8GB (3680455198).jpg); Licensed under Creative Commons-Share Alike 3.0, via Wikimedia Commons: 16T (Photograph by Fletcher 6, File: Salem_Town_Hall_in_

Old_Town_Hall_Historic_District.jpg), 72 (Photograph by Дар Ветер, File:Maslo maslo.jpg); Licensed under the Creative Commons Attribution-Share Alike 4.0 International license, via Wikimedia Commons: 81B (Photograph by Jacek Halicki, File:2023 Agrafki.jpg), 129B (Photograph by Draconichiaro, File:PlayerPianoRoll.jpg), 131T (Image of Scott Joplin, File:Scott Joplin 1907 (transparent).png), 191R (Photograph originally published by the *Los Angeles Times*, restored by Adam Cuerden, File:Ruth Handler in 1961.jpg), 210B (Apple 1 computer formerly owned by Joey Copson, File:CopsonApple1 2k.jpg) 258T (Image of Saul Perlmutter by Christopher Michel, File:Saul Perlmutter in 2024 at Berkeley Lab 02.jpg); Licensed under the GNU Free Documentation License, via Wikimedia Commons: 214 (Photograph by Fred Hsu, File:First Rollerblade brand inline skate from early 1980s-Rockered wheel setup-IMG 5651-FRD.png); Image courtesy of the LuminAID team: 252; Metropolitan Museum of Art: Gift of Mrs. Emily Winthrop Miles, 1946 56B, The Jefferson R. Burdick Collection, Gift of Jefferson R. Burdick 137; Mr. Mixer of Augusta, Kansas. Photo courtesy of Zach Dinicola: 150; NASA, ESA: Allison Loll/ Jeff Hester (Arizona State University). Acknowledgement: Davide De Martin (ESA/Hubble) 227TL, M. Livio and the Hubble 20th Anniversary Team (STScI) 227TR, the Hubble Heritage Team (STScI/ AURA), A. Nota (ESA/STScI), and the Westerlund 2 Science Team 227B; National Museum of American History, Smithsonian Institution: 220T; Courtesy of the New Hampshire Historical Society: 39; Photo Courtesy Orange County Archives 187B; Photofest: 201; Public Domain via Wikimedia Commons: 12 (Logo by the American Academy of Arts and Sciences, File:American Academy of Arts and Sciences (ancient logo).png), 14 (Engraving of a flatboat by Alfred Waud, File:Traveling-by-flatboat-engraving-by-Alfred-R-Waud.png), 19R (Portrait of Alexander Hamilton by John Trumbull, File: Alexander Hamilton.jpg), 22T (Anonymous, File:Henry Curlhair Sr. in the 1850 United States census living in Wright, Schoharie, New York.jpg), 26B (Wood engraving of Eli Whitney's cotton gin, File:Eli Whitney's cotton gin.jpg), 34 ("The Retribution Comet" editorial cartoon by George Frederick Keller, File:Mussel Slough editorial cartoon - The Wasp 8 July 1881.jpg), 36–37 (Painting of the Lewis and Clark Expedition by Charles Marion Russell, File:Lewis and Clark Reach Shoshone Camp Led by Sacajawea.jpg), 61B (Photograph by Evan-Amos, File:Graham-Cracker-Stack.jpg), 53 (Portrait of Thomas L. Jennings, File:Thomas-L-Jennings.jpg), 54L (Colored drawing of actor James Hewlett, engraved by I. Scoles from a sketch by Parisen, File:Houghton TCS 44 - James Hewlett as Richard the Third.jpg), 54R (Henry Box Brown, File:Henry Box Brown.jpg), 64 (Postcard of Peterborough Town Library, File:Town Library, Peterborough, NH.jpg), 65 (US Patent of the seed planter by Henry Blair, File:US Patent 8447X figures.png), 70R (US Patent of the seed planter by Henry Blair, File:US Patent 8447X figures.png), 71B (Patent illustration of W. S. Otis' excavator, Public Domain, File:Otis Excavator 1841.jpg), 78B (Photograph by Seth C. Landon, File:First known photograph of a baseball game in progress.jpg), 80 (Design of Richard March Hoe's printing press by Nathaniel Orr, File:Hoe's six-cylinder press.png), 83T (The Bloomer Polka, File:The Bloomer Polka (BM 1925,0404.63).jpg), 83B (Portrait of Amelia Bloomer by Elizabeth Cady Stanton, Susan B. Anthony, and Matilda Joslyn Gage, File:HOWS V1 D536 Amelia Bloomer.jpg), 86B (*Walden* by Henry David Thoreau, File:Walden Thoreau.jpg), 87L (*Leaves of Grass* by Walt Whitman, File:Leaves of Grass (1855) cover.jpg), 90 (Scan by CodeName47km, File:03. Signal box for demonstration.jpg), 100B (*Little Women* by Louisa May Alcott, File:Houghton AC85.Aℓ194L.1869 pt.2aa - Little Women, spines.jpg), 101 (Photograph by Andrew J. Russell, File:East and West Shaking hands at the laying of last rail Union Pacific Railroad - Restoration.jpg), 102 (Advertisement by Black Jack advertising, File:Black Jack gum advertisement 1925.jpg), 103B (Advertisement by Levi Strauss, File:Advertising for Levi Strauss—It's no use they can't be ripped.jpg), 109 (Burpee's Farm Annual 1886 by W. Atlee Burpee & Co., File:Burpee's Farm Annual 1886 front cover.jpg), 111T (Maria E. Beasley, File:Maria E. Beasley.png), 124 (The Allegheny Athletic Association football team, File:Allegheny Athletic Association football team 1896.jpg), 125B (Portrait of Pudge Heffelfinger, File:Pudge heffelfinger.jpg), 125T (The Pittsburgh Athletic Club football team, File:Pittsburgh Athletic Club football team 1892.jpg), 126B (Portrait of Josephine Garis Cochrane, File:Josephine Garis Cochrane.png), 136 (Photograph by David Fletcher, File:Holt75pk.jpg), 156 (Postcard by Tichnor Brothers, File:Motel Inn, San Luis Obispo, Cal., U.S. 101, North City Limits (80459).jpg), 163R (Diagrams of Rickenbacker lap steel guitar by George D. Beauchamp, File:RickenbackerFryingpanPatentDiagram.png), 173R (Photograph of Lynda Carter by ABC Television, File:Lynda Carter Wonder Woman.JPG), 176 (Smokey Bear poster by the National Agricultural Library, File:Uncle Sam style Smokey Bear Only You.jpg), 189 (Photograph by Petey21, File:Frisbee 090719.jpg), 192B (First-edition cover of *To Kill a Mockingbird* by Harper Lee, File:To Kill a Mockingbird (first edition cover).jpg), 202T (Photograph by Neil Armstrong, File:Buzz Aldrin and Apollo 11 Lunar Lander, AS11-40-5927.jpg), 223 (Photograph by the National Institutes of Health, File:Aids Quilt.jpg); REGENT: 259; Ryan Herd "The Smart Guy": 256; Courtesy of Science History Institute: 47; Shutterstock: Andrea Izzotti 8, Andrew Angelov 239, Artem Sokolov 174T, berkah jaya material 99B, BrandonKleinPhoto 250, CEW 207R, Dennis MacDonald 68T, Dimitrios Karamitros 36B, gfdunt 195T, Ian Dewar Photography 97T, Ichpochmak 139B, IgorGolovniov 215T, IVY PHOTOS 229, Julie Clopper 183B, Keith Homan 88B, Kletr 248T, Mamun Sheikh 241, OttoPles 174B, Photology1971 238T, Pierre Dahl 195B, puha dorin 208B, Rawpixel.com 226, SFCphotos 196; National Portrait Gallery, Smithsonian Institution: 19L, 19M, 84L, Gift of Jean McGinley Draper 28, Gift of Wendell G. Sykes and Betsy J. Constantine 71B; Smithsonian Institution, American History Museum: 168T, Mr. and Mrs. Kermit Roosevelt: 133; Smithsonian Institution, National History Museum: 70L; Collection of the Smithsonian National Museum of African American History and Culture, Gift of the Lyles Station Historic Preservation Corporation: 20B; Smithsonian Institution, National Museum of American History: 25; Smithsonian Institution, National Postage Museum: 85; University of Virginia Library: 91; © 2025 ViiV Healthcare group of companies or its licensor. Used with permission: 258; Yale Law Library: 19T.

All best efforts have been made to identify and secure permissions for the images in this book.

weldon**owen**
an imprint of Insight Editions
P.O. Box 3088
San Rafael, CA 94912
www.weldonowen.com

CEO Raoul Goff
SVP Group Publisher Jeff McLaughlin
VP Publisher Roger Shaw
Executive Editor Karyn Gerhard
Editorial Assistant Jon Ellis
Art Director Megan Sinead Bingham
Production Designer Jean Hwang
VP Manufacturing Alix Nicholaeff
Senior Production Manager Joshua Smith
Strategic Production Planner Lina s Palma-Temena

Layout Design by Edwin Kuo
Cover Art by Chui Kanela

Weldon Owen would also like to thank Ellen Foreman and Crystal Coppock.

ISBN: 979-8-88674-264-0

Manufactured in China by Insight Editions
10 9 8 7 6 5 4 3 2 1

ROOTS of PEACE REPLANTED PAPER

Insight Editions, in association with Roots of Peace, will plant two trees for each tree used in the manufacturing of this book. Roots of Peace is an internationally renowned humanitarian organization dedicated to eradicating land mines worldwide and converting war-torn lands into productive farms and wildlife habitats. Roots of Peace will plant two million fruit and nut trees in Afghanistan and provide farmers there with the skills and support necessary for sustainable land use.

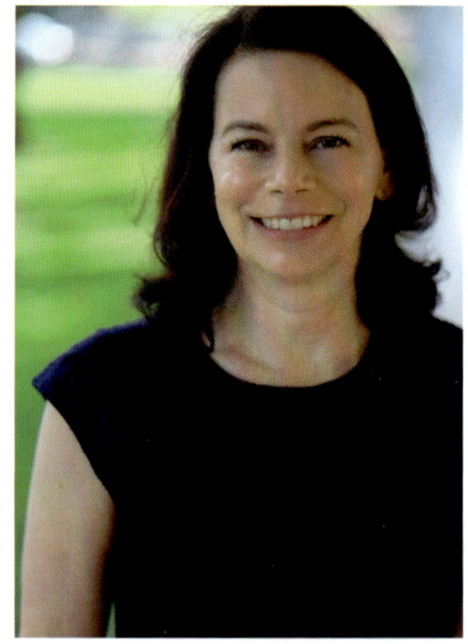

ABOUT THE AUTHOR

Jennifer Newens is an author, editor, literary agent, and former publisher who has been developing book content for more than twenty-five years. She has written for, edited, acquired, and represented everything from Williams Sonoma cookbooks to cocktail books to YA graphic novels. Her most recent work is *Walt Disney Treasures: Art and Artifacts from the Walt Disney Family Museum*. She lives in the San Francisco Bay Area.